Groundwater Geochemistry

Groundwater Geochemistry

Pollution and Remediation Methods

Edited by

Sughosh Madhav
Shaheed Bhagat Singh College,
Delhi, India

and

Pardeep Singh
PGDAV College, University of Delhi,
Delhi, India

WILEY Blackwell

This edition first published 2021

Registered Offices
John Wiley & Sons, Inc., 111 River Street, Hoboken, NJ 07030, USA
John Wiley & Sons Ltd, The Atrium, Southern Gate, Chichester, West Sussex, PO19 8SQ, UK

Editorial Office
9600 Garsington Road, Oxford, OX4 2DQ, UK

For details of our global editorial offices, customer services, and more information about Wiley products visit us at www.wiley.com.

Wiley also publishes its books in a variety of electronic formats and by print-on-demand. Some content that appears in standard print versions of this book may not be available in other formats.

Library of Congress Cataloging-in-Publication Data

Names: Madhav, Sughosh, editor. | Singh, Pardeep, editor.
Title: Groundwater geochemistry : pollution and remediation methods / edited by Sughosh Madhav and Pardeep Singh.
Description: Hoboken, NJ : Wiley-Blackwell, 2021. | Includes index.
Identifiers: LCCN 2021001375 (print) | LCCN 2021001376 (ebook) | ISBN 9781119709695 (hardback) | ISBN 9781119709718 (adobe pdf) | ISBN 9781119709701 (epub)
Subjects: LCSH: Groundwater–Pollution. | Groundwater–Purification. | Groundwater–Quality.
Classification: LCC TD426 .G7145 2021 (print) | LCC TD426 (ebook) | DDC 628.1/14–dc23
LC record available at https://lccn.loc.gov/2021001375
LC ebook record available at https://lccn.loc.gov/2021001376

Cover Design: Wiley
Cover Image: Courtesy of Sughosh Madhav

Set in 9.5/12.5pt STIXTwoText by Straive, Pondicherry, India
Printed and bound by CPI Group (UK) Ltd, Croydon, CR0 4YY

C9781119709695_240521

Contents

Preface

Water is the most significant universal component which is necessary for all living organisms, and this is why it is known as the matrix of life. Water has a finite presence on this planet, and freshwater is even more limited. The entire freshwater supply is only 2.7% of all water present on the Earth's surface. Out of the total freshwater, 68% is frozen in the form of ice and glaciers, 30% of it is groundwater, and a very minimal amount of about 1.3% is found on the Earth's surface in lakes and rivers. This is the reason why dependence for portable and usable water is met through groundwater resources in most parts of the world. In the past few decades, this scant resource has arrived at a crisis position because of over-exploitation and intense irrigation. Groundwater scarcity and pollution remain among the most significant challenges faced by developing countries. Freshwater demand tremendously increased with population, urbanization, and intense irrigation. Because of the insufficient supply of surface water, which is aggravated by easy contamination through sources like untreated wastewater, urbanization, and industrialization, groundwater is preferred for many people in developing countries like India, where the majority of the population uses groundwater resources for household consumption, manufacturing, and farming purposes. Depletion of the water table owing to overexploitation for irrigation and urbanization, groundwater pollution, drying of aquifers, waterlogging, and saltwater intrusion are some of the major issues which need immediate attention. Water scarcity is further aggravated by a looming threat of climate change. With only minimal water present as groundwater, it requires greater attention paid to its management and conservation.

It is essential to recognize the hydrogeochemical features of groundwater and the evolution of the hydrogeochemical process for sustainable development and effective groundwater management. Groundwater quality of any area is a function of physicochemical factors that are significantly governed by geological configurations and human conduct. Natural aspects that have direct control over water composition incorporate rainfall model and quantity, geological characteristics of the basin and aquifer, climatic features, and various water–rock interface procedures in the subsurface environment. Human behaviors that manipulate the water composition comprise management of domestic and industrial effluents and mining and farming actions. Groundwater quality plays a vital role in groundwater protection and quality conservation. Hence, assessment of the groundwater quality is essential not only for use by the present generation but also for future consumption. Therefore, knowledge of the hydrochemistry of water is necessary for assessing groundwater quality in any basin or population area, which affects the suitability of water for domestic, industrial, and irrigation purposes.

Through this book, we have tried to address all the impending problems with a certain degree of solutions to manage and conserve this scarce resource. The book covers holistic issues related to groundwater resources associated with its quality and quantity. The various chapters will help to understand the hydrogeochemistry of groundwater resources with the impact of agriculture, domestic, municipal, and industrial pollution in altering the geochemistry. We will also try to track the quantitative and qualitative challenges faced by water scarcity in the era of rapid urbanization and industrial development. The book also talks about different organic, inorganic, and emerging contaminants causing potential threats to groundwater quality. The focus is also on saltwater intrusion in coastal areas and salinity issues, as well as waterlogging due to overdrafting and pumping. New trends in groundwater contamination remediation measures will help researchers working in water conservation and management.

It will be useful for not only researchers but stakeholders from all sectors, and along with students, industries and governmental agencies who are directly or indirectly associated with groundwater research and management. As for the courses, based on the major topics listed in the features and contents section of this document, a critical discussion can be done in academic and research fields covering broad areas like hydrogeochemistry, groundwater pollution, modelling aspects, remediation, and management. It focuses on groundwater resources challenges, including different sources of ions in groundwater, new remediation measures, and case studies for better understanding of groundwater resources.

This book contains both practical and theoretical latest and broad aspects of groundwater contaminants, their sources, and remediation. An emphasis has been made on the recent research of groundwater resources and their impact across the globe. This book will be useful for undergraduates, university students, and researchers, mostly working in hydrogeochemistry, water pollution, and the effects of anthropogenic activities on groundwater systems. The book is a humble attempt to reflect upon the various aspects of groundwater hydrogeochemistry, pollution, and remediation measures, hoping that it would be a significant addition to the available literature on the topic. The contributors to the book, having different backgrounds, provide holistic literature on the topic, imbibing diverse approaches and perspectives. We express our sincere gratitude to all the contributors and publishers to produce a remarkable and meaningful edited volume on an important issue.

About the Editors

Dr. Sughosh Madhav

Dr. Sughosh Madhav is presently working as Assistant Professor on a guest basis at Shaheed Bhagat Singh College, University of Delhi. He has obtained his master's degree in Environmental Sciences from the Department of Environmental Science, Banaras Hindu University, Varanasi, India. He earned his doctorate from Jawaharlal Nehru University, New Delhi, India. The area of his doctoral research is the environmental impact of textile effluents on groundwater and soil quality. He has published various research papers and book chapters in environmental geochemistry, water pollution, and wastewater remediation.

Dr. Pardeep Singh

Dr. Pardeep Singh is presently working as Assistant Professor in the Department of Environmental Science, PGDAV College University of Delhi, New Delhi, India. He has obtained his master's degree from the Department of Environmental Science, Banaras Hindu University, Varanasi, India. He has obtained his doctorate from the Indian Institute of Technology, Banaras Hindu University, Varanasi, India. The area of his doctoral research is the degradation of organic pollutants through various indigenous isolated microbes and by using multiple types of photocatalysts. He has published more than 45 papers in international journals.

List of Contributors

Narsimha Adimalla
School of Environmental Science and Engineering
Chang'an University
Xi'an, China

Pinki Rani Agrawal
Academy of Scientific and Innovative Research (AcSIR)
CSIR-National Physical Laboratory campus
New Delhi
India

Deeksha Aithani
School of Environmental Sciences
Jawaharlal Nehru University
New Delhi
India

Zeenat Arif
Department of Chemical Engineering and Technology
Indian Institute of Technology (BHU)
Varanasi, Uttar Pradesh
India

Rishikesh Bharti
Department of Civil Engineering
Indian Institute of Technology
Guwahati
Assam
India

Rabin Chakrabortty
Department of Geography
The University of Burdwan
Bardhaman
West Bengal
India

Vinod Kumar Chaudhary
Department of Environmental Sciences
Dr. Ram Manohar Lohia Awadh University
Ayodhya
Uttar Pradesh
India

Indrajit Chowdhuri
Department of Geography
The University of Burdwan
Bardhaman
West Bengal
India

Abhishek Dawar
Department of Biotechnology
Panjab University
Chandigarh
India

Pooja Devi
Academy of Scientific and Innovative Research (AcSIR)
Ghaziabad, Uttar Pradesh
India

Central Scientific Instruments
Organisation
Chandigarh
India

Chandra Shekhar Dwivedi
Department of Geoinformatics, Central
University of Jharkhand
Ranchi, Jharkhand
India

Balaji Etikala
Department of Geology
Sri Venkateswara University
Tirupati, Andhra Pradesh
India

Neera Garg
Academy of Scientific and Innovative
Research (AcSIR)
Ghaziabad, Uttar Pradesh
India
Central Scientific Instruments Organisation
Chandigarh
India

Ajay Giri
Department of Botany
Banaras Hindu University
Varanasi, Uttar Pradesh
India

Deen Dayal Giri
Department of Botany
Maharaj Singh College Saharanpur
Saharanpur, Uttar Pradesh
India

R. N. Jadeja
Department of Environmental Studies
Faculty of Science
The Maharaja Sayajirao University
of Baroda
Vadodara, Gujarat
India

Rishabh Jain
Academy of Scientific and Innovative
Research (AcSIR)
Ghaziabad
Uttar Pradesh
India
Central Scientific Instruments
Organisation
Chandigarh
India

Santosh Kumar Jha
Department of Bioengineering
Birla Institute of Technology
Mesra, Ranchi
Jharkhand
India

Paramdeep Kaur
Department of Biotechnology
Panjab University
Chandigarh
India

P. L. Keshava Kiran Kumar
Department of Geology
Yogi Vemana University
Kadapa
Andhra Pradesh
India

Juhi Khan
Department of Botany
IFTM University
Moradabad
Uttar Pradesh
India

Amrish Kumar
Pollution Ecology Research Laboratory
Department of Botany
Hindu College
Moradabad
Uttar Pradesh
India

Pradeep Kumar
Department of Chemical Engineering & Technology
Indian Insitute of Technology (BHU)
Varanasi
Uttar Pradesh
India

Jyoti Kushawaha
School of Environmental Sciences
Jawaharlal Nehru University
New Delhi
India

Sughosh Madhav
School of Environmental Sciences
Jawaharlal Nehru University
New Delhi
India

P. K. Mishra
Department of Chemical Engineering and Technology
Indian Institute of Technology (BHU) Varanasi
Uttar Pradesh
India

Virendra Kumar Mishra
Institute of Environment and Sustainable Development
Banaras Hindu University
Varanasi, Uttar Pradesh
India

Saumitra Mukherjee
School of Environmental Sciences
Jawaharlal Nehru University
New Delhi
India

Parisa Nikpour
Department of Chemical Engineering
Babol Noshirvani University of Technology
Babol
Iran

Ankita Ojha
Department of Chemistry
IIT (BHU)
Varanasi
Uttar Pradesh
India

Department of Chemistry
Maharaja College
VKSU
Arrah, Bihar
India

Dan Bahadur Pal
Department of Chemical Engineering
Birla Institute of Technology
Mesra
Ranchi
Jharkhand
India

Subodh Chandra Pal
Department of Geography
The University of Burdwan
Bardhaman
West Bengal
India

Manvendra Patel
School of Environmental Sciences
Jawaharlal Nehru University
New Delhi
India

Naveen Patel
Department of Civil Engineering
Institute of Engineering & Technology
Dr. Ram Manohar Lohia Awadh University
Ayodhya
Uttar Pradesh
India

Majid Peyravi
Department of Chemical Engineering
Babol Noshirvani University of Technology
Babol
Iran

Uday Bhan Prajapati
Patanjali Herbal Research Department
Patanjali Research Institute
Haridwar
Uttarakhand
India

Anita Punia
Department of Civil Engineering
Indian Institute of Technology
Guwahati
Assam
India

Vikas Rena
School of Environmental Sciences
Jawaharlal Nehru University
New Delhi
India

Paramita Roy
Department of Geography
The University of Burdwan
Bardhaman
West Bengal
India

Rajesh Saha
Department of Chemical Engineering & Technology
Indian Institute of Technology (BHU) Varanasi
Uttar Pradesh
India

Naresh K. Sethy
Department of Chemical Engineering and Technology
Indian Institute of Technology (BHU) Varanasi
Uttar Pradesh
India

Rahul Sharma
Academy of Scientific and Innovative Research (AcSIR)
CSIR-National Physical Laboratory campus
New Delhi
India

Sushil Kumar Shukla
Department of Transport Science and Technology
Central University of Jharkhand
Ranchi
Jharkhand
India

Anubhuti Singh
Institute of Environment and Sustainable Development
Banaras Hindu University
Varanasi
Uttar Pradesh
India

Baljinder Singh
Department of Biotechnology
Panjab University
Chandigarh
India

Deepali Singh
School of Environmental Sciences
Jawaharlal Nehru University
New Delhi
India

Gurudatta Singh
Institute of Environment and Sustainable Development
Banaras Hindu University
Varanasi
Uttar Pradesh
India

Priyanka Singh
Institute of Environment and Sustainable Development
Banaras Hindu University
Varanasi
Uttar Pradesh
India

Sanchita Singhal
Academy of Scientific and Innovative Research (AcSIR)
CSIR-National Physical Laboratory campus
New Delhi
India

Saurabh Kumar Singh
School of Environmental Sciences
Jawaharlal Nehru University
New Delhi
India

K. S. Sista
Research and Development
Tata Steel, Jamshedpur
Jharkhand
India

Srinivasa Gowd Somagouni
Department of Geology
Yogi Vemana University
Kadapa
Andhra Pradesh
India

Arun Lal Srivastav
Chitkara University School of Engineering and Technology
Chitkara University
Solan, Himachal Pradesh
India

Anupma Thakur
Academy of Scientific and Innovative Research (AcSIR)
Ghaziabad
Uttar Pradesh
India

Central Scientific Instruments Organisation
Chandigarh
India

Amit Kumar Tiwari
Department of Chemical Engineering
Birla Institute of Technology
Mesra, Ranchi
Jharkhand
India

Dhanesh Tiwary
Department of Chemistry
IIT (BHU)
Varanasi
Uttar Pradesh
India

Anamika Tripathi
Pollution Ecology Research Laboratory
Department of Botany
Hindu College
Moradabad
Uttar Pradesh
India

Ashutosh Tripathi
Amity Institute of Environmental Sciences
Amity University
Noida, Uttar Pradesh
India

Vinod Kumar Tripathi
Department of Farm Engineering
Institute of Agricultural Sciences
Banaras Hindu University
Varanasi
Uttar Pradesh
India

B. Verma
Department of Chemical Engineering and Technology
Indian Institute of Technology (BHU)
Varanasi
Uttar Pradesh
India

Anusha Vishwakarma
Department of Environmental Science
Central University of Jharkhand
Ranchi
Jharkhand
India

Chandrashekhar Azad Vishwakarma
TERI School of Advanced Studies
New Delhi
India

Monika Yadav
Department of Environmental Studies
Faculty of Science
The Maharaja Sayajirao University of Baroda
Vadodara
Gujarat
India

1

Geogenic Pollutants in Groundwater and Their Removal Techniques

Jyoti Kushawaha and Deeksha Aithani

School of Environmental Sciences, Jawaharlal Nehru University, New Delhi, India

1.1 Introduction

Water resources are one of the essential resources of nature. Being fluid, water's nature allows it to flow into the low-pressure zone from the high pressure. On the surface, it is found in the geographical form of rivers and streams, and it flows with varying pace depending on the gravity, pressure, and geography of the area. The contact time of water with its geographical space is lower due to pace in surface water flow, and this is the reason it has far fewer or negligible geogenic contaminants compared to anthropogenic contaminants. In contrast, the subsurface water moves at a very slow pace (a few millimetres a day) through the pore spaces or cracks of the soil and rocks. Sometimes the water becomes stagnant, like in perched aquifers and in the impermeable hard rock terrains. This very low movement of aquifer waters (groundwater) allows more time for water to interact with the surrounding natural environment, which may be hard rock, soft rock, or soil and enriched with the geogenic constituents. The groundwater contamination largely depends on the soil geochemistry through which water travelled before reaching the aquifers (Achary 2014a). Hydrological processes are important in governing groundwater contamination. Minerals mobilize in the aquifer system in response to the constituents and minerals present in the rock matrix and their depositional history, along with geochemical conditions (Garduño et al. 2011).

Approximately 1.5 billion people are dependent on underground sources of water (Mukherjee et al. 2012); this is why geogenic pollution has become a major threat to groundwater contamination. Groundwater contaminated by the rock–water interaction has resulted in geologically induced constituents such as As, F, Fe, Mn, Se, Cr, etc. Long-term intake of F-contaminated groundwater leads to severe fluorosis, both dental and skeletal, as well as a range of non-skeletal effects. As-contaminated water caused severe health effects such as arsenicosis, skin cancer, respiratory problems, and other cancers.

Groundwater Geochemistry: Pollution and Remediation Methods, First Edition.
Edited by Sughosh Madhav and Pardeep Singh.

The commonly used remedial techniques for the removal of As and F from groundwater are membrane separation, ion exchange resins, coagulation-precipitation (also known as Nalgonda technique), and adsorption filter beds. It is essential to understand the operational parameters before adopting these potential techniques for remediation such as the local water demand, water quality parameters, initial concentration, and daily basis water use patterns. As-contamination is highest in the Indo-Gangetic plain in the eastern and northeastern parts of India.

Geogenic contaminants, including arsenic, fluoride, and iron are commonly observed in nature. In India, the foremost geogenic contamination in aquifers is Arsenic and Fluoride (Garduño et al. 2011). Other contaminants include nitrate, phosphate, heavy metals and trace metals which may result of the human activities including domestic savage, septic tank, industrial effluents and agricultural practises (Madhav et al. 2018). In India, there are several states and districts which are affected by geogenic contamination such as Arsenic (10 states and 68 districts), Fluoride (20 states and 276 districts), and Iron (24 states and 294 districts) (CGWB 2014).

1.2 Arsenic

Arsenic is overarching with a variable amount in the Earth's crust, mostly in the form of arsenate and arsenite (WHO 2011). In water, generally, it is present as arsenate (As V) in oxidizing condition and as arsenite (As III) form in reducing condition. The fresh biomass, rainfall infringement in the recent alluvium may be the reason for the evacuation of As and Fe from the sediments or soils (Aulakh et al. 2009; Raju et al. 2012). It is introduced in water via different sources such as the dissolution of minerals, rocks, and soil, mining activity (mining waste), mineral smelting, coal combustion, industrial effluents, and through the dry and wet deposition of atmosphere's dust. It is used commercially as the alloy in the formation of semiconductors and in other electricals. Moreover, it used as a preservative for wood, and industrially in the textile, paper, and glass industries for processing. It is also used in pharmaceuticals, food additives, and pesticides to a small extent. The average concentration ranged from 1–2 mg/kg in the continental crust, in which 1.5–3.0 mg/kg was in igneous rock followed by 1.7–400 mg/kg in sedimentary (CGWB 2014).

Arsenic contamination is distinguished in important basin river basins of India, the Ganga and Brahmaputra river basins. These two rivers originated from the Himalayas and create the two significant river basins of India. The As contaminants are prevalent in the lowland area rich with either organic or clayey deltaic sediments in the Bengal basin, as well as sites of the entrenched river channels having the similar facies but sometimes in small pockets in the middle Ganga plain (MGP), i.e. Uttar Pradesh, Bihar, Jharkhand, and West Bengal (Chakraborti et al. 2002; Raju et al. 2012; Saha and Sahu 2016; Tirkey et al. 2017; Kumar et al. 2019). Arsenic induced groundwater deterioration in MGP to the lower Ganga Plain including Uttar Pradesh, Bihar, Jharkhand, and West Bengal and consistently created an adverse impact on human health.

Arsenic is more concentrated in the alluvial aquifers, small lenses in the MGP, low-lying basin of Bengal, and localized contamination through the gneissic aquifers in

Chhattisgarh. Geographically, the large area comes under arsenic contamination from the 89 blocks and 57 blocks in West Bengal and Bihar, respectively. Arsenic contamination is also pervasive sequentially in Uttar Pradesh, Jharkhand, Haryana, Punjab, Manipur, and Assam (Aulakh et al. 2009; Saha and Sahu 2016; Lapworth et al. 2017; Tirkey et al. 2017; Kumar et al. 2019). Most of the cases were reported from the younger alluvial aquifer systems (CGWB 2014).

1.2.1 Health Impact

As poses more adverse health impact to human than animals due to different gastrointestinal absorption. The consumption of As-contaminated water damages human health, including respiratory distress resulting into laryngitis, bronchitis, or rhinitis, cardiovascular effects, and gastrointestinal effects like lips burning, pain while swallowing, abdominal pain, thirst, and nausea. Consumption of the inorganic As increases the risk of lung cancer, including the side effects of it such as headache, lethargy, hallucination, keratosis and hyperpigmentation in skins, seizures, and mental confusion (Mandal et al. 1996; CGWB 2014).

The organic and inorganic arsenic compound has been introduced in the water system through geological and anthropogenic sources. It is available in all geological material in variable concentration. Arsenic mobilization depends on three mechanisms in groundwater, which have been proposed:

i) Oxidation of arsenic-bearing pyrite minerals.
   $FeAsS + 13Fe^{3+} + 8H_2O \rightarrow 14Fe^{2+} + SO42- + 13H^+ + H_3AsO_4(aq)$
ii) Dissolution of As-rich iron oxyhydroxides (FeOOH) under reducing conditions:
   $8FeOOH\text{-}As(s) + CH3COOH + 14H_2CO_3 \rightarrow 8Fe^{2+} + As(d)+ + 16HCO_3^- + 12H_2O$
iii) Release of As present in the aquifer media exchanged with phosphate (H_2PO^-) linked with percolation of phosphate ions into the aquifer when excess application is done in farming practices (Acharya et al. 1999; Pokhrel et al. 2009).

The dissolution of FeOOH under reducing conditions reflected the possible reason for the elevated concentration of arsenic in subsurface water (Harvey et al. 2002). Apart from these, arsenic shows a strong affinity for protein; the biological sources also contribute arsenic though the soil and water ecosystem. Arsenic exhibits a strong affinity for proteins, lipids, and other cellular components and as such, accumulates readily in living tissues (Ferguson and Gavis 1972). Besides this, arsenic concentration was observed high in the aquatic organism through the processes of biomagnification.

1.2.2 Remediation

Several conventional methods are available for arsenic removal from the water system, and the most commonly applied technologies include coprecipitation with adsorption onto coagulated flocs and sorptive media, oxidation method, treatment with lime, membrane filter technique, and ion exchange process. In the oxidation method, the dissolved oxygen converts the mobile arsenite into the less mobile arsenate and dissolved ferrous into ferric ion lowered the arsenic concentration in groundwater (Ahmed 2001; Ayoob et al. 2007).

1.3 Fluoride

Fluorine is highly electronegative in nature and cannot stand isolated in the environment due to its high reactivity. It occurs in the form of oxides in the natural system with the oxidation state of −1. It presents in water as fluoride. Fluoride occurrence in groundwater is predominantly geogenic in nature. Fluoride comes in groundwater by the dissolution of F-bearing rocks, as well as anthropogenic pollution. Fluorite (CaF_2), fluore-apetite ($Ca_5(PO_4)_3F$ and apatite $Ca(PO_4)_3(F/OH/Cl)$, hornblende $Ca_2(Mg,Fe,Al)_5(Al,Si)_8O_{22}(O.H.,F)_2$, and biotite $K(Mg,Fe)_3(AlSi_3\ O_{10})\ (F,OH)_2$ in gneisses are important fluoride-bearing minerals. Moreover, fluorine may also present as the constituent of clay minerals and through rock–water interaction it liberates as fluoride into the subsurface water (Raju et al. 2009; Banerjee et al. 2011). Moreover, leaching of fluoride depends on the alkalinity of the groundwater (Brindha and Elango 2011). In the alkaline environment, the leaching of fluoride will be higher; the alkaline environment is attributed to the dissolution of silicate minerals and the leaching of organic matters from the soil layer (Hoque et al. 2000). Some amount of fluorides may occur in groundwater because of mineral fluorite (CaF_2) dissolution. The reaction is given below (Helgeson 1969):

$$CaF_2 \leftrightarrow Ca + 2F^-$$

If the aquifer system has a high mineral content of calcite, it also supports fluoride dissolution from the fluoride-rich minerals. Therefore F^- will release in water if soil and groundwater have lower calcium content (Kundu and Mandal 2009; Brindha and Elango 2011). The factors that influence the fluoride concentration in groundwater include fluoride-bearing minerals, pH, temperature, anion exchange capacity of aquifer media ($O.H.^-$ for F^-), residence time, porosity, soil structure, depth, groundwater age, and bicarbonates (Grützmacher et al. 2013). The fluorine concentration varies in different rock types such as the igneous rocks (100–>1000 ppm), sedimentary rocks (100 ppm in limestone to 1000 ppm in shales), and in metamorphic rocks (up to 5000 ppm) (International Groundwater Resources Assessment Centre, Report nr. SP 2004-2). Apart from this it leaches from the agricultural activities through phosphatic fertilizers and from the effluents of the ceramic industries in which cay has been used, as well as is present in high amounts in the flying ash from the burning of coal. In geological material, the median fluoride concentration present in the sequence of metamorphic rocks ≥ granitoid ≥ complex rock (Manikandan et al. 2014). Besides geogenic sources, phosphatic fertilizer, cow dung, industrial effluents, and other urban waste are responsible for the fluoride in groundwater (CGWB 2014).

1.3.1 Health Impact

Fluoride upholds healthy teeth and bone development in ranges of 0.7–1.2 mg/L. In developed countries, over 50% of the populations fluoridate water up to this range (Alfredo et al. 2014). However, at higher concentration, i.e. above 1.5 mg/L, it can have disadvantageous health effects as it incorporates into budding enamel crystals and substitutes the hydroxyl ions in the apatite structure. Extended consumption of highly fluoride-contaminated water during the budding phases of life can cause fluorosis problems linked to mottled or brittle teeth, or even more dangerous in the form of extreme skeletal fluorosis linked with porous bone structures. The World Health Organization (WHO) has reported that the consumption of highly fluoridated water (>1.5 mg/L) is a health concern.

1.3.2 Remediation

The generally applied methods for defluoridation are membrane separation, coagulation-precipitation, adsorption, lime softening, and activated alumina (Grützmacher et al. 2013). Hydrous bismuth oxides (HBOs) have been examined as a potential adsorbent for the removal of F^- from the contaminated water or aqueous solution (Srivastav et al. 2013). Reverse osmosis is comparatively insensitive to pH because it is a membrane separation technique (Arora and Evans 2011). However, it necessitates a cautious assessment of water characteristics and pretreatment to avert fouling. In addition to this, resistance and fouling of membrane increases due to the accumulation of rejected species and particles. In poor and rural areas of developing countries, the membrane is seldom considered a suitable technology because of issues like expense, fouling, operational sophistication, and difficulty of intermittent operation. Adoption of these advanced technologies in rural, resource-limited areas of the world is not practical. Hence, a technique called the Nalgonda technique, which was first established by the National Environmental Research Institute in India, is applied as a household treatment. This technique is based on the coagulation-flocculation-sedimentation process of lime and aluminium sulphate (alum)– for fluoride elimination. Adsorption methods for fluoride removal are primarily based on clay (Mahramanlioglu et al. 2002; Chidambaram et al. 2003), charcoal (Mjengera and Mkongo 2002; Medellin-Castillo et al. 2007), and aluminium-based adsorbents (Ghorai and Pant 2004; Sarkar et al. 2007; Alfredo et al. 2014).

1.4 Salinity (Na and Cl)

Commonly salinity problems in groundwater are very prominent in the coastal region, followed by arid and semiarid regions. The coastal areas of Gujarat, Maharashtra, Goa, Kerala, Tamilnadu, Odisha, and West Bengal are facing the problem of saltwater intrusion termed as coastal salinity, and inland salinity problems have been reported in the states of Haryana, Rajasthan, Punjab, and Gujarat, with some limited problems in other states also. Mainly seawater intrusion is responsible for the salinity in groundwater in coastal areas, whereas agricultural wastes, agriculture runoff, heavy uses of fertilizers, and industrial effluents have caused the salinity in arid and semiarid regions. Ion exchange processes, rock–water interaction within subsurface, and the surface water with urban and semi-urban wastes percolates through the soil and enters into the aquifer system and leads to salinity problems.

The Indian subcontinent has a coastline stretched about 7500 km long. Saltwater can intrude laterally or by coming up from the deeper layer when the groundwater level has dropped below the sea level. Moreover, tides and coastal floods may contribute to salinity in water by infiltration. Seawater intrusion incidents are common and have been observed in several states, including in Tamil Nadu, Pondicherry, and Saurashtra in Gujarat (Mondal et al. 2010; Garduño et al. 2011). There is no uptake of sodium salt by the plants.

Only evaporation eliminates the sodium salts from the solution. The most significant source of sodium and chloride in groundwater, particularly in arid and semiarid expanses, is the precipitation of this salt permeating the soil in the shallow water tracts. Na^+ and Cl^- concentration was reported higher in the coastal zone due to saltwater intrusion. Na^+ concentration increases with Cl^- concentration, which resulted in an increase in the weathering of halite minerals in the groundwater. Other sources of Cl^- includes natural weathering of bedrocks, volcanic activity, natural brines, saline intrusions, and atmospheric deposition along with the geographical locations, i.e. coastal/inland areas (Grützmacher et al. 2013).

1.4.1 Health Impacts

No health-based guideline value is suggested by WHO (2011), however Cl^- concentration exceeding the 250 mg/L may lead to a noticeable taste (WHO 2011). But it has been reported that high salinity in irrigation water can damage the crops, affect the plant growth, reduce the soil fertility, and deteriorate the water quality. High chloride may harm the aquatic life through leaf burn, defoliation in sensitive crops, and disturbing the oxygen distribution in water, as well as increasing the metal concentration in water, although health-related issues are not critically observed yet. However, high sodium water can pose heart disease and high blood pressure, predominantly in vulnerable entities.

1.4.2 Remediation

Distillation, membrane technology, including reverse osmosis and microfiltration, ion exchange, and treatment using hydrotalcite are generally applied to remediate the high salinity (Grützmacher et al. 2013).

1.5 Sulphate

Geogenic sources of sulphate are sulphur-bearing minerals (e.g. gypsum, anhydrite, and pyrite (Grützmacher et al. 2013). Naturally occurring sedimentary rocks containing sulphur are pyrite (FeS_2) and gypsum ($CaSO_4.2H_2O$) (Berner 1987; Gourcy et al. 2000). Oxidative weathering of pyrite occurring in alluvial sediments is also one of the causes of high SO_4^{2-} in water (Raju et al. 2011).

$$2FeS_2 + 7O_2 + 2H_2O = 2Fe^{2+} + 4SO_4^{2-} + 4H^+$$

Sulphuric acid reacts with calcium carbonate found in weathered zones and produces calcium sulphate, given by the following reaction:

$$H_2SO_4 + CaCO_3 = H_2CO_3 + CaSO_4$$

Factors influencing the concentration of sulphate are sulphate-rich minerals and reducing condition in the aquifer material (Grützmacher et al. 2013)

1.5.1 Health Impact

SO_4 concentration above 1000 mg/L can cause purgative effects (WHO 2011).

1.5.2 Remediation

A natural method for removing sulphate is an anaerobic reduction in constructed wetlands. Other technical methods include reverse osmosis, N.F., and ion exchange (Grützmacher et al. 2013).

1.6 Heavy Metals

Heavy metal is a term given to a division of metals and metalloids that are differentiated by high atomic weight and specific density, five times higher to water or greater than 4000 kg/m^3 (Hashim et al. 2011; Kura et al. 2018). Heavy metals are toxic even at deficient concentrations and harmful to human health (Madhav et al. 2020). The following heavy metals are addressed as geogenic pollutants in this chapter.

1.6.1 Iron

Iron is the second-most abundant metallic element in the Earth's outer crust, but its concentration in water is usually low. It occurs in many oxidation states, such as 0, +2, +3, and +6. Many mixed-valence compounds, like magnetite and prussian blue, have both Fe (II) and Fe (III) centres (Hashim et al. 2011). Igneous rock minerals such as pyroxenes, the amphiboles, biotite, magnetite, and especially the nesosilicate olivine are rich in iron content. In these minerals, iron mostly exists in the ferrous form (Fe^{3+}) but ferric (Fe^{3+}) form also occurs, as in magnetite, Fe_3O_4. The ferrous polysulfides like pyrite, marcasite, and the less constant species such as mackinawite and greigite might exist in the presence of sulphur and when reducing conditions prevail. Siderite ($FeCO_3$) may form when sulphur is less plentiful, whereas under oxidizing environments, ferric oxides or oxyhydroxides species like haematite, Fe_2O_3, goethite, FeOOH, or other minerals having these compositions generally occur (CGWB 2014).

In soil, it is commonly present in organic waste and as plant debris. The oxidation intensity and pH firmly control the chemical nature of iron and its solubility in water. The interaction amid oxidized iron minerals and organic matter or dissolution of $FeCO_3$ leads to a higher concentration of iron in groundwater. This type of water is clear when withdrawn, but soon it becomes cloudy and then browns due to precipitation of Fe $(OH)_3$ (Hashim et al. 2011; Achary 2014a). Organic matter removes the dissolved oxygen within the sediments which creates reduced conditions in which iron-bearing minerals (siderite/marcasite) has higher solubility causing enriching of the dissolved iron in the groundwater (Applin and Zhao 1989).

The dissolution of Fe in groundwater strongly depends on the concentration of dissolved oxygen and also on the pH of the water to a lesser extent. In groundwater, it generally occurs in two forms as Fe^{2+} and as Fe^{3+}. Iron occurs as Fe^{3+} when the concentration of dissolved oxygen in groundwater is greater than 1–2 mg/L and as Fe^{2+} when dissolved oxygen is in low concentration. At normal pH of water, Fe^{2+} is soluble, and Fe^{3+} is insoluble.

At low pH and dissolved oxygen, iron (also manganese) will dissolve readily in groundwater. Aquifers having higher depth and rich in organic matter typically contain less dissolved oxygen. Decomposition of the organic matter leads to depletion of the oxygen in the water, which results in the dissolution of the iron as Fe^{2+}. In the oxygen-deprived water, when pumped to the surface, a rust-coloured iron mineral forms due to the reaction of dissolved iron and the oxygen (CGWB 2014).

In India, groundwater from a large number of areas is highly contaminated with iron. These areas include west Bengal, Rajasthan, Orissa, Goa, Haryana, Jammu and Kashmir, Andhra Pradesh, Karnataka, and Kerala (Garduño et al. 2011). Higher Fe constituent in the coastal aquifers may be due to the interface of oxidized Fe minerals and organic matter and subsequent dissolution of Fe_2CO_3 at a moderately lower pH.

$$FeCO_3(s) + CO_2 + H_2O = Fe^{+2} + 2HCO_3^-$$

The brown colour of groundwater after extraction from the aquifers is the primary sign of iron presence. The high concentration of Fe in water causes awful taste, staining, turbidity, and equipment problem in the water supply.

1.6.1.1 Health Impact

Permissible limit of Fe in drinking water prescribed by WHO (2011) and BIS (2012) is 300 μg/L. The use of contaminated groundwater having iron concentration above the permissible limit is highly objectionable as it causes health disorders. The reported health disorders due to Fe contaminated groundwater are disorders of the skin, digestive, respiratory and nervous system, kidney, spinal cord, and heart, as well as mental imbalance, miscarriage, and cancer (Achary 2014b). Deficiency of iron causes anaemia, whereas prolonged consumption of drinking water with a high concentration of iron may lead to a liver disease called hemosiderosis. The water with high iron concentration may seem brownish because of the precipitation of ferric hydroxide and taste astringent. The USEPA (US Environmental Protection Agency) maintains that though drinking water having iron may be consumed safely, iron-bearing sediments may comprise trace impurities or harbour bacteria that might be damaging. Iron has nutritional value for human beings as it plays an important role in the formation of haemoglobin protein, enzymes, and also used in cellular metabolism. Lesser storage of iron in the body causes the iron deficit, anaemia, fatigue, and affects the immune system. Iron deficiency in children negatively disturbs mental growth, resulting in irritability and concentration ailment. Chronic consumption of surplus amounts of iron results in an ailment termed iron overload, which occurs due to gene mutation. Iron overload, if untreated, can lead to haemochromatosis, a severe illness that could harm the body's organs. Early symptoms of haemochromatosis are fatigue, weight loss, and joint pain; when not cured appropriately it can cause heart disease, liver complications, and diabetes (Garduño et al. 2011; CGWB 2014; Duggal et al. 2017).

1.6.2 Manganese

Manganese is the 12th most abundant element in the environment, and it is a crucial component of plants and animals. It has a strong positive correlation with iron, and its chemistry is also somewhat like iron as both metals take part in redox reactions in weathering conditions. Manganese substitutes the iron, magnesium, or calcium in silicate structures as it is not a crucial component of any of the common silicate rock minerals. Manganese exists in divalent form in several igneous and metamorphic minerals as a small component, but it is an essential component of basalt. It has occurred in minerals like olivine, pyroxene, and amphibole and a minor amount generally occurred in dolomite and limestone as a substitute for calcium (CGWB 2014).

In soil, manganese arises from mineral weathering and atmospheric deposition instigating from both natural and anthropogenic sources. In soil solution, only the divalent ion form is stable, whereas Mn (III) and Mn (IV) are steady in the solid phase of soil. Soil parameters like acidity, wetness, organic matter content, biological activity, etc., affect the mobility of manganese. The solubility of manganese in the soil is controlled by redox potential and soil pH. Its mobility rises at lower pH or lower redox potential because these favour the reduction of insoluble manganese oxides. Manganese forms bonds with organic matter, oxides, and silicates, which lead to a decrease in solubility at soil pH above 6. Hence high pH and higher organic matter content usually lower the availability and solubility of manganese, whereas in acidic soils with low organic matter, it is readily available. The anaerobic conditions (at pH above 6) and aerobic conditions pH below 5.5 both resulted in an increase of solubility of manganese (CGWB 2014).

The existence of different minerals at the aquifer influence the levels of Mn found in groundwater from natural leaching processes. It usually occurs in deeper wells where the water normally interacts with the rock for an extended time. In groundwater, manganese often occurs together with iron as it originated from ferromagnesian, but its concentration is usually lower than iron. Though it is crucial for humans and other living beings, at higher concentration, it is lethal. The higher concentration of manganese is stated primarily from West Bengal, Tamil Nadu, Orissa, UP, and Bihar (Garduño et al. 2011).

1.6.2.1 Health Impact

The permissible limit of Mn in drinking water given by WHO (2011) and BIS (2012) is 300 and 100 μg/L, respectively. Inhalation or contact with a higher concentration of Mn can cause damage to the central nervous system (Singh et al. 2011). It can readily accumulate in the brain, particularly in the basal ganglia, and can result in an irretrievable neurological syndrome like Parkinson's disease. Comparatively higher doses of manganese affect DNA replication and lead to mutations in the microorganism and mammalian cells. In mammalian cells, manganese causes DNA impairment and chromosome abnormalities. The higher concentration of it affect fertility in mammals and are deadly to the embryo and foetus (Singh et al. 2011).

1.6.3 Chromium

Chromium mostly occurs in trivalent and hexavalent forms depending on pH. In shallow aquifers, Cr (VI) is the dominant and toxic form, and its major species include chromate CrO_4^{-2} and dichromate $Cr_2O_7^{-2}$ (especially Ba^{2+}, Pb^{+2}, and Ag^{+}) (Hashim et al. 2011). At low pH (<4), Cr (III) is the dominant form of Cr. In soil organic matter,

S^{2-} and Fe^{2+} ions under anaerobic conditions lead to the reduction of Cr (VI) to Cr (III) (Hashim et al. 2011). The Cr (VI) has higher mobility and toxicity than Cr (III), and it is more problematic to eliminate Cr (VI) in water. In water, chromium (III) is typically present as the free-ion state, but it is also related to hydroxide ions depending on pH and forming chromium $(OH)^{2+}$, chromium $(OH)^{2+}$, chromium (OH) 3, and chromium $(OH)^{4-}$. The major Cr species found in aqueous solution are $SO_4{}^{2-}$, $NH_4{}^{+}$, and CN; these are formed by complexes of Cr^{3+} with organic and inorganic ligands. The equilibrium of Cr among the two oxidation forms has relied on oxidation kinetics which further depends on biochemical settings, such as pH, redox, and nutrient levels that direct microbial activity (CGWB 2014).

Chromium has been found in diverse parts of India. Its source has been generally attributed to anthropogenic inputs. However, chromite deposits in the Sukinda area of Orissa are identified as geogenic sources. In the Sukinda area, the lateralization progress including oxidation and change of the serpentines generates alkaline pore water which enables the formation of Cr (VI) from the inactive chromites and leads to chromium pollution of water in and around the area (CGWB 2014).

1.6.3.1 Health Impact

Chromium (III) has a particular transportation mechanism; because of this, only limited quantities of it pass in the cells. However, numerous *in vitro* studies specified that higher levels of Cr (III) in the cell could cause DNA impairment. Reasonable uptake of Cr (III) via dietary supplements causes no risk, whereas its severe oral toxicity varies between 1.5 and 3.3 mg/kg.

The severe toxicity of chromium (VI) is because of its strong oxidation properties. Once it enters the bloodstream, it harms the kidneys, the liver, and blood cells by oxidation reactions (CGWB 2014). Excess Cr (VI) can cause damage to the nervous system, fatigue, irritability, skin allergies, dermatitis, dermal necrosis, and dermal corrosion (Singh et al. 2011).

1.6.4 Remediation of Heavy Metals

Heavy metal can be eliminated from groundwater by applying various chemical, biological/biochemical, and physicochemical remediation technologies. Chemical methods include a reduction (by using dithionites, H_2S gas, iron-based technologies such as zerovalent ion, iron salts), chemical flushing (using soil washing, chelate flushing, ion exchange) and in situ chemical fixation. Biological/biochemical remediation has been carried out by biological activity in the subsurface, biosorption (by using biosurfactant, metal uptake by the organism, cellulosic material and agricultural wastes) and enhanced bio restoration. Physico-chemical remediation applied by using adsorption/absorption/filtration (using surfactants, membrane separation, synthetic activated carbon, industrial byproducts and waste, ferrous materials, iron-based soil amendments, mineral products), electrokinetic treatment and permeable reactive barriers, which include sorption (in red mud, activated carbon, zeolite, iron sorbent), precipitation via zero-valent ion, complexing agents, atomized slag, caustic magnesia, and biological barriers (denitrification and sulphate reduction, microbes) (Hashim et al. 2011).

1.7 Selenium

Selenium is a crucial element, but at higher concentrations, it is toxic (Sharma et al. 2015). Natural sources of Se account for 50–65% of total emissions, globally. Natural sources of atmospheric selenium are crustal erosion, which causes its suspension in soil and dust, biologically originated volatile organo-selenium compounds, volcanoes, and aerosol particles related with sea spray (Nriagu and Pacyna 1988; Sharma et al. 2015). Sedimentary rocks have a higher concentration of Se than the igneous rocks. Its concentrations in underlying rocks are strongly correlated with its concentration in soils. Soils rich in Se are frequently originated from marine deposits. In the oxic environment of surface soils, the main species of Se are selenite (Se (IV)) and selenate (Se (VI)) (Sharma et al. 2015).

It is a minor constituent of natural waters, and its level in water is generally found to be less than 10 mg/L (Sohrin and Bruland 2011). Laboratory studies suggested that a nonvolatile dimethyl selenonium ion can probably be changed into volatile dimethyl selenide at neutral pH (Cooke and Bruland 1987). This can be attributed as its pathway for the in-situ origin of dimethyl selenide in natural waters. Selenium chemistry in groundwater is controlled by four crucial variables. These are the prime source, the transportation in solution, the courses of mobilization, and retaining at the water–solid interface and its level in groundwater under a semiarid and arid environment (Nicolli et al. 2012). Selenium in groundwater is restricted to particular locations and has been reported from dispersed areas in Himachal Pradesh and Punjab (CGWB 2014; Lapworth et al. 2017).

1.7.1 Health Impact

It has been observed that acute oral doses of selenite and other selenium compounds cause symptoms such as nausea, diarrhoea, abdominal pain, chills, tremor, numbness in limbs, irregular menstrual bleeding, and marked hair loss. The higher level of selenium in humans leads to hair loss and nail drop, whereas in plants, it causes snow-white chlorosis which appears on juvenile leaves and sheaths. Some of these leaves changed to light pink or developed purple-white tips having green mid-veins (Aulakh et al. 2009). In northwestern India, its toxicity caused snow-white chlorosis in wheat and sugar cane while chronic selenosis in faunas and humans (CGWB 2014).

It accumulates in the tissues of organisms and shows the biomagnification at a higher trophic level. Se accumulates in the animals when they ingest large quantities of plants that are uptaking higher concentrations of selenium. In many areas, Se concentration is found to be very high in aquatic organisms because of irrigation runoff. Selenium could cause reproductive failure and birth imperfections at an elevated concentration in animals. The most reliable clinical appearance indicated by animals is losing body state and hair loss, necrosis of the tip of the tail, reluctance to move, stiff gate, and overgrowth of hoof followed by progress of cracks and irregularities in horn development. Food or water are common exposure pathways of selenium in human beings. Typical symptoms of selenium toxicity in humans are hair and finger- and toenail loss, continuous worsening of health, queasiness, headache, tooth deterioration, tint of teeth, and nails with fragility and longitudinal strips (CGWB 2014).

Table 1.1 Geogenic contaminants in different parts of the country (India).

Study area	As	F	SO_4	Na	Cl	Se	Cr	Mn	Fe	References
Cuttack City, Odisha, India									0.03–1.5 ppm	Achary (2014a)
Bhuvaneshwar City, Odisha, India									0.32–7.7 ppm	Achary (2014a)
Gangetic Plain, Jharkhand, India	1–133 ppb in premonsoon, 2–98 ppb in monsoon and 7–115 ppb in postmonsoon									Alam et al. (2016)
Punjab, India	4–688 ppb	0–10.09 ppm				0.2–69.5 ppb				Aulakh et al. (2009)
Tripura, India	0.025 ppm		58.53 ppm					2.68 ppm	14.67 ppm	Banerjee et al. (2011)
Golaghat District, Assam, India	0.074–0.128 ppm							0.02–2.5 ppm	3.5–5.9 ppm	Chetia et al. (2011)
Kanpur, India	Not detected						0.02–0.04 ppm			Dotaniya et al. (2017)
Northeast Rajasthan, India	0.62–1.60 ppb						2.04–3.48 ppm	0.004–0.072 ppm	0.045–0.151 ppm	Duggal et al. (2017)
Uranium Mining Site, Jharkhand, India								0.01–1.3 ppm	0.06–5.3 ppm	Giri et al. (2010)

Northern Maharashtra, India								0.026–0.361 ppm	0–0.096 ppm	Golekar et al. (2013)
Jhajjar district, Haryana, India		1.1 ppm			214.2 ppm					Gupta and Misra (2018)
Sehore District, Madhya Pradesh, India							0–27 ppb	0–121 ppb	0–3330 ppb	Jinwal et al. (2009)
Himanchal Pradesh, India	2.25–117.5 ppb						1–322.58 ppm	0.02–0.041 ppm	0.12–0.60 ppm	Kashyap et al. (2018)
Medak District, Andhra Pradesh, India	2.9–1257 ppb						2.2–74.7 ppb	2.9–14 971 ppb	38.8–576	Krishna et al. (2009)
Upper Gangetic basin, India	1–36 ppb							0.0011–0.415 ppm		Kumar et al. (2019)
Ghaziabad, India	0.04–0.54 mg/L			9.20–3887 ppm			4.51–7.09 ppm	0.0–0.350 ppm	BDL (below detection limit)	Kumari et al. (2014)
Bist-Doab region of Punjab, northwest India,						10–40 ppb				Lapworth et al. (2017)
East Bokaro Coalfield, Jharkhand, India	0.01–4.0 ppb						2.5–25.6 ppb	11.0–397.1 ppb	194–1685 ppb	Mahato et al. (2016)

(*Continued*)

Table 1.1 (Continued)

Study area	As	F	SO_4	Na	Cl	Se	Cr	Mn	Fe	References
Krishnagiri district Tamilnadu, India		0.5–5.45 ppm	10.76 ppm	31.4–220.6 ppm	49.59–292.7 ppm					Manikandan et al. (2014)
West Bengal, India	0.2–1219 ppb			2.9–64.6 mg/L		0.2–7 ppb	0.1–1.6 ppb	25.1–1538 ppb	0.06–18.5 ppm	Rahman et al. (2015)
Hakinaka Taluk, India	BDL						0.01–0.05 ppm		0–0.175 ppm	Rajappa et al. (2010)
Sonbhadra District, Uttar Pradesh, India		0.4–5.6 ppm in the pre-monsoon, 0.1–6.7 ppm in the post-monsoon	10–190 ppm in premonsoon, 25–195 ppm in postmonsoon	55.2 ppm in pre and 62.9 ppm in postmonsoon						Raju et al. (2012)
Sonbhadra, Uttar Pradesh, India		0.3–11 ppm								Raju (2017)
Telangana State, India		0.99	51.53 ppm	117.87 ppm	119.27 ppm					Rao et al. (2018)
Andhra Pradesh, India	0.011 and 0.081 ppm					0.008–0.084 ppm	0.010–0.091 ppm			Reddy et al. (2012)
North district of Delhi, India	0.0003–0.1 ppm	0.05–7.14 ppm	25–1202 ppm	20–2000 ppm	11–4433 ppm					Sarkar and Shekhar (2013)
Vaishali and Bhagalpur District, Bihar, India	20 ppb in Vaishali and 143 ppb in Bhagalpur									Singh et al. (2014)

Bhalswa landfill, Delhi, India								0.01–1.90 ppm	0.64–6.34 ppm	Srivastava and Ramanathan (2008)
Ballia and Ghazipur, Uttar Pradesh, India	43.75–620.75 ppb									Srivastava and Sharma (2013)
Ranchi city, Jharkhand, India	0.0–200 ppb	0–2.19 ppm	0–268 ppm	18.2–303 ppm	2–200 ppm	0–140 ppb		0.0–4200 ppb		Tirkey et al. (2017)
Parichha, Jhansi, India							0.036–0.061 ppm	0.013–0.178 ppm	0.186–11.98 ppm	Verma et al. (2016)
Puri city, Odisha, India		0.01–2.04 ppm	7–320 ppm		3–11 222 ppm				0.04–3.5 ppm	Vijay et al. (2011)
Kadava River basin, Nashik, India							34–3516 ppb	0.013–0.142 ppm	0.092–3.558 ppm	Wagh et al. (2018)

1.7.2 Remediation

Various techniques might be adopted to lessen the concentration of Se in drinking water. The commonly applied techniques are ion exchange, activated alumina (AA), reverse osmosis (RO), and distillation. Ion (anion) exchange can decrease selenium by 90–95%, as selenate ion is highly favoured. Though oxidation of Se (IV) is more problematic, it can be done easily with the help of free chlorine. Se (IV) can be transformed to Se (VI) within five minutes by maintaining the concentration of free chlorine to 2 mg/L. The pH between 6.5 and 8 is observed to be an ideal rate of oxidation. Removal rates for other methods used for selenium elimination are distillation (>98% removal), activated alumina (85–95% removal), and reverse osmosis (RO) (>90% removal) (CGWB 2014).

1.8 Conclusion

India has different biogeographical regions including the Himalayan regions, desert, semiarid areas, Western Ghats, Deccan Plateau, Gangetic plain, northeast India, islands and coastal regions. Based on this study, it can be seen that most of the geographical areas of India are facing geogenic contamination in groundwater. Arsenic and fluoride contamination have been reported in bulk in the Gangetic Plain. Salinity problems are more concentrated in the coastal, arid, and semiarid regions, while heavy metal contamination has been observed more or less everywhere in India. There is a need for continuous monitoring of the groundwater parameters to check the concentration of the geogenic contaminant in the water system. The remediation method should be applied in the regions of the geogenic contaminants to cope with the present and future worst situations of health-related problems.

References

Achary, G.S. (2014a). Studies on ground water pollution due to iron content in Cuttack city, Odisha, India. *International Journal of Current Engineering and Technology* 2: 86–89.

Achary, G.S. (2014b). Studies on ground water pollution due to iron content in Bhubaneswar, Odisha, India. *International Journal of Current Engineering and Technology* 4 (1): 88–93.

Acharya, S.K., Chakraborty, P., Lahiri, S. et al. (1999). Arsenic poisoning in the Ganges Delta. *Nature* 401: 545.

Ahmed, M.F. (2001, May). An overview of arsenic removal technologies in Bangladesh and India. Proceedings of BUET-UNU international workshop on technologies for arsenic removal from drinking water, Dhaka (pp. 5–7).

Alfredo, K.A., Lawler, D.F., and Katz, L.E. (2014). Fluoride contamination in the Bongo District of Ghana, West Africa: geogenic contamination and cultural complexities. *Water International* 39 (4): 486–503.

Alam, M.O., Shaikh, W.A., Chakraborty, S. et al. (2016). Groundwater arsenic contamination and potential health risk assessment of Gangetic Plains of Jharkhand, India. *Exposure and Health* 8 (1): 125–142.

Applin, K.R. and Zhao, N. (1989). The kinetics of Fe (II) oxidation and well screen encrustation. *Groundwater* 27 (2): 168–174.

Arora, A. and Evans, R.W. (2011). Dental caries in children: a comparison of one non-fluoridated and two fluoridated communities in NSW. *New South Wales Public Health Bulletin* 21 (12): 257–262.

Aulakh, M.S., Khurana, M.P.S., and Singh, D. (2009). Water pollution related to agricultural, industrial, and urban activities, and its effects on the food chain: case studies from Punjab. *Journal of New Seeds* 10 (2): 112–137.

Ayoob, S., Gupta, A.K., and Bhakat, P.B. (2007). Performance evaluation of modified calcined bauxite in the sorptive removal of Arsenic (III) from aqueous environment. *Colloids and Surfaces A: Physicochemical and Engineering Aspects* 293 (1–3): 247–254.

Banerjee, S., Das, B., Umlong, I.M. et al. (2011). Heavy metal contaminants of underground water in Indo Bangla border districts of Tripura, India. *International Journal of ChemTech Research* 3 (1): 516–522.

Berner, R.A. (1987). Models for carbon and sulfur cycles and atmospheric oxygen; application to Paleozoic geologic history. *American Journal of Science* 287 (3): 177–196.

Brindha, K. and Elango, L. (2011). Fluoride in groundwater: causes, implications and mitigation measures. *Fluoride Properties, Applications and Environmental Management* 1: 111–136.

Central Ground Water Board (2014). *Concept Note on Geogenic Contamination of Ground Water in India*. Faridabad, Haryana: Bujal Bhawan, NH-IV www.cgwb.gov.in.

Chakraborti, D., Rahman, M.M., Paul, K. et al. (2002). Arsenic calamity in the Indian subcontinent: what lessons have been learned? *Talanta* 58 (1): 3–22.

Chidambaram, S., Ramanathan, A.L., and Vasudevan, S. (2003). Fluoride removal studies in water using natural materials. *Water SA* 29 (3): 339–344.

Chetia, M., Chatterjee, S., Banerjee, S. et al. (2011). Groundwater arsenic contamination in Brahmaputra river basin: a water quality assessment in Golaghat (Assam), India. *Environmental Monitoring and Assessment* 173 (1–4): 371–385.

Cooke, T.D. and Bruland, K.W. (1987). Aquatic chemistry of selenium: evidence of biomethylation 1. *Environmental Science and Technology* 21: 1214–1219.

Dotaniya, M.L., Meena, V.D., Rajendiran, S. et al. (2017). Geo-accumulation indices of heavy metals in soil and groundwater of Kanpur, India under long term irrigation of tannery effluent. *Bulletin of Environmental Contamination and Toxicology* 98 (5): 706–711.

Duggal, V., Rani, A., Mehra, R., and Balaram, V. (2017). Risk assessment of metals from groundwater in Northeast Rajasthan. *Journal Geological Society of India* 90: 77–84. https://doi.org/10.1007/s12594-017-0666-z.

Ferguson, J.F. and Gavis, J. (1972). A review of the arsenic cycle in natural waters. *Water Research* 6: 1259–1274.

Garduño, H., Romani, S., Sengupta, B. et al. (2011). *India Groundwater Governance Case Study. Technical Report*. Washington, DC: World Bank.

Ghorai, S. and Pant, K.K. (2004). Investigations on the column performance of fluoride adsorption by activated alumina in a fixed-bed. *Chemical Engineering Journal* 98 (1–2): 165–173.

Giri, S., Singh, G., Gupta, S.K. et al. (2010). An evaluation of metal contamination in surface and groundwater around a proposed uranium mining site, Jharkhand, India. *Mine Water and the Environment* 29 (3): 225–234.

Golekar, R.B., Patil, S.N., and Baride, M.V. (2013). Human health risk due to trace element contamination in groundwater from the Anjani and Jhiri river catchment area in northern Maharashtra, India. *Earth Sciences Research Journal* 17 (1): 17–23.

Gourcy, L., de Paulet, F.C., and Laurent, A. (2000). Sulfur origin and influences of water level variation on SO4 concentration in groundwater of the transboundary carboniferous limestone aquifer (Belgium, France). *Procedia Earth and Planetary Science* 7: 309–312.

Grützmacher, G., Kumar, P.S., Rustler, M. et al. (2013). Geogenic groundwater contamination – definition, occurrence and relevance for drinking water production. *Zentralblatt für Geologie und Paläontologie, Teil I* 1: 69–75.

Gupta, R. and Misra, A.K. (2018). Groundwater quality analysis of quaternary aquifers in Jhajjar district, Haryana, India: focus on groundwater fluoride and health implications. *Alexandria Engineering Journal* 57: 375–381.

Harvey, C.F., Swartz, C.H., Badruzzaman, A.B.M. et al. (2002). Arsenic mobility and groundwater extraction in Bangladesh. *Science* 298 (5598): 1602–1606.

Hashim, M.A., Mukhopadhyay, S., Sahu, J.N., and Sengupta, B. (2011). Remediation technologies for heavy metal contaminated groundwater. *Journal of Environmental Management* 92 (10): 2355–2388.

Helgeson, H.C. (1969). Thermodynamics of hydrothermal systems at elevated temperatures and pressures. *American Journal of Science* 267: 729–780.

Hoque, B.A., Mahmood, A.A., Quadiruzzaman, M. et al. (2000). Recommendations for water supply in arsenic mitigation: a case study from Bangladesh. *Public Health* 114 (6): 488–494.

Jinwal, A., Dixit, S., and Malik, S. (2009). Some trace elements investigation in ground water of Bhopal and Sehore district in Madhya Pradesh: India. *Journal of Applied Sciences and Environmental Management* 13 (4): 47–50.

Kashyap, R., Verma, K.S., Uniyal, S.K., and Bhardwaj, S.K. (2018). Geospatial distribution of metal(loid)s and human health risk assessment due to intake of contaminated groundwater around an industrial hub of Northern India. *Environmental Monitoring and Assessment* 190: 136.

Krishna, A.K., Satyanarayanan, M., and Govil, P.K. (2009). Assessment of heavy metal pollution in water using multivariate statistical techniques in an industrial area: a case study from Patancheru, Medak District, Andhra Pradesh, India. *Journal of Hazardous Materials* 167 (1–3): 366–373.

Kumar, M., Nagdev, R., Tripathi, R. et al. (2019). Geospatial and multivariate analysis of trace metals in tubewell water using for drinking purpose in the upper Gangetic basin, India: heavy metal pollution index. *Groundwater for Sustainable Development* 8: 122–133.

Kumari, S., Singh, A.K., Verma, A.K., and Yaduvanshi, N.P.S. (2014). Assessment and spatial distribution of groundwater quality in industrial areas of Ghaziabad, India. *Environmental Monitoring and Assessment* 186 (1): 501–514.

Kundu, M.C. and Mandal, B. (2009). Agricultural activities influence nitrate and fluoride contamination in drinking groundwater of an intensively cultivated district in India. *Water, Air, and Soil Pollution* 198 (1–4): 243–252.

Kura, N.U., Ramli, M.F., Sulaiman, W.N.A. et al. (2018). An overview of groundwater chemistry studies in Malaysia. *Environmental Science and Pollution Research* 25 (8): 7231–7249.

Lapworth, D.J., Krishan, G., MacDonald, A.M., and Rao, M.S. (2017). Groundwater quality in the alluvial aquifer system of Northwest India: new evidence of the extent of anthropogenic and geogenic contamination. *Science of the Total Environment* 599: 1433–1444.

Madhav, S., Ahamad, A., Kumar, A. et al. (2018). Geochemical assessment of groundwater quality for its suitability for drinking and irrigation purpose in rural areas of Sant Ravidas Nagar (Bhadohi), Uttar Pradesh. *Geology, Ecology, and Landscapes* 2 (2): 127–136.

Madhav, S., Ahamad, A., Singh, A.K. et al. (2020). Water pollutants: sources and impact on the environment and human health. In: *Sensors in Water Pollutants Monitoring: Role of Material* (eds. D. Pooja, P. Kumar, P. Singh and S. Patil), 43–62. Singapore: Springer.

Mahato, M.K., Singh, P.K., Tiwari, A.K., and Singh, A.K. (2016). Risk assessment due to intake of metals in groundwater of East Bokaro coalfield, Jharkhand, India. *Exposure and Health* 8 (2): 265–275.

Mahramanlioglu, M., Kizilcikli, I., and Bicer, I.O. (2002). Adsorption of fluoride from aqueous solution by acid treated spent bleaching Earth. *Journal of Fluorine Chemistry* 115 (1): 41–47.

Manikandan, S., Chidambaram, S., Ramanathan, A.L. et al. (2014). A study on the high fluoride concentration in the magnesium-rich waters of hard rock aquifer in Krishnagiri district, Tamilnadu, India. *Arabian Journal of Geosciences* 7 (1): 273–285.

Mandal, B.K., Chowdhury, T.R., Samanta, G. et al. (1996). Arsenic in groundwater in seven districts of West Bengal, India–the biggest arsenic calamity in the world. *Current Science* 70: 976–986.

Medellin-Castillo, N.A., Leyva-Ramos, R., Ocampo-Perez, R. et al. (2007). Adsorption of fluoride from water solution on bone char. *Industrial and Engineering Chemistry Research* 46 (26): 9205–9212.

Mjengera, H., and Mkongo, G. (2002). Appropriate technology for use in fluoritic areas in Tanzania, 3rd Waternet. In WARFSA Symposium on water demand management for sustainable use of water resources, University of Dar Es Salaam.

Mondal, N.C., Singh, V.P., Singh, V.S., and Saxena, V.K. (2010). Determining the interaction between groundwater and saline water through groundwater major ions chemistry. *Journal of Hydrology* 388 (1–2): 100–111.

Mukherjee, P., Singh, C.K., and Mukherjee, S. (2012). Delineation of groundwater potential zones in arid region of India – a remote sensing and GIS approach. *Water Resources Management* 26 (9): 2643–2672.

Nicolli, H.B., Bundschuh, J., Blanco, M.D.C. et al. (2012). Arsenic and associated trace-elements in groundwater from the Chaco-Pampean plain, Argentina: results from 100 years of research. *Science of the Total Environment* 429: 36–56.

Nriagu, I.O. and Pacyna, I.M. (1988). Quantitative assessment of worldwide contamination of air, water and soils by trace metals. *Nature (London)* 333: 134–139.

Pokhrel, D., Bhandari, B.S., and Viraraghavan, T. (2009). Arsenic contamination of groundwater in the Terai region of Nepal: an overview of health concerns and treatment options. *Environment International* 35 (1): 157–161.

Rahman, M.M., Dong, Z., and Naidu, R. (2015). Concentrations of Arsenic and other elements in groundwater of Bangladesh and West Bengal, India: potential cancer risk. *Chemosphere* 139: 54–64.

Rajappa, B., Manjappa, S., and Puttaiah, E.T. (2010). Monitoring of heavy metal concentration in groundwater of Hakinaka Taluk, India. *Contemporary Engineering Sciences* 3 (4): 183–190.

Raju, N.J. (2017). Prevalence of fluorosis in the fluoride enriched groundwater in semiarid parts of eastern India: geochemistry and health implications. *Quaternary International* 443: 265–278.

Raju, N.J., Dey, S., and Das, K. (2009). Fluoride contamination in groundwaters of Sonbhadra district, Uttar Pradesh, India. *Current Science* 96: 979–985.

Raju, N.J., Shukla, U.K., and Ram, P. (2011). Hydrogeochemistry for the assessment of groundwater quality in Varanasi: a fast-urbanizing center in Uttar Pradesh, India. *Environmental Monitoring and Assessment* 173 (1–4): 279–300.

Raju, N.J., Dey, S., Gossel, W., and Wycisk, P. (2012). Fluoride hazard and assessment of groundwater quality in the semiarid Upper Panda River basin, Sonbhadra district, Uttar Pradesh, India. *Hydrological Sciences Journal* 57 (7): 1433–1452.

Rao, N.S., Sunitha, B., Rambabu, R. et al. (2018). Quality and degree of pollution of groundwater, using PIG from a rural part of Telangana State, India. *Applied Water Science* 8 (8): 227.

Reddy, V.H., Prasad, P.M.N., Reddy, A.R., and Reddy, Y.R. (2012). Determination of heavy metals in surface and groundwater in and around Tirupati, Chittoor (Di), Andhra Pradesh, India. *Der Pharma Chemica* 4 (6): 2442–2448.

Saha, D. and Sahu, S. (2016). A decade of investigations on groundwater arsenic contamination in Middle Ganga Plain, India. *Environmental Geochemistry and Health* 38 (2): 315–337.

Sarkar, A. and Shekhar, S. (2013). An assessment of groundwater quality of lesser contaminated aquifers in North District of Delhi. *Proceeding of the Indian National Science Acadamy* 79: 235–243.

Sarkar, M., Banerjee, A., Pramanick, P.P., and Sarkar, A.R. (2007). Design and operation of fixed bed laterite column for the removal of Fluoride from water. *Chemical Engineering Journal* 131 (1–3): 329–335.

Singh, R., Gautam, N., Mishra, A., and Gupta, R. (2011). Heavy metals and living systems: an overview. *Indian Journal of Pharmacy* 43 (3): 246–253.

Singh, S.K., Ghosh, A.K., Kumar, A. et al. (2014). Groundwater arsenic contamination and associated health risks in Bihar, India. *International Journal of Environmental Research* 8 (1): 49–60.

Sharma, V.K., McDonald, T.J., Sohn, M. et al. (2015). Biogeochemistry of selenium. A review. *Environmental Chemistry Letters* 13 (1): 49–58.

Sohrin, Y. and Bruland, K.W. (2011). Global status of trace elements in the ocean. *TrAC Trends in Analytical Chemistry* 30 (8): 1291–1307.

Srivastav, A.L., Singh, P.K., Srivastava, V., and Sharma, Y.C. (2013). Application of a new adsorbent for fluoride removal from aqueous solutions. *Journal of Hazardous Materials* 263: 342–352.

Srivastava, S.K. and Ramanathan, A.L. (2008). Geochemical assessment of groundwater quality in vicinity of Bhalswa landfill, Delhi, India, using graphical and multivariate statistical methods. *Environmental Geology* 53 (7): 1509–1528.

Srivastava, S. and Sharma, Y.K. (2013). Arsenic occurrence and accumulation in soil and water of eastern districts of Uttar Pradesh, India. *Environmental Monitoring and Assessment* 185 (6): 4995–5002.

The Bureau of Indian Standards (BIS) (2012). *Indian Standard Drinking Water – Specification (Second Revision)*. NewDelhi: Publication Unit, BIS; 2012.

Tirkey, P., Bhattacharya, T., Chakraborty, S., and Baraik, S. (2017). Assessment of groundwater quality and associated health risks: a case study of Ranchi city, Jharkhand, India. *Groundwater for Sustainable Development* 5: 85–100.

Verma, C., Madan, S., and Hussain, A. (2016). Heavy metal contamination of groundwater due to fly ash disposal of coal-fired thermal power plant, Parichha, Jhansi, India. *Cogent Engineering* 3 (1): 1179243.

Vijay, R., Khobragade, P., and Mohapatra, P.K. (2011). Assessment of groundwater quality in Puri City, India: an impact of anthropogenic activities. *Environmental Monitoring and Assessment* 177 (1–4): 409–418.

Wagh, V.M., Panaskar, D.B., Mukate, S.V. et al. (2018). Health risk assessment of heavy metal contamination in groundwater of Kadava River Basin, Nashik, India. *Modeling Earth Systems and Environment* 4 (3): 969–980.

WHO Guidelines for Drinking Water Quality (2011). 4th Edition Vol. 2 Health Criteria and other supporting information.

2

Fluoride Contamination in Groundwater, Impacts, and Their Potential Remediation Techniques

Monika Yadav[1], *Gurudatta Singh*[2], *and R.N. Jadeja*[1]

[1] *Department of Environmental Studies, Faculty of Science, The Maharaja Sayajirao University of Baroda, Vadodara, Gujarat, India*
[2] *Institute of Environment and Sustainable Development, Banaras Hindu University, Varanasi, Uttar Pradesh, India*

2.1 Introduction

Groundwater is the major source for drinking and irrigation purposes in Asian countries like Nepal, Bangladesh, China, India, and Pakistan (Al-Hatim et al. 2015; Raj and Shaji 2017). Annual consumption of groundwater by these countries is around 300 km^3, equivalent to 50% of the total consumption of the world (Shah et al. 2003). Asian countries have a population size of 4.5 billion which constitutes more than half of the total human population of world (WPR 2018). Thus, the large population size and overexploitation of water results in enhanced requirement, as well as demand of groundwater for the purpose of irrigation and drinking that ultimately results in reduction of the level of groundwater in Asia (Gleeson et al. 2012; Gupta et al. 2013; Alhababy and Al 2015). Irrigation utilizes around 85% of Asia's groundwater and is considered to be the main reason behind groundwater depletion (FAO 2013). This groundwater contains various geochemical components such as iron, fluoride, arsenic concentration, etc., which may lead to some serious health ailments. Presence of fluoride anion at higher concentrations, i.e. more than 1.5 mg/L in drinking water, might result in dental and skeletal fluorosis (Vithanage and Bhattacharya 2015). However, at lower concentration, F^- in drinking water is proven to be beneficial for consumers as it reduces dental cavities (Guissouma et al. 2017; Raj and Shaji 2017; Yadav et al. 2015). The consumption of fluoride-contaminated groundwater leads to endemic fluorosis, which causes dangerous health issues among a large percentage of population in India and thus becomes a very challenging and extensively studied national health problem (Rudra 2012).

Fluoride is one of the extensively present contaminants dispersed either through natural or artificial means. Naturally, fluoride contamination occurs due to its dissolution from rocks and soils into the groundwater (Podgorski et al. 2018; Barathi et al. 2019). Anthropogenic activities like industrial effluents released from aluminum and steel

Groundwater Geochemistry: Pollution and Remediation Methods, First Edition.
Edited by Sughosh Madhav and Pardeep Singh.

industries, glass and semiconductor manufacturing industries, fertilizer, and electroplating industries also lead to fluoride contamination in groundwater (Bhatnagar et al. 2011). Some other sources, like wastewater released from electronics, toothpaste, and insecticide manufacturing industries also result in enhanced levels of fluoride in groundwater (Jagtap et al. 2012).

The fluoride ion occurs naturally in groundwater, however numerous anthropogenic activities like fluorinated industrial waste (glass, aluminum, iron, steel), and also agricultural activities like application of fertilizers (phosphate, potassium), pesticides, etc. increase its concentration in the environment and cause detrimental effects (Gupta et al. 2015, 2019; Srivastav et al. 2018; Maurya et al. 2019). The fluoride concentration varies from region to region daily. The population consuming a dietary habit of fish and tea are much more prone to fluoride, because fluoride is present in higher concentration in them (Kanduti et al. 2016). The problem of fluoride contaminants has become a worldwide problem in groundwater because it is natural as well as uncontrollable (Brunt et al. 2004). According to reports, the geogenic sources are the main cause for exposure of fluoride and its adverse effect increases from consuming fluoride-polluted groundwater (Sharma et al. 2015; Mukherjee and Singh 2018). The threatening concentration of fluoride ions present in groundwater is higher in different countries in South and Southeastern Asia, in addition to other toxic and infectious substances (WHO 2000).

The southern part of Asia is represented by South Asia, including countries like Bhutan, Bangladesh, Afghanistan, Maldives, India, Nepal, Sri Lanka, and Pakistan. These countries inhabit around 39.5% of the Asian population and also constitute 24% of the world's population (Sikdar 2019). A higher concentration of fluoride in the groundwater of countries like China, India, and Pakistan put a large population's health under risk. The problem of skeletal fluorosis in adults and children occurs due to consumption of groundwater containing high fluoride concentration (3 mg/L). Endemism of fluorosis is found in at least 25 countries of all continents including North America, South America, Europe, Africa, and Asia (Fawell et al. 2006). The large intake of fluoride causes numerous health issues like anorexia, muscle weakness, sweating, dyspnea, severe gastroenteritis, salivation, restlessness, etc. (Sahu et al. 2017). Many chemical factors like dissolution, weathering, ion exchange, precipitation, and several biological processes occurring under the Earth's surface influence the concentration of fluoride in groundwater. Since only a limited number of studies have been done on fluoride impact on human health, and considering its detrimental effects, human health issues were investigated in Iran, mainly in the Sistan and Baluchestan provinces.

Numerous problems ought to be considered while assessing fluoride contamination and its effects:

1) A higher concentration of fluoride in groundwater is limited to arid and semiarid regions of North Africa and Asia.
2) Rocks and sediments having fluorine-containing minerals are the main source of fluoride for geogenic origin, whereas pesticide and industrial wastes contribute to anthropogenic sources of fluoride.
3) The fluoride ion in water binds with different ions such as chlorides and bicarbonates (Kanchan et al. 2015; Ali et al. 2016). Various studies have been done on defluoridation, although no proper solution to this problem could be found.

2.2 Source of Fluoride in the Environment

The fluoride-containing minerals and rocks naturally introduce fluoride and are also considered the largest fluoride reserve. Figure 2.1 represents the expected sources of fluoride in environment.

2.2.1 Natural Sources

2.2.1.1 Rocks and Minerals

One of the most abundant trace elements present in the Earth's crust is fluoride, having 625 mg/kg average concentration in different kinds of rock (Edmunds and Smedley 2005; Tavener and Clark 2006). The rocks containing fluoride-rich minerals are the largest fluoride reserve. According to some studies, the highest fluoride concentrations are combined with quartz, felsic, gneisses, syenites, granites, and alkaline volcanic (Robinson Jr and Kapo 2003; Rosi et al. 2003; Moore 2004; Chae et al. 2006, 2007; Ozsvath 2006). Rocks found in the Coimbatore district of Tamil Nadu contain 180–2600 mg/kg fluoride (Table 2.1). Among various fluoride-rich minerals, some of them are fluorite (CaF_2), micas, amphiboles, villiaumite (NaF), and topaz ($Al_2[SiO_4]F_2$). These chemicals are abundantly found minerals in sediments and rocks (Cronin et al. 2000; Saxena and Ahmed 2003; Edmunds and Smedley 2005; Chae et al. 2007).

2.2.1.2 Groundwater

There are several factors affecting the fluoride presence like granite, gneissic rocks, and volcanic as well as mountainous area sediments of marine origin. The above-mentioned rocks are rich in fluoride and are often found beneath the Earth's surface, ultimately

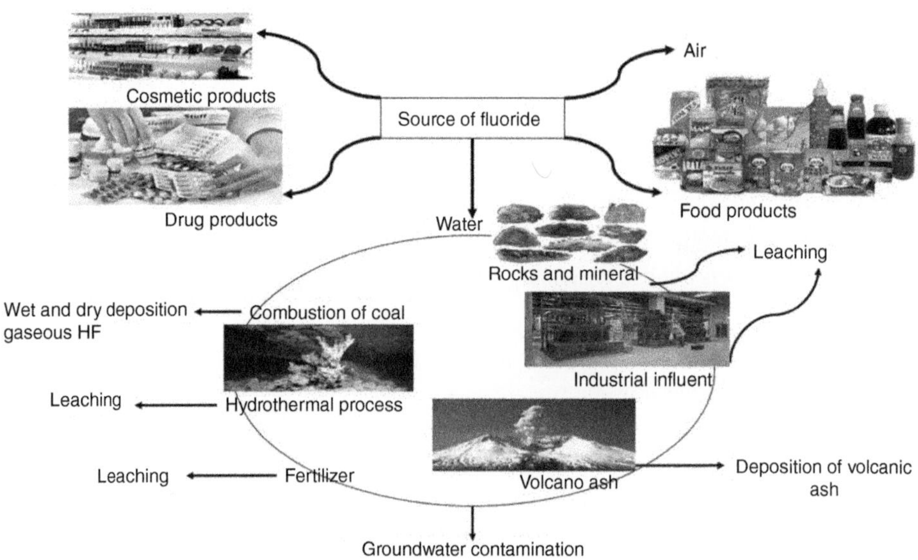

Figure 2.1 Sources of fluoride contamination in the environment.

Table 2.1 Fluoride levels reported in different regions.

Country	Location	Fluoride concentration (mg/L)	References
India	Hyderabad, Andhra Pradesh	0.38–4.0	Sreedevi et al. (2006)
	Ranga Reddy, Andhra Pradesh	0.4–4.8	Sujatha (2003)
	Karbi Anglong, Assam	0.4–20.6	Chakraborti et al. (2000)
	Bihar Shallow	0.1–2.5	Ray et al. (2000)
	Delhi	0.2–32.5	Raju et al. (2009)
	Gujarat	0.1–40	Raju et al. (2009)
	Palghat, Kerala	0.2–5.75	Shaji et al. (2007)
	Chandidongri, Madhya Pradesh	1.5–4.0	Chatterjee and Mohabey (1998)
	Shivpuri, Madhya Pradesh	0.2–6.4	Ayoob and Gupta (2006)
	Orissa	0.1–10.1	Kundu et al. (2001)
	Churu/Dungarpur, Rajasthan	0.1–14	Muralidharan et al. (2002); Choubisa (2001)
	Kancheepuram, Tamil Nadu	1–3.24	Dar et al. (2011)
	Tamil Nadu	0.5–4.0	Raju et al. (2009)
	Cambay, North Gujarat	0–10	Gupta et al. (2005)
	Varanasi, Uttar Pradesh	0.2–2.1	Raju et al. (2009)
	Sonbhadra, Uttar Pradesh	0.48–6.7	Raju et al. (2009)
	Mathura, Uttar Pradesh	0.6–2.5	Misra et al. (2006)
Canada	Gaspe, Quebec	0.05–10.9	Boyle and Chagnon (1995)
Ghana	Nathenje and Lilongwe	0.5–7.02	Msonda et al. (2007)
Pakistan	Nagar Parkar	1.13–7.85	Naseem et al. (2010)
Sri Lanka	Dry Zone	0.02–5.30	Chandrajith et al. (2011)
Iran	Posht-e-Kooh-e-Dashtestan	0.7–6.6	Battaleb-Looie and Moore (2010)
China	Taiyuan Basin	0.4–2.4	Li et al. (2011)
Germany	Muenster Region	0.01–8.8	Queste et al. (2001)
Mexico	Hermosillo city, Sonara	0–7.59	Valenzuela-Vasquez et al. (2006)

leading to groundwater contamination. Alarming high fluoride ion concentrations together with various other toxic and infectious substances present in the groundwater of South and Southeastern Asia are of utmost concern (Ghosh et al. 2013).

2.2.2 Anthropogenic Sources

In many developing countries as well as developed countries, the fluoride concentration present in the environment has been majorly altered by industrial discharge. The main sources of fluoride input into the environment via anthropogenic activities are the aluminum and zinc industry, coal-burning, brick/clay burning, steel production, oil refining, chemical production, uranium trifluoride, magnesium smelting, ceramic glass and uranium hexafluoride production, enamel manufacturing, and fluoride-containing fertilizers or pesticide industries (Sujatha 2003). Discharge from industries heavily pollutes the soil and water, as well as vegetation cover around the industry and far away from it. Other than industrial emissions, agriculture runoffs having fluoride-containing fumigants, fertilizers, and pesticides are some of the other predominant causes of fluoride pollution (Kundu and Mandal 2009; Borah and Saikia 2011). Industries involving coal burning pollute the atmosphere in a small area, and the extent of the pollution depends on the origin and type of coal. It has been estimated that burning of biomass releases 76 Gg fluoride into the air annually (Jayarathne et al. 2014).

2.3 Occurrence of Fluoride in the World and India

2.3.1 World

The quantity of fluoride in drinking water varies around the world as well as region to region depending on the geographical location. Contamination of fluoride ions has been largely illustrated in the groundwater of mainly humid, tropical parts of the world. The countries having this type of climatic condition include China, countries in South Asia, and countries in Africa (Ayoob and Gupta 2006). According to WHO 2002, the recommended fluoride concentration of drinking water is 1.5 mg/L. However, estimations show that 200 million or more people are consuming fluoride-contaminated drinking water worldwide. According to some studies, it has been revealed that over 5 million people are exposed to fluoride-contaminated groundwater in Mexico (Ayoob and Gupta 2006).

2.3.1.1 America

A high level of fluoride in groundwater has been reported from the USA, mainly in facility wells of industries in Pennsylvania, having 3.2 and 6.5 mg/L of fluoride; Lakeland, Southern California, having 3.6–5.3 mg/L of fluoride; and deep aquifers present in the western US comprising 5–15 mg/L of fluoride (Cohen and Conrad 1998). The presence of fluorosis has also been reported in different states of the USA, such as Oklahoma, Nevada, South Carolina, North Dakota, Texas, Oregon, California, Utah, Colorado, New Mexico, Virginia, and North Carolina. Five million people in Mexico (around 6% of the total population of the country) have been affected by pollution of fluoride in groundwater. In Canada, there are several communities whose natural drinking water sources contain higher levels of fluoride (as high as 4.3 mg/L). Industrial discharges are the main reason behind fluoride contamination reported in the USA and Canada (Rose and Marier 1977). A few parts of Argentina contain fluoride concentrations of about 5 mg/L in groundwater (Kruse and Ainchil 2003).

2.3.1.2 Indian Scenario

Twelve million tons of fluoride deposits are found in India from the Earth crust out of 85 million tons of total deposition (Teotia and Teotia 1984). Therefore, the contamination of fluoride has broadly spread at an alarming rate in the Indian scenario. In the capital city of India, Delhi, the percentage of groundwater crossing the maximum permissible limit of fluoride present in drinking water is around 50% (Datta et al. 1996). According to the report of Jacks et al. (2005), the main reason behind the higher concentration of fluoride ion found in the groundwater of many parts of India was due to the evapotranspiration of groundwater with residual alkalinity. In the southern parts of India, major proportion of fluoride contamination in groundwater is due to fluoride enriched rocks for instance, groundwater of Andhra Pradesh, precisely Nalgonda district has high fluoride level, i.e. 320–3100 mg/kg due to presence of fluoride-rich granitic rocks. The average fluoride level found in the granites of Hyderabad is 910–920 mg/kg (Ramamohana Rao et al. 1993).

According to the investigation the two sites of Uttar Pradesh and Madhya Pradesh have considerably higher concentrations of fluoride, i.e. 0.1–0.3 mg/L (Das et al. 1981; Satsangi et al. 1998; Singh et al. 2001). The main cause of the increasing fluoride contamination in this region was predicted as deposition of soil dust. According to the study of Jain et al. (2000), in Haryana, wet deposition of crustal material increases the fluoride load. Thirteen sites in Madhya Pradesh show about 0.05–0.22 mg/L concentration of fluoride as reported by Chandrawanshi and Patel (1999) and the area is considerably close to the industrial aluminum plant. A recently reported evaluation of dry deposition near Agra, reported by Satsangi et al. (2002) shows higher amounts of fluoride due to atmospheric deposition. Several authors have claimed that the atmospheric deposition is mainly from the crustal source. Concentrations of fluoride in different regions are presented in Table 2.1.

2.4 Effects of Fluoride on Human Health

The effect of fluoride contamination on human health has been studied by researchers from all over the world for more than a century. Fluoride causes both good and bad effects on the human body depending on the level of exposure. According to the study of Ozsvath (2009), ingestion of a moderate amount of fluoride can actively decrease the risk of occurrence of dental caries as well as promote the growth of strong bones under certain conditions. Chronic exposure to fluoride can cause various ill effects on human health such as dental fluorosis and skeletal fluorosis; it can also increase the rate of urolithias, and decrease natality and IQ level of children. In some cases, chronic exposure might also lead to a number of defects such as genetic mutations, birth defects, and Alzheimer's disease; the scientific data at present are inconclusive (Ozsvath 2009). According to the World Health Organization (WHO), the maximum intake of fluoride in drinking water is recommended as 1.5 mg/L (Edition 2011). Among various other adverse health impacts on the body, fluorosis remains the major problem in affected populations and is categorized as dental fluorosis and skeletal fluorosis, as discussed in Sections 2.4.1 and 2.4.2.

2.4.1 Dental Fluorosis

Dental fluorosis is a developmental disturbance of tooth enamel or tooth surface that occurs due to systemic overexposure of fluoride during enamel formation (Kanduti et al. 2016). During the first six years of life, the development of enamel occurs over the tooth and increased mineralization is accompanied by reduction of matrix protein. Exposure to fluoride causes dose relationship disruption in the process of amelogenesis and dentinogenesis that ultimately results in deformity in the crystalline structure of teeth (Swarup and Dwivedi 2002). Dental fluorosis is characterized by yellow to brownish mottling of enamel with narrow white horizontal striations (DenBesten and Li 2011). The dental fluorosis severity depends upon the degree of exposure of fluoride on humans. In India, 70% of adolescents consuming fluoride-contaminated drinking water with more than the recommended value of the World Health Organization are affected by dental fluorosis (Chaudhry et al. 2017). Enamel opacities also occur due to malnutrition and deficiency of Vitamin A and D as well as due to low protein-energy intake. Therefore, fluoride is not the only cause of dental enamel defects (Zohoori and Duckworth 2017).

2.4.2 Skeletal Fluorosis

Skeletal fluorosis is a somewhat more severe adverse health impact, characterized by increased bone mass and bone density (osteosclerosis) that occurs due to a prolonged period of exposure to fluoride at the time of bone modeling and/or remodeling. Fluoride exposure of more than 4 mg/L concentration may lead to skeletal fluorosis; the exposure may be either direct (ingestion) or indirect (inhalation) (Yadav et al. 2019). Skeletal fluorosis occurs in three stages: initial, intermediate, and final:

1) The initial stage – The initial stage shows various mild symptoms like joint pain, stiffness of bones and joints, muscle weakness, periodic pain, and chronic fatigue.
2) The intermediate stage – The intermediate stage is characterized by calcification of bone followed by hardening and stiffening of joints as well as calcification of ligaments in the body. Patients may develop a "Poker back" situation in severe cases of skeletal fluorosis. **Poker back**- a condition in which the whole spine becomes a fixed column due to increasing bone stiffness.
3) The final stage (crippling skeletal fluorosis) – In this stage, joint movements become very limited and skeletal bone deformities, acute calcification of ligaments, muscle wasting, and neurological defects can be seen in patients (Itai et al. 2010).

According to some literature, it has been revealed that maximum ingestion fluoride ions can cause other health effects such as headache, deformities in red blood cells, rash over skin, gastrointestinal problems, depression, low haemoglobin levels, nausea, pain in abdomen, fingers and toes with tingling sensations, and reduced immunity, as well as neurological manifestations which are quite similar to pathological changes occurring in patients with Alzheimer's. These effects of fluoride have received less attention in comparison to the dental and skeletal fluorosis typical of high fluoride-contaminated areas (Thole 2013).

2.5 Remediation Techniques for Fluoride Contamination

2.5.1 Remediation of Fluoride

As we all know, clean water is one of our basic needs. The increase of fluoride content in drinking water due to various sources ultimately increases the above-mentioned detrimental effects, so a temporary solution may be that it's better to avoid that particular source. The methods used for the treatment of fluoride-contaminated groundwater are in-situ and

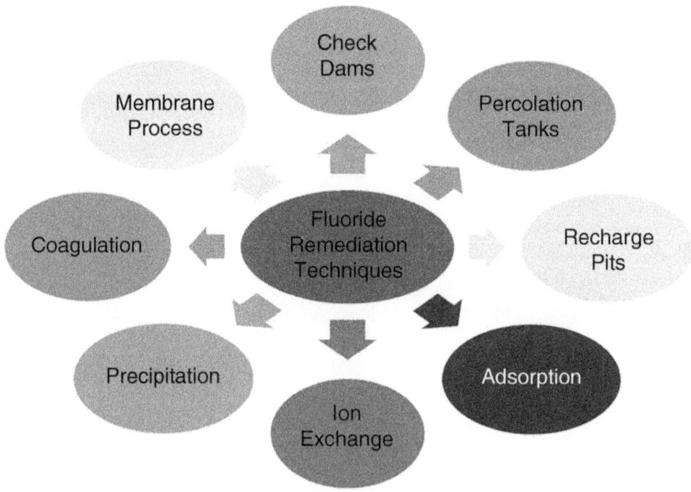

Figure 2.2 Remediation techniques for fluoride removal.

ex-situ methods that reduce fluoride content and bring it to a usable form. Several types of remediation techniques have been shown in Figure 2.2.

2.5.2 *In-situ* Treatment Methods for Fluoride Removal

These methods are focused on straight dilution of fluoride concentration of groundwater in aquifer only, which can be accomplished by artificial recharge. In order to evaluate the efficacy of managed aquifer recharge (MAR), measures like check dams, percolation tanks, and recharge pots came to light as part of the in-situ treatment methods.

2.5.2.1 Check Dams

Check dam construction reduces fluoride concentrations and also improves the quality of groundwater in accordance with domestic and agricultural use. The fact that groundwater quality has been improved by recharge using the dams has also been confirmed by spatial variation in the saturation index of minerals. The content of fluoride in wells of groundwater reduces near the check dam whereas areas away from the dam have higher fluoride content. Higher groundwater level and reduced fluoride content near the check dam represent the major advantage of recharging from the dam. The fluoride concentration becomes lower than the permissible limit of BIS (Bureau of Indian Standards) around the area of 4 km^2 near a dam present in Krishnagiri District of Tamil Nadu (Gowrisankar et al. 2017). One of the known examples of in-situ method in India is the Anantapur district check dam construction in Andhra Pradesh, which aided in reduction of fluoride content of groundwater (Bhagavan and Raghu 2005).

2.5.2.2 Percolation Tank and Recharge Pits

Artificial recharging structures built at suitable places decrease fluoride concentration. The method of rainwater harvesting also known as rainwater recharge could be adopted

Figure 2.3 *In-situ* remediation techniques: (A and B) Check dams; (C) Recharge pits; (D and E) Percolation tanks.

using percolation tanks and recharge pits that may prove to be very helpful. Utilizing existing wells to recharge rainwater after filtration is also an important and useful method to improve groundwater quality. Artificial recharging structures can be built in suitable places which will decrease concentration. See Figure 2.3 for in-situ remediation techniques.

2.5.3 *Ex-situ* Treatment Methods for Fluoride Removal

Several ex-situ treatment methods are available for defluoridation of water at both domestic and community levels.

2.5.3.1 Adsorption

Adsorption is one of the most feasible and cost-effective methods for removal of fluoride (Figure 2.4). In this process, water is passed via a contact bed adsorbing fluoride on the matrix. The effective removal depends upon the concentration of fluoride, time of contact, pH, size, and type of adsorbent. After a period of time, the saturated column or bed should be refilled or regenerated. The first defluoridation was done around 1930 using activated alumina. It has a large surface area and porous aluminum oxide; the continuous aluminum lattice provides a localized positive charge area, creating a good adsorbent for many anionic adsorbates. Since it has a greater preference for fluoride ion in comparison to other ions, it is thus used widely in defluoridation (George et al. 2010). Activated charcoal, activated aluminum, fly ash, serpentine, brick, charfines, bone char, waste mud, red mud, kaolinite, rice husk, ceramic, bentonite, bioadsorbents, etc. are several capable adsorbents used for removal of fluoride-contaminated water.

2.5.3.2 Ion Exchange

The ion exchange method involves the passage of water through a column containing ion exchange resin, and calcium ions are replaced by fluoride ions in resin. Once the saturation is obtained, the resin is backwashed with a chloride-containing solution like sodium chloride, in which chloride replaces back the fluoride ions in resin so that it can be reused. The backwashing should be done with proper care.

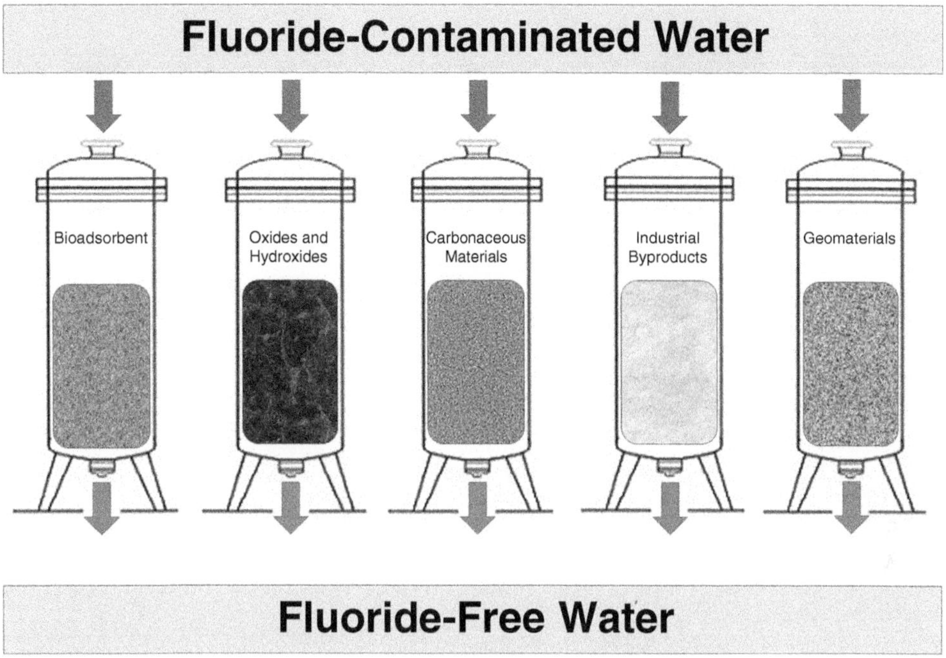

Figure 2.4 Adsorption technique for fluoride removal using various adsorbents.

2.5.3.3 Coagulation-Precipitation

The precipitation method involves certain chemical additions like calcium resulting in precipitation of fluorite; in the place of calcium, one can also use aluminum. One of the widely used Nalgonda techniques uses alum, lime, and bleaching powder with vigorous mixing, flocculation, sedimentation, and filtration. The method was initiated and developed by NEERI (National Environmental Engineering Research Institute) for community and household levels under Rajiv Gandhi Drinking Water Mission (Figure 2.5). The sludge obtained by this process has high aluminum and fluoride, disposal of which is also problematic.

2.5.3.4 Membrane Process

The membrane process is one of the ex-situ techniques also known as reverse osmosis and electrodialysis. This method requires financial input as it uses a semipermeable membrane to remove dissolved solutes from water (Brindha and Elango 2011). This process is utilized as advanced water treatment technology mainly in the treatment of pure and ultrapure water.

The research of Somnath Rudra (2012) pointed out some important holistic alternatives to mitigate the problem of fluorosis:

- Utilization of fluoride removal filters for treatment of excess fluoride at home.
- River and stream water based piped water supply.
- Harvesting rainwater.
- Utilizing traditional sources of water (bunds/ponds) after treatment.

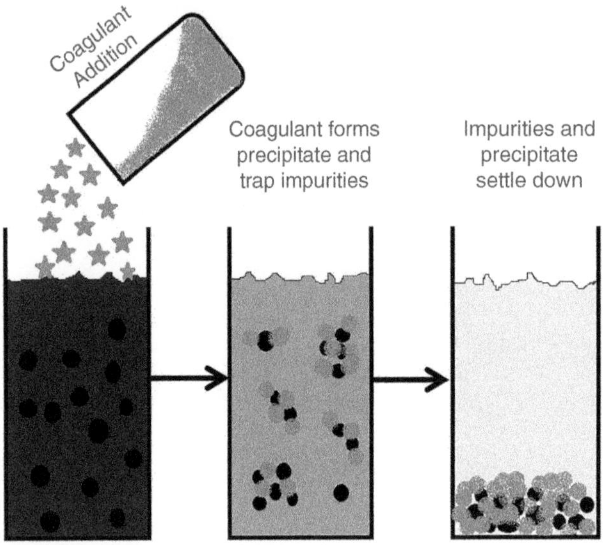

Figure 2.5 Coagulation and precipitation method.

- Substantial water quality monitoring program.
- Diagnosis and remediation of fluorosis
- Providing nutritional supplement of calcium and other vitamins (like C, E, and D)
- Addition of calcium-, vitamin C-, and vitamin D-rich food to diet and avoid chewing supari, gutka, etc.
- Awareness, motivation and training of the community.

2.5.3.5 Reverse Osmosis

A tank divided into two parts with the help of a semipermeable membrane comprises a reverse osmosis system. Fluoride-contaminated water passes via semipermeable membrane with the help of hydraulic pressure on a side of tank. Along with water, few impurities also cross the semipermeable membrane, whereas salts and other impurities cannot cross the membrane (Dubey et al. 2018). Successful filtration of various impurities with different sizes has been reported while crossing the semipermeable membranes by exerting pressure. In the process of reverse osmosis oman-induced pressure toward the contaminated water side of the membrane proved to be advantageous to overcome the naturally occurring osmotic pressure that flows in the opposite direction. In order to enhance the reaction speed of the process some more osmotic pressure has been added by forcing water across the semipermeable membrane toward the opposite, clean side (Demeuse 2009; Water Professionals 2018). According to reports, reverse osmosis results in accessing a better quality of drinking water, having salinity less than 0.1 g salts/L (Mazighi et al. 2015). Ions residing in water could be carried via membranes with the help of electric currents in the electrodialysis process. The two major parameters affecting the performance of membrane are pH and temperature. The ability to remove fluoride via the reverse osmosis process varies from 45 to 90% by increasing the pH level from 5.5 to 7. Some problems associated with reverse osmosis are chemical attacks, fouling due to particulate matter, plugging of membrane, and producing large quantities of waste. The waste generated by this process is much more than in the ion exchange process. Pretreatment of raw water must be carried out before introduction into the production unit if the raw water condition is not optimal. Another demerit is eliminating all the ions of contaminated water. As we all know, humans require minerals for proper growth and metabolism thus remineralization of reverse osmosis-treated water is required, hence increasing the cost of the process in comparison to other options. This method involves more financial input as the purification process makes water more acidic, thus several pH corrections are required, generating lots of brine waste (Kumar and Gopal 2000). The important points of consideration for the reverse osmosis membrane selection method are recovery of purified water, raw water composition, pretreatment steps, cost, rejection properties, etc. (Jagtap et al. 2012; Yadav et al. 2017; Yadav et al. 2018).

2.5.3.6 Nanofiltration

One more removal method for fluoride is nanofiltration which decreases hardness of water with the help of membranes having high retention capacities for charged particles like bivalent ions. Nanofiltration has been considered as the optimum membrane process for eliminating fluoride having this inherent property of specific membrane selectivity (Tahaikt et al. 2007). Many researchers have shown successful elimination of fluoride with the help of the nanofiltration method. Reports suggest that nanofiltration membranes have better success in elimination of fluoride from polluted drinking water in comparison to LPRO (low-pressure reverse osmosis) membranes. One study reported two commercially available nanofiltration membranes: NF-90 and NF-270, in which NF-270 reduces the concentration of fluoride from 10 to 1.5 mg/L whereas NF-90 decreases the concentration from 20 to 0.5 mg/L. The presence of anions like bicarbonates exert no noticeable negative effects on the purification process whereas the elimination of fluoride reduces under acidic condition (Hoinkis et al. 2011). According to the study of Bejaoui et al. (2014), successful elimination of fluoride has been reported utilizing reverse osmosis comparing against NF-90 considering different parameters like pH, ionic strength, feed pressure, and fluoride concentration, as well as nature of cations present along with fluoride. The results revealed that optimization of fluoride removal was done at higher pH as enhancing overall negative charges of membrane has been tested (Bejaoui et al. 2014). The study of Emamjomeh et al. (2018) shows a lab-scale study of nanofiltration membrane (FILMTEC-NF90-4040) using a pilot plant: fluoride removal from contaminated water with concentration lying between 1.50 and 2.17 mg/L. The effectively considered parameters were pressure (between 4 and 12 bars) and temperature (between 10 and 30 °C). Results revealed minimum and maximum removal percentage of fluoride, i.e. 30 and 70%, respectively. One other point brought into light was that increasing pressure and temperature enhanced the performance of fluoride removal and membrane permeate flow rate (Emamjomeh et al. 2018).

According to the study of Van der Bruggen et al. (2008), membrane fouling, chemical resistance, and insufficient separation limited lifetimes, as well as rejection, are some significant disadvantages of the nanofiltration method that increase the financial input of the method. The production of fouling that should be collected and disposed off is also considered as one main disadvantage of this method. One study shows the comparison between NF5 and NF9, two commercial nanofiltration membranes for removing fluoride from groundwater (Nasr et al. 2013). The concentration of fluoride in purified water using NF5 and NF9 nanofiltration membrane was found to be 1.45 and 0.38 mg/L, respectively. Mainly, NF5 and NF9 membranes successfully removed higher amounts of divalent anions from water but smaller ions could be removed more effectively in comparison to others. The reason behind this could be the salvation energy of smaller ions. Results also revealed that chlorine ions penetrate the NF5 and NF9 nanofiltration membrane faster than fluoride. BW30 and NF90 nanofiltration membrane can successfully eliminate fluoride, reducing the concentration from 417.9 to 1.5 mg/L (Shen and Schäfer 2014). According to the reports, calcium carbonate is the main fouling component on membranes of nanofiltration. Recovery of nanofiltration membranes could be achieved by using the citric acid and ammonia cleaning method (Wei-fang et al. 2009).

2.5.3.7 Electrocoagulation

Electrocoagulation (EC) is another filtration method removing suspended solids (fluorides) to μm level from water (Noling 2004). Electrocoagulation is a type of electrolytic method in which metallic cation synthesis occurs at sacrificial anodes (Kobya et al. 2016). The electrocoagulation utilization has been enhanced in the last decade. This method is introduced as one of the suitable methods for elimination of fluoride from contaminated drinking water. It can effectively remove a wide range of pollutants such as oil, heavy metals, dye, and fluoride (Hu et al. 2005; Malakootian et al. 2011). The electrocoagulation method does not release secondary pollutants and retain beneficial components of raw water during defluoridation. According to the report of Sinha et al. (2012), the electrocoagulation successfully eliminates fluoride and aluminum simultaneously under the condition of 230 V DC using aluminum electrodes. Better fluoride removal from drinking water can be achieved by providing longer retention time. One of the most important parameters in this method is charge loading for controlling EC reaction rates that ultimately decides coagulation rates (Kobya et al. 2016). The fluoride depletion and charge loading performance do not have a linear relationship. Despite being the important parameter for EC, charge loading is not considered a critical parameter in fluoride removal from drinking water. Enhanced charge loading reduces fluoride concentration initially in treated water, whereas after a critical point the fluoride concentration decrease was not significant (Sinha et al. 2012). The electrocoagulation method releases less aluminum from water in comparison to the active aluminum process and the Nalgonda technique. Data revealed that aluminum concentration in treated water is enhanced when input energy has been enhanced (Sinha et al. 2012). From steel industry wastewater, successful removal of fluoride has been done using the electrocoagulation method. The fluoride removal has been optimized under several conditions like "hydraulic retention time, pH, temperature, voltage, number of aluminum plates between anode and cathode to assess the performance of this method. Results revealed that increase in hydraulic retention time by 5 min shows enhanced fluoride removal performance". The concentration of fluoride reduces from 4 to 6 mg/L to less than 0.5 mg/L (Khatibikamal et al. 2010). According to the study of Emamjomeh and Sivakumar (2006) the performance of fluoride removal depends upon various parameters like current density, initial concentration of fluoride, flow rate of wastewater, and pH. The fluoride removal is directly related to aluminum F− hydroxide complex [AlnFm(OH)3n−m] formation during electrocoagulation (Emamjomeh et al. 2011).

2.6 Conclusion and Future Perspective

The contamination of fluoride in drinking water and groundwater has reached an alarming level already. This needs the immediate concern and involvement of people to alleviate this problem. Suitable remedial measures should be explored, meeting the geohydrological, sociocultural, ecopolitical, and environmental aspects of the area and also the people. Several defluoridation techniques have been employed when the level of fluoride increases in potable water. The fluoride concentration limit of potable water for any region is based on its dietary habits, annual daily temperature, nature, and level of exposure in the area. The remediation technology should consider the facts of availability of material, cost-effectiveness, and level of fluoride removal, as well as technical complexity.

References

Alhababy, A.M. and Al, A.J. (2015). Groundwater quality assessment in Jazan region, Saudi Arabia. *Current World Environment* 10 (1): 22.

Al-Hatim, H.Y., Alrajhi, D., and Al-Rajab, A.J. (2015). Detection of pesticide residue in dams and well water in Jazan area, Saudi Arabia. *American Journal of Environmental Sciences* 11 (5): 358.

Ali, S., Thakur, S.K., Sarkar, A., and Shekhar, S. (2016). Worldwide contamination of water by fluoride. *Environmental Chemistry Letters* 14 (3): 291–315.

Ayoob, S. and Gupta, A.K. (2006). Fluoride in drinking water: a review on the status and stress effects. *Critical Reviews in Environmental Science and Technology* 36 (6): 433–487.

Barathi, M., Kumar, A.S.K., and Rajesh, N. (2019). Impact of fluoride in potable water–an outlook on the existing defluoridation strategies and the road ahead. *Coordination Chemistry Reviews* 387: 121–128.

Battaleb-Looie, S. and Moore, F. (2010). A study of fluoride groundwater occurrence in Posht-e-Kooh-e-Dashtestan, South of Iran. *World Applied Sciences Journal* 8 (11): 1317–1321.

Bejaoui, I., Mnif, A., and Hamrouni, B. (2014). Performance of reverse osmosis and nanofiltration in the removal of fluoride from model water and metal packaging industrial effluent. *Separation Science and Technology* 49 (8): 1135–1145.

Bhagavan, S.V.B.K. and Raghu, V. (2005). Utility of check dams in dilution of fluoride concentration in ground water and the resultant analysis of blood serum and urine of villagers, Anantapur District, Andhra Pradesh, India. *Environmental Geochemistry and Health* 27 (1): 97–108.

Bhatnagar, A., Kumar, E., and Sillanpää, M. (2011). Fluoride removal from water by adsorption – a review. *Chemical Engineering Journal* 171 (3): 811–840.

Borah, J. and Saikia, D. (2011). Estimation of the concentration of fluoride in the ground water of Tinsukia town master plan area of the Tinsukia district, Assam, India. *Archives of Applied Science Research* 3 (3): 202–206.

Boyle, D.R. and Chagnon, M. (1995). An incidence of skeletal fluorosis associated with groundwaters of the maritime carboniferous basin, Gaspé region, Quebec, Canada. *Environmental Geochemistry and Health* 17 (1): 5–12.

Brindha, K. and Elango, L. (2011). Fluoride in Groundwater: Causes, Implications and Mitigation Measures. In: *Fluoride Properties, Applications and Environmental Management* (ed. S.D. Monroy), 111–136. New York: Nova Science Publishers.

Brunt, R., Vasak, L. and Griffioen, J., 2004. Fluoride in groundwater: Probability of occurrence of excessive concentration on global scale. International Groundwater Resources Assessment Centre (IGRAC), Utrecht, The Netherlands. 1–12.

Chae, G.T., Yun, S.T., Kim, K., and Mayer, B. (2006). Hydrogeochemistry of sodium-bicarbonate type bedrock groundwater in the Pocheon spa area, South Korea: water–rock interaction and hydrologic mixing. *Journal of Hydrology* 321 (1–4): 326–343.

Chae, G.T., Yun, S.T., Mayer, B. et al. (2007). Fluorine geochemistry in bedrock groundwater of South Korea. *Science of the Total Environment* 385 (1–3): 272–283.

Chakraborti, D., Chanda, C.R., Samanta, G. et al. (2000). Fluorosis in Assam India. *CurrSci* 78: 1421–1423.

Chandrajith, R., Dissanayake, C.B., Ariyarathna, T. et al. (2011). Dose-dependent Na and Ca in fluoride-rich drinking water – another major cause of chronic renal failure in tropical arid regions. *Science of the Total Environment* 409 (4): 671–675.

Chandrawanshi, C.K. and Patel, K.S. (1999). Fluoride deposition in central India. *Environmental Monitoring and Assessment* 55 (2): 251–265.

Chatterjee, M.K. and Mohabey, N.K. (1998). Potential fluorosis problems around Chandidongri, Madhya Pradesh, India. *Environmental Geochemistry and Health* 20 (1): 1–4.

Chaudhry, M., Prabhakar, I., Gupta, B. et al. (2017). Prevalence of dental fluorosis among adolescents in schools of Greater Noida, Uttar Pradesh. *Journal of Indian Association of Public Health Dentistry* 15 (1): 36.

Choubisa, S.L. (2001). Endemic fluorosis in southern Rajasthan, India. *Fluoride* 34 (1): 61–70.

Cohen, D. and Conrad, H.M. (1998). 65,000 GPD fluoride removal membrane system in Lakeland, California, USA. *Desalination* 117 (1–3): 19–35.

Cronin, S.J., Manoharan, V., Hedley, M.J., and Loganathan, P. (2000). Fluoride: a review of its fate, bioavailability, and risks of fluorosis in grazed-pasture systems in New Zealand. *New Zealand Journal of Agricultural Research* 43 (3): 295–321.

Dar, M.A., Sankar, K., and Dar, I.A. (2011). Fluorine contamination in groundwater: a major challenge. *Environmental Monitoring and Assessment* 173 (1–4): 955–968.

Das, D.K., Burman, G.K., and Kidwai, A.L. (1981). Chemical composition of monsoon rainwater over Bhopal, Madhya Pradesh (India) during 1977 and 1978. *Indian Journal of Meteorology, Hydrology, and Geophysics (at present Mousam)* 32 (3): 221–228.

Datta, P.S., Deb, D.L., and Tyagi, S.K. (1996). Stable isotope (18O) investigations on the processes controlling fluoride contamination of groundwater. *Journal of Contaminant Hydrology* 24 (1): 85–96.

Demeuse, M.T. (2009). Production and applications of hollow fibers. In: *Handbook of Textile Fibre Structure* (eds. S.J. Eichhorn, J.W.S. Hearle and T. Kikutani), 485–499. Woodhead Publishing.

DenBesten, P. and Li, W. (2011). Chronic fluoride toxicity: dental fluorosis. In: *Fluoride and the Oral Environment*, vol. 22 (ed. M.A.R. Buzalaf), 81–96. Karger Publishers.

Dubey, S., Agarwal, M., and Gupta, A.B. (2018). Recent developments in defluoridation of drinking water in India. In: *Environmental Pollution* (eds. V.P. Singh, S. Yadav and R.N. Yadava), 345–356. Singapore: Springer.

Edmunds, W.M. and Smedley, P.L. (2005). *Fluoride in Natural Waters Essentials of Medical Geology* (eds. B.J. Alloway and O. Selinus). Science and Education Publishing, Elsevier.

Emamjomeh, M.M. and Sivakumar, M. (2006). An empirical model for defluoridation by batch monopolar electrocoagulation/flotation (ECF) process. *Journal of Hazardous Materials* 131 (1–3): 118–125.

Emamjomeh, M.M., Sivakumar, M., and Varyani, A.S. (2011). Analysis and the understanding of fluoride removal mechanisms by an electrocoagulation/flotation (ECF) process. *Desalination* 275 (1–3): 102–106.

Emamjomeh, M.M., Safari Varyani, A., Alijani, M.H. et al. (2018). Effect of temperature and pressure on removal of fluoride from groundwater using nanofiltration. *Journal of Mazandaran University of Medical Sciences* 27 (156): 166–176.

Edition, F. (2011). Guidelines for drinking-water quality. *WHO Chronicle* 38 (4): 104–108.

FAO (2013). *Statistical Yearbook. World Food and Agriculture*, 289. Rome: Food and Agriculture Organization of the United Nations.

Fawell, J., Bailey, K., Chilton, J. et al. (2006). *Fluoride in Drinking-Water*. IWA publishing.

George, S., Pandit, P., and Gupta, A.B. (2010). Residual aluminium in water defluoridated using activated alumina adsorption–Modeling and simulation studies. *Water Research* 44 (10): 3055–3064.

Gleeson, T., Wada, Y., Bierkens, M.F., and Van Beek, L.P. (2012). Water balance of global aquifers revealed by groundwater footprint. *Nature* 488 (7410): 197–200.

Ghosh, A., Mukherjee, K., Ghosh, S.K., and Saha, B. (2013). Sources and toxicity of fluoride in the environment. *Research on Chemical Intermediates* 39 (7): 2881–2915.

Gowrisankar, G., Jagadeshan, G., and Elango, L. (2017). Managed aquifer recharge by a check dam to improve the quality of fluoride-rich groundwater: a case study from southern India. *Environmental Monitoring and Assessment* 189 (4): 200.

Guissouma, W., Hakami, O., Al-Rajab, A.J., and Tarhouni, J. (2017). Risk assessment of fluoride exposure in drinking water of Tunisia. *Chemosphere* 177: 102–108.

Gupta, S.K., Deshpande, R.D., Agarwal, M., and Raval, B.R. (2005). Origin of high fluoride in groundwater in the North Gujarat-Cambay region, India. *Hydrogeology Journal* 13 (4): 596–605.

Gupta, N., Yadav, K.K., Kumar, V., and Singh, D. (2013). Assessment of physicochemical properties of Yamuna River in Agra City. *International Journal of ChemTech Research* 5 (1): 528–531.

Gupta, N., Yadav, K.K., and Kumar, V. (2015). A review on current status of municipal solid waste management in India. *Journal of Environmental Sciences* 37: 206–217.

Gupta, N., Yadav, K.K., Kumar, V. et al. (2019). Trace elements in soil-vegetables interface: translocation, bioaccumulation, toxicity and amelioration-a review. *Science of the Total Environment* 651: 2927–2942.

Hoinkis, J., Valero-Freitag, S., Caporgno, M.P., and Pätzold, C. (2011). Removal of nitrate and fluoride by nanofiltration–a comparative study. *Desalination and Water Treatment* 30 (1–3): 278–288.

Hu, C.Y., Lo, S.L., and Kuan, W.H. (2005). Effects of the molar ratio of hydroxide and fluoride to Al (III) on fluoride removal by coagulation and electrocoagulation. *Journal of Colloid and Interface Science* 283 (2): 472–476.

Itai, K., Onoda, T., Nohara, M. et al. (2010). Serum ionic fluoride concentrations are related to renal function and menopause status but not to age in a Japanese general population. *ClinicaChimicaActa* 411 (3–4): 263–266.

Jain, M., Kulshrestha, U.C., Sarkar, A.K., and Parashar, D.C. (2000). Influence of crustal aerosols on wet deposition at urban and rural sites in India. *Atmospheric Environment* 34 (29–30): 5129–5137.

Jacks, G., Bhattacharya, P., Chaudhary, V., and Singh, K.P. (2005). Controls on the genesis of some high-fluoride groundwaters in India. *Applied Geochemistry* 20 (2): 221–228.

Jagtap, S., Yenkie, M.K., Labhsetwar, N., and Rayalu, S. (2012). Fluoride in drinking water and defluoridation of water. *Chemical Reviews* 112 (4): 2454–2466.

Jayarathne, T., Stockwell, C.E., Yokelson, R.J. et al. (2014). Emissions of fine particle fluoride from biomass burning. *Environmental Science & Technology* 48 (21): 12636–12644.

Kanchan, S., Kumar, V., Yadav, K.K. et al. (2015). Effect of fly ash disposal on ground water quality near Parichha thermal power plant, Jhansi-a case study. *Current World Environment* 10 (2): 572.

Kanduti, D., Sterbenk, P., and Artnik, B. (2016). Fluoride: a review of use and effects on health. *Materia Socio-Medica* 28 (2): 133.

Khatibikamal, V., Torabian, A., Janpoor, F., and Hoshyaripour, G. (2010). Fluoride removal from industrial wastewater using electrocoagulation and its adsorption kinetics. *Journal of Hazardous Materials* 179 (1–3): 276–280.

Kobya, M., Demirbas, E., and Ulu, F. (2016). Evaluation of operating parameters with respect to charge loading on the removal efficiency of arsenic from potable water by electrocoagulation. *Journal of Environmental Chemical Engineering* 4 (2): 1484–1494.

Kruse, E. and Ainchil, J. (2003). Fluoride variations in groundwater of an area in Buenos Aires Province, Argentina. *Environmental Geology* 44 (1): 86–89.

Kumar, S. and Gopal, K. (2000). A review on fluorosis and its preventive strategies. *Indian Journal of Environmental Protection* 20 (6): 430–440.

Kundu, M.C. and Mandal, B. (2009). Assessment of potential hazards of fluoride contamination in drinking groundwater of an intensively cultivated district in West Bengal, India. *Environmental Monitoring and Assessment* 152 (1–4): 97.

Kundu, N., Panigrahi, M., Tripathy, S. et al. (2001). Geochemical appraisal of fluoride contamination of groundwater in the Nayagarh District of Orissa, India. *Environmental Geology* 41 (3–4): 451–460.

Li, X., Hou, X., Zhou, Z., and Liu, L. (2011). Geochemical provenance and spatial distribution of fluoride in groundwater of Taiyuan basin, China. *Environmental Earth Sciences* 62 (8): 1635–1642.

Madhnure, P., Sirsikar, D.Y., Tiwari, A.N. et al. (2007). Occurrence of fluoride in the groundwaters of Pandharkawada area, Yavatmal district, Maharashtra, India. *Current Science* 92 (5): 675–679.

Malakootian, M., Yousefi, N., and Fatehizadeh, A. (2011). Survey efficiency of electrocoagulation on nitrate removal from aqueous solution. *International Journal of Environmental Science and Technology* 8 (1): 107–114.

Maurya, P.K., Malik, D.S., Yadav, K.K. et al. (2019). Bioaccumulation and potential sources of heavy metal contamination in fish species in river Ganga basin: possible human health risks evaluation. *Toxicology Reports* 6: 472–481.

Mazighi, A., Lounici, H., Drouiche, N. et al. (2015). Economic study of groundwater defluoridation of the North African Sahara. *Desalination and Water Treatment* 54 (10): 2681–2691.

Misra, A.K., Mishra, A., and Premraj (2006). Escalation of groundwater fluoride in the Ganga alluvial plain of India. *Fluoride* 39 (1): 35–38.

Moore, R.B., 2004. Quality of water in the fractured-bedrock aquifer of New Hampshire (No. 2004). US Department of the Interior, US Geological Survey.

Msonda, K.W.M., Masamba, W.R.L., and Fabiano, E. (2007). A study of fluoride groundwater occurrence in Nathenje, Lilongwe, Malawi. *Physics and Chemistry of the Earth, Parts A/B/C* 32 (15–18): 1178–1184.

Mukherjee, I. and Singh, U.K. (2018). Groundwater fluoride contamination, probable release, and containment mechanisms: a review on Indian context. *Environmental Geochemistry and Health* 40 (6): 2259–2301.

Muralidharan, D., Nair, A.P., and Sathyanarayana, U. (2002). Fluoride in shallow aquifers in Rajgarh Tehsil of Churu District, Rajasthan–an arid environment. *Current Science* 83 (6): 699–702.

Naseem, S., Rafique, T., Bashir, E. et al. (2010). Lithological influences on occurrence of high-fluoride groundwater in Nagar Parkar area, Thar Desert, Pakistan. *Chemosphere* 78 (11): 1313–1321.

Nasr, A.B., Charcosset, C., Amar, R.B., and Walha, K. (2013). Defluoridation of water by nanofiltration. *Journal of Fluorine Chemistry* 150: 92–97.

Noling, C. (2004). New electrocoagulation system addresses challenges of industrial storm, wash water. *Industrial WaterWorld* 5 (4): 22.

Ozsvath, D.L. (2006). Fluoride concentrations in a crystalline bedrock aquifer Marathon County, Wisconsin. *Environmental Geology* 50 (1): 132–138.

Ozsvath, D.L. (2009). Fluoride and environmental health: a review. *Reviews in Environmental Science and Bio/Technology* 8 (1): 59–79.

Podgorski, J.E., Labhasetwar, P., Saha, D., and Berg, M. (2018). Prediction modeling and mapping of groundwater fluoride contamination throughout India. *Environmental Science & Technology* 52 (17): 9889–9898.

Queste, A., Lacombe, M., Hellmeier, W. et al. (2001). High concentrations of fluoride and boron in drinking water wells in the Muenster region-results of a preliminary investigation. *International Journal of Hygiene and Environmental Health* 203 (3): 221–224.

Raj, D. and Shaji, E. (2017). Fluoride contamination in groundwater resources of Alleppey, Southern India. *Geoscience Frontiers* 8 (1): 117–124.

Raju, N.J., Dey, S., and Das, K. (2009). Fluoride contamination in groundwaters of Sonbhadra district, Uttar Pradesh, India. *Current Science* 96 (7): 979–985.

Rao, N.R., Rao, N., Rao, K.S.P., and Schuiling, R.D. (1993). Fluorine distribution in waters of Nalgonda district, Andhra Pradesh, India. *Environmental Geology* 21 (1–2): 84–89.

Ray, D., Rao, R.R., Bhoi, A.V. et al. (2000). Physico-chemical quality of drinking water in Rohtas district of Bihar. *Environmental Monitoring and Assessment* 61 (3): 387–398.

Robinson, G.R. Jr. and Kapo, K.E. (2003). *Generalized Lithology and Lithogeochemical Character of Near-Surface Bedrock in the New England Region*. United States Geological Survey.

Rose, D. and Marier, J.R., 1977. Environmental Fluoride, 1977 (p. 151). National Research Council of Canada, Associate Committee on Scientific Criteria for Environmental Quality.

Rosi, M., Papale, P., Lupi, L., and Stoppato, M. (2003). *Volcanoes: A Firefly Guide*, 235. New Yark: Baffalo.

Rudra, S. (2012). Fluoride contamination of ground water: a geographical analysis Purulia Block-1 of Purulia District, West Bengal, India. *Analyst* 2 (1): 1–6.

Sahu, B.L., Banjare, G.R., Ramteke, S. et al. (2017). Fluoride contamination of groundwater and toxicities in dongargaon block, Chhattisgarh, India. *Exposure and Health* 9 (2): 143–156.

Satsangi, G.S., Lakhani, A., Khare, P. et al. (1998). Composition of rain water at a semi-arid rural site in India. *Atmospheric Environment* 32 (21): 3783–3793.

Satsangi, G.S., Lakhani, A., Khare, P. et al. (2002). Measurements of major ion concentration in settled coarse particles and aerosols at a semiarid rural site in India. *Environment International* 28 (1–2): 1–7.

Saxena, V. and Ahmed, S. (2003). Inferring the chemical parameters for the dissolution of fluoride in groundwater. *Environmental Geology* 43 (6): 731–736.

Singh, S.P., Khare, P., Satsangi, G.S. et al. (2001). Rainwater composition at a regional representative site of a semi-arid region of India. *Water, Air, and Soil Pollution* 127 (1–4): 93–108.

Sinha, R., Khazanchi, I. and Mathur, S., 2012. Fluoride Removal By A Continuous Flow Electrocoagulation Reactor From Groundwater Of Shivdaspura.

Shah, T., Roy, A.D., Qureshi, A.S., and Wang, J. (2003). Sustaining Asia's groundwater boom: an overview of issues and evidence. In: *Natural Resources Forum*, vol. 27, No. 2, 130–141. Oxford, UK: Blackwell Publishing Ltd.

Shaji, E., Viju, J., and Thambi, D.S. (2007). High fluoride in groundwater of Palghat District, Kerala. *Current Science* 92 (2): 240–245.

Sharma, S., Yadav, K.K., Gupta, N. et al. (2015). Effects of seasonal variation on major ion chemistry of Pahuj reservoir, Jhansi, Uttar Pradesh, India. *Universal Journal of Environmental Research and Technology* 5: 79–89.

Shen, J. and Schäfer, A. (2014). Removal of fluoride and uranium by nanofiltration and reverse osmosis: a review. *Chemosphere* 117: 679–691.

Sikdar, P.K. (2019). Problems and challenges for groundwater management in South Asia. In: *Groundwater Development and Management* (ed. P.K. Sikdar), 1–18. Cham: Springer.

Sreedevi, P.D., Ahmed, S., Madé, B. et al. (2006). Association of hydrogeological factors in temporal variations of fluoride concentration in a crystalline aquifer in India. *Environmental Geology* 50 (1): 1–11.

Srivastav, A., Yadav, K.K., Yadav, S. et al. (2018). Nano-phytoremediation of pollutants from contaminated soil environment: current scenario and future prospects. In: *Phytoremediation* (eds. A.A. Ansari, S. Singh Gill, R. Gill, et al.), 383–401. Cham: Springer.

Sujatha, D. (2003). Fluoride levels in the groundwater of the south-eastern part of Ranga Reddy district, Andhra Pradesh, India. *Environmental Geology* 44 (5): 587–591.

Swarup, D. and Dwivedi, S.K. (2002). *Environmental Pollution and Effects of Lead and Fluoride on Animal Health.* Food and Agricultural Organization of the United Nations.

Tahaikt, M., El Habbani, R., Haddou, A.A. et al. (2007). Fluoride removal from groundwater by nanofiltration. *Desalination* 212 (1–3): 46–53.

Tavener, S.J. and Clark, J.H. (2006). Fluorine: friend or foe? A green chemist's perspective. *Advances in Fluorine Science* 2: 177–202.

Teotia, S.P. and Teotia, M. (1984). Endemic fluorosis in India: a challenging national health problem. *The Journal of the Association of Physicians of India* 32 (4): 347–352.

Thole, B. (2013). Ground water contamination with fluoride and potential fluoride removal technologies for east and southern Africa. *Perspectives in Water Pollution*: 66–90.

Van der Bruggen, B., Mänttäri, M., and Nyström, M. (2008). Drawbacks of applying nanofiltration and how to avoid them: a review. *Separation and Purification Technology* 63 (2): 251–263.

Valenzuela-Vasquez, L., Ramirez-Hernandez, J., Reyes-Lopez, J. et al. (2006). The origin of fluoride in groundwater supply to Hermosillo City, Sonora, Mexico. *Environmental Geology* 51 (1): 17–27.

Vithanage, M. and Bhattacharya, P. (2015). Fluoride in the environment: sources, distribution and defluoridation. *Environmental Chemistry Letters* 13 (2): 131–147.

WaterProfessionals®, q, 2018. Reverse Osmosis. WaterProfessionals®, Learning Center. http://www.waterprofessionals.com/learning-center/reverse.

Wei-fang, M.A., Wen-jun, L. and Guo-wei, C., 2009, June. Factors influencing the removal of fluoride from groundwater by nanofiltration. In 2009 3rd International Conference on Bioinformatics and Biomedical Engineering (pp. 1–5). IEEE.

WHO (2002). Fluorides, Geneva, World Health Organization. *Environmental Health Criteria* 227: 268.

World Health Organization (WHO) (2000). Fluorides. In: *Air Quality Guideline for Europe*, 2e (ed. F. Theakston), 1–273. Copenhagen: World Health Organization, Regional Office for Europe.

WPR, 2018. http://worldpopulationreview.com, Accessed date: 14 April 2018.

Yadav, K.K., Gupta, N., Kumar, V. et al. (2015). Water quality assessment of Pahuj River using water quality index at UnnaoBalaji, MP, India. *International Journal of Science: Basic Applied Research* 19 (1): 241–250.

Yadav, K.K., Singh, J.K., Gupta, N., and Kumar, V.J.J.M.E.S. (2017). A review of nanobioremediation technologies for environmental cleanup: a novel biological approach. *Journal of Material and Environmental Science* 8 (2): 740–757.

Yadav, K.K., Gupta, N., Kumar, A. et al. (2018). Mechanistic understanding and holistic approach of phytoremediation: a review on application and future prospects. *Ecological Engineering* 120: 274–298.

Yadav, K.K., Kumar, S., Pham, Q.B. et al. (2019). Fluoride contamination, health problems and remediation methods in Asian groundwater: a comprehensive review. *Ecotoxicology and Environmental Safety* 182: 109362.

Zohoori, F.V. and Duckworth, R.M. (2017). Fluoride: intake and metabolism, therapeutic and toxicological consequences. In: *Molecular, Genetic, and Nutritional Aspects of Major and Trace Minerals* (ed. J.F. Collins), 539–550. Academic Press.

3

Salinity Problems in Groundwater and Management Strategies in Arid and Semi-arid Regions

Balaji Etikala[1], *Narsimha Adimalla*[2], *Sughosh Madhav*[3], *Srinivasa Gowd Somagouni*[4], *and P.L. Keshava Kiran Kumar*[4]

[1]*Department of Geology, Sri Venkateswara University, Tirupati, Andhra Pradesh, India*
[2]*School of Environmental Science and Engineering, Chang'an University, Xi'an, China*
[3]*School of Environmental Sciences, Jawaharlal Nehru University, New Delhi, India*
[4]*Department of Geology, Yogi Vemana University, Kadapa, Andhra Pradesh, India*

3.1 Introduction

Salinity is a major social, economic, and environmental menace in climates with low rainfall and high evapotranspiration (Jabbar and Chen 2008; Abuelgasim and Ammad 2018). In general, the surface water supply in arid and semi-arid climates is scarce, which shifts the attention of policymakers to groundwater. Groundwater in these regions is very limited and considered a major resource for sustaining terrestrial ecosystems (Balaji et al. 2019a, b; Huang et al. 2019). In addition, intense urbanisation, demand for freshwater due to an increase in population, and poor management strategies have generated additional stress to this limited resource which leads to lowering groundwater levels (Cosgrove and Loucks 2015; Nagaraju et al. 2016). Therefore, people dig bore wells to great depths for groundwater that are basically rich in soluble salts (Miglietta et al. 2017; Akinlalu and Afolabi 2018). Among complex environmental issues such as droughts, heavy blowing winds, heatwaves, and floods, salinity is also a major issue. It turns soils and irrigated land more saline, which impairs crop growth and leads to low production and land degradation (Shrivastava and Kumar 2015). Moreover, salinity is a serious public health concern and its consequences are seen mostly in coastal drylands. Consuming a higher amount of salt increases blood pressure, which increases the risk of cardiovascular disease that induces heart stroke and attack. Nowadays, it accounts for a large number of deaths worldwide. Among various direct and indirect sources, salinity is one of many serious issues that affect the hydrological cycle in terms of water quality deterioration (Pulido-Bosch et al. 2018). In urban landscapes (non-agricultural lands) it affects the structures by subsidence, corrosion, and water quality deterioration. All these practices result in the loss of arable lands that affects terrestrial habitats, particularly in

Groundwater Geochemistry: Pollution and Remediation Methods, First Edition.
Edited by Sughosh Madhav and Pardeep Singh.

drylands. It is estimated that the annual loss of arable land due to salinity is 20–50% (Pitman and Läuchli 2002). It is projected that about 50% of the world's arable land that is to be lost by 2050 will be attributed to the salinity problem (FAO 2009; Hussain et al. 2020).

Therefore, some knowledge on salinity-related health issues, management strategies, and reclamation techniques are needed to tackle this exacerbated situation. Hence, the sources, implications, and reclamation strategies are discussed in this chapter.

3.2 Problem of Salinity

Salinity refers to the amount of soluble salts in water and soils, mostly of ions such as Na^+, Cl^-, SO_4^{2-}, HCO_3^-, K^+, NO_3^{2-}, and F^- (Imadi et al. 2016; Artiola et al. 2019). It is likely to be one of the major issues that affect the world economy in the near future, especially in drylands. Figure 3.1 represents the world map showing countries with salinity issues. Salinity could be either natural or human-induced. Weathering of minerals, sea breeze (mainly in coastal areas), and capillary rise of saline water from lands of low water tables are the natural factors. This can be further accentuated by irrigated agriculture, intense fertilization, and seawater intrusion due to critical groundwater overdrafts, which are human induced. It is a serious issue affecting crop production worldwide. Abrol et al. (1988) reported that more than 932.2 million hectares of the world's fertile land are at risk due to salinization which makes the arable lands unusable for farming. According to UNEP (1992) (cf. FAO-ITPS-GSP 2015; Shahid et al. 2018) 1030.1 million hectares of the world's arable land were affected by salinity. About 20–50% of the world's arable land is salt-affected and degraded (Pitman and Läuchli 2002; Glick et al. 2007). The current scenario of land degradation could be much higher than previously thought.

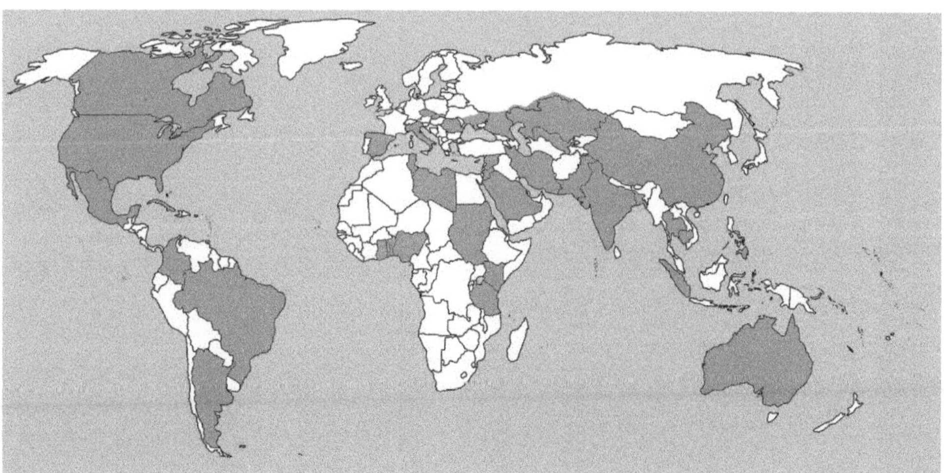

Figure 3.1 World map showing countries with salinity issues.

3.3 Sources of Aquifer Salinity

Salinity in aquifers can be either inland or coastal induced. The most common sources of salinity in aquifers is given in the following sections.

3.3.1 Inland Aquifer Salinity

Inland salinity is one of the main concerns linked to groundwater pollution and other associated issues worldwide (Greene et al. 2016). It is a daunting challenge for policymakers to handle water resources effectively. Inland salinity may result from a number of mechanisms like weathering of native rocks, erratic water supply due to low rainfall or high evaporation, leaching of poor quality irrigation water and untreated sewage effluents, recharge water quality, the inflow of paleo-saline water from adjacent formations due to critical pumping of groundwater, rise in water level due to removal of vegetation and poor drainage system, and use of excess fertilizers (Figure 3.2).

3.3.2 Coastal Aquifer Salinity

Saline water intrusion is a global threat to coastal aquifers that leads freshwater ecosystems to be contaminated due to excessive groundwater pumping (Badaruddin et al. 2015). This excess pumping of groundwater near coastal aquifers reduces the hydraulic head of inland groundwater that allows the saltwater to enter into inland aquifers, which leads to aquifer salinization. In fact, it is not only the cause, but it is also affected by global warming, which increases the opportunity for the intrusion of saltwater into coastal aquifers (Figure 3.3).

3.4 Types of Salinity

Based on the causes, salinity has been divided into four types. They are given in the following sections.

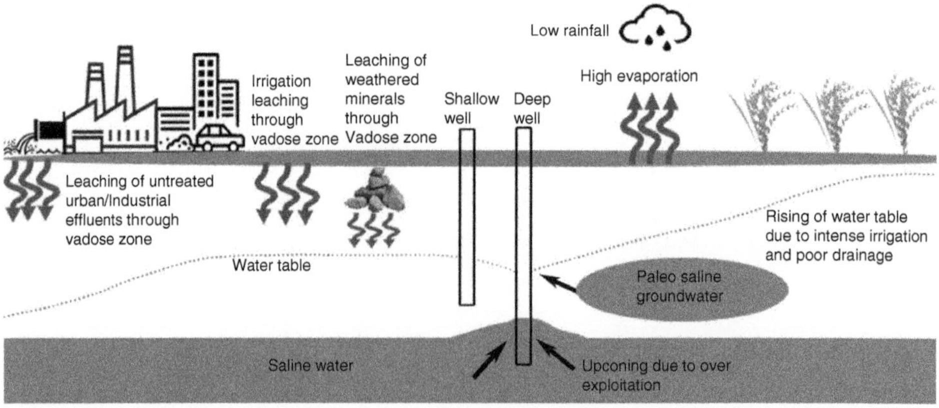

Figure 3.2 Salinization of inland aquifers. *Source:* Brindha and Schneider (2019), Elsevier.

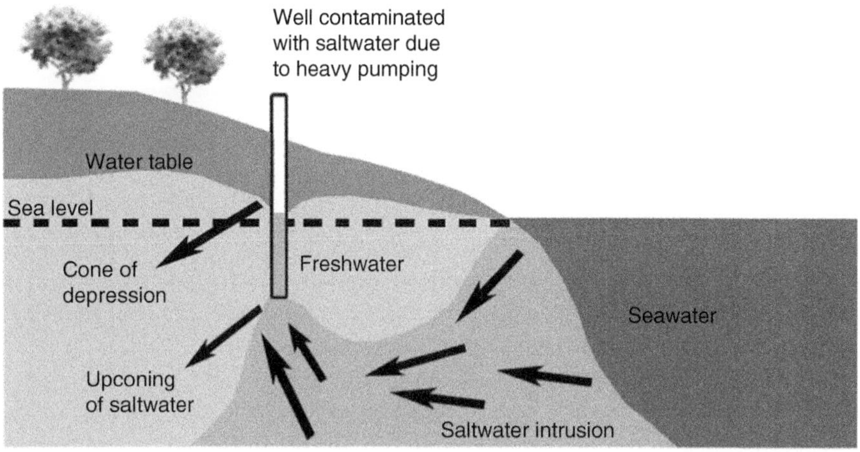

Figure 3.3 Salinization of groundwater in coastal areas.

3.4.1 Primary Salinity

Primary salinity is also called natural salinity. The most common sources of primary salinity are the rainfall, the characteristics of the parent rock, and seawater intrusion (Podmore 2009). In general, the rain leaves a certain amount of salt in the soils through evaporation. Over many cycles, these salts in the soil reach elevated levels. Rocks such as granites, rhyolites, and marine sediments left by the retreating of seas can contain high salts, which may release into the soil and mobilise into groundwater through weathering. Moreover, salts may be brought into the lands by strong winds and some salts may enter into the coastal aquifer by seawater intrusion.

3.4.2 Secondary Salinity

Secondary salinity is also referred to as dryland salinity and it is a major problem in the world. It is caused by the rising of the water table due to the evaporation of water from the soil. In drylands, the water loss is reduced due to the removal of vegetation and change in land use patterns that allows accumulation of more salts into the soil and groundwater, which have adverse effects on plant growth and lead to low crop yield (Zaman et al. 2018). The other causes that induce water table rise are restricted drainage systems, excessive recharge of groundwater due to heavy rains and floods, and replacement of deep-rooted perennial plants with shallow-rooted annual plants.

3.4.3 Tertiary Salinity

Tertiary salinity is also called irrigated salinity. It is characterised by the rise in the local groundwater level due to repetitive irrigation with large quantities of water over many cycles. This process can add some salts to the soil profile, which may also be mobilised into the groundwater. Each successive irrigation or reuse of saline groundwater keeps adding more salts to the groundwater, which progressively becomes more saline, resulting in higher levels of salinity over several cycles (Zaman et al. 2018). It gets even worse when irrigating from poor quality water or saltwater.

3.4.4 Urban Salinity

Urban salinity is a combination of both dryland salinity and irrigation salinity that has the potential to affect valuable assets. It is characterized by the rise in groundwater level that is possibly due to blocking or changing natural drainage patterns due to urban developmental activities like laying roads, constructing buildings and other infrastructures, and leakage of pipes and drains. The most common sources of urban salinity are untreated urban effluents/industrial wastewater, building materials, fertilizers, and chemicals (Brindha and Schneider 2019).

3.5 Effects on Agriculture

3.5.1 Soil Structure

The bivalent cations such as Ca^{2+} and Mg^{2+} tend to flocculate, which facilitates penetration of water into roots, whereas the monovalent cations such as Na^+ and K^+ disperse the clay particles, which reduces the porosity of the soil. Hence the excess amount of Na^+ and K^+ has a profound impact on the relationship between soil and water, resulting in soil erosion and crop failure (Chibowski 2011).

3.5.2 Oxidative and Alkaline Stress

In general, the osmotic gradient helps in taking water from the soil by roots. Salinity in soil diminishes the osmotic gradient, which reduces the intake of water by roots and hinders cellular activities. This leads to the loss of vacuole fluid and the plant starts to wilt (Litalien and Zeeb 2019). The alkaline soils, usually saline/saline-sodic with pH above 8, tend to reduce the absorption of nutrients that due to the redox potential of major nutrients (Husson 2013).

3.5.3 Ion Toxicity

Long-term saline stress in terms of excess Cl^- and Na^+ ions in soils induces the accumulation of ions into the plant that leads to ion toxicity. A high concentration of Cl^- ions in the soils affect the plant; further, it can affect photosynthesis, which leads to leaf burn and necrosis. Whereas the excess amount of Na^+ ions in soil reduces the intake of K^+ ions, which is highly desirable (White and Broadley 2001; Barhoumie et al. 2007; Machado and Serralheiro 2017). Boron toxicity is a common issue in the soils of the arid and semi-arid regions. It affects various aspects of the plant growth, such as metabolism alteration, lowering chlorophyll content in leaves, and decreasing root growth (Nable et al. 1990, 1997).

3.5.4 Nutrient Deficiencies

Continuous salt accumulation in soils over a period of time can cause an ionic imbalance in plant cells that inhibits the absorption of core elements like Ca^{2+}, K^{+}, and NO_3^{2-}. Accumulation of Na^{+} and Cl^{-} ions in soil induces nutritional deficiencies in plant tissues, which results in Na^{+} induced Ca^{2+} and K^{+}, Ca^{2+} induced Mg^{2+} and Cl^{-} induced NO_3^{2-} deficiencies (Grattan and Grieve 1992). Excess boron in soils results in deficiencies of Ca^{2+} in plants that cause necrosis of the leaf (Abdulnour et al. 2000).

3.6 Effects on Non-Agricultural Lands and Other Natural Resources

3.6.1 Subsidence of Land

Salinity has a profound effect on land subsidence, especially in clay-dominated coastal soils. Higher salinity in water reduces the interconnectivity of the pores by converting clay fabrics into parallel alignments that induce the fast dissipation of pore water. Hence, consolidation is more pronounced (Sarah et al. 2018).

3.6.2 Corrosive Risk

In general, the corrosive risk of freshwater is lower than that of saline water. The water containing a high percentage of dissolved oxygen and sodium and other chlorides makes metals like steel and low-alloy steels more susceptible to metal corrosion. In fact, these are not only the causes but are also affected by the other dependant factors such as pH, temperature, amount of soluble gases, and pollutants (Zakowski et al. 2014).

3.6.3 Deterioration of Water Quality

Salinity is one of the major issues that affect water resources in various forms. It is possibly due to both natural and anthropogenic activities. Seawater intrusion, rise in the water table due to poor irrigation and drainage, disturbance in existing groundwater salinity stratification by digging bore wells, and industrial effluents are the notable natural and human-induced groundwater salinity sources that can deteriorate the water quality by acidification and release of toxic ions into it. (Greene et al. 2016).

3.7 Effects of Saline Water on Human Health

Salinity is a serious environmental issue worldwide, especially in drylands and coastal regions. Dryland salinity is a major environmental degradation problem observed in Australia (Lambers 2003). Seawater intrusion is a major concern in coastal areas of Bangladesh, Brazil, California, China, India, Indonesia, Netherlands, and Vietnam (Chakraborty et al. 2019; Rahaman et al. 2020). The sea-level rise is a major issue for coastal cities such as Chennai, Cochin, Kolkata, and Mumbai. The major fertile river deltas in

India such as Cauvery, Indus, and Krishna are vulnerable to floods and seawater intrusion (Rahaman et al. 2020). The salinity issues in terms of seawater intrusion and sea-level rise may increase in the future due to climate change, and human activities like an increase in groundwater overdrafts and shrimp culture along the seacoasts, which may affect the coastal ecosystem to a greater extent (Akib Jabed et al. 2018).

Drinking saline water is a global health issue notably in coastal areas. Earlier, a number of studies reported that the people drinking large quantities of saline water may suffer from cardiovascular disease, diarrhoea, rise in blood pressure, hypertension, infant mortality, and skin and respiratory diseases (Dasgupta et al. 2016; Akib Jabed et al. 2018; Chakraborty et al. 2019). Though the salinity is a global issue, its health effects are often seen in low-income countries where water is poorly treated or totally untreated (Vineis et al. 2011). Health issues such as chronic malnutrition, low-calorie intake and hypertension were reported in coastal peoples of Bangladesh (Nahian et al. 2017; Rahaman et al. 2020). Mental and respiratory diseases were reported in Australia due to the inland salinity issue (Jardine et al. 2007). Studies in Arizona, Illinois, and Massachusetts, USA, suggested that high intake of salts would lead to raising blood pressure (Tuthill and Calabrese 1981; Welty et al. 1986; Rahaman et al. 2020). A study from Vietnam reported that high salt intake is highly associated with a rise in blood pressure that increases the risk of cardiovascular disease (Do et al. 2016). Moreover, the salinity shows some considerable impacts on soil microbial species which lowers the crop productivity (Dasgupta et al. 2017).

3.8 Management Strategies

Proper management strategies have to be followed to reduce the salinity effects, especially in arid and semi-arid areas. They are given in the following sections.

3.8.1 Lowering of the Groundwater Table

Extensive withdrawal of groundwater in the upstream side of the river may reduce the groundwater level in the coastal areas that allows the intrusion of seawater. Hence, crop cultivation is quite difficult in coastal areas due to the presence of saltwater. In order to reduce the salinity effects in coastal areas, leaching of salts has to be reduced. Hence, it is necessary to maintain a proper drainage system to lower the water table at least 1.5 m from the surface of the soil to prevent salt accumulation (Alam et al. 2017).

3.8.2 Construction of Water Harvesting Structures

In order to reduce the effect of salinity, groundwater consumption has to be substituted with freshwater (rainfall) for irrigation. Proper rainwater harvesting structures have to be constructed to cope with salinity for sustaining agricultural livelihood in drylands. Floodwater harvesting such as spat irrigation and runoff farming, macro-catchments such as highlands, and micro-catchment structures such as bunds, pits holes, and basins are the rainwater harvesting structures that can reduce the groundwater consumption for irrigation especially in drylands (Gebreyess and Abayineh 2019).

3.8.3 Reclamation of Saline Soils

Soil reclamation is one of the best ways to reduce the impact of salinity. In general, it refers to the strategies to extract soluble salts from the roots of the crops. Some of the best practices to reduce the impact of salinity are leaching, enhanced water management techniques, establishing surface (through ditches) and subsurface drainages (through open ditches, mole drains, crop rotations), use of organic or chemical fertilizers and use of salt-tolerant cultivars (Esenov and Redjepbaev 1999; Shrivastava and Kumar 2015; Kaledhonkar et al. 2019).

3.8.4 Leaching

Soil salinity is one of the important factors that influence plant physical and biological activities thereby reducing the crop yields. Hence, it is necessary to remove the excess salts from the root zone of the soil for improving crop production. Leaching is one of the important processes to remove salts from the root zone by applying a large amount of freshwater into the field and allowing the water to infiltrate. During infiltration, the excess salts from the root zone are washed away into deep soil layers. This process is effective when it is to be done in soils with low moisture and deep groundwater water tables (Zaman et al. 2018).

3.8.5 Surface and Subsurface Drainage Systems

Drainage refers to the removal of surface or subsurface water by natural or by installing artificial drainage systems. Drainage helps in lowering the water table and reducing the risk of rising groundwater table and accumulation of salts through the capillary rise. Hence, it is necessary to have a proper drainage system in order to reduce the negative impacts of rising water table and accumulation of salts. It can be accomplished by establishing surface (through ditches) and subsurface drainages (through open ditches, mole drains) (Shahid et al. 2018).

3.8.6 Possible Strategies and Practices to Reduce Salinity-Related Health Issues

Generally, the people from drylands and coastal areas are more vulnerable to salinity-related health issues to a large extent. Hence, it is necessary to create awareness among the communities to find alternative freshwater supplies to cope with salinity related health issues. Though there are several desalination processes which are of high cost, some low-cost techniques such as rainwater harvesting structures and ponds should be constructed. Optimal use and reutilisation of rainwater for various purposes is needed to reduce the risk of salinity to some extent.

3.8.7 Organic or Chemical Fertilizers

Organic fertilizers help in releasing various elements like Ca^{2+} and Mg^{2+} into the soil through decomposition and help in increasing soil water holding capacity. It also helps in reducing Na^{+} toxicity through cation exchange capacity (Machado and Serralheiro 2017).

Application of potash fertilizers can reduce the salinity effect of soils on crops. It reduces the uptake of Na^+, which helps in increasing crop production. It also helps in the uptake of elements such as Ca^{2+}, Mg^{2+}, K^+, and P. The boron toxicity can be overcome by adding nitrogen fertilizers (Koohkan and Maftoun 2016). The effects of Cl^- toxicity in soils can be ameliorated by adding nitrogen fertilizers (Bar et al. 1997; Karaivazoglou et al. 2005). The toxic effects of sodic soils on crops can be rectified by adding gypsum to the soil (Zaman et al. 2018).

3.8.8 Salt-Tolerant Cultivars

Salt accumulation in soils is one of the major issues worldwide. An uptaking of a large amount of soluble salts through roots has a significant impact on plant physiological and metabolic processes, which reduces crop yields. Usage of salt-tolerant cultivars is one of the major mitigation processes to cope with salinity problems. To overcome this issue, high tolerant crops have to be grown. Table 3.1 shows the list of salt-tolerant cultivars.

3.8.9 Water Management

Efficient water management techniques in saline soils can reduce the risk of salt accumulation in root zones and increase the crop yield. They are given in the following sections.

Table 3.1 List of salt-tolerant cultivars.

Sensitive	Moderately tolerant	Tolerant	Highly tolerant
Pea (*Pisum sativum*)	Wheat (*Triticum aestivum*)	Cabbage (*Brassica oleracea var. capitata*)	Asparagus (*Asparagus officinalis*)
Soybean (*Glycine max*)	Sunflower (*Helianthus annus*)	Olive tree (*Olea europaea*)	Beetroot (*Beta vulgaris*)
Gram (*Cicer arietium*)	Onion (*Allium cepa*)	Tomato (*Solanum lycopersicum*)	Rye (*Secale cereale*)
Groundnut (*Arachis hypogea*)	Barley (*Hordeum vulgare*)	Rice (*Oryza sativa*)	Date palm (*Phoenix dactylifera*)
Peach (*Prunus persica*)	Lucerne (*Medicago sativa*)	Spinach (*Spinacia oleracea*)	
Sesamum (*Sesamum oriental*)	Garlic (*Allium sativum*)	Sugarbeet (*Beta vulgaris*)	
Mung (*Phaseolus aureus*)	Oat (*Avena sativa*)	Dhaincha (*Sesbania aculeata*)	
Maize (*Zea mays*)	Pearl millet (*Pennisetum typhoides*)	Carrot (*Daucus carota subsp. sativus*)	
Lime (*Citrus aurantiifolia*)			

Source: Abrol et al. (1988), Singh (2009), Galvani (2006).

3.8.9.1 Irrigation Methods

To alleviate the saline stress coupled with low soil moisture, the crops grown under saline conditions should be more frequently irrigated than the non-saline conditions (Shrivastava and Kumar 2015). Sprinkler irrigation is the best method for frequently irrigated lands. It has the advantage of releasing small amounts of water for the infiltration process that reduces the leaching of salts (Minhas 1996). Drip irrigation is one of the best irrigation methods in lands irrigated with saline water. It keeps the plant root zone hydrated, which maintains low salt levels (Alhammadi and Al-Shrouf 2013).

3.8.9.2 Mulching

Mulching is the prominent process of soil moisture conservation from evaporation by placing polyethylene sheets, grass, and crop residues at the top of the soil. It helps to improve the quality of soil by reducing soil erosion and weed growth, regulating soil temperature, improving aeration, and supplying nutrients to the roots. Moreover, it also reduces the upward movement and accumulation of salts in the root zone, which helps to increase crop yields (Abd El-Mageed et al. 2016).

3.8.9.3 Crop Rotation

Crop rotation is the widely used cropping system in combating salinity which gives better results when the crop rotation is accompanied by good quality water and salt-tolerant cultivars. Growing of crops that rely on long fallowing for soil moisture conservation may favour rising of the groundwater table. It brings salts to the surface, which inhibits crop growth. In order to reduce the salinity perennial crops to be grown in rotation with annual crops. The earlier studies show that growing of Lucerne in rotation with wheat has a significant impact in combating salinity (Jobbágy and Jackson 2007).

3.9 Conclusions

Salinity is a serious threat to the environment which reduces agricultural yields, economic outcomes, and soil erosion that eventually leads to land deterioration, particularly in drylands. Moreover, it affects public health to a greater extent. Therefore, it is necessary to have basic knowledge of crop response to salt tolerance and proper management strategies such as constructions of artificial recharge structures, reclamation of soils, and water management methods to boost the global economy by increasing food production and reducing the risk of exposure to salinity-associated health issues. Earlier literature showed that some cropping patterns, management strategies, and methods of water management have been successfully adopted by the farming community and governing bodies and yielded good results in managing dryland salinity and other salinity-related problems. Considering the earlier literature as a reference, some future research is required in this field for sustainable development of agricultural livelihood and human ecology in combating salinity.

References

Abd El-Mageed, T.A., Semida, W.M., and Abd El-Wahed, M.H. (2016). Effect of mulching on plant water status, soil salinity and yield of squash under summer-fall deficit irrigation in salt affected soil. *Agricultural Water Management* 173 (1): 12. https://doi.org/10.1016/j.agwat.2016.04.025.

Abdulnour, J.E., Donnelly, D.J., and Barthakur, N.N. (2000). The effect of boron on calcium uptake and growth in micro propagated potato plantlets. *Potato Research* 43: 287–295. https://doi.org/10.1007/BF02358088.

Abrol, I.P., Yadav, J.S.P., and Massoud, F.I. (1988). *Salt-Affected Soils and Their Management.* FAO Soils Bulletin 39. Rome, Italy: Food and Agriculture Organization of the United Nations 131 pp.

Abuelgasim, A. and Ammad, R. (2018). Mapping soil salinity in arid and semi-arid regions using Landsat 8 OLI satellite data. *Remote Sensing Applications: Society and Environment.* https://doi.org/10.1016/j.rsase.2018.12.010.

Akib Jabed, M., Paul, A., and Nath, T.K. (2018). Peoples' perception of the water salinity impacts on human health: a case study in south-eastern coastal region of Bangladesh. *Exposure and Health* https://doi.org/10.1007/s12403-018-0283-0.

Akinlalu, A.A. and Afolabi, D.O. (2018). Borehole depth determination to freshwater and well design using geophysical logs in coastal regions of Lagos, southwestern Nigeria. *Applied Water Science* 8: 152. https://doi.org/10.1007/s13201-018-0798-3.

Alhammadi, M.S. and Al-Shrouf, A.M. (2013). Irrigation of Sandy soils, basics and scheduling. *Crop Production.* https://doi.org/10.5772/55117.

Alam, M.Z., Carpenter-Boggs, L., Mitra, S. et al. (2017). Effect of salinity intrusion on food crops, livestock, and fish species at Kalapara Coastal Belt in Bangladesh. *Journal of Food Quality* 2017: 1–23. https://doi.org/10.1155/2017/2045157.

Artiola, J.F., Walworth, J.L., Musil, S.A., and Crimmins, M.A. (2019). Soil and land pollution. *Environmental and Pollution Science*: 219–235. https://doi.org/10.1016/b978-0-12-814719-1.00014-8.

Balaji, E., Veeraswamy, G., Peiyue, L., and SiddiRaju, R. (2019a). Deciphering groundwater potential zones using MIF technique and GIS: a study from Tirupati area, Chittoor District, Andhra Pradesh, India. *HydroResearch* 1: 1–7. https://doi.org/10.1016/j.hydres.2019.04.001.

Balaji, E., Veeraswamy, G., Narsimha, A., and Subba Rao, M. (2019b). Factors controlling groundwater chemistry of Renigunta area, Chittoor District, Andhra Pradesh, South India. A multivariate statistical approach. *HydroResearch* 1: 57–62. https://doi.org/10.1016/j.hydres.2019.06.002.

Bar, Y., Apelbaum, A., Kafkafi, U., and Goren, R. (1997). Relationship between chloride and nitrate and its effect on growth and mineral composition of avocado and citrus plants. *Journal of Plant Nutrition* 20 (6): 715–731. https://doi.org/10.1080/01904169709365288.

Barhoumie, Z., Djebali, W., Smaoui, A. et al. (2007). Contribution of NaCl excretion to salt resistance of Aeluropus littoralis (Willd) Parl. *Plant Physiology* 164: 842–850.

Badaruddin, S., Werner, A.D., and Morgan, L.K. (2015). Water table salinization due to intrusion. *Water Resources Research* 51 (10): 8397–8408. https://doi.org/10.1002/2015wr017098.

Brindha, K. and Schneider, M. (2019). Impact of urbanization on groundwater quality. GIS and geostatistical techniques for groundwater. *Science*: 179–196. https://doi.org/10.1016/b978-0-12-815413-7.00013-4.

Chakraborty, R., Khan, K.M., Dibaba, D.T. et al. (2019). Health implications of drinking water salinity in coastal areas of Bangladesh. *International Journal of Environmental Research and Public Health* 16 (19): 3746. https://doi.org/10.3390/ijerph16193746.

Chibowski, E. (2011). Flocculation and dispersion phenomena in soils. *Encyclopedia of Earth Sciences Series*: 301–304. https://doi.org/10.1007/978-90-481-3585-1_59.

Cosgrove, W.J. and Loucks, D.P. (2015). Water management: current and future challenges and research directions. *Water Resources Research* 51 (6): 4823–4839. https://doi.org/10.1002/2014wr016869.

Dasgupta, S., Huq, M., Mustafa, M.G., et al. (2016). The impact of climate change and aquatic salinization on fish species and poor communities in the Bangladesh sundarbans. World Bank Policy Research Working Paper # 7593.

Dasgupta, S., Huq, M., Mustafa, M.G. et al. (2017). The impact of aquatic salinization on fish habitats and poor communities in a changing climate: evidence from southwest coastal Bangladesh. *Ecological Economics* 139: 128–139. https://doi.org/10.1016/j.ecolecon.2017.04.009.

Do, H.T.P., Santos, J.A., Trieu, K. et al. (2016). Effectiveness of a Communication for Behavioral Impact (COMBI) intervention to reduce salt intake in a Vietnamese Province based on estimations from spot urine samples. *The Journal of Clinical Hypertension* 18 (11): 1135–1142. https://doi.org/10.1111/jch.12884.

Esenov, P.E. and Redjepbaev, K.R. (1999). The reclamation of saline soils. In: *Desert Problems and Desertification in Central Asia* (ed. A.G. Babaev). Berlin, Heidelberg: Springer https://doi.org/10.1007/978-3-642-60128-6_15.

FAO (2009). *The State of Food Insecurity in the World*. Rome: Food and Agriculture Organization of the United Nations.

FAO-ITPS-GSP (2015) Status of the world's soil resources. FAO-ITPS-GSP Main Report, Food and Agriculture Organization of the United Nations, Rome, Italy, pp 125–127.

Gebreyess, B.F. and Abayineh, A. (2019). Water harvesting technologies in semi-arid and arid areas. *Journal of Degraded and Mining Lands Management* 7 (1): 1921–1928, ISSN 2502-2458. doi:https://doi.org/10.15243/jdmlm.2019.071.1921.

Galvani, A. (2006). The challenge of the food sufficiency through salt tolerant crops. *Reviews in Environmental Science and Bio/Technology* 6 (1–3): 3–16. https://doi.org/10.1007/s11157-006-0010-3.

Glick, B.R., Cheng, Z., Czarny, J., and Duan, J. (2007). Promotion of plant growth by ACC deaminase-producing soil bacteria. *European Journal of Plant Pathology* 119: 329–339.

Grattan, S.R. and Grieve, C.M. (1992). Mineral element acquisition and growth response of plants grown in saline environments. *Agriculture, Ecosystems and Environment* 38 (4): 275–300. https://doi.org/10.1016/0167-8809(92)90151-z.

Greene, R., Timms, W., Rengasamy, P. et al. (2016). Soil and aquifer salinization: toward an integrated approach for salinity management of groundwater. In: *Integrated Groundwater Management*, vol. 2016 (eds. A.J. Jakeman, O. Barreteau, R.J. Hunt, et al.), 377–412. Switzerland: Springer International Publishing AG https://doi.org/10.1007/978-3-319-23576-9_15.

Huang, F., Zhang, Y., Zhang, D., and Chen, X. (2019). Environmental groundwater depth for groundwater-dependent terrestrial ecosystems in arid/semi-arid regions: a review. *International Journal of Environmental Research and Public Health* 16 (5): 763. https://doi.org/10.3390/ijerph16050763.

Hussain, M.I., Farooq, M., Muscolo, A., and Rehman, A. (2020). Crop diversification and saline water irrigation as potential strategies to save freshwater resources and reclamation of marginal soils—a review. *Environmental Science and Pollution Research*. https://doi.org/10.1007/s11356-020-09111-6.

Husson, O. (2013). Redox potential (Eh) and pH as drivers of soil/plant/microorganism systems: a transdisciplinary overview pointing to integrative opportunities for agronomy. *Plant and Soil* 362 (1–2): 389–417.

Imadi, S.R., Shah, S.W., Kazi, A.G. et al. (2016). Phytoremediation of saline soils for sustainable agricultural productivity. *Plant Metal Interaction*: 455–468. https://doi.org/10.1016/b978-0-12-803158-2.00018-7.

Jabbar, M. and Chen, X. (2008). Land degradation due to salinization in arid and semi-arid regions with the aid of geo-information techniques. *Geo-Spatial Information Science* 11 (2): 112–120. https://doi.org/10.1007/s11806-008-0013-z.

Jardine, A., Speldewinde, P., Carver, S., and Weinstein, P. (2007). Dryland salinity ecosystem distress syndrome: human health implications. *EcoHealth* 4 (1): 10–17.

Jobbágy, E.G. and Jackson, R.B. (2007). Groundwater and soil chemical changes under phreatophytic tree plantations. *Journal of Geophysical Research* 112 (G2) https://doi.org/10.1029/2006jg000246.

Karaivazoglou, N.A., Papakosta, D.K., and Divanidis, S. (2005). Effect of chloride in irrigation water and form of nitrogen fertilizer on Virginia (flue-cured) tobacco. *Field Crops Research* 92 (1): 61–74. https://doi.org/10.1016/j.fcr.2004.09.006.

Kaledhonkar, M.J., Meena, B.L., and Sharma, P.C. (2019). Reclamation and nutrient management for salt-affected soils. *Indian Journal of Fertilisers* 15 (5): 566–575.

Koohkan, H. and Maftoun, M. (2016). Effect of nitrogen– boron interaction on plant growth and tissue nutrient concentration of canola (Brassica napus L.). *Journal of Plant Nutrition* 39 (7): 922–931. https://doi.org/10.1080/01904167.2016.1143492.

Lambers, H. (2003). Introduction, dryland salinity: a key environmental issue in southern Australia. *Plant and Soil* 257: 5–7. https://doi.org/10.1023/B:PLSO.0000003909.80658.d8.

Litalien, A. and Zeeb, B. (2019). Curing the earth: a review of anthropogenic soil salinization and plant-based strategies for sustainable mitigation. *Science of the Total Environment*: 134235. https://doi.org/10.1016/j.scitotenv.2019.134235.

Machado, R. and Serralheiro, R. (2017). Soil salinity: effect on vegetable crop growth. Management practices to prevent and mitigate soil salinization. *Horticulturae* 3 (2): 30. https://doi.org/10.3390/horticulturae3020030.

Minhas, P.S. (1996). Saline water management for irrigation in India. *Agricultural Water Management* 30 (1): 1–24. https://doi.org/10.1016/0378-3774(95)01211-7.

Miglietta, P., Toma, P., Fanizzi, F. et al. (2017). A Grey water footprint assessment of groundwater chemical pollution: case study in Salento (Southern Italy). *Sustainability* 9 (5): 799. https://doi.org/10.3390/su9050799.

Nable, R.O., Cartwright, B., and Lance, R.C. (1990). Genotypic differences in boron accumulation in barley: relative susceptibilities to boron deficiency and toxicity. In: *Genetic*

Aspects of Plant Mineral Nutrition (eds. N. El Bassam, M. Dambroth and B. Loughman), 243–251. Dordrecht, The Netherlands: Kluwer Academic Publishers.

Nable, R.O., Banuelos, G.S., and Paull, J.G. (1997). Boron toxicity. *Plant and Soil* 198: 181–198.

Nagaraju, A., Balaji, E., and Dash, P. (2016). Land use/land cover analysis based on various comprehensive geospatial data sets: a case study from Tirupati Area, South India. *Advanced Remote Sensing* 5: 73–82. https://doi.org/10.4236/ars.2016.52006.

Nahian, M.A., Ahmed, A., Lazar, A.N. et al. (2017). Drinking water salinity associated health crisis in coastal Bangladesh. *Elementa: Science of the Anthropocene* 6 https://doi.org/10.1525/elementa.143.

Pitman, M.G. and Läuchli, A. (2002). Global impact of salinity and agricultural ecosystems. Salinity: environment - plants. *Molecules*: 3–20. https://doi.org/10.1007/0-306-48155-3_1.

Podmore, C. (2009). Irrigation salinity-causes and impacts. *Primefacts* 937: 1–4.

Pulido-Bosch, A., Rigol-Sanchez, J.P., Vallejos, A. et al. (2018). Impacts of agricultural irrigation on groundwater salinity. *Environmental Earth Sciences* 77 (5) https://doi.org/10.1007/s12665-018-7386-6.

Rahaman, M.A., Rahman, M.M., and Nazimuzzaman, M. (2020). Impact of salinity on infectious disease outbreaks: experiences from the global coastal region. In: *Good Health and Well-Being. Encyclopedia of the UN Sustainable Development Goals* (eds. F.W. Leal, T. Wall, A.M. Azul, et al.). Cham: Springer https://doi.org/10.1007/978-3-319-95681-7.

Sarah, D., Hutasoit, L., Delinom, R. et al. (2018). A physical study of the effect of groundwater salinity on the compressibility of the Semarang-DemakAquitard, Java Island. *Geosciences* 8 (4): 130. https://doi.org/10.3390/geosciences8040130.

Singh, G. (2009). Salinity-related desertification and management strategies: Indian experience. *Land Degradation and Development* 20 (4): 367–385. https://doi.org/10.1002/ldr.933.

Shahid, S.A., Zaman, M., and Heng, L. (2018). Soil salinity: historical perspectives and a world overview of the problem. In: *Guideline for Salinity Assessment, Mitigation and Adaptation Using Nuclear and Related Techniques*. Cham: Springer https://doi.org/10.1007/978-3-319-96190-3_2.

Shrivastava, P. and Kumar, R. (2015). Soil salinity: a serious environmental issue and plant growth promoting bacteria as one of the tools for its alleviation. *Saudi Journal of Biological Sciences* 22 (2): 123–131. https://doi.org/10.1016/j.sjbs.2014.12.001.

Tuthill, R.W. and Calabrese, E.J. (1981). Drinking water sodium and blood pressure in children: a second look. *American Journal of Public Health* 71 (7): 722–729.

UNEP (1992) Proceedings of the Ad-hoc expert group meeting to discuss global soil database and appraisals of GLASOD/SOTER, 24–28 February 1992, Nairobi, Kenya, 39 pp

Vineis, P., Chan, Q., and Khan, A. (2011). Climate change impacts on water salinity and health. *Journal of Epidemiology and Global Health* 1 (1): 5–10. https://doi.org/10.1016/j.jegh.2011.09.001.

White, P.J. and Broadley, M.R. (2001). Chloride in soils and its uptake and movement within the plant: a review. *Annals of Botany* 88: 967–988.

Welty, T.K., Freni, L.W., Zack, M.M. et al. (1986). Effects of exposure to salty drinking water in an Arizona community. Cardiovascular mortality, hypertension prevalence, and relationships between blood pressure and sodium intake. *Journal of the American Medical Association* 255: 622–626.

Zakowski, K., Narozny, M., Szocinski, M., and Darowicki, K. (2014). Influence of water salinity on corrosion risk—the case of the southern Baltic Sea coast. *Environmental Monitoring and Assessment* 186 (8): 4871–4879. https://doi.org/10.1007/s10661-014-3744-3.
Zaman, M., Shahid, S.A., and Heng, L. (2018). Guideline for salinity assessment. *Mitigation and Adaptation Using Nuclear and Related Techniques* https://doi.org/10.1007/978-3-319-96190-3.

4

Heavy Metal Contamination in Groundwater Sources

Pinki Rani Agrawal, Sanchita Singhal, and Rahul Sharma

Academy of Scientific and Innovative Research (AcSIR), CSIR-National Physical Laboratory campus, New Delhi, India

4.1 Introduction

"We drink water but are always hesitant about the quality." This seems to be miserable, that though we have plenty of water around, the supply of fresh potable water is still a heavy burden on our pockets (Vanloon and Duffy 2017). Being a key supportive system in daily livelihood and industrial development, ensuring the continuous supply of fresh water is an essential aspect. As per the United Nations Human Right Council (UNHRC), safe drinking water and sanitation are the fundamental right of every citizen for a sustainable healthy environment. The Council commits that by 2030 all people in the world should benefit from clean water assistance so as to allow equity for social and economic development. Such reforms and targets are already getting positive results in some developed countries but it seems that the situation is much more challenging for developing nations. In fact, it has been reported that developing countries like China, India, Sri Lanka, Egypt, Malaysia, Nigeria, Indonesia, Philippines, Bangladesh, etc. account for around 90% of debris disposed in oceans (Tran et al. 2020). Every day around two million tons of sewage and other waste get into water bodies, which corresponds to around eight million deaths every year where untreated water consumption is a direct cause (WHO, 14 May 2019).

Considering the water in the world, only ~2.5% is available as non-saline freshwater but unfortunately, this is continuously undergoing contamination and thus a water crisis arises. From big quantities of garbage to some nano-sized chemicals, a wide range of pollutants indulge in these water bodies and make them unfit for potable purposes. Along with these sources, the exploding population, lack of sustainable usage of water, climate change, rising industrial demand, farming and domestic sectors, and changing consumption patterns have also contributed in the freshwater crisis. It has been reported that water use has been increasing by about 1% per year since the 1980s and if this persists, then by 2050 an increase of ~20–30% above the current level of water use is expected (Islam and Karim 2019). Recently, the World Economic Forum in 2019 has categorized water pollution

Groundwater Geochemistry: Pollution and Remediation Methods, First Edition.
Edited by Sughosh Madhav and Pardeep Singh.

in one of the largest global risks due to human interventions (Băhnăreanu 2019). It has been reported that one-sixth of the world's population is suffering from freshwater unavailability (Elimelech 2006).

Water contamination could be natural or man-made. Natural contaminants include geological materials from sedimentary rocks, floods, volcanic eruption, etc. with a wide range of elements like magnesium (Mg), calcium (Ca), copper (Cu), lead (Pb), iron (Fe), etc. These are essential for human health at some acceptable limits, but beyond that cause acute health effects (Vörösmarty et al. 2005; Ghrefat et al. 2014). Artificial contaminants include by-products of petroleum, dyes, chemical, oil, batteries, and foodstuff industries. The Centres for Disease Control and Prevention reports Hepatitis A virus, norovirus, *Escherichia coli*, Salmonella, fluoride ions, heavy metals like As, Pb, Cd, Hg, Cr, Zn, perfluoroalkyl substances (PFAS), chlorinated solvents, pesticides, nitrates, and carbonates as the leading causes of these waterborne diseases (Cramer et al. 2008).

In addition, the use of pesticides, insecticides, and fertilizers in agricultural practices also add to this section (Kass et al. 2005). Pathogens and parasites, spread by human and animal waste, also pollute water to some extent; however, their proportion is comparatively less than natural and artificial contaminants.

Volatile organic compounds (VOCs) including chemicals like toluene, styrene, phosgene, gasoline additives, and adhesives, and microorganisms like Salmonella, Vibrio cholera, and Shigella, also contribute to chronic diseases like typhoid, fever, vomiting, diarrhoea, dysentery, gastroenteritis, and cholera (Council 2009; Soni et al. 2018).

Generally, water spoiled with pollutants causes different health problems to humans, plants, and aquatic life (which directly/indirectly affects humans). Exposure to these noxious wastes lead to chronic poisoning characterized by neurological, gastrointestinal, and cardiovascular problems, and liver damage (Dada et al. 2016).

Fluoride in water is essential for protection against bone weakening but concentration above 0.5 mg/L leads to fluorosis. Ingesting large quantities of Cu-contaminated water causes diarrhoea, fever, liver, and kidney damage. Exposure to PFAS can lead to fertility problems, cholesterol rise, cancer, and thyroid problems. According to the Environmental Protection Agency, a person's exposure to PFAS should be less than 70 parts per trillion and in Hoosick Falls, it was detected to be 130 000 parts per trillion (Michaels 2017).

Long-term Cd exposure by consuming the food crops irrigated by effluents may cause kidney and skeletal damage (Friberg and Vahter 1983). Infants and pregnant ladies consuming Lead (Pb) contaminated water can experience developmental delay, abdominal pain, birth defects, hypertension, and preeclampsia. Pb contaminates in drinking water from pipes, solder, and household plumbing systems and affects the blood, central nervous system, and kidneys. Flint, Michigan, is an example in this regard. The city officials decided to start using Flint River as an alternative point water source for short period of time in 2014. The new water pipeline was built but not treated with any anti-corrosive agent to deter Pb contamination. Studies revealed that this negligence led to increase in the blood lead levels by a factor of two, three, or more, which actually causes many skin-related problems (Grossman and Slutsky 2017).

Metallic Hg is an allergen which may cause oral lichen and damage to the nervous system. The Minamata case is the best known example, where the Minamata Bay was contaminated by high content of Hg discharge. Around 2000 were poisoned; hundreds of deaths were reported from Minamata disease, which arises due to the consumption of fish containing methyl mercury (Mishima 1992). Intake of large quantities of arsenic (As) metal for long periods of time leads to haemolysis, melanosis, polyneuropathy, bone marrow depression, etc.

Accumulation of heavy metals also affects the aquatic flora and fauna and constitutes public health problem when contaminated species are used for food. It has been reported by World Health Organisation (WHO) in 2019 that transmitted diseases from polluted water like diarrhoea, pneumonia, typhoid, cholera, polio, etc. cause more than 485 000 deaths per year (WHO, 14 May 2019). Despite affecting human health, water pollution also has an adverse effect on plants. Nutrient deficiency in water ecosystem, excess use of agricultural chemicals, etc. hinders plant growth and also poisons the plant.

Though we have made technological advancements in the previous decades, the lack of sustainable usage and pretreatment solutions have made thousands of rural towns unavailable of safe potable water. As per the WHO report, two million people received groundwater comprising contaminants like arsenic, radium, and fluoride ions. It is not surprising that thousands of communities could not afford to filter these pollutants off and intake the same polluted water (Shannon et al. 2010). Here, it should be mentioned that heavy metal sources are the major source in water deterioration.

The chapter discusses all the above mentioned aspects and is divided in sections. The first section addresses the general introduction of water quality. The second section emphasizes the different sources of the major pollutant, i.e. heavy metal contamination in groundwater for detailed study. The third part of this chapter contributes to the type of pollution. This section emphasizes heavy metal pollution. In the fourth section, we have discussed the effect of heavy metal pollution on human body. The fifth and sixth sections throw light on recent strategies to control water pollution and remediation of heavy metals. We have discussed the future strategies adopted by different governments and suggest some measures to remove these exudates in the water. At last, this chapter ends with future expects and conclusions drawn.

4.2 Sources of Heavy Metal Contamination

Pollutants can enter the water bodies from various sources and then get transported along the streamline. Legal and illegal discharges from factories, spills and leakages in underground oil pipelines or during transportation, hydraulic fracturing operations, and sewer overflows pollute various water discharge bodies. Other potential sources include radiation leaks from nuclear power plants and drinking water disinfection processes.

This can be broadly categorized under pointed and non-pointed sources. Pointed sources refer to pollutants that enter the waterway through traceable sources like industrial outlet pipes, domestic sewage outlets, etc. It should be noted that most of these pointed sources are due to human intervention and untreated waste discharge. Non-pointed sources refer to diffused contamination of water bodies through activities like leaching, littering garbage, etc., which cannot be monitored and controlled owing to the difficulty of tracing them. The sources of heavy metal pollution in water can also be categorized as natural as well as anthropogenic. These are discussed in brief in Figure 4.1.

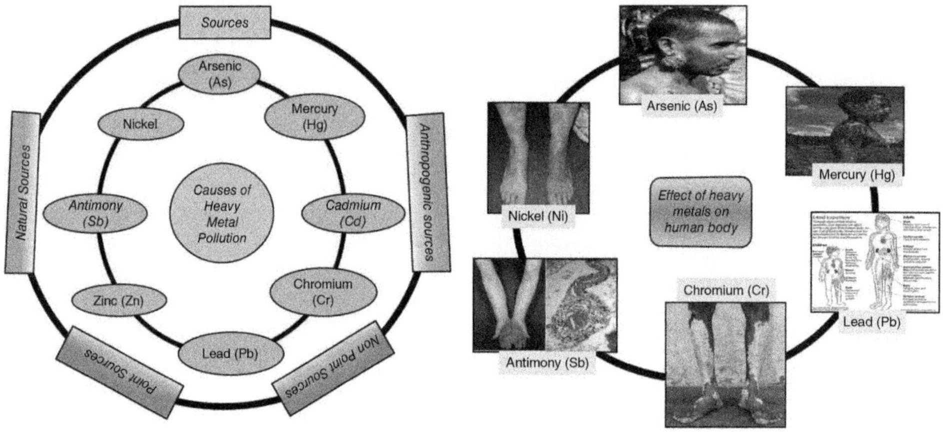

Figure 4.1 Sources of heavy metal pollution in ground water and their effects on human health.

4.2.1 Natural Sources

Natural sources like geothermal and volcanic activity and rock weathering are some natural sources known to play their part in contamination of water sources. Rainwater is also a common source of water contamination as it dissolves air pollutants and transports them in groundwater, as is the case in acid rain. Moreover, the rock oxidation, sea intrusion, decay of radioisotopes (uranium-enriched bedrock), and surrounding vegetation add on to these contaminants in water sources. Rock oxidation and rock weathering are also some natural activities that are major causes of heavy metal pollution (Talabi and Kayode 2019).

4.2.2 Anthropogenic Sources

Anthropogenic sources (caused by human activity) are the main source of heavy metal pollution. Industries and chemical companies directly discharge contaminated water without any pretreatment in groundwater and surface water sources. On the other side, agricultural irrigation is a huge reason for the degradation of drinking water quality owing to excess use of various types of harmful pesticides, fertilizers, and insecticides for crop production (Talabi and Kayode 2019). Other reasons for water pollution are discharge of sewage waste (wastewater, waste dump, and solid waste), exoneration of the by-products and wastes produced by mining activities, dissociation of radioactive elements, over pumping, etc. The discharges of waste chemicals (inorganic and organic) by laboratories, experimental wastes which are directly thrown in water, are some types of anthropogenic sources which deteriorate the water quality. These anthropogenic activities are more toxic and harmful for humans as well as animals.

4.2.3 Point Sources

Point sources are recognizable, quantifiable, and controllable sources. The drainage pipes of industries and factories that directly break out into the water, oil spilling by oil ship and tankers, and discharge of municipal and sewage waste are all major point sources of water pollution. The other point sources are domestic, hospital, industrial, and mining wastes.

The domestic wastes that occur from residence, faculties, institutions, and laboratories are major threats to human health because these sources directly pollute the drinking water. While the industrials and landfills are vital sources, heavy metal pollution is a huge cause of contamination of groundwater as well as surface water (Talabi and Kayode 2019). Consequently, it has an adverse effect on human health, animals, and marine life owing to its carcinogenic nature. Most likely, the hospital waste discharged knowingly and unknowingly into water bodies diminishes the water quality day by day (Kümmerer 2001; Emmanuel et al. 2005; Emmanuel et al. 2009). Over the last decades, mining wastes have also drawn more attention for water pollution. The mining waste from minerals, coal mines, and heavy metal ores are major origins for heavy metal pollution (Pal et al. 2010; Lapworth et al. 2012).

4.2.4 Non-Point Sources

Generally, the natural, agricultural and anthropogenic sources of water pollution are classified as non-point sources, which are non-recognizable and non-controllable (Lapworth et al. 2012; Talabi and Kayode 2019). These sources are huge contributors to the contamination of water bodies as well as in diminishing the water resource quality. Naturally occurring volcanic activity, rain, rock dissociation, storms, and so on also produce heavy metal contamination in water. Soil erosion also contributes to water contamination. The chemical contamination includes heavy metals, organic impurities, and inorganic impurities (WHO 2011; Russoniello et al. 2013; Megremi et al. 2013; Werner et al. 2013). In the case of industrial waste, it directly involves contaminating the ground as well as surface water and is threatening to water bodies all over the world. Climate change and water scarcity are other causes of water pollution. Owing to water scarcity, the excessive use of groundwater is the other source of origin of heavy metal contamination. Nowadays, agricultural sources, for example pesticides, fertilizers, and insecticides, originate the pollution in surface and groundwater also (Velthof et al. 2009; Savci 2012; Cruz et al. 2013).

4.3 Types of Water Pollution

4.3.1 Surface Water Pollution

Surface water, such as rivers, streams, and lakes are directly consumed in rural areas. As we know, the surface water is a huge resource of drinking water for rural and urban areas. Mainly, anthropogenic sources lead to the pollution in surface water. However, industrial contaminants like heavy metals are directly released into surface water. The chemical companies which are situated in the bank of rivers, lakes and streams are a major cause of heavy metal and chemical pollution of surface water. The other reason for surface water pollution is discharge of municipal waste in water. Additionally, soil erosion is another cause of generating heavy metal contamination in surface water bodies.

4.3.2 Groundwater Pollution

Over the last decades, groundwater has become one of the important sources of water for human and animals provided by nature. Unfortunately; the groundwater gets contaminated both knowingly and unknowingly by human carelessness. The natural sources explained in Section 4.2.1 are major causes of groundwater contamination. These natural sources are playing a part in contaminating the groundwater by heavy metal. The excavation of rock, which is a rich source of heavy metal, creates the heavy metal pollution in groundwater because the heavy metals are easily soluble in water, and can also be suspended in water. Additionally, anthropogenic sources are a big contributor to the heavy metal pollution in groundwater.

4.3.3 Heavy Metal Pollution

The heavy metals, which are denser than water, are easily sustained in water and create heavy metal pollution in water. As documented in the literature, the presence of heavy metal contamination in water is more threatening to human health, as well as animals and marine life, owing to their carcinogenic nature. These include arsenic (As), mercury (Hg), lead (Pb), cadmium (Cd), chromium (Cr), zinc (Zn), antimony (Sb), and nickel (Ni). The exposure to heavy metals by human beings is lethal and toxic (Mohankumar et al. 2016; Talabi and Kayode 2019; Ahamad et al. 2020; Pooja et al. 2020). To address the carcinogenic nature of the elements, various foundations like the WHO, Bureau of Indian Standard (BIS), United States Environmental Protection Agency (US-EPA), European Commission Environment (ECE), Australian Drinking Water Guidelines (ADWG), and Norma Official Mexicana (NOM)-127-SSA1-1994 prescribe the limit (Fernandez-Luqueno et al. 2013) of contaminants in water as described in Table 4.1.

4.3.3.1 Arsenic

Arsenic (As), the twentieth-most abundant element in the Earth's crust, is a well-known deadly poison from time immemorial (Mandal and Suzuki 2002). The compounds of arsenic in water are found in both organic and inorganic moieties. The organic moieties of arsenic existing in compound form are monomethylarsonicacid (MMA[V]), dimethylarsinic acid (DMA[V]), monomethylarsonous acid (MMA[III]), dimethylarsinous acid (DMA[III]), thiomethylarsonic acid (Thio MMA), thiodimethylarsinic acid (Thio DMA),

Table 4.1 Acceptable limits of heavy metals in ppb in drinking water reported by different foundations.

Heavy metal	WHO	ECE	USEPA	ADWG	NOM-127	BIS
Arsenic	10	10	10	10	25	10
Mercury	6	1	2	1	1	–
Cadmium	3	5	5	2	5	3
Chromium	50	100	50	50	50	50
Antimony	20	5	6	3	–	–
Lead	10	10	15	10	10	10
Nickel	70	–	20	20	–	20
Zinc	–	–	500	3000	5000	5000

arsenobetaine, and arsenocholine (Gomez-Caminero et al. 2001). Apart from this, the inorganic compounds of As are about 100 times more toxic than the organic ones and are more prevalent in water (Jain and Ali 2000). The major inorganic compounds of arsenic as pentavalent arsenate ions [As(V)] like H_3AsO_4, $H_2AsO_4^-$, $HAsO_4^{2-}$, and AsO_4^{3-} are found in surface water and trivalent arsenite ions [As(III)] like H_3AsO_3, $H_2AsO_3^-$, and $HAsO_3^{2-}$ are found in groundwater (Jain and Ali 2000). As(V) is found as an oxidizing agent in surface water. Arsenic exists as H_3AsO_4 at $pH > 2$ and in the form like $H_2AsO_4^-$ and $HAsO_4^{2-}$ at pH 2–11. However, As(III) is found under reducing conditions at low pH. Considering this, As(III) is more toxic than As(V). Although As(III) is more lethal, the metabolism of As(V) plays an important role in its toxicity in the human body. Arsenate has the same structure as the phosphate ion and it can replace the phosphate ion from enzymatic reactions, which occur in the human body (Fan et al. 2018). The mechanism of As(V) in human body is completed in the following steps (Thomas et al. 2001; Hughes 2002; Fan et al. 2018):

- As(V) involves to reduce As(III) in presence of glutathionine (GSH). GSH takes part in reduction reaction as an electron donor.
- As(III) takes place oxidative methylation to pentavalent state (As[V]) in presence of S-adenosylmethyl.

4.3.3.2 Mercury

Mercury (Hg) has received more attention in groundwater owing to their carcinogenic nature. The different path of mercury contributes to groundwater pollution. For instance, coal-fired plants, smelting, alkali processing, and other industrial actions cause mercury pollution in groundwater. Natural activities are another cause of mercury contamination in groundwater. Apart from this, over the last decades, metallic mercury has been utilized in various fields like medical fields (thermometers, barometers, and other instruments for measuring blood pressure), which are causes of mercury pollution in groundwater, including the consumption of calomel and mercury amalgam to healing teeth (feeling and diuretics) in the field of dentistry contributing to Hg groundwater pollution (Barringer et al. 2013). The utilization of Hg as voltametric sensor to detect the trace metals in water is other reason for the contribution of Hg to water contamination. The inorganic Hg is less toxic than organic mercury. However, inorganic Hg is easily transformed into methyl mercury as organic compound, which is more stable and exposed to fish. Humans consume the organic Hg through the food chain like fish consumption and dental amalgam (Järup 2003; Hashim et al. 2011).

4.3.3.3 Cadmium

Cadmium is symbolized by Cd and belongs to 3D block elements. Cd is introduced as a toxic element and the sources of Cd are rock, coal, and petroleum. Cd is often found in combination with zinc. Cd is found in two forms: metallic form and cationic form (Cd^{+2}) (Smith 1995). Both natural and human sources contribute to the cadmium impurities in groundwater. The industrial activities, like manufacturing of batteries (NiCd), pigments, plastic, and electroplating, directly discharge into the water and contaminate the water bodies. Other activities like mining, seepage of hazardous waste materials from sites, and discharge of waste industrial water cause the increasing concentration of cadmium in water day by day. On the other side, agricultural sources like cadmium-containing phosphate fertilizers are also producing cadmium contamination in water resources (Ryan et al. 2000; Järup 2003; Hashim et al. 2011).

4.3.3.4 Lead

Lead (Pb) pollution in groundwater is a cumbersome and daunting affair for the global population. The manufacturing plants of lead acid batteries contribute to lead pollution in groundwater. Old pipes that contain lead are direct sources of lead contamination in water. The industrial sources and vehicle exhaust also produce Pb contamination in water. Pb is addressed as a normal constituent of The Earth's crust and it is found into two oxidation states as 0 and 2. It exists in water as Pb^{2+}, $PbOH^{+}$, and $PbHCO3^{+}$ at normal pH (Raviraja et al. 2008). The exposure to a small amount of Pb by humans is more lethal than other heavy metal contamination (Hashim et al. 2011).

4.3.3.5 Antimony

Intake of antimony (Sb) in lower quantities is good for human health. But the acute exposure to Sb (more than prescribed limit as given in Table 4.1) is lethal for the human body. The Sb contamination is released in water by anthropogenic activities, heavy commercial activities, and natural activities. Sb has similar chemical and physical properties as arsenic because Sb and As belong to same family in the periodic table (Willis et al. 2011). As arsenic, Sb(III) is also more toxic than Sb(V) (WHO 2003).

4.3.3.6 Zinc

Zinc (Zn) is a 3D block element which is found in the Earth's crust. Generally, Zn is found in +2 oxidation states and it forms compounds with the anions, amino acid, and organic acid. Zn is soluble in water bodies at neutral and acidic pH values. At basic pH, it makes the carbonate and hydroxide compounds, which are insoluble in water. The anthropogenic and natural sources release the Zn contamination in ground as well as surface water. Zn is an essential element for the human body in limited quantities. The exposure to high concentrations of Zn is harmful for many age groups (Hashim et al. 2011).

4.3.3.7 Chromium

Chromium (Cr) belongs to 3D block elements and it exists in three valence states as metallic state Cr(0), Cr(III), and Cr(VI). Cr(III) is insoluble in water; it only forms hydroxide and oxy-hydroxide and solid solution with iron (Fe), while Cr(VI) is resolvable in water under environmental conditions. Cr occurs in the form of chromate ($CrO_4{}^{2-}$) and dichromate ions ($Cr_2O_7{}^{2-}$). Cr(VI) is more lethal than Cr(III). Chromium pollution in water is accumulated by natural sources as well as anthropogenic sources. However, industrial sources as chrome plating, steel production, corrosion inhibition, wood preservative, well drilling, and paint and primer pigments lead to chromium contamination in water. Nevertheless, natural sources like rock ores also lead to chromium contamination in groundwater (Zhitkovich 2011).

4.4 Effects on the Human Body

The various diseases like nausea, vomiting, diarrhoea, cyanosis, cardiac arrhythmia, confusion and hallucinations can arise in the human body owing to short- term consumption of arsenic. The long-term consumption of arsenic through food, water, and other sources is associated with various types of health problems like skin disease (hyperpigmentation), skin cancer, tumours in lungs, bladder, kidneys, and liver, spontaneous abortion, depression, numbness, sleeping disorders, and headaches (Ali et al. 2012; Singh et al. 2015). Poisonous levels diverge between various compounds; for example, monomethyl arsenic acid and inorganic arsenide have higher levels of deadliness than arsenic choline. Acute toxicity is generally higher for inorganic arsenic compounds than for organic arsenic compounds (Hughes 2002). Arsenic can also cause low birth weight. Owing to the serious effect of arsenic on people's health, WHO recommended 10 μgL^{-1} As in drinking water in 2001 (WHO 2001). The US Environmental Protection Agency and European Commission have also revised the maximum contaminant level (MCL) for arsenic in drinking water to 50–10 μgL^{-1} (EPA 2002; European Commission Directive 1998).

As discussed in Section 4.3.3.2, Hg is found in three forms as metallic, inorganic, and organic Hg. The exposure of organic mercury is more lethal than inorganic mercury. The organic Hg is consumed through fish, water, and food in the body and it can cause different human health issues. The long-term consumption of Hg causes various symptoms in the human body like damage to the nervous system and circulatory system, skin problems, and kidney disease. It also causes rheumatoid arthritis, which shows symptoms over a long period (Järup 2003, Azimi et al. 2017; Filter 2020). Owing to their poisonous nature, WHO and BIS reported the acceptable limit of mercury in water as summarized in Table 4.1.

The impact of Cd on human health is more dangerous. When Cd is exposed in very low quantities, it causes nausea with vomiting, headache, and cough. However, the intake of cadmium in high quantities affects kidney and liver. Cd can replace calcium from bones, which is the reason for different bone diseases like painful bone disorders, renal failure, and insinuation in human hypertension. The long period exposure of Cd can cause chronic anaemia (Järup 2003; Burke et al. 2016).

Intake of Pb over the acceptable limit is carcinogenic for humans. The short-term as well as long-term exposure to lead affects children more than adults. The intake of Pb through water, food and other resources causes various diseases like damage to the foetal brain, kidney disease, and circulatory and nervous system damage. Pregnant people exposed to lead through food and water can experience miscarriage, stillbirth, premature delivery, and low birth weight (Järup 2003).

Generally, the toxicity of Sb is similar to the toxicity of arsenic. The Sb(III) is more carcinogenic than Sb(V). The inhalation of antimony leads to carcinogenic disease and respiratory system problems. The intake of Sb in higher quantities leads to premature delivery, miscarriage, liver damage, etc. (WHO 2003; Sundar and Chakravarty 2010).

Zn is as an essential element to intake up to an acceptable limit; it helps in various metabolic processes, embryonic development, cellular differentiation, and cell proliferation. Moreover, it also delivers the substrates for the manifestation of genetic potential like optimum growth, health, reproduction, and development. However, the consumption of high doses of Zn is hazardous and leads to depression, lethargy, neurological signs, and thirst increment diseases (Plum et al. 2010).

Chromium (Cr) is also a more toxic element. Both forms, Cr(III) and Cr(VI), are carcinogenic for humans. But Cr(VI) is more lethal than Cr(III) because of its more powerful oxidizing agent, which has a tendency to be irritating and corrosive and it is easily inhaled by human cells, which have seen the carcino-toxicity. The long-term consumption of Cr affects the respiratory system, cardiovascular system, haptic system, gastrointestinal system, etc. It can also cause carcinogenic diseases (Zhitkovich 2011; Mcneill et al. 2012; Wilbur et al. 2012).

Overall, the consumption of heavy metal/metalloids over the permissible limit by humans through drinking water leads to skin problems, stomach problems, and carcinogenic problems. It produces different types of cancer like lung, kidney, skin, etc. The intake of heavy metals causes brain-related diseases like Parkinson's disease, disturbances in nerve cells, brain tumours, depression, and more (Momodu and Anyakora 2010; Mebrahtu and Zerabruk 2011; Mohod and Dhote 2013). Day by day, human health is deteriorated by consuming the contaminated water and toxic food. Keeping in mind their toxicity, several researchers are working to overcome this problem and demonstrate the methods to decontaminate the water bodies. The government has propounded some strategies to stop water pollution and water problems, which are discussed in Section 4.5.

4.4.1 Impact of Heavy Metal Pollution in Groundwater: Some Case Studies

Groundwater resources are world's largest freshwater sources. They provide adequate water supply for domestic, agricultural and industrial practices. Uniquely, their role in sustainability of ecosystems is immensely vital. Therefore, the sustainability of groundwater quality and quantity itself is very important for the above-mentioned phenomena. However, owing to various anthropogenic activities in the recent past, groundwater resources are undergoing catastrophic depletion and deterioration. Particularly, the overexploitation and continuous groundwater pollution works in a synergic fashion and cause non-availability of groundwater and making it unfit for any use. Here, in this section we are putting forward few global examples that illustrates the devastating impacts of polluted groundwater on smaller and bigger scales.

Boateng et al. have studied the extent of heavy metal contamination in Oti landfill site, Kumasi, and further evaluated the effect of this contamination on human health. Through the standard methods and procedures for the examination of water and wastewater, authors reported that the concentration of many metals like Pb, Cd, Fe, and Cr was above the acceptable limits as set by the WHO for drinking water. It was reported that only the concentration of Zn and Cu was within the permissible limit. Further, it was revealed that this high level of heavy metal contamination in groundwaters of Oti has grave health consequences and water needs to be pretreated before its consumption (Boateng et al. 2019).

In a recent report, a meta-analysis on the impact of consumption of polluted groundwater on health of children was executed in 10 developing countries including India. The research shows that blood of the children from these countries have relatively higher levels of heavy metals than normal cases and this problem is only attributed to the polluted groundwater by the researchers Horton et al. (2013); (Singaraja et al. 2015; Mohankumar et al. 2016). The groundwater pollution problem is a very serious concern across the world, however, the situation in developing economies, especially heavily populated ones, is very critical. In these countries, industrial growth and agricultural practices are rapidly increasing and consequently, offering more and complex hurdles in the sustainable management of groundwater resources. For instance, the Kurichi Industrial Cluster near Coimbatore city does not have agricultural activities and waste dumping nearby it, however, it has given a comprehensive Environmental Pollution Index (CEPI) score of 58.75 by the Central Pollution Control Board (CPCB, India) in 2009, which implies that the water was critically polluted (Action plan for critically polluted area, Tamilnadu Pollution Control Board, 2010, www.tnpcb.gov.in/pdf/Action_plan_cbe.PDF). This is because of the release of unchecked wastewater streams of local industries in groundwater. This transformed the local groundwater and nearby groundwater tables into highly toxic systems. The groundwater sources are not suitable for domestic and agricultural practices. Further, many ill effects on human health, soil fertility, bioaccumulation, and biomagnification of the toxic heavy metals are spreading with a higher pace in the region. To mitigate the issues, the local authorities had imposed stronger legislations on the industries and wastewater treatment plants were commissioned in many industries.

Many such studies are now published by many research groups, NGOs (nongovernmental organizations), environmentalists, and local governments. However, the real scenario is even worse than that reported in these studies, especially in the areas where infrastructure is not that strong. Moreover, many examples of the hazardous effects of consumption of groundwater contaminated with heavy metals are presented throughout the chapter.

Therefore, a comprehensive review on such aspects and illustration of the true picture of the problem is required for making the sustainable policies and strategies to propagate the sustainability of groundwater resources and their uses.

4.5 Recent Strategies to Control Heavy Metals

As discussed, the main sources of heavy metal pollution are anthropogenic sources, which are the major concern for contaminants in groundwater as well as surface water resources. To overcome these problems, government must take appropriate action to solve these issues. In this regard, government has taken a decision to prevent and overcome heavy metal pollution in water.

NEERI (National Environmental Engineering Research Institute) has addressed the technique to decontaminate the water from heavy metals as well as for waste management of the land sector. The kit is also receommended to check the quality of drinking water (Marg 2011).

For preventing Hg pollution, the major Hg pollution sources like medical devices and CFL (compact fluorescent light) bulbs should be replaced by non-mercury containing products (Marg 2011). The industries should set up treatment technology for spontaneous remediation of the heavy metal-based wastewater before discharging the wastewater into water bodies.

Domestic wastage like municipal and sewage waste should be prevented from discharging in water. It should be monitored and made a rule to stop the disposal of waste in water. To cure the drinking water, the metal-based pesticides and insecticides should be bonded within the agricultural field. Government must give proper attention to organic farming. Phytoremediation is a best tool to control the heavy metal contamination. Other spontaneous chemical and physical treatments have been utilized to overcome heavy metal pollution in water (Marg 2011; Talabi and Kayode 2019).

4.6 Remediation Methods of Heavy Metals

To solve this problem, over the last decades various types of methods are used for the elimination of heavy metals from water (see Figure 4.2). The literature reveals that some methods, such as membrane filtration, oxidation, ion exchange, coagulation-flocculation, phytoremediation, electro-kinetics, and adsorption are carried out to eliminate the heavy metals from contaminated water (Barakat 2011; Agarwal and Singh 2017; Azimi et al. 2017). All these methods with their advantages and limitation are explained in brief.

4.6.1 Oxidation

The oxidation method includes the use of an oxidizing agent (Cl_2, ClO_2, O_3, H_2O_2, NH_2Cl, MnO_4^-, FeO_4^{2-}, etc.) to eliminate heavy metals from water (Sharma et al. 2007; Mondal et al. 2013; Bora and Dutta 2019). Investigation of the photochemical and photocatalytic oxidation process includes the oxidation of heavy metals using UV radiation, O_2, etc. for the removal of dissolved pollutants. UV/solar radiation assists to develop the hydroxyl radicals during the photolysis of iron and iron hydroxide (fe$(OH)_3$), hydroxyl radicals and O_2 oxidizes the toxic metals like As(III) to As(V). The presence of these radicals, the oxidation reaction becomes faster (Yoon and Lee 2005; Sharma et al. 2007). On the other hand, the photocatalytic reaction is carried out in the presence of TiO_2 for oxidation of the heavy metals (Miller et al. 2011).

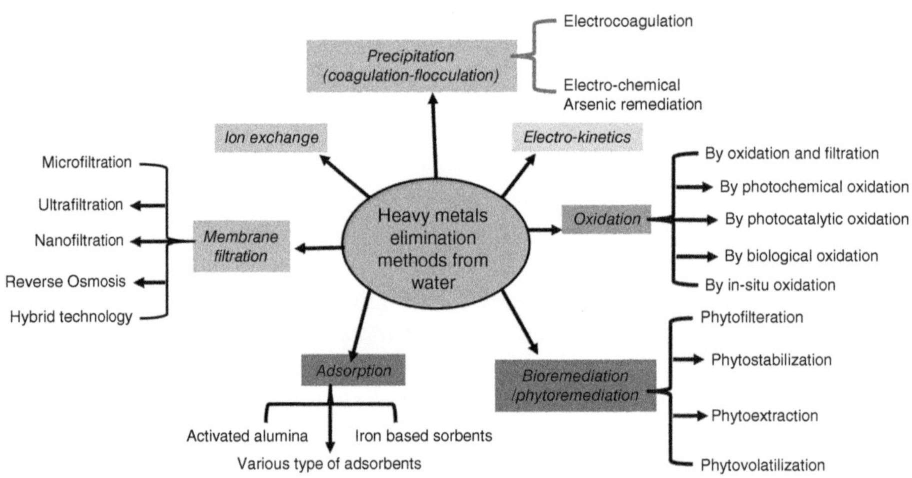

Figure 4.2 Schematic of conventional different conventional methods to decontaminate water from heavy metals.

4.6.2 Coagulation-Flocculation

The coagulation and flocculation method has been investigated to eliminate the toxic metal pollution from contaminated water (Abouri et al. 2019; Bora and Dutta 2019). In this process, the coagulant is incorporated into the contaminated water and then forms the floc, which has the potential to eliminate heavy metals from the water. The positively charged coagulants such as aluminium and iron salts, which are widely used for heavy metal like As removal, help to decrease the negative charge of colloids. The larger particles (floc) form due to agglomeration of the particles, which settle down in water due to the influence of gravity (Choong et al. 2007). The formed floc helps to precipitate out the soluble heavy metals from the water and the solution can be filtered rapidly. Several reported coagulants like aluminium, ferrous sulphate, and ferric chloride result in the formation of amorphous metal hydroxide precipitates that are more suitable to eliminate toxic metals from contaminated water (Pallier et al. 2010). At lower pH, the colloidal substances with negative charges can be coagulated, but the cationic ion cannot be eliminated and at higher pH, the turbidity removal decreases and the cationic removal is favoured.

4.6.3 Phytoremediation

The phytoremediation process is based on the use of plants for the removal of toxic metals from the environment and it is even used to clean up contaminated air, soil, and water. It consists of several steps like phytoextraction, phytoming, phytostabilisation, rhizofiltration, and phytovolatilization (Dickinson et al. 2009; Behera 2014). Over the last decades, several researchers have been using various types of plants like *Pteris Vittata*, Vetiver grass, for remediation of toxic metals through the phytoremediation process (Hosamane 2012). The disadvantage of this process is that the plants adsorb high levels of toxic metals, which contaminate food crops and require more time to eliminate heavy metals from soil and water.

4.6.4 Membrane Filtration

This technique shows great potential for decontamination of heavy metals for their high efficiency and easy operation (Hao et al. 2018; Khulbe and Matsuura 2018). The membrane process has been explored with various types of techniques like microfiltration (MF), ultrafiltration (UF), reverse osmosis (RO), and electrodialysis. In the UF technique, a low-pressure driven membrane process including pore size (10–10 000 Å) is used for the removal of dissolved and colloidal materials. High removal efficiency of metal ions could be achieved by using micellar-enhanced ultrafiltration (MEUF) and polymer-enhanced ultrafiltration (PEUF). The micelles containing membrane are having smaller pore size than a UF retaining contaminants, where the untapped species readily pass through the UF membrane. An anionic surfactant, sodium do-decyl sulphate (SDS), is often selected for the effective elimination of toxic metal ions in MEUF. PEUF uses a water-soluble polymer to complex metallic ions and form a macromolecule which will be retained, when pumped through UF membrane. Thereafter, the retained material can be recycled in order to recover metallic ions and to reuse the polymeric agent. In PEUF, the main complexing agents like polyacrylic acid (PAA) (Labanda et al. 2009) and polyethylene imine (PEI) (Aroua et al. 2007) used to achieve selective separation and recovery of heavy metals with low energy requirements.

The electrodialysis (ED) process carries out the separation of ions across charged membranes from one solution to another solution following an electric field as the driving force. Mostly, ion-exchange membranes (cationic and anionic) are used in ED processes. It has been widely used for the production of drinking water, process water from brackish water and seawater, treatment of industrial effluents, recovery of useful materials from effluents, salt production, and for heavy metal wastewater treatment.

According to the literature, the RO (reverse osmosis) process is a very old and famous method thought of as the best method to remove arsenic from water. The RO membrane has extremely small pores ($<0.001\,\mu m$). The RO membrane process can achieve a high rejection of low molecular mass compound and ions. The cellulose acetate RO membrane was been investigated for the first time in the 1980s, and has a 90% As removal efficiency with the RO system with 400 psi high pressure. However, Akin et al. have investigated the RO operational parameters. A semi-permeable membrane is used, which allows the fluid that is being purified to pass through it, while rejecting the contaminants. It accounts for more than 20% of the world's desalination capacity (Shahalam et al. 2002). RO is an alternative option for wastewater treatment in chemical and environmental engineering. The major problem with RO is the high power consumption due to the pumping pressures, and the restoration of the membranes.

4.6.5 Ion Exchange

The ion exchange method has been employed to eliminate both cationic and anionic impurities from water. It has many advantages, such as high treatment capacity, high removal efficiency, and fast kinetics (Rahimizadeh and Liaghatb 2015). The resin used may be either synthetic or natural solid resin (synthetic resins are comparatively more effective) (Alyüz and Veli 2009). As the solution containing heavy metal passes through the cations column, metal ions are exchanged for the hydrogen ions on the resin. The uptake of heavy metal ions by ion-exchange resins is affected by certain variables such as pH, temperature, initial metal concentration, and contact time (Gode and Pehlivan 2006). Ionic charge also plays an important role in the ion-exchange process. In contrast to resin, natural zeolites also exhibit good cation exchange capacities for heavy metal ions under different experimental conditions. Clinoptilolite is one of the most frequently studied natural zeolites that has received extensive attention due to its selectivity for heavy metals (Doula 2009). Although there are many reports on the use of zeolites and montmorillonites as ion-exchange resin to decontaminate heavy metal, they are limited at present compared with the synthetic resins and the application of zeolites is on the laboratory experimental scale.

4.6.6 Electrokinetics Remediation

Electrokinetics remediation (ER) has been evaluated for heavy metal removal from contaminated fine grain soils (Chen and Zhang 2012; Hussein and Alatabe 2019). The ER process contains three phenomena: electroosmosis (EO), electromigration (EM) and electrophoresis (EP). The ER process works on low-level direct current like a "cleaning agent" including several transport mechanisms as EO, EM, EP, and electrochemical reactions (electrolysis and electrodeposition). It is an easily controllable method both in-situ and ex-situ. This method is applicable for low permeability soil treatment and removal of organic and inorganic pollutants from soil also. The EO process occurs due to fluid flow from anode to cathode. On the other hand, the EP process exhibits the movement of the charged particle under applied electric field. The last phenomenon in the ER process is EM, where cations move towards the negatively charged cathode and anions moves towards the anode under an applied electric field. The electrokinetics method has been implemented to remove various types of contaminants like polychlorinated biphenyls, pesticides, and heavy metals. The major disadvantage of this method is that it is not appropriate for removing nonpolar pollutants and this method takes more time to complete the remediation process.

4.6.7 Adsorption Method

The adsorption method is considered to be the best method in comparison to the other methods discussed here because of their limitations like low efficiency, high cost, etc. The key advantages of the adsorption method are low cost, sludge free, high efficiency, easy recovery of metals and possibility to reuse the adsorbent (Sharma et al. 2015; Agrawal et al. 2018; Agrawal et al. 2019). Generally, adsorption is a mass transfer process, in which a substance transfers from the liquid phase to the solid phase and bounds by physical and chemical interactions (Babel and Kurniawan 2003). The adsorption process can be accomplished either by electrostatic interaction (ionic interaction between positively charged metal ions and negatively charged matrices) or by chelation (donation of the lone-pair electrons of the matrices to metal ions to form coordinate bonds). More effective adsorption of toxic metals depends on the properties of the heavy metal solutions such as temperature, pH of the solution, and specification of heavy metal; for example, at neutral or lower pH, As(V) shows better adsorption efficiency than As(III). The use of activated carbon (Li et al. 2018; Hashim et al. 2019), activated alumina (Szatyłowicz and Skoczko 2018), carbon nanotubes (CNTs) (Zhu et al. 2019), porous carbon (Agrawal et al. 2019), graphene (Fausey et al. 2019; Dai et al. 2020), and ferric oxide particles (Majumder et al. 2019) have generated much interest, and novel metal modified adsorbents have demonstrated superior performance towards heavy metal decontamination. Carbonaceous materials have comparatively good efficiency of adsorption, large surface area, tuneable pores, high porosity, and good sorption sites for adsorbate, and are easy to synthesize and cost effective.

4.7 Conclusion and Future Aspects

Being the largest freshwater sources, groundwaters not only provide adequate supply for domestic, agricultural, and industrial activities but also ensure sustainability of the ecohydrological phenomenon. Their continuous depletion (due to overexploitation) and deterioration (due to pollution) are causing grave environmental and health consequences. Particularly, over the past few decades, release of unchecked industrial wastewaters into streams of heavy metal ions into groundwater resources has been found to be catastrophic and the devastating impacts of the use of heavy metal-contaminated groundwater can be seen across the globe despite strong and rigid legislations. Therefore, the major challenge now is to maintain the sustainability of groundwater resources in conjunction with socio-economic-industrial development. The development of efficient methodologies that can be used to monitor groundwater sources in real time from the perspective of heavy metal pollution are highly desirable. Further, technologically feasible and commercial viable treatment processes need to be designed and developed for the pretreatment of industrial waste before its discharge to groundwater sources and other water bodies. Moreover, attention needs to be paid to the sustainability of new-generation advanced treatment technologies like membrane filtration and adsorption so that treatment of contaminated groundwater can be achieved on a bulk scale at the point of source prior to its consumption. These technologies not only offer purified water from contaminated water but also provide an opportunity for recovery of heavy metals as resource materials from polluted waters. Therefore, in a nutshell, a long-term planning solution is required to optimize groundwater resources by considering the future projections of industrial expansions, irrigation demands, and drinking water supplies.

References

Abouri, M., Souabi, S., and Jada, A. (2019). Optimization of coagulation flocculation process for the removal of heavy metals from real textile wastewater. *Advanced Intelligent Systems for Sustainable Development (AI2SD'2018)* 3: 257.

Agarwal, M. and Singh, K. (2017). Heavy metal removal from wastewater using various adsorbents: a review. *Journal of Water Reuse and Desalination* 7: 387–419.

Agrawal, P.R., Singh, N., Kumari, S., and Dhakate, S.R. (2018). Multiwall carbon nanotube embedded phenolic resin-based carbon foam for the removal of As (V) from contaminated water. *Materials Research Express* 5: 035601.

Agrawal, P.R., Singh, N., Kumari, S., and Dhakate, S.R. (2019). The removal of pentavalent arsenic by graphite intercalation compound functionalized carbon foam from contaminated water. *Journal of Hazardous Materials* 377: 274–283.

Ahamad, A., Madhav, S., Singh, A.K. et al. (2020). Types of water pollutants: conventional and emerging. In: *Sensors in Water Pollutants Monitoring: Role of Material* (eds. D. Pooja, P. Kumar, P. Singh, et al.), 21–41. Singapore: Springer.

Ali, I., Gupta, V.K., Khan, T.A., and Asim, M. (2012). Removal of arsenate from aqueous solution by electro-coagulation method using Al-Fe electrodes. *International Journal of Electrochemical Science* 7: 1898–1907.

Alyüz, B. and Veli, S. (2009). Kinetics and equilibrium studies for the removal of nickel and zinc from aqueous solutions by ion exchange resins. *Journal of Hazardous Materials* 167: 482–488.

Aroua, M.K., Zuki, F.M., and Sulaiman, N.M. (2007). Removal of chromium ions from aqueous solutions by polymer-enhanced ultrafiltration. *Journal of Hazardous Materials* 147: 752–758.

Azimi, A., Azari, A., Rezakazemi, M., and Ansarpour, M. (2017). Removal of heavy metals from industrial wastewaters: a review. *ChemBioEng Reviews* 4: 37–59.

Babel, S. and Kurniawan, T.A. (2003). Low-cost adsorbents for heavy metals uptake from contaminated water: a review. *Journal of Hazardous Materials* 97: 219–243.

Băhnăreanu, C. 2019. World Economic Forum 2019: Globalization 4.0–A Better Version. *Strategic Impact*, 79–82.

Barakat, M. (2011). New trends in removing heavy metals from industrial wastewater. *Arabian Journal of Chemistry* 4: 361–377.

Barringer, J.L., Szabo, Z. and Reilly, P.A. 2013. Occurrence and mobility of mercury in groundwater. Current perspectives in contaminant hydrology and water resources sustainability, 117–149.

Behera, K.K. (2014). Phytoremediation, transgenic plants and microbes. In: *Sustainable Agriculture Reviews*, vol. 13 (ed. E. Lichtfouse). Springer https://doi.org/10.1007/978-3-319-00915-5_4.

Boateng, T.K., Opoku, F., and Akoto, O. (2019). Heavy metal contamination assessment of groundwater quality: a case study of Oti landfill site, Kumasi. *Applied Water Science* 9: 33.

Bora, A.J. and Dutta, R.K. (2019). Removal of metals (Pb, Cd, Cu, Cr, Ni, and Co) from drinking water by oxidation-coagulation-absorption at optimized pH. *Journal of Water Process Engineering* 31: 100839.

Burke, F., Hamza, S., Naseem, S. et al. (2016). Impact of cadmium polluted groundwater on human health: winder, Balochistan. *SAGE Open* 6: 2158244016634409.

Chen, J. and Zhang, W.-L. (2012). An experiment on removing heavy metals from sewage sludge by electrokinetic technology. *Environmental Science Survey* 21: 3–21.

Choong, T.S., Chuah, T., Robiah, Y. et al. (2007). Arsenic toxicity, health hazards and removal techniques from water: an overview. *Desalination* 217: 139–166.

Council, N.R. (2009). *Systemic Exposures to Volatile Organic Compounds and Factors Influencing Susceptibility to their Effects. Contaminated Water Supplies at Camp Lejeune: Assessing Potential Health Effects.* National Academies Press (US).

Cramer, E.H., Blanton, C.J., Otto, C., and Team, V.S.P.E.H.I. (2008). Shipshape: sanitation inspections on cruise ships, 1990–2005, vessel sanitation program, centers for disease control and prevention. *Journal of Environmental Health* 70: 15–21.

Cruz, J.V., Silva, M., Dias, M.I., and Prudêncio, M.I. (2013). Groundwater composition and pollution due to agricultural practices at Sete Cidades volcano (Azores, Portugal). *Applied Geochemistry* 29: 162–173.

Dada, O.A., Adekola, F.A., and Odebunmi, E.O. (2016). Kinetics and equilibrium models for sorption of Cu (II) onto a novel manganese nano-adsorbent. *Journal of Dispersion Science and Technology* 37: 119–133.

Dai, K., Liu, G., Xu, W. et al. (2020). Judicious fabrication of bifunctionalized graphene oxide/ MnFe2O4 magnetic nanohybrids for enhanced removal of Pb (II) from water. *Journal of Colloid and Interface Science* 579: 815–822.

Dickinson, N.M., Baker, A.J., Doronila, A. et al. (2009). Phytoremediation of inorganics: realism and synergies. *International Journal of Phytoremediation* 11: 97–114.

Doula, M.K. (2009). Simultaneous removal of Cu, Mn and Zn from drinking water with the use of clinoptilolite and its Fe-modified form. *Water Research* 43: 3659–3672.

Elimelech, M. (2006). The global challenge for adequate and safe water. *Journal of Water Supply: Research and Technology—AQUA* 55: 3–10.

Emmanuel, E., Perrodin, Y., Keck, G. et al. (2005). Ecotoxicological risk assessment of hospital wastewater: a proposed framework for raw effluents discharging into urban sewer network. *Journal of Hazardous Materials* 117: 1–11.

Emmanuel, E., Pierre, M.G., and Perrodin, Y. (2009). Groundwater contamination by microbiological and chemical substances released from hospital wastewater: health risk assessment for drinking water consumers. *Environment International* 35: 718–726.

EPA, U.S. Minor clarification of National Primary Drinking Water Regulation for arsenic proposed; rule Fed. Regist. 67(2002) 78202–78209.

European Commission Directive EC, related with drinking water quality intended for human consumption Brussels, Belgium, 1998.

Fan, C., Liu, G., Long, Y. et al. (2018). Thiolation in arsenic metabolism: a chemical perspective. *Metallomics* 10: 1368–1382.

Fausey, C.L., Zucker, I., Shaulsky, E. et al. (2019). Removal of arsenic with reduced graphene oxide-TiO2-enabled nanofibrous mats. *Chemical Engineering Journal* 375: 122040.

Fernandez-Luqueno, F., López-Valdez, F., Gamero-Melo, P. et al. (2013). Heavy metal pollution in drinking water-a global risk for human health: a review. *African Journal of Environmental Science and Technology* 7: 567–584.

Filter, T.B.W. 29 March 2020. *Health Effects of Mercury in Drinking Water* [Online]. Berkey Water Filter. Available: https://theberkey.com/blogs/water-filter/160104135-health-effects-of-mercury-in-drinking-water.

Friberg, L. and Vahter, M. (1983). Assessment of exposure to lead and cadmium through biological monitoring: results of a UNEP/WHO global study. *Environmental Research* 30: 95–128.

Ghrefat, H., Nazzal, Y., Batayneh, A. et al. (2014). Geochemical assessment of groundwater contamination with special emphasizes on fluoride, a case study from Midyan Basin, northwestern Saudi Arabia. *Environmental Earth Sciences* 71: 1495–1505.

Gode, F. and Pehlivan, E. (2006). Removal of chromium (III) from aqueous solutions using Lewatit S 100: the effect of pH, time, metal concentration and temperature. *Journal of Hazardous Materials* 136: 330–337.

Gomez-Caminero, A., Howe, P.D. et al. (2001). *Arsenic and Arsenic Compounds*. World Health Organization.

Grossman, D. and Slutsky, D.J. 2017. The effect of an increase in lead in the water system on fertility and birth outcomes: The case of Flint, Michigan.

Hao, L., Wang, N., Wang, C., and Li, G. (2018). Arsenic removal from water and river water by the combined adsorption-UF membrane process. *Chemosphere* 202: 768–776.

Hashim, M.A., Mukhopadhyay, S., Sahu, J.N., and Sengupta, B. (2011). Remediation technologies for heavy metal contaminated groundwater. *Journal of Environmental Management* 92: 2355–2388.

Hashim, M.A., Kundu, A., Mukherjee, S. et al. (2019). Arsenic removal by adsorption on activated carbon in a rotating packed bed. *Journal of Water Process Engineering* 30: 100591.

Horton, L.M., Mortensen, M.E., Iossifova, Y. et al. (2013). What Do We Know of Childhood Exposures to Metals (Arsenic, Cadmium, Lead, and Mercury) in Emerging Market Countries? *International Journal of Pediatrics.* 2013: 872596. https://doi.org/10.1155/2013/872596.

Hosamane, S.N. (2012). Removal of arsenic by phytoremediation-a study of two plant spices. *International Journal of Scientific Engineering and Technology* 1: 218–224.

Hughes, M.F. (2002). Arsenic toxicity and potential mechanisms of action. *Toxicology Letters* 133: 1–16.

Hussein, A. and Alatabe, M. (2019). Remediation of lead-contaminated soil, using clean energy in combination with electro-kinetic methods. *Pollution* 5: 859–869.

Islam, S.M.F. and Karim, Z. (2019). *World's Demand for Food and Water: The Consequences of Climate Change. Desalination-Challenges and Opportunities*. IntechOpen.

Jain, C. and Ali, I. (2000). Arsenic: occurrence, toxicity and speciation techniques. *Water Research* 34: 4304–4312.

Järup, L. (2003). Hazards of heavy metal contamination. *British Medical Bulletin* 68: 167–182.

Kass, A., Gavrieli, I., Yechieli, Y. et al. (2005). The impact of freshwater and wastewater irrigation on the chemistry of shallow groundwater: a case study from the Israeli Coastal Aquifer. *Journal of Hydrology* 300: 314–331.

Khulbe, K. and Matsuura, T. (2018). Removal of heavy metals and pollutants by membrane adsorption techniques. *Applied Water Science* 8: 19.

Kümmerer, K. (2001). Drugs in the environment: emission of drugs, diagnostic aids and disinfectants into wastewater by hospitals in relation to other sources–a review. *Chemosphere* 45: 957–969.

Labanda, J., Khaidar, M.S., and Llorens, J. (2009). Feasibility study on the recovery of chromium (III) by polymer enhanced ultrafiltration. *Desalination* 249: 577–581.

Lapworth, D., Baran, N., Stuart, M., and Ward, R. (2012). Emerging organic contaminants in groundwater: a review of sources, fate and occurrence. *Environmental Pollution* 163: 287–303.

Li, J., Xing, X., Li, J. et al. (2018). Preparation of thiol-functionalized activated carbon from sewage sludge with coal blending for heavy metal removal from contaminated water. *Environmental Pollution* 234: 677–683.

Majumder, A., Ramrakhiani, L., Mukherjee, D. et al. (2019). Green synthesis of iron oxide nanoparticles for arsenic remediation in water and sludge utilization. *Clean Technologies and Environmental Policy* 21: 795–813.

Mandal, B.K. and Suzuki, K.T. (2002). Arsenic round the world: a review. *Talanta* 58: 201–235.

Marg, B.Z. 2011. Hazardous metals and minerals pollution in India: Sources, toxicity and management. A Position Paper, Indian National Science Academy, New Delhi.

Mcneill, L., Mclean, J., Edwards, M., and Parks, J. (2012). State of the science of hexavalent chromium in drinking water. *Pollution Engineering* 44: 6666.

Mebrahtu, G. and Zerabruk, S. (2011). Concentration and health implication of heavy metals in drinking water from urban areas of Tigray region, Northern Ethiopia. *Momona Ethiopian Journal of Science* 3: 105–121.

Megremi, I., Vasilatos, C., Atsarou, A. et al. (2013). Geochemical evidences for the sources of the Cr (VI) contamination in groundwater in Central Euboea and Assopos-Thiva basins, Greece: natural versus anthropogenic origin. *European Water* 41: 23–34.

Michaels, R.A. (2017). Lessons learned from legacy contaminants of emerging concern: perfluoroalkyl compounds (PFCs) in the village of Hoosick Falls, Rensselaer County, New York. *Environmental Claims Journal* 29: 4–48.

Miller, S.M., Spaulding, M.L., and Zimmerman, J.B. (2011). Optimization of capacity and kinetics for a novel bio-based arsenic sorbent, TiO2-impregnated chitosan bead. *Water Research* 45: 5745–5754.

Mishima, A. (1992). *Bitter Sea: The Human Cost of Minamata Disease.* Kosei Publishing Company.

Mohankumar, K., Hariharan, V., and Rao, N.P. (2016). Heavy metal contamination in groundwater around industrial estate vs residential areas in Coimbatore, India. *Journal of Clinical and Diagnostic Research: JCDR* 10: BC05.

Mohod, C.V. and Dhote, J. (2013). Review of heavy metals in drinking water and their effect on human health. *International Journal of Innovative Research in Science, Engineering and Technology* 2: 2992–2996.

Momodu, M. and Anyakora, C. (2010). Heavy metal contamination of ground water: the Surulere case study. *Research Journal Environmental and Earth Sciences* 2: 39–43.

Mondal, P., Bhowmick, S., Chatterjee, D. et al. (2013). Remediation of inorganic arsenic in groundwater for safe water supply: a critical assessment of technological solutions. *Chemosphere* 92: 157–170.

Pal, A., Gin, K.Y.-H., Lin, A.Y.-C., and Reinhard, M. (2010). Impacts of emerging organic contaminants on freshwater resources: review of recent occurrences, sources, fate and effects. *Science of the Total Environment* 408: 6062–6069.

Pallier, V., Feuillade-Cathalifaud, G., Serpaud, B., and Bollinger, J.-C. (2010). Effect of organic matter on arsenic removal during coagulation/flocculation treatment. *Journal of Colloid and Interface Science* 342: 26–32.

Plum, L.M., Rink, L., and Haase, H. (2010). The essential toxin: impact of zinc on human health. *International Journal of Environmental Research and Public Health* 7: 1342–1365.

Pooja, D., Kumar, P., Singh, P., and Patil, S. (2020). *Sensors in Water Pollutants Monitoring: Role of Material.* Springer.

Rahimizadeh, M. and Liaghatb, A. 2015. Biosorbents for adsorption of heavy metals: A review. International Confenerence on Environmental Science, Engineering and Technologies, CESET, 1–13.

Raviraja, A., Babu, V., Narayanamurthy, G. et al. (2008). Lead toxicity in a family as a result of occupational exposure. *Arhiv za Higijenu Rada i Toksikologiju* 59: 127–133.

Russoniello, C.J., Fernandez, C., Bratton, J.F. et al. (2013). Geologic effects on groundwater salinity and discharge into an estuary. *Journal of Hydrology* 498: 1–12.

Ryan, P.B., Huet, N., and Macintosh, D.L. (2000). Longitudinal investigation of exposure to arsenic, cadmium, and lead in drinking water. *Environmental Health Perspectives* 108: 731–735.

Savci, S. (2012). An agricultural pollutant: chemical fertilizer. *International Journal of Environmental Science and Development* 3: 73.

Singaraja, C., Chidambaram, S., Srinivasamoorthy, K. et al. (2015). A study on assessment of credible sources of heavy metal pollution vulnerability in groundwater of Thoothukudi districts, Tamilnadu, India. *Water Quality, Exposure and Health* 7: 459–467.

Singh, R., Singh, S., Parihar, P. et al. (2015). Arsenic contamination, consequences and remediation techniques: a review. *Ecotoxicology and Environmental Safety* 112: 247–270.

Shahalam, A.M., Al-Harthy, A., and Al-Zawhry, A. (2002). Feed water pretreatment in RO systems: unit processes in the Middle East. *Desalination* 150: 235–245.

Shannon, M.A., Bohn, P.W., Elimelech, M., et al. 2010. Science and technology for water purification in the coming decades. Nanoscience and technology: a collection of reviews from nature Journals. World Scientific.

Sharma, V.K., Dutta, P.K., and Ray, A.K. (2007). Review of kinetics of chemical and photocatalytical oxidation of arsenic (III) as influenced by pH. *Journal of Environmental Science and Health, Part A* 42: 997–1004.

Sharma, R., Singh, N., Tiwari, S. et al. (2015). Cerium functionalized PVA–chitosan composite nanofibers for effective remediation of ultra-low concentrations of Hg (II) in water. *RSC Advances* 5: 16622–16630.

Smith, L.A. (1995). *Remedial Options for Metals-Contaminated Sites*. Lewis Publ.

Soni, V., Singh, P., Shree, V., and Goel, V. (2018). Effects of VOCs on human health. In: *Air Pollution and Control* (eds. N. Sharma, A.K. Agarwal, P. Eastwood, et al.). Springer.

Sundar, S. and Chakravarty, J. (2010). Antimony toxicity. *International Journal of Environmental Research and Public Health* 7: 4267–4277.

Szatyłowicz, E. and Skoczko, I. (2018). The use of activated alumina and magnetic field for the removal heavy metals from water. *Journal of Ecological Engineering* 19: 61–67.

Talabi, A.O. and Kayode, T.J. (2019). Groundwater pollution and remediation. *Journal of Water Resource and Protection* 11: 1–19.

Thomas, D.J., Styblo, M., and Lin, S. (2001). The cellular metabolism and systemic toxicity of arsenic. *Toxicology and Applied Pharmacology* 176: 127–144.

Tran, D.A., Tsujimura, M., Kambuku, D., and Dang, T.D. (2020). Hydrogeochemical characteristics of a multi-layered coastal aquifer system in the Mekong Delta, Vietnam. *Environmental Geochemistry and Health* 42: 661–680.

Vanloon, G.W. and Duffy, S.J. (2017). *Environmental Chemistry: A Global Perspective*. Oxford university press.

Vörösmarty, C.J., Douglas, E.M., Green, P.A., and Revenga, C. (2005). Geospatial indicators of emerging water stress: an application to Africa. *Ambio: A Journal of the Human Environment* 34: 230–236.

Velthof, G., Oudendag, D., Witzke, H. et al. (2009). Integrated assessment of nitrogen losses from agriculture in EU-27 using MITERRA-EUROPE. *Journal of Environmental Quality* 38: 402–417.

Werner, A.D., Bakker, M., Post, V.E. et al. (2013). Seawater intrusion processes, investigation and management: recent advances and future challenges. *Advances in Water Resources* 51: 3–26.

WHO 2001. Arsenic in drinking water. World Health Organization. http://www.who.int/mediacentre/factsheets/fs210/EN.

WHO 2003. Antimony in drinking water. Background document for development of WHO Guidelines for Drinking-water Quality. WHO Geneva.

WHO (2011). Guidelines for drinking-water quality. *World Health Organization* 216: 303–304.

WHO. 14 May 2019. *Drinking water* [Online]. Available: https://www.who.int/news-room/fact-sheets/detail/drinking-water.

Wilbur, S., Abadin, H., Fay, M. et al. (2012). *Toxicological Profile for Chromium*. Atlanta (GA): US Department of Health and Human Services. Public Health Service, Agency for Toxic Substances and Disease Registry, 24049864.

Willis, S.S., Haque, S.E., and Johannesson, K.H. (2011). Arsenic and antimony in groundwater flow systems: a comparative study. *Aquatic Geochemistry* 17: 775–807.

Yoon, S.-H. and Lee, J.H. (2005). Oxidation mechanism of As (III) in the UV/TiO2 system: evidence for a direct hole oxidation mechanism. *Environmental Science and Technology* 39: 9695–9701.

Zhitkovich, A. (2011). Chromium in drinking water: sources, metabolism, and cancer risks. *Chemical Research in Toxicology* 24: 1617–1629.

Zhu, S., Xing, C., Wu, F. et al. (2019). Cake-like flexible carbon nanotubes/graphene composite prepared via a facile method for high-performance electromagnetic interference shielding. *Carbon* 145: 259–265.

5

Source, Assessment, and Remediation of Metals in Groundwater

Anita Punia[1], *Saurabh Kumar Singh*[2], *and Rishikesh Bharti*[1]

[1]*Department of Civil Engineering, Indian Institute of Technology, Guwahati, Assam, India*
[2]*School of Environmental Sciences, Jawaharlal Nehru University, New Delhi, India*

5.1 Introduction

Groundwater is a renewable resource and a major source of freshwater for domestic, agricultural, and industrial purposes (Golchin and Moghaddam 2016). It has inherent advantages over surface water: purity, lower evaporation, and wider distribution. Compared to groundwater, surface water is more easily contaminated due to its high probability of direct exposure to contamination sources. The surface water resources are easily contaminated through the atmospheric deposition of pollutants and mixing of contaminated runoff. In the case of groundwater, leaching and infiltration of heavy metal from upper earth surface are complex processes and soil layers interrupt the direct contact of heavy metals with groundwater. The groundwater resources are supposed to be less contaminated compared to surface water. But once the groundwater gets contaminated it is very difficult to remediate the groundwater resources due to its existence deep underground. The self-purification or rejuvenation capacity of groundwater is also very slow.

Heavy metals are natural components of earth and some (Cu, Zn, Fe, and Mn) are essential micronutrient for living things. They have atomic density greater than 4000 kg/m^3 or 5 times more than water (Garbarino et al. 1995). Heavy metals have tendency to accumulate in living things, persistent for a long duration and non-degradable in nature. The source of heavy metals could be geogenic (natural) or anthropogenic (manmade). Industries, mineral exploration or processing units, urban settlements, and fertilizers are the major contributors of heavy metal enrichment into the environment and groundwater resources.

The wastewater discharged from industries are enriched in heavy metals and increase the load of heavy metals in soil and surface water bodies in the vicinity. The heavy metals percolate to groundwater from the surface water bodies and soil. The groundwater arising from the industrial areas have high concentrations of heavy metals, and particular industries discharge specific heavy metals, i.e. Pb and Cd from textile industries (Noreen

Groundwater Geochemistry: Pollution and Remediation Methods, First Edition.
Edited by Sughosh Madhav and Pardeep Singh.

et al. 2017) and Cr from tanneries (Khan et al. 2020). The chemical fertilizers are also enriched in heavy metals and extensive use of fertilizers in agricultural fields results in accumulation of heavy metals in soil and further in groundwater through infiltration (Atafar et al. 2010). Another important anthropogenic source of heavy metals is the mining sector. Both abandoned and active mines are potential sources of heavy metal contamination in the surrounding environment (Akcil and Koldas 2006).

Apart from anthropogenic activities, several geogenic factors such as geology of the area, rate and degree of weathering, rock–water interaction during recharge, rate, and residence time of groundwater flow control the ion concentration and quality of groundwater (Andre et al. 2005; Singh et al. 2008). The degree of contamination or dissolution of minerals depends upon the infiltration capacity of the aquifer and depth of groundwater along with the intensity of rainfall. The chances of contamination are high at shallow aquifers compared to the highly deep aquifers. The impermeable or comparatively non-reactive rocks (silicates) act as obstacle in the leaching of contaminants into groundwater, while highly permeable rocks such as carbonates, shale, or sulphides promote the percolation of heavy metals or contaminants into the groundwater. Hence, natural factors are also responsible for the enrichment of heavy metals and the water quality assessment is important prior to its use for domestic purposes.

The contamination of heavy metals is a major problem by virtue of their non-degradable nature and persistence in groundwater for longer duration. The contamination of available water resources decreases its availability for drinking purposes. The water quality assessment and hydrogeochemical characterization are also essential for effective management of groundwater resources and sustainable development of a region (Sethy et al. 2016). Some of the heavy metals (e.g. Fe, Cu, Zn, etc.) are micronutrients for living things and some are only harmful (e.g. Pb, Cr, Hg, etc.) (Table 5.1). The excessive inhalation and ingestion of heavy metals in human beings cause many diseases. Co leads to excessive erythrocytosis and chronic mountain sickness (Jefferson et al. 2002), and Cobalt Lung (Lison et al. 1996). The overload of Fe in the human body can lead to colorectal cancer and liver cancer (Huang 2003). High accumulation of Cd and Pb in kidneys leads to their dysfunction (Baldwin and Marshall 1999).

The probability of occurrence of carcinogenic effects from consuming heavy metal-contaminated groundwater can be determined via health risk assessment methods using parameters of daily intake, exposure duration, and body weight. The assessment of heavy metal accumulation in the body can be done by measuring its concentration in hair or nails. The presence of heavy metals even in low concentration is also a cause for concern as

Table 5.1 List of essential and non-essential metals..

Essential metals	Cobalt (Co), Copper (Cu), Chromium (Cr), Iron (Fe), Magnesium (Mg), Manganese (Mn), Molybdenum (Mo), Nickel (Ni), Selenium (Se), and Zinc (Zn)
Nonessential metals	Aluminium (Al), Antimony (Sb), Arsenic (As), Beryllium (Ba), Bismuth (Bi), Cadmium (Cd), Gallium (Ga), Germanium (Ge), Gold (Au), Indium (In), Lead (Pb), Lithium (Li), Mercury (Hg), Nickel (Ni), Platinum (Pt), Silver (Ag),Strontium (Sr), Tellurium (Te), Tin (Sn), Titanium (Ti), Vanadium (V), and Uranium (U).

Source: (Tchounwou et al. 2012)

heavy metals have high accumulation capacity in living organism. Heavy metals accumulate into human beings through three major routes: inhalation, ingestion, and dermal adsorption. The accumulation of high concentrations of carcinogenic elements, specifically Arsenic (As), from groundwater to fingernails and hair of humans is observed in Thailand (Wongsasuluk et al. 2018) and Bangladesh (Rakib et al. 2013). The health hazards caused by non-carcinogenic elements can lead to many fatal diseases or even organ failure. Ogunlaja et al. (2019) observed that gasoline spills due to pipeline vandalisation resulted in elevation of Cr, Cu, Ni, and Zn in soil and water sources at the Agaye community of Lagos State, Nigeria. The residents are vulnerable to risk of non-carcinogenic hazards if they consumed groundwater with elevated levels of Cr, Cu, Ni, and Zn.

The above studies confirm that the abundance of heavy metals in groundwater leads to health hazards in human beings. The implementation of remediation technologies by the concerned authorities are important to reduce the health hazards. The implementation of remediation technologies at the contamination site is more feasible and cheaper than at the consumption point. The implementation of remediation technologies at the source would save large populations from health risks of heavy metals. Thus, it is important to discuss the sources of heavy metals and remediation technologies for safe use of water. In the present chapter we will be discussing sources, pathways, monitoring methods, and the existing remediation technologies to understand the factors influencing the heavy metal contamination of groundwater.

5.2 Sources of Contamination

5.2.1 Geogenic Sources

The source of heavy metals in groundwater could be geogenic or anthropogenic. The geogenic sources of heavy metals are weathering and leaching of heavy metals from parent rocks. The abundance of heavy metals in the case of geogenic sources are usually lower or within the permissible limits prescribed by the global or local environmental protection agencies. However, some studies have reported high concentrations of heavy metals originating from parent rocks. Arslan et al. (2017) observed that As, B, and Fe concentrations exceeded the Turkish drinking water guideline values in groundwater at Northern Develi Closed Basin, Kayseri, Turkey. The source of metals (As, B, Fe, Pb, Zn, Fe, and Cu) is geogenic and attributed to the interaction of highly altered volcanic and pyroclastic rocks with water.

The major geogenic sources of dissolved ions in surface and groundwater is weathering of silicate, carbonate, and sulphide minerals, and the dissolution of evaporates (Singh et al. 2013). The weathering and dissolution of minerals depend on many factors such as climate and chemical composition. It is observed that sulphides react fastest followed by carbonates and silicates (Sherlock et al. 1995). Among carbonates, calcite dissolves more compared to other minerals. In silicates, olivine, wollastonite, and serpentine phases also offer more dissolution (Jambor et al. 2002, 2007).

In sulphide-rich regions, the acid neutralization capacity of carbonate minerals and pH of groundwater are important factors in influencing the dissolution of minerals. The acid generated by the oxidation of sulphide can be neutralized by the dissolution of carbonate minerals. The dissolution of carbonates maintains the pH of groundwater within the alkaline range and prevents further dissolution of sulphides. In addition, it is observed that the neutralization capacity of clay is less than that of calcite (Lottermoser 2010), indicating the possibility of acidic conditions in aquifers.

The naturally occurring carcinogenic contaminants such as uranium, radon, and arsenic are recently reported in the aquifers and lead to health hazards in human beings. The geogenic source of uranium (U) in groundwater is reported worldwide and causes health-related issues in the community (Stalder et al. 2012; Liesch et al. 2015; Shin et al. 2016; Burow et al. 2017; Sharma et al. 2019). Naturally, uranium is found in the forms of carbonates (Bayleyite, Andersonite, and Grimselite), phosphates (Autunite and Torbernite), oxides (Metastudtite, Metaschoepite and Uraninite), and silicates (Uraninite and Swamboite). Radon is also observed to be high in the groundwater contaminated with U. Under high pressure, radon is readily dissolved in aquifer water from the soil and rocks (Srinivasa et al. 2019).

Under natural conditions, Uraninite, a U-enriched silicate, is immobile in nature and readily oxidized to soluble uranyl ions ($U(VI)O_2^{2+}$) on rock–water interactions. Under anoxic conditions, the reduced sulphur and iron species may reduce immobile uranium [U(VI)] to mobile [$U_3O_8/U_4O_9/UO_2$] (Hua and Deng 2008) (Eqs. 5.1–5.4).

$$UO_2^{2+} + \equiv FeS \leftrightarrow \equiv S^{2-} - UO_2^{2+} + Fe^{2+} \tag{5.1}$$

$$\equiv S^{2-} - UO_2^{2+} \leftrightarrow S^0(s) - UO_2(s) \tag{5.2}$$

$$FeS(s) + H_2O \leftrightarrow Fe^{2+} + HS^- + OH^- \tag{5.3}$$

$$UO_2^{2+} + HS^- \leftrightarrow UO_2(s) - S^0(s) + H^+ \tag{5.4}$$

Arsenic (As) concentration is observed to be high in the Gangetic plain and Bengal, India, and the sources of high concentration of As is unweathered alluvium sediments originated from Himalayas (Kumar et al. 2016). The sediments and organic matter create a reducing environment, which results in mobilization of As. The calcium arsenate ($Ca_3(AsO_4)_2$), coal-bearing rocks, and iron oxides (Fe- and Mn-oxi/hydroxides) release As during natural weathering. The Ganges-Brahmaputra plains are also prone to high concentrations of As in the groundwater (Jain et al. 2018). Weathering of silicate minerals in the host granitic rocks are a source of high content of Fe and Mn in the groundwater and the geochemical processes related with natural dissolution and precipitation govern the heavy metal species in the groundwater (Chung et al. 2019) (Eq. 5.5).

$$FeAsS + 3.5O_2 + 4H_2O \rightarrow Fe(OH)_3 + H_3AsO_4 + 2H^+ + SO_4^{2-} \tag{5.5}$$

The presence of iron and phosphate enhance the mobility of As into the groundwater. Phosphorous has a stronger affinity to mineral surface complexes compared to As, which influences the desorption and sorption processes. Phosphorous substitutes As leading to the desorption of As (V) from the mineral surface and contaminates the groundwater (Maier et al. 2017). The enrichment of As in groundwater of Assam, India, is attributed to geogenic source, i.e. the reductive dissolution of iron oxyhydroxides (Nath et al. 2018).

5.2.2 Anthropogenic Sources

Anthropogenic sources of heavy metal contamination are industries, mines, urban areas, and fertilizers. The anthropogenic groundwater extraction affects the chemical composition of local shallow aquifers (Neidhardt et al. 2013). The anthropogenic can be classified two categories, i.e. point (mines and industries) and non-point (fertilizers).

5.2.2.1 Industries

The rapid industrialization and gross negligence of pollution abatement measures by industrial units resulted in heavy metal (i.e. Cr, Cyanide) contamination of groundwater due to industrial effluents (Kakar and Bhatnagar 1981). Industries (pharmaceutical products, agrochemicals, electronics, textiles, paints, leather processing, electroplating, etc.) discharge untreated wastewater into groundwater and contaminate the water resources. The discharged wastewater accumulates on the soil surface and poses a risk to groundwater quality even after the closure of the industry. The Hazaribagh industrial area in Dhaka city, Bangladesh, is a hotspot for Cr pollution for soil and groundwater due to discharge from 150 tanneries. Even after the closure of the tanneries a high concentration of Cr was observed in groundwater near the sites (Khan et al. 2020).

Infiltration of illegal wastewater discharged from factories, such as chemical plants and electronics plants, into groundwater results in an increase in concentration of Hg and Co (Huang et al. 2018). Similarly, the textile industries are also potential contaminators of groundwater. It is observed that textile industries at Faisalabad, Pakistan, are sources of Pb and Cd in groundwater (Noreen et al. 2017). The heavy metals Cr, Cd, Ni, and Pb were observed high in the soil and groundwater of Hattar, Pakistan, near the textile industry (Manzoor et al. 2006). The change in industrial wastewater treatment procedure, i.e. waste crushing and volume reduction, does not have significant influence on the concentration of Zn and Cd in groundwater indicating the plant was the main source of contamination (Ozaki et al. 2019). Other industrial steps such as edible oil, battery fabricating, metal plating, metal amalgams, plastic items, synthetic substances, and so forth are also observed as sources of As, Cr, Cd, Ni, Pb, and Zn in the groundwater of Hyderabad, India (Krishna et al. 2019).

5.2.2.2 Urbanization

In urban areas, leakage of sewage is an important cause for the deterioration of groundwater quality. Sewage contains nutrients and it is mainly applied on agricultural fields for irrigation purposes and to act as a fertilizer. The mixing of industrial effluents with sewage wastewater enhances the problem and concentration of heavy metal content. The soil irrigated with heavy metal enriched sewage wastewater also contains a high concentration of heavy metals such as Pb, Cr, Cd, and Ni and further contaminates the groundwater resources (Dheri et al. 2007). The soil texture is an important parameter and coarse texture soil supports the leaching of heavy metals into the groundwater. In the locations where soil texture is coarse, the shallow aquifers are more contaminated with heavy metals compared to clay and deep aquifers.

The waste generated from urban areas is dumped in landfills with limited or no application of proper scientific methodologies in developing countries. The filling of landfill sites without sorting degradable and non-degradable waste leads to huge heaps of waste. Sorting of waste segregates toxic waste from the less toxic waste and reduces the burden on landfill sites. It helps in reducing the maintenance and infrastructural cost for proper disposal and treatment of toxic waste. The leaching of heavy metals from landfills into groundwater nowadays is considered one of the major sources of heavy metals in groundwater. Landfill sites maintained below the standard limits and lack of sorting of waste generate leachates and toxic gases into the environment. Krčmar et al. (2018) observed that the leachates from landfill sites contaminate the groundwater resources with emerging contaminants, i.e. Cr and Cu. The percolation of leachates from municipal solid waste resulted in an increase in concentration of Pb, Fe, Cd, and Cr that was above the acceptable limits of the World Health Organization for drinking water (Boateng et al. 2019).

The groundwater is contaminated with heavy metals and trace element concentration near urban areas. The concentration of B and Se are observed high in the urban area of Hong Kong, China, and it was observed that the corrosion or leakage of stormwater drains in the region was a source of B and Se in groundwater (Leung and Jiao 2006). Electronic waste generated from households is a source of heavy metals (Cr and Hg) and even the e-waste recycling sites are also contaminating groundwater. The shallow groundwater is more prone to contamination due to landfills compared to deep aquifers. The shallow groundwater (within 30 m depth) around the Zhoukou landfill is not suitable for drinking (Han et al. 2014). The shallow groundwater in the surrounding area of the e-waste recycling area of southeast China was seriously contaminated by Cd (Wu et al. 2014). The source of Cd is e-waste and justified by its presence in the topsoil. The reconstruction of heavy metal (Pb, Mn V, Cu, Cd, and Al) pollution in shallow groundwater using the lead isotopic composition of urban speleothems in the Paris (France) from last three centuries indicates the anthropogenic sources are feeding the speleothems (Pons-Branchu et al. 2015).

5.2.2.3 Mines and Mineral Exploration

Mines (metallic and non-metallic) are also one of the major sources of heavy metal contamination for both surface and groundwater resources. Groundwater quality is degrading in many parts of the world such as in the United States (Miller et al. 1982); China (Zhang and Zhao 1996); Chile (Leybourne and Cameron 2008); Iran (Aghazadeh and Mogaddam 2011); and India (Punia and Siddaiah 2017) due to mining. Mines generate sulphide-rich mining waste, i.e. tailings and overburden rocks during extraction and mineral processing. Sulphide generates acid mine drainage (AMD) enriched in heavy metals on exposure to water and oxygen which contaminate the surroundings. Acid generation lowers the pH which further enhances the release of heavy metals. The heavy metal contamination is observed high in alluvium aquifers compared to hard rock aquifers indicating leaching of heavy metals. In the mineral exploration area, the presence of carbonate and silicate minerals reduces the probability of acidification of groundwater or generation of AMD (Ahmadi et al. 2018).

During the rainy season, the runoff from mining waste piles (tailings and overburden rocks) generate acid enrich drainage and mixes with surface water resources such as river, lakes, and ponds. Runoff from mine waste piles and AMD from both active and abandoned mines are accounted significantly for the heavy metal contamination of surface water streams (Schaider et al. 2014). Similarly, the literature survey suggests heavy metal contamination of groundwater due to mines along with changes in hydrogeochemical processes (Lei et al. 2010; Das et al. 2013; Corcho Alvarado et al. 2014).

Some studies reported negligible impact of mining activities on the groundwater quality. A study conducted by Amari et al. (2014) shows concentration of heavy metals (Cu, Fe, Pb, S, Cd, Zn, Co, Cr, Ni, Se, and As) in groundwater within acceptable limits of Italian standards, despite of their higher concentration in AMD of tailing (Kettara). The possible reason for this may be the dry climatic condition which resists the metal mobility. Tailings enriched in heavy metals could easily contaminate the water bodies because of mobility of heavy metals (Jang et al. 2005).

5.2.2.4 Fertilizers

The fertilizers are blended with trace metals and contaminate soil on application. The fertilizers manufactured from carbonatites have higher contents of REE (Sr, Ba, and Th) and fertilizers from phosphorites have higher contents of heavy metals (Cd, U, and As) (Otero et al. 2005). The phosphate fertilizers have Cd as impurity and their application leads to change in soil geochemistry and further leads to contamination of groundwater (Taylor et al. 2016; Kubier et al. 2019). The mobility of Se in sediments is enhanced by the infiltration of NO_3^-, and finally results in the high concentration of Se in groundwater (Huang et al. 2018). The agricultural irrigation, ammonium fertilizers, and seawater intrusion enhanced the mobility and toxicity of Cr by direct and indirect processes in the Central Evia (Greece) (Remoundaki et al. 2016). Soil acts as a barrier in the transportation of heavy metal into the groundwater. The mobility of heavy metal depends on the nature of the soil (clay and sand) and the contaminant. The risk of groundwater contamination in sandy soil is as follows: lead < nickel < phosphate < nitrate; for clay soil, it is phosphates < lead < nickel < nitrate indicating lowest retention of nitrate irrespective of sand and clay content (Abdelwaheb et al. 2019). The clay cannot constitute a barrier for the groundwater metal content movement at neutral pH.

5.2.2.5 Intrusion of Seawater in Coastal Regions

Understanding of heavy metal contamination of groundwater due to the intrusion of saline ocean water in coastal region is very limited. The intrusion of saline water or extensive extraction leads to the salinization of groundwater resources in the coastal regions. A study carried out by Wen et al. (2019) shows that seawater intrusion into the shallow aquifer results in high risk of heavy metal contamination in the coastal region. The high concentrations of As, Co, Cr, Cu, Fe, Mn, and Ni in the shallow aquifers are influenced by the groundwater salinization.

5.3 Pathways of Contamination

Mobility of heavy metal into the groundwater follows chemical processes such as ion exchange, complexation, adsorption, absorption, precipitation, and evaporation. Groundwater resources are less prone to pollution compared to surface water resources as the latter are in direct contact with contaminants. The heavy metals reach groundwater following two paths, i.e. leaching or direct contact with groundwater.

5.3.1 Leaching of Heavy Metals

The process of leaching is a complex phenomenon and depends on many factors such as geology, soil texture, runoff flow rate, depth of groundwater, etc. Soil properties play a major role in the percolation and filtration of water to groundwater. Soil pH is an important parameter in understanding the solubility and mobility of heavy metals to groundwater. Uddh et al. (2019) observed that despite the high concentration of heavy metals in the waste zone and pore water of glass landfills, the concentration of Cd and Pb at the transition of landfills and natural soil decreases abruptly. They observed no groundwater contamination suggesting high cation retention efficiency due to high pH.

Percolation of leachate from landfill sites contains high concentration of heavy metals and contaminates the groundwater (Boateng et al. 2019) (Figure 5.1). The leachate content depends on physical composition of landfill, e.g. landfill sites with a majority of electronic waste have a high concentration of Hg and Pb. Similarly, there is little or no impact of heavy metal contamination in the downstream groundwater of municipal landfill sites (Mor et al. 2006). Thus, apart from the soil properties the leaching of heavy metals in groundwater also depends on the composition of the upper surface.

The upper earth surface or soil acts as a hurdle and retain heavy metals through the adsorption and precipitation process. The seepage and leaching of heavy metals in groundwater depend on the geology, flow rate of infiltrate, rock–water interaction and climate of the region. The shallow aquifers are more prone to heavy metal contamination compared to deep aquifers (Ravindra and Mor 2019). Fewer contaminants percolate in the deep aquifers. The dissolution of minerals in shallow aquifer is more due to closeness of earth surface. Leaching of water dissolves the minerals and increase solute concentration of groundwater. Along with silicates other elements also dissolve and increase the concentration of heavy metals.

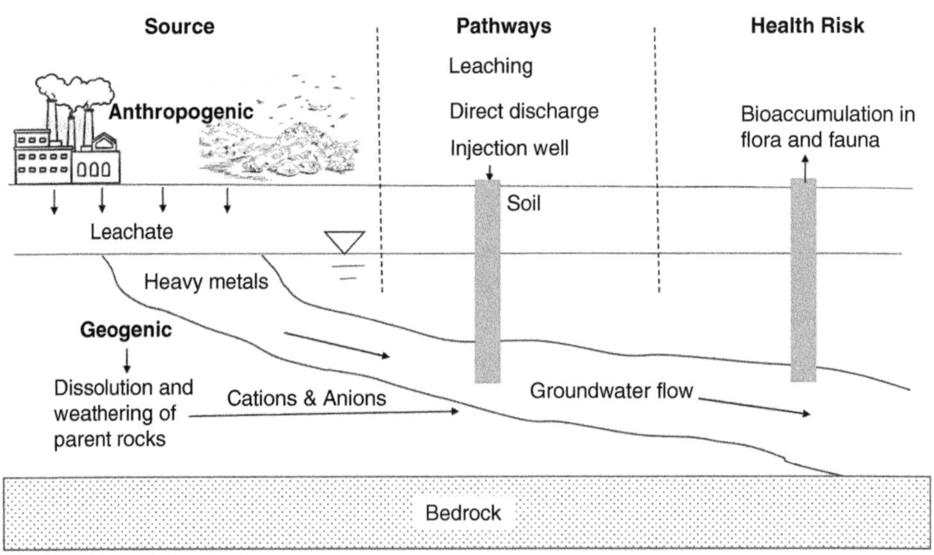

Figure 5.1 The fate of heavy metals from source to living beings.

Atmospheric deposition of heavy metals on the surfaces of plants, water bodies, and soil emitted from anthropogenic sources such as mines and industries are one of the main polluters. Vehicular emissions lead to the enrichment and accumulation of heavy metals such as Cu, Zn, Pb, Cd, and Sb to the roadside soils (Yan et al. 2018). Vehicular movement for industrial purposes could lead to the enrichment of heavy metals in roadside soil. In the mining sector, dispersal and atmospheric deposition of ore particles enrich in heavy metals during ore transportation is also a source of contamination. It is observed that areas having high concentration of heavy metals in soil also have high concentration of heavy metals in the groundwater (Manzoor et al. 2006; Khan et al. 2020).

5.3.2 Direct Contact with Contaminates

The industrial discharge contains high concentrations of heavy metals as discussed earlier in the chapter. The implementation of wastewater treatment plants and proper discharge is an expensive process and increases the production cost of material. Industrial wastewater discharged directly into the groundwater resources through injection wells is commonly used in developing countries to reduce the production cost. The direct discharge of wastewater into groundwater is a major concern as the self-purification capacity of groundwater is very low.

In the mining sector, the direct contact of sulphide ores exposed during the extraction process with the groundwater table is a common phenomenon. In the case of underground or surface mining the ore-bearing rocks are present near to groundwater level. The continued extraction of ore-bearing rocks by mining authorities till the groundwater level results in oxidation of exposed sulphides (Punia and Siddaiah 2017). The direct contact of groundwater with the exposed ores would lead to heavy metal contamination of whole aquifer.

5.4 Monitoring and Assessment Tools

The monitoring and assessment of concentration of heavy metals is important to evaluate the level of pollution. The advanced spectroscopic-based technologies are frequently being used for the quantification of heavy metals. The inductively coupled plasma mass spectrometry (ICPMS) detects the heavy metals in parts per billion (ppb) concentration. Matrix effect is a major problem faced during spectroscopic analysis. It may increase or decrease the actual values and sometimes it is not detected (Thomsen et al. 2006). The matrix effect can be grouped into three classes such as physical, chemical, and spectral (Thomsen and Schatzlein 2006). The physical effect is due to viscosity, surface tension, and associated elemental concentrations. The chemical effect is associated with the formation of complex compound which may get precipitated or may remain intact in the sample solution. The high temperature of plasma is capable of dissociating most such compounds, which minimizes the effect, but the precipitated compound remains in a storage sample bottle and thus is never analysed (Jiang 1996). The spectral effect is associated with the overlap of emission lines, which enhances the observed signals and gives false readings (Thomsen et al. 2006).

Instruments use different procedures to deal with the matrix effect. In atomic absorption spectroscopy, the use of mini-flow condition is efficient in dealing the matrix problem

(L'vov and Frech 1993). The matrix effect in ICP-AES (inductively coupled plasma atomic emission spectroscopy) can be dealt by using single-point standard and by optimisation of plasma. Among these two, single-point standard addition is most efficient during multi-elemental analysis (Sadler et al. 1997). In ICPMS, the use of cyclonic spray chamber for sample introduction, addition of extra electrons between the skimmer and lens, and the use of robust plasma can solve the inaccuracy problems created by the matrix effect (Agatemor and Beauchemin 2011).

The monitoring of heavy metal concentration is not feasible for each site at the regional scale, hence geospatial tools are explored for the easy and fast prediction of concentration of heavy metal. Additionally, the multivariate statistical techniques (principal component analysis [PCA], fuzzy comprehensive assessment [FCA], cluster analysis [CA], etc.) are beneficial for assessing the source of contamination. The multivariate techniques (PCA and CA) alter the normal clustering pattern of environmental media and reveal the significant enrichments of heavy metals (Ogunlaja et al. 2019). In the following sections, indices and geospatial tools used for assessing and evaluating the environmental contamination are discussed.

5.4.1 Indices

Indices assess the pollution level of heavy metals irrespective of their source, i.e. anthropogenic or geogenic. Indices are used to assess the enrichment of heavy metal and predict the probable adverse environmental impact. The maximum permissible limits set by the WHO or the national agencies such as the Bureau of Indian Standards (BIS) are used for the calculation of indices. Permissible limits for some of heavy metals are given in Table 5.2.

Indices such as groundwater quality index (GWQI) (Vasanthavigar et al. 2010), heavy metal pollution index (HPI) (Mohan et al. 1996), contamination index (CI) (Adamu et al. 2015), index of environmental risk (IER) (Rapant and Kordík 2003), heavy metal evaluation index (HEI) (Edet and Offiong 2002) and heavy metal contamination index (HCI) (Rajkumar et al. 2020) are commonly used for evaluations of pollution level. The indices classified water samples in different categories according to their potential toxicity. High value of index indicates more harm to living beings upon consumption of groundwater. Some of the equations used for calculating the indices are given:

Table 5.2 The permissible limits of the WHO (2011) and the BIS (2012) for some heavy metals.

Parameters	WHO limit (2011)	BIS (2012) Desirable Limit	BIS (2012) Permissible Limit
Cu	2	0.05	1.5
Zn	3	5	15
Fe	–	0.3	No relaxation
Co	–	0.03	0.2
Ni	0.07	0.02	No relaxation
Mn	0.4	0.1	0.3
Pb	0.01	0.01	No relaxation

Groundwater Quality Index (GWQI) (Eqs. 5.6 and 5.7)

$$GQWI = \Sigma SI_i = \Sigma\left(w_i \times q_i\right) \tag{5.6}$$

$$= \Sigma\left(\left(\frac{w_i}{\sum_{i=1}^{n} w_i}\right) \times \left(\frac{Ci}{Si} \times 100\right)\right) \tag{5.7}$$

where w_i is unit weight factor, q_i is water quality rating, C_i is measured concentration of ith metal, and S_i is WHO standard for ith metal.

HPI (Eqs. 5.8 and 5.9)

$$HPI = \frac{\sum_{i=1}^{n} W_i Q_i}{\sum_{i=1}^{n} W_i} \tag{5.8}$$

$$Q_i = \sum_{i=1}^{n} \frac{C_i - L_i}{M_i - L_i} \times 100 \tag{5.9}$$

where W_i is unit weight, Q_i is quality rating, C_i is measured concentration of ith metal, M_i is maximum permissible limit, and L_i is lower desirable limit for ith metal.

Metal Index (MI) (Eq. 5.10)

$$MI = \sum_{i=1}^{n} \frac{Ci}{Si} \times 100 \tag{5.10}$$

where C_i is concentration of ith metal and S_i is permissible limit.

HEI (Eqs. 5.11 and 5.12)

$$HEI = \sum_{i=1}^{n} HEI_i \tag{5.11}$$

$$HEI_i = \frac{M_i}{MAC_i} \tag{5.12}$$

where HEI_i is heavy metal evaluation of ith metal, M_i is concentration of ith metal, MAC_i is maximum allowable concentration of ith metal.

CI (Eqs. 5.13 and 5.14)

$$CI = \sum_{i=1}^{n} C_{fi} \tag{5.13}$$

$$C_{fi} = \frac{C_a}{C_p} - 1 \tag{5.14}$$

where, C_a is analysed concentration and C_p is permissible limit.

HPI represents the composite influence of all heavy metals on the quality of water. Kumar et al. (2012) used HPI model for evaluating the heavy metal (Cd, Cr, Cu, Pb, and Zn) contamination in the groundwater of Chennai city, India. They observed that the major cause for the accumulation of metals in the groundwater is due to industrialization and improper waste dumping and the groundwater is not safe for drinking purposes. MI is also used for estimating the water quality and it is simple to calculate using measured concentration of heavy metals and drinking water standards. On comparing the HPI and MI, Ravindra and Mor (2019) observed that HPI is more reliable than the MI for quality assessment. HPI considers the unit weightage of each heavy metal and relative importance is given to concentration of each measured heavy metal, including drinking water quality standards of a nation. HPI offers a better association of its outcomes (water quality) with the sources of contamination, i.e. municipal dumping sites and industrial area.

Recently, a new index named HCI was developed by Rajkumar et al. (2020) for the assessment of heavy metal (Cd, Cr, Cu, Fe, Mn, Ni, Pb, and Zn) contamination. The index is developed after the Tiwari and Mishra (1985) method, which is water quality evaluation methods using the physio-chemical parameters pH, electrical conductivity, total dissolved solids, total hardness, and major cations. It is observed that HCI is more efficient compared to HPI, HEI, and CI in assessing the water pollution and it classifies the water samples using mean deviation and percent deviation.

The ecological risk or potential ecological hazards due to heavy metal contamination in groundwater are assessed by computing the mathematical models (Sharifi et al. 2016; Wen et al. 2019; Mukherjee et al. 2020). The ecological risk index (ERI) is classified into different pollution or risk levels based on ranking criteria, namely low risk ($ERI < 110$), moderate risk ($110 \leq ERI < 200$), considerable risk ($200 \leq ERI < 400$), and very high risk ($ERI \geq 400$) (Sharifi et al. 2016). The groundwater quality of the semiarid region of Birbhum district, India, was evaluated using the modified HPI, Nemerow index, and HEI methods and indicate medium to high heavy metal contamination in groundwater. The ecological risk is also observed to be moderate to very high risk for the groundwater-dependent ecosystems in the region (Mukherjee et al. 2020). Similarly, the low and moderate or higher groundwater heavy metal pollution based on HPI evaluation results in low ecological risks to groundwater systems from Laizhou Bay, a typical coastal bay located in the northern Shandong Peninsula of China (Wen et al. 2019). Thus, it can be stated that the high abundances of heavy metals in groundwater are posing significant potential risk to ecological system and indices are good tools for assessing the potential risk.

5.4.2 Geospatial Tools

The Geographic Information System (GIS) based mapping of pollutants are extensively used for the spatial representation of contaminates. The analysed concentrations can be plotted using GIS techniques to represent the distribution and visualization of pollutants. The interpolation models are widely used to predict or estimate the heavy metal concentrations at unsampled sites. The various interpolation methods (inverse distance weighting [IDW], local polynomial, ordinary kriging and radial basis functions) are algorithms featured in GIS software are frequently used nowadays to map the meteorological data on a daily (Garen and Marks 2005), monthly, or annual basis (Mutua and Kuria 2012). It is used to predict the unrecorded climatic data at each station in a catchment. These tools help in extracting the new information from available spatially variable data, suggesting its possible use for predicating the heavy metal concentration in groundwater at unsampled locations.

Recently, the interpolation methods were used to predict the concentration of heavy metals in the sediment (Delgado et al. 2010), soil (Xie et al. 2011), and groundwater (Arslan and Turan 2015). Increase in sampling percentage gradually increases the stability and accuracy of different interpolation models (Liao et al. 2018). Assessing the heavy metal concentration in groundwater (ground truth points) will enhance the accuracy of interpolation techniques. It reduces the time and labour. It helps in easy identification of most affected sites and the highly impacted areas at regional scale. The calculated indices data should be interpolated for the generation of spatial distribution of heavy metals at regional scale.

Different interpolation methods available in ArcGIS software are being used for the representation of spatiotemporal variation of heavy metal contamination. In the IDW technique, the attribute values of unsampled point are assumed by a weighted average of measured values within the fixed distance surrounding the unsampled point (Burrough and McDonnell 1998). The Kriging method considers spatial correlation between the measured point to assume the values of unsampled locations (Goovaerts 1999). Thus, the attribute values of unsampled points are predicted using the information related to other attributes. Spline interpolation is a polynomial interpolation that consists of the approximation of a polynomial over adjacent intervals. The continuous derivatives from the end point of the intervals are used.

Achieving good accuracy is important for better results. The accuracies of six spatial interpolation methods (IDW, NN [nearest-neighbour], Regularized Spline, Tension Spline, Ordinary Kriging, Universal Kriging) were compared for the Zayandeh-Rud River basin, Iran, and suggest that the accuracy of each method depends on geographical location (Amini et al. 2019). Mukherjee et al. (2020) also identified the spline method as the best method for interpolation of groundwater pollutants. Luque-Espinar et al. (2018) observed that the Kriging interpolation method is a suitable method for the representation of spatial variation in relation to environmental factors (geogenic and anthropogenic). Local scale of variability (5 km) represents local anomalies related to natural or anthropogenic factors. Longer range of variability (135 km) represents clearly the geochemical background.

5.5 Remediation Methods

The removal or extraction of heavy metals from groundwater is very difficult process to implement, and is also costly. Many techniques, namely membrane filtration, ion exchange, coagulation, adsorption, reduction or oxidation, and chemical precipitation are used to remove heavy metals from contaminated soil and water (Rajasulochana and Preethy 2016). The classification of remediation technologies is difficult to categorize on the basis of processes followed, as most of the technologies are based on coupled processes. But to simplify the understanding of available technologies, the remediation technologies could be divided into three categories, based on chemical, physical, and biological/biochemical processes.

5.5.1 Chemical Processes

The oxidizing or reducing chemicals are used for the treatment of heavy metal in water resources by transforming or changing the oxidation number of toxic metals into the less toxic. The chemical treatment follows the processes of adsorption, precipitation, coagulation, ion exchange, and complexation to reduce the mobility and toxicity of heavy metals. In the following sections different types of methods based on chemical processes such as coagulation, oxidation, and adsorption for the removal of heavy metals will be discussed.

The low-cost method of oxidation-coagulation-adsorption at optimized pH (OCOP) is capable of efficiently removing heavy metals Cd, Co, Ni, Cu, Cr, and Pb along with As, Fe, and Mn from the groundwater (Bora and Dutta 2019). The removal efficiency of heavy metal increases in the order of Cd (79.0%) < Co (94.8%) < Ni (94.4%) < Cu (98.0%) < Cr (98.3%) < Pb (99.5%) from initial concentration of 2 mg/L of groundwater. In the method, $NaHCO_3$, $KMnO_4$, and $FeCl_3$ are used as pH conditioner, oxidant, and coagulant, respectively. The method is tested in laboratory or artificial conditions and its in situ application is debatable.

Coagulants such as lime and calcium carbonate are effective for the treatment of heavy metal-contaminated groundwater. Mixture of granulated lime ($Ca(OH)_2$) and calcium carbonate ($CaCO_3$) in 1 : 1 (v/v) ratio removes 98% of total content of As and Ni from the contaminated groundwater (Lee et al. 2007). Independently, lime achieved 98% removal efficiency for As and Ni; removal efficiency of calcium carbonate is low for As, i.e. 50%, and Ni with 97%. The mobility of heavy metals is comparatively higher in an acidic environment than an alkaline environment. The coagulants such as lime and calcium carbonate have high acid neutralization capacity and are less toxic to the environment.

Iron as Fe (0) and Fe (II) have reducing capability to reduce the redox sensitive elements, so it is coupled with other adsorbent materials to enhance the adsorption efficiency. The excessive application of iron could result in high concentration of Fe in the groundwater. Ferrous salts and ferrous sulphates are used to reduce the chromate and Cr(VI), respectively (Brown et al. 1998; Puls et al. 1999) (Eq. 5.15).

$$Fe^{2+} + CrO_4^{2-} + 4H_2O \rightarrow \left(Fe_{x_2}Cr_{1-x}\right)\left(OH\right)_3 + 5OH^- \tag{5.15}$$

A novel material of activated carbon/nZVI (C-Fe(0)) composite prepared by Wu et al. (2013) shows high efficiency of Cr reducibility from Cr (VI) to insoluble Cr (III) and the removal capacity is up to 99.5%. Further, Fe-C micro-electrolysis process is investigated by Jia et al. (2020) for the mitigation of Cr and Pb along with nitrate from the groundwater. The nZVI/palladium (Z-Fe/Pd) loaded on zeolite found suitable for the remediation of Cr (VI) from the groundwater following the adsorption and co-precipitation process (He et al. 2020). The Cr removal efficiency is higher under anoxic conditions compared to open atmosphere suggesting its possible use for in situ remediation of groundwater.

Biochar is used for the treatment and remediation of heavy metal and organic contaminants from soil and groundwater by virtue of its porous nature and high surface area. It is produced from different types of biomasses at low pyrolysis temperatures and oxygen-limited conditions. Biochar is combined with functional groups such as graphene, iron oxides, and goethite to enhance the adsorption capacity for heavy metals. The iron scraps and biochar were incorporated into constructed wetlands (CWs) shows the effective purification performance with removal rate of 75–97% for Cr and Pb (Jia et al. 2020). The composites of biochar/iron (BC/Fe), iron sulphide/BC, nano zero-valent iron (nZVI)/BC and iron oxide/BC enhance the properties of biochar such as functional groups, surface area, and electron transfer efficiency to maximize removal potential (Lyu et al. 2020).

5.5.2 Physical Processes

A permeable reactive barrier (PRB) created by placing high permeability materials to treat groundwater is a well-known method. It is cost effective, stops or converts pollutants, and minimizes the plume contamination (Khan et al. 2004; Rad and Fazlali 2020) (Figure 5.2). The height, length, and width are important parameters for PRB design for optimisation the treatment of pollutants and cost (Rad and Fazlali 2020). Additionally, the reduction in the concentration of contaminates depend on the material used for PRB design. Materials such as a primarily zero-valent iron (ZVI) (Simon and Meggyes 2000) and nZVI (Eljamal et al. 2018) are significant reducers and are found suitable for PRB.

The PRB technology can be successfully used for the removal of radioactive elements from the groundwater. The PRB loaded with hydroxyapatite and quartz sand is an effective treatment technology for the remediation of U from the aqueous solutions (Zhang et al. 2019). The in situ reactive zone (IRZ) derived from PRB immobilizes the heavy metals from groundwater. The application of $Mg(OH)_2$ on IRZ increases the alkalinity of groundwater and decreases the mobility and toxicity of Pb(II), Cd(II), and Cr(III) through the process of adsorption and precipitation (Dong et al. 2017).

The filtration technologies made up of different types of materials are frequently used. The efficiency of filtration units particularly depends on the number of pores and pore size. Nano filtration (NF) and reverse osmosis (RO) are found to be effective in removing As from the groundwater (Figoli et al. 2020). The modification in membranes of NF increases

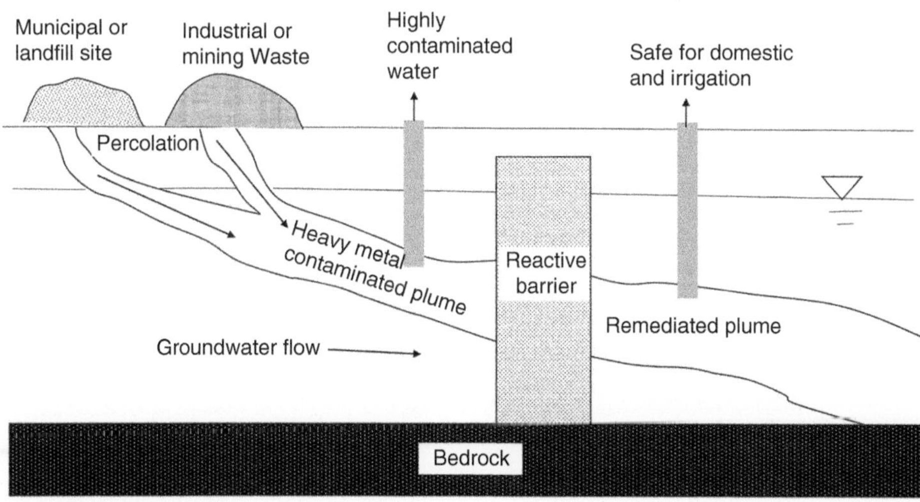

Figure 5.2 The remediation of heavy metal by permeable reactive barrier (PRB). *Source:* Khan et al.,(2004) and Rad et al.,(2020).

the removal efficiency. The polyamide composite NF flat sheet membranes are found to be successful in removing As (V) from the contaminated groundwater at alkaline pH (Sen et al. 2010). Additionally, As (III) can be removed after its oxidation into As (V) using $KMnO_4$ as oxidizing agent.

5.5.3 Biological/Biochemical Processes

The microbial species such as *Flavobacterium*, *Pseudomonas*, *Azotobacter*, *Bacillus*, *Arthrobacter*, *Corynebacterium*, *Aspergilus niger*, *Rhodococcus*, *Methosinus*, *Mycobacterium*, *Stereum hirsutum*, *Nocardia*, *Methanogens*, *Rhizopus arrhizus*, *Alcaligenes*, *Pleurotus ostreatus*, *Phormidium valderium*, and *Ganoderma applantus* are helpful in bioremediation of heavy metals (Verma and Kuila 2019). The removal or treatment of heavy metals at the source site has many advantages over the treatment of water at community level. The in situ bioremediation is a cost-effective method and could be applied at a large scale for the treatment of groundwater.

Microorganisms have a capability to metabolize or transform heavy metals present in the groundwater into less toxic forms (Figure 5.3). The sulphate-reducing bacteria or microorganisms generate sulphides and they strongly bind with cations or heavy metals present in the groundwater. Bacterial species such as *Wolinella, Bacillus, Clostridium, Citrobacter, Desulfomicrobiumetc*, and *Sulfurospirillum*, are identified as potential arsenate reducers (Silver and Phung 2005). Biswas et al. (2019) isolates indigenous *Bacillus Sp.* BAR1 from the highly As contaminated aquifer of Bhojpur, Bihar, and suggest its possible use for in situ bioremediation of contaminated groundwater. The anthropogenic originated heavy metals get diluted in the downstream direction away from the source. In contrast, the heavy metals originated from geogenic sources spread in large area with respect to parent rock and affect

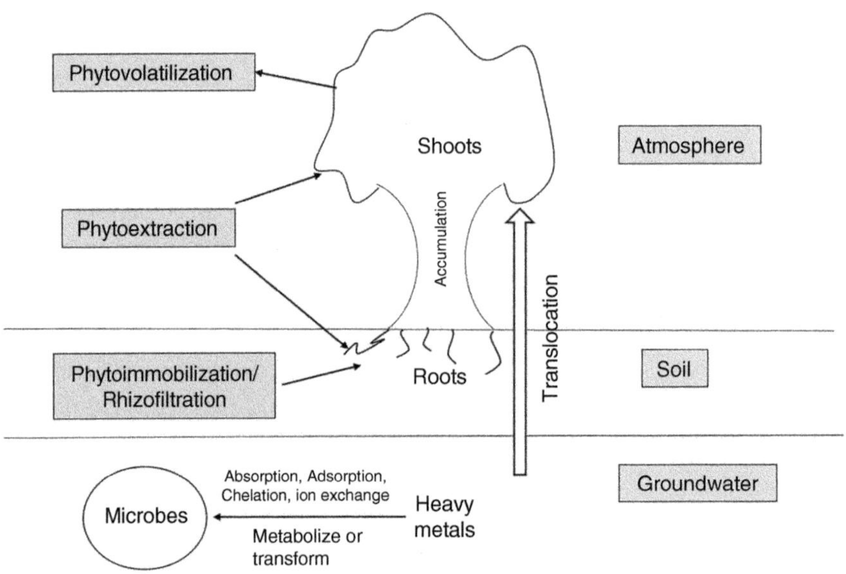

Figure 5.3 The phytoremediation of heavy metals from shallow groundwater.

living things at the regional level. The application of microbial-based remediation technology is the potential solution to remove the geogenic originated heavy metals from the groundwater.

Huang et al. (2017) studied the removal of toluene and copper from contaminated groundwater using microcosm experiments. The microcosm are the experiments designed for creating the natural conditions for the study. The microcosm was inoculated with toluene (17.5 mg/L) and copper (12 mg/L) and after 40 days of operation 99% of both toluene and copper were removed following sulphate-reducing and heavy metal bio-precipitation mechanisms, respectively. They detected 12 dominant sulphate-reducing bacteria and petroleum hydrocarbon degraders. The presence of toluene facilitates the bio-precipitation mechanism for the removal of Cu from water.

Bioaccumulation and biosorption of heavy metals from shallow aquifers into plant tissues is a potential groundwater treatment method (Wang and Zhao 2009). Phytoremediation is a cost-effective and sustainable approach for the removal of heavy metals from shallow aquifers. The absorption of heavy metals by plant roots is known as rhizofiltration. In the phytoextraction process, heavy metals are up taken by plant roots and translocated or accumulated into the aboveground biomass, i.e. shoots and leaves (Rafati et al. 2011). Phytostabilization is the stabilization and immobilization of heavy metals in soils through sorption, complexation, and precipitation processes. It prevents the migration of heavy metals from soil to groundwater (Wuana and Okieimen 2011). Phytovolatilization is a conversion of pollutants into volatile form and then a release into the atmosphere by plants (Padmavathiamma and Li 2007).

Biosorption of heavy metals by plants follow processes such as chelation, ion exchange, complexation, and adsorption by physical forces. Apart from the tolerance level of the plant for heavy metals, the presence of sources for P and N are also important in phytoremediation. Yang et al. (2017) used *Pteris vittata* for the remediation of arsenic-contaminated groundwater and observed that phosphate rock (PR) facilitated As removal. A bio-electrochemical system (BES) is an interaction system between microbes and solid electron acceptor/donors for the generation of electricity from wastewater. Electrodes act as an electron acceptor/donor to promote the microbial oxidation and reduction processes for the remediation of wastewater or groundwater enriched in organic waste and heavy metals (Jain and He 2018; Beretta et al. 2019). The remediation technologies based on chemical, physical, and biological/biochemical processes have some advantages and disadvantages, and some of them are given in Table 5.3.

The economic feasibility of remediation technologies is a major concern for authorities, especially in developing countries. Additionally, implementation of remediation is also a big challenge due to lack of awareness and enthusiasm regarding the environmental ethics. The innovation of cost-effective technologies is needed for successful implementation of remediation methods in developing countries. Many approaches are studied by researchers from developing countries (Biswas et al. 2019; Rad and Fazlali 2020). Chemical and nanofiltration membrane methods are found to be efficient in removing heavy metals from groundwater in laboratory conditions (Sen et al. 2010; Bora and Dutta 2019), but their economic feasibility is a major issue. As contamination is a big problem in the states of Bihar and Assam, India, and recently it was observed that *Bacillus sp.* BAR1 isolated from the aquifer of Bhojpur, Bihar, is capable of reducing arsenate to arsenite or As bio-transformations (Biswas et al. 2019).

Table 5.3 The advantages and disadvantages of physical, chemical, and remediation technologies.

	Advantages		Disadvantages
	Chemical Processes		
1.	High efficiency in removing the heavy metals	1.	Generate toxic gases or chemical
2.	Easy to implement through injection wells	2.	High cost of chemicals
	Physical Processes		
1.	The reactive barrier is effective in remediating the groundwater flowing downgradient of the barrier	1.	Costly and disturbs the natural hydrogeological systems
2.	High adsorption capacity of reactive media for the contaminates	2.	Difficult to implement and clogging of barrier is prominent
	Biological/Biochemical Processes		
1.	No harmful by-products	1.	Generate high quantity of sludge
2.	Low implementation cost	2.	Slow process and only remove particular heavy metals

5.6 Summary

Industries, mines, and urban expansion result in the accumulation and enrichment of heavy metals in the soil and water resources. The geogenic contribution of heavy metals is low and the contribution from anthropogenic sources accelerates the concentration of heavy metals in the water resources. The extensive monitoring and assessment of heavy metals in the groundwater is costly and needs extensive labour. Recently, new technologies based on GIS are being used for the interpolation of missing data from the available data. The developed indices such HPI, MI, and CI for groundwater quality assessment help in identifying the level of groundwater pollution. Pollution of surface resources increases the pressure on groundwater resources and further leads to overpumping of groundwater resources. The overexploitation of groundwater resources increases ionic and heavy metal concentration.

Interdisciplinary research is needed for the development of cost-effective remediation technology considering physical, chemical, and biological processes. The new remediation approaches based on coupling of different processes should be promoted to maximize their efficiency. The groundwater in the industrial and mining regions are contaminated with a mixture of different heavy metals in excessive quantities. The remediation technologies are developed for particular heavy metals and technologies for the removal of a wide range of heavy metals need to be developed. It is more difficult to treat the contaminated aquifers than to remove or reduce the heavy metal contamination at source. The experimental setups are tested at microscale while implementation of developed technology is at the regional level. The environmental conditions at the regional level vary from the artificial to laboratory-based experimental setup. Considering all the above issues, it can be stated that further research is desirable for the development of low-cost and effective remediation technology for the treatment of heavy metal contamination in groundwater.

References

Abdelwaheb, M., Jebali, K., Dhaouadi, H., and Dridi-Dhaouadi, S. (2019). Adsorption of nitrate, phosphate, nickel and lead on soils: risk of groundwater contamination. *Ecotoxicology and Environmental Safety* 179: 182–187.

Adamu, C.I., Nganje, T.N., and Edet, A. (2015). Heavy metal contamination and health risk assessment associated with abandoned barite mines in Cross River State, Southeastern Nigeria. *Environmental Nanotechnology, Monitoring and Management* 3: 10–21.

Agatemor, C. and Beauchemin, D. (2011). Matrix effects in inductively coupled plasma mass spectrometry: a review. *Analytica Chimica Acta* 706 (1): 66–83.

Aghazadeh, N. and Mogaddam, A.A. (2011). Investigation of hydrochemical characteristics of groundwater in the Harzandat aquifer, Northwest of Iran. *Environmental Monitoring and Assessment* 176: 183–195.

Ahmadi, S., Jahanshahi, R., Moeini, V., and Mali, S. (2018). Assessment of hydrochemistry and heavy metals pollution in the groundwater of Ardestan mineral exploration area, Iran. *Environmental Earth Sciences* 77: 212.

Akcil, A. and Koldas, S. (2006). Acid Mine Drainage (AMD): causes, treatment and case studies. *Journal of Cleaner Production* 14: 1139–1145.

Amari, K.E., Valera, P., Hibti, M. et al. (2014). Impact of mine tailings on surrounding soils and ground water: case of Kettara old mine, Morocco. *Journal of African Earth Science* 100: 43–449.

Amini, M.A., Torkan, G., Eslamian, S. et al. (2019). Analysis of deterministic and geostatistical interpolation techniques for mapping meteorological variables at large watershed scales. *Acta Geophysica* 67: 191–203.

Andre, L., Franceschi, M., Pouchan, P., and Atteia, O. (2005). Using geochemical data and modelling to enhance the understanding of groundwater flow in a regional deep aquifer, Aquitaine Basin, south-west of France. *Journal of Hydrology* 305: 40–62.

Arslan, H. and Turan, N.A. (2015). Estimation of spatial distribution of heavy metals in groundwater using interpolation methods and multivariate statistical techniques; its suitability for drinking and irrigation purposes in the Middle Black Sea Region of Turkey. *Environmental Monitoring and Assessment* 187: 516.

Arslan, Ş., Yücel, Ç., Çallı, S.S., and Çelik, M. (2017). Assessment of heavy metal pollution in the groundwater of the northern Develi Closed Basin, Kayseri, Turkey. *Bulletin of Environmental Contamination and Toxicology* 99: 244–252.

Atafar, Z., Mesdaghinia, A., Nouri, J. et al. (2010). Effect of fertilizer application on soil heavy metal concentration. *Environmental Monitoring and Assessment* 160: 83.

Baldwin, D.R. and Marshall, W.J. (1999). Heavy metal poisoning and its laboratory investigation. *Annals of Clinical Biochemistry* 36: 267–300.

Beretta, G., Daghio, M., Espinoza Tofalos, A. et al. (2019). Progress towards bioelectrochemical remediation of hexavalent chromium. *Water* 11: 2336.

Biswas, R., Majhi, A.K., and Sarkar, A. (2019). The role of arsenate reducing bacteria for their prospective application in arsenic contaminated groundwater aquifer system. *Biocatalysis and Agricultural Biotechnology* 20: 101218.

Boateng, T.K., Opoku, F., and Akoto, O. (2019). Heavy metal contamination assessment of groundwater quality: a case study of Oti landfill site, Kumasi. *Applied Water Science* 9: 33.

Bora, A.J. and Dutta, R.K. (2019). Removal of metals (Pb, Cd, Cu, Cr, Ni, and Co) from drinking water by oxidation-coagulation-absorption at optimized pH. *Journal of Water Process Engineering* 31: 100839.

Brown, R.A., Leaby, M.C., and Pyrih, R.Z. (1998). In situ remediation of metals comes of age. *Remediation Journal* 8: 81–96.

Bureau of Indian Standards (BIS) (2012) Indian standard drinking water- specifications (second revision) IS 10500:2012, New Delhi.

Burow, K.R., Belitz, K., Dubrovsky, N.M., and Jurgens, B.C. (2017). Large decadal-scale changes in uranium and bicarbonate in groundwater of the irrigated western U.S. *Science of the Total Environment* 586: 87–95.

Burrough, P.A. and McDonnell, R.A. (1998). Creating continuous surfaces from point data. In: *Principles of Geographic Information Systems* (eds. P.A. Burrough, M.F. Goodchild, R.A. McDonnell, et al.), 17–34. Oxford: Oxford University Press.

Chung, S.Y., Rajesh, R., and Venkatramanan, S. (2019). Evaluation of heavy-metal contamination in groundwater using hydrogeochemical and multivariate statistical analyses. GIS and geostatistical techniques for groundwater. *Science*: 331–346. https://doi.org/10.1016/B978-0-12-815413-7.00024-9.

Corcho Alvarado, J.A., Balsiger, B., Röllin, S. et al. (2014). Radioactive and chemical contamination of the water resources in the former uranium mining and milling sites of Mailuu Suu (Kyrgyzstan). *Journal of Environmental Radioactivity* 138: 1–10.

Das, S., Patnaik, S.C., Sahu, H.K. et al. (2013). Heavy metal contamination, physico-chemical and microbial evaluation of water samples collected from chromite mine environment of Sukinda, India. *Transactions of Nonferrous Metals Society of China* 23: 484–493.

Delgado, J., Nieto, J.M., and Boski, T. (2010). Analysis of the spatial variation of heavy metals in the Guadiana Estuary sediments (SW Iberian Peninsula) based on GIS-mapping techniques. *Estuarine, Coastal and Shelf Science* 88: 71–83.

Dheri, G.S., Brar, M.S., and Malhi, S.S. (2007). Heavy-metal concentration of sewage-contaminated water and its impact on underground water, soil, and crop plants in alluvial soils of northwestern India. *Communications in Soil Science and Plant Analysis* 38: 1353–1370.

Dong, J., Li, B., and Bao, Q. (2017). In situ reactive zone with modified Mg(OH)2 for remediation of heavy metal polluted groundwater: immobilization and interaction of Cr(III), Pb(II) and Cd(II). *Journal of Contaminant Hydrology* 199: 50–57.

Edet, A.E. and Offiong, O.E. (2002). Evaluation of water quality pollution indices for heavy metal contamination monitoring. A study case from Akpabuyo-Odukpani area, lower cross river basin (southeastern Nigeria). *Geojournal* 57: 295–304.

Eljamal, O., Mokete, R., Matsunaga, N., and Sugihara, Y. (2018). Chemical pathways of Nanoscale Zero-Valent Iron (nZVI) during its transformation in aqueous solutions. *Journal of Environmental Chemical Engineering* 6 (5): 6207–6220.

Figoli, A., Fuoco, I., Apollaro, C. et al. (2020). Arsenic-contaminated groundwaters remediation by nanofiltration. *Separation and Purification Technology* 238: 116461.

Garbarino, J.R., Hayes, H., Roth, D. et al. (1995). *Contaminants in the Mississippi River*, 1133. Virginia, USA: U. S. Geological Survey Circular.

Garen, D.C. and Marks, D. (2005). Spatially distributed energy balance snowmelt modelling in a mountainous river basin: estimation of meteorological inputs and verification of model results. *Journal of Hydrology* 315 (1–4): 126–153.

Golchin, I. and Moghaddam, M.A. (2016). Hydro-geochemical characteristics and groundwater quality assessment in Iranshahr plain aquifer, Iran. *Environmental Earth Science* 75: 317.

Goovaerts, P. (1999). Geostatistics in soil science: state-of-the-art and perspectives. *Geoderma* 89: 1–45.

Han, D., Tong, X., Currell, M.J. et al. (2014). Evaluation of the impact of an uncontrolled landfill on surrounding groundwater quality, Zhoukou, China. *Journal of Geochemical Exploration* 136: 24–39.

He, Y., Lin, H., Luo, M. et al. (2020). Highly efficient remediation of groundwater co-contaminated with Cr(VI) and nitrate by using nano-Fe/Pd bimetal-loaded zeolite: process product and interaction mechanism. *Environmental Pollution* 263: 114479.

Hua, B. and Deng, B. (2008). Reductive immobilization of uranium(VI) by amorphous iron sulfide. *Environmental Science and Technology* 42 (23): 8703–8708.

Huang, X. (2003). Iron overload and its association with cancer risk in humans: evidence for iron as a carcinogenic metal. *Mutation Research* 533 (1–2): 153–171.

Huang, W., Dong, C., Chen, C. et al. (2017). Application of sulfate reduction mechanisms for the simultaneous bioremediation of toluene and copper contaminated groundwater. *International Biodeterioration and Biodegradation* 124: 215–222.

Huang, G., Zhang, M., Liu, C. et al. (2018). Heavy metal(loid)s and organic contaminants in groundwater in the Pearl River Delta that has undergone three decades of urbanization and industrialization: distributions, sources, and driving forces. *Science of the Total Environment* 635: 913–925.

Jain, A. and He, Z. (2018). Cathode-enhanced wastewater treatment in bioelectrochemical systems. *npj Clean Water* 1: 23.

Jain, C.K., Sharma, S.K., and Singh, S. (2018). Physico-chemical characteristics and hydrogeological mechanisms in groundwater with special reference to arsenic contamination in Barpeta District, Assam (India). *Environmental Monitoring and Assessment* 190: 417.

Jambor, J.L., Dutrizac, J.E., Groat, L., and Raudsepp, M. (2002). Static tests of neutrailization potentials of silicate and aluminiosilicates minerals. *Environmental Geology* 43: 1–17.

Jambor, J.L., Dutrizac, J.E., and Raudsepp, M. (2007). Measured and computed neutralization potentials from static tests of diverse rock types. *Environmental Geology* 52: 1019–1031.

Jang, M., Hwang, J.S., Choi, S.I., and Park, J.K. (2005). Remediation of arsenic-contaminated soils and washing effluents. *Chemosphere* 60: 344–345.

Jefferson, J.A., Escudero, E., Hurtado, M.A. et al. (2002). Excessive erythrocytosis, chronic mountain sickness, and serum cobalt levels. *Lancet* 359 (9304): 407–408.

Jia, L., Liu, H., Kong, Q. et al. (2020). Interactions of high-rate nitrate reduction and heavy metal mitigation in iron-carbon-based constructed wetlands for purifying contaminated groundwater. *Water Research* 169: 115285.

Jiang, C. (1996). Solubility and solubility constant of barium sulfate in aqueous sodium sulfate solutions between 0 and 80° C. *Journal of Solution Chemistry* 25: 105–111.

Kakar, Y.P. and Bhatnagar, N.C. (1981). Ground water pollution due to industrial effluents in Ludhiana, India. *Studies in Environmental Science* 17: 265–272.

Khan, F., Husain, T., and Hejazi, R. (2004). An overview and analysis of site remediation technologies. *Journal of Environmental Management* 71 (2): 95–122.

Khan, A., Michelsen, N., Marandi, A. et al. (2020). Processes controlling the extent of groundwater pollution with chromium from tanneries in the Hazaribagh area, Dhaka, Bangladesh. *Science of the Total Environment* 710: 136213.

Krčmar, D., Tenodi, S., Grba, N. et al. (2018). Preremedial assessment of themunicipal landfill pollution impact on soil and shallow groundwater in Subotica, Serbia. *Science of the Total Environment* 615: 1341–1354.

Krishna, A.K., Mohan, K.R., and Dasaram, B. (2019). Assessment of groundwater quality, toxicity and health risk in an industrial area using multivariate statistical methods. *Environmental Systems Research* 8: 26.

Kubier, A., Wilkin, R.T., and Pichler, T. (2019). Cadmium in soils and groundwater: a review. *Applied Geochemistry* 108: 104388.

Kumar, P.J.S., Delson, P.D., and Babu, P.T. (2012). Appraisal of heavy metals in groundwater in Chennai City using a HPI model. *Bulletin of Environmental Contamination and Toxicology* 89 (4): 793–798.

Kumar, M., Rahman, M.M., Ramanathan, A.L., and Naidu, R. (2016). Arsenic and other elements in drinking water and dietary components from the middle Gangetic plain of Bihar, India: health risk index. *Science of the Total Environment* 539: 125–134.

Lee, M., Paik, I.S., Kim, I. et al. (2007). Remediation of heavy metal contaminated groundwater originated from abandoned mine using lime and calcium carbonate. *Journal of Hazardous Materials* 144: 208–214.

Lei, L., Song, C., Xie, X. et al. (2010). Acid mine drainage and heavy metal contamination in groundwater of metal sulfide mine at arid territory (BS mine, Western Australia). *Transactions of Nonferrous Metals Society of China* 20 (8): 1488–1493.

Leung, C. and Jiao, J.J. (2006). Heavy metal and trace element distributions in groundwater in natural slopes and highly urbanized spaces in mid-levels area, Hong Kong. *Water Research* 40: 753–767.

Leybourne, M.I. and Cameron, E.M. (2008). Source, transport, and fate of rhenium, selenium, molybdenum, arsenic, and copper in groundwater associated with porphyry–Cu deposits, Atacama Desert, Chile. *Chemical Geology* 247: 208–228.

Liao, Y., Li, D., and Zhang, N. (2018). Comparison of interpolation models for estimating heavy metals in soils under various spatial characteristics and sampling methods. *Transactions in GIS* 22 (2): 409–434.

Liesch, T., Hinrichsen, S., and Goldscheider, N. (2015). Uranium in groundwater – fertilizers versus geogenic sources. *Science of the Total Environment* 536: 981–995.

Lison, D., Lauwerys, R., Demedts, M., and Nemery, B. (1996). Experimental research into the pathogenesis of cobalt/hard metal lung disease. *European Respiratory Journal* 9: 1024–1028.

Lottermoser, B.G. (2010). *Mien Wastes: Characterization, Treatment and Environmental Impacts*, 3e, 400. Berlin, Heidelberg: Springer-Velag.

Luque-Espinar, J.A., Pardo-Igúzquiza, E., Grima-Olmedo, J., and Grima-Olmedo, C. (2018). Multiscale analysis of the spatial variability of heavy metals and organic matter in soils and groundwater across Spain. *Journal of Hydrology* 561: 348–371.

L'vov, B.V. and Frech, W. (1993). Matrix vapours and physical interference effects in graphite furnace atomic absorption spectrometry – I. End-heated tubes. *Spectrochimica Acta Part B: Atomic Spectroscopy* 48 (3): 425–433.

Lyu, H., Tang, J., Cui, M. et al. (2020). Biochar/iron (BC/Fe) composites for soil and groundwater remediation: synthesis, applications, and mechanisms. *Chemosphere* 246: 125609.

Maier, M.V., Isenbeck-Schröter, M., Klose, L.B. et al. (2017). In situ-mobilization of arsenic in groundwater – an innovative remediation approach? *Procedia Earth and Planetary Science* 17: 452–455.

Manzoor, S., Shah, M.H., Shaheen, N. et al. (2006). Multivariate analysis of trace metals in textile effluents in relation to soil and groundwater. *Journal of Hazardous Materials* A137: 31–37.

Miller, W.R., Ficklin, W.H., and Learned, R.E. (1982). Hydrogeochemical prospecting for porphyry copper deposits in the tropical-marine climate of Puerto Rico. *Journal of Geochemical Exploration* 16 (3): 217–233.

Mohan, S.V., Nithila, P., and Reddy, S.J. (1996). Estimation of heavy metal in drinking water and development of heavy metal pollution index. *Journal of Environmental Science and Health* A31: 283–289.

Mor, S., Ravindra, K., Dahiya, R.P., and Chandra, A. (2006). Leachate characterization and assessment of groundwater pollution near municipal solid waste land fill site. *Environmental Monitoring and Assessment* 118: 435–456.

Mukherjee, I., Singh, U.K., Singh, R.P. et al. (2020). Characterization of heavy metal pollution in an anthropogenically and geologically influenced semi-arid region of east India and assessment of ecological and human health risks. *Science of the Total Environment* 705: 135801.

Mutua, F. and Kuria, D. (2012). A comparison of spatial rainfall estimation techniques: a case study of Nyando River Basin Kenya. *Journal of Agricultural Science and Technology* 14 (2): 149–165.

Nath, B.K., Chaliha, C., Bhuyan, B. et al. (2018). GIS mapping-based impact assessment of groundwater contamination by arsenic and other heavy metal contaminants in the Brahmaputra River valley: a water quality assessment study. *Journal of Cleaner Production* 201: 1001–1011.

Neidhardt, H., Berner, Z., Freikowski, D. et al. (2013). Influences of groundwater extraction on the distribution of dissolved as inshallow aquifers of West Bengal, India. *Journal of Hazardous Materials* 262: 941–950.

Noreen, M., Shahid, M., Iqbal, M. et al. (2017). Measurement of cytotoxicity and heavy metal load in drains water receiving textile effluents and drinking water in vicinity of drains. *Measurement* 109: 88–99.

Ogunlaja, A., Ogunlaja, O.O., Okewole, D.M., and Morenikeji, O.A. (2019). Risk assessment and source identification of heavy metal contamination by multivariate and hazard index analyses of a pipeline vandalised area in Lagos State, Nigeria. *Science of the Total Environment* 651: 2943–2952.

Otero, N., Vitòria, L., Soler, A., and Canals, A. (2005). Fertiliser characterisation: major, trace and rare earth elements. *Applied Geochemistry* 20 (8): 1473–1488.

Ozaki, H., Ichise, H., Kitaura, E. et al. (2019). Immutable heavy metal pollution before and after change in industrial waste treatment procedure. *Scientific Reports* 9: 4499.

Padmavathiamma, P.K. and Li, L.Y. (2007). Phytoremediation technology: hyperaccumulation metals in plants. *Water Air and Soil Pollution* 184: 105–126.

Pons-Branchu, E., Ayrault, S., Roy-Barman, M. et al. (2015). Three centuries of heavy metal pollution in Paris (France) recorded by urban speleothems. *Science of the Total Environment* 518–519: 86–96.

Puls, R.W., Paul, C.J., and Powell, R.M. (1999). The application of in situ permeable reactive (zero-valent iron) barrier technology for the remediation of chromate contaminated groundwater: a field test. *Applied Geochemistry* 14: 989–1000.

Punia, A. and Siddaiah, N.S. (2017). Assessment of heavy metal contamination in groundwater of Khetri copper mine region, India and health risk assessment. *Asian Journal of Water Environment and Pollution* 14 (4): 9–19.

Rad, P.R. and Fazlali, A. (2020). Optimization of permeable reactive barrier dimensions and location in groundwater remediation contaminated by landfill pollution. *Journal of Water Process Engineering* 35: 101196.

Rafati, M., Khorasani, N., Moattar, F. et al. (2011). Phytoremediation potential of Populus alba and Morus alba for cadmium, chromuim and nickel absorption from polluted soil. *International Journal of Environmental Research* 5: 961–970.

Rajasulochana, P. and Preethy, V. (2016). Comparison on efficiency of various techniques in treatment of waste and sewage water – a comprehensive review. *Resource-Efficient Technologies* 2 (4): 175–184.

Rajkumar, H., Naik, P.K., and Rishi, M.S. (2020). A new indexing approach for evaluating heavy metal contamination in groundwater. *Chemosphere* 245: 125598.

Rakib, M.A., Huda, M.E., Hossain, S.M. et al. (2013). Arsenic content in inactive tissue: human hair and nail. *Journal of Scientific Research and Reports* 2 (2): 522–535.

Rapant, S. and Kordík, J. (2003). An environmental risk assessment map of the Slovak Republic: application of data from geochemical atlases. *Environmental Geology* 44: 400–407.

Ravindra, K. and Mor, S. (2019). Distribution and health risk assessment of arsenic and selected heavy metals in groundwater of Chandigarh, India. *Environmental Pollution* 250: 820–830.

Remoundaki, E., Vasileiou, E., Philippou, A. et al. (2016). Groundwater deterioration: the simultaneous effects of intense agricultural activity and heavy metals in soil. *Procedia Engineering* 162: 545–552.

Sadler, D.A., Sun, F., Howe, S.E., and Littlejohn, D. (1997). Comparison of procedures for correction of matrix interferences in the multi-element analysis of soils by ICP-AES with a CCD detection system. *Microchimica Acta* 126: 301–311.

Schaider, L.A., Senn, D.B., Estes, E.R. et al. (2014). Sources and fates of heavy metals in a mining-impacted stream: temporal variability and the role of iron oxides. *Science of the Total Environment* 490: 456–466.

Sen, M., Manna, A., and Pal, P. (2010). Removal of arsenic from contaminated groundwater by membrane-integrated hybrid treatment system. *Journal of Membrane Science* 354: 108–113.

Sethy, S.N., Syed, T.H., Kumar, A., and Sinha, D. (2016). Hydrogeochemical characterization and quality assessment of groundwater in parts of Southern Gangetic Plain. *Environmental Earth Science* 75: 232.

Sharifi, Z., Hossaini, S.M.T., and Renella, G. (2016). Risk assessment for sediment and stream water polluted by heavy metals released by a municipal solid waste composting plant. *Journal of Geochemical Exploration* 169: 202–210.

Sharma, T., Sharma, A., Kaur, I. et al. (2019). Uranium distribution in groundwater and assessment of age dependent radiation dose in Amritsar, Gurdaspur and Pathankot districts of Punjab, India. *Chemosphere* 219: 607–616.

Sherlock, E.J., Lawrence, R.W., and Poulin, R. (1995). On the neutralization of acid rock drainage by carbonate and silicate minerals. *Environmental Geology* 25: 43–54.

Shin, W., Oh, J., Choung, S. et al. (2016). Distribution and potential health risk of groundwater uranium in Korea. *Chemosphere* 163: 108–115.

Silver, S. and Phung, L.T. (2005). Genes and enzymes involved in bacterial oxidation and reduction of inorganic arsenic. *Applied and Environmental Microbiology* 71: 599–608.

Simon, F.G. and Meggyes, T. (2000). Removal of organic and inorganic pollutants from groundwater using permeable reactive barriers. *Land Contamination and Reclamation* 8 (2): 103–116.

Singh, A.K., Mondal, G.C., Kumar, S. et al. (2008). Major ion chemistry, weathering processes and water quality assessment in upper catchment of Damodar River basin, India. *Environmental Geology* 54: 45–758.

Singh, A.K., Raj, B., Tiwari, A.K., and Mahato, M.K. (2013). Evaluation of hydrogeochemical processes and groundwater quality in the Jhansi district of Bundelkhand region, India. *Environmental Earth Science* 70: 1225–1247.

Srinivasa, E., Rangaswamy, D.R., and Sannappa, J. (2019). Measurement of radon concentration and evaluation of total dose in drinking water of Chikmagalur city, Karnataka. *Journal of Geological Society of India* 94: 100–104.

Stalder, E., Blanc, A., Haldimann, M., and Dudler, V. (2012). Occurrence of uranium in Swiss drinking water. *Chemosphere* 86: 672–679.

Taylor, M., Kim, N., Smidt, G. et al. (2016). Trace element contaminants and radioactivity from phosphate fertiliser. In: *Phosphorus in Agriculture: 100% Zero* (eds. E. Schnug and L.J. De Kok), 231–266. Dordrecht: Springer.

Tchounwou, P.B., Yedjou, C.G., Patlolla, A.K., and Sutton, D.J. (2012). Heavy metal toxicity and the environment. In: *Molecular, Clinical and Environmental Toxicology. Experientia Supplementum*, vol. 101 (ed. A. Luch). Basel: Springer.

Thomsen, V. and Schatzlein, D. (2006). The Laws of Spectrochemistry. *Spectroscopy* 21 (5): 44–48.

Thomsen, V., Schatzlein, D., and Mercuro, D. (2006). Interelement corrections in spectrochemistry. *Spectroscopy – Springfield then Eugene then Duluth* 21 (7): 32–40.

Tiwari, T.N. and Mishra, M. (1985). A preliminary assignment of water quality index of major Indian rivers. *Indian Journal of Environmental Protection* 5 (4): 276–279.

Uddh, S.T., Berggren, K.D., Åström, M. et al. (2019). Metal solubility and transport at a contaminated landfill site – from the source zone into the groundwater. *Science of the Total Environment* 668: 1064–1076.

Vasanthavigar, M., Srinivasamoorthy, K., Vijayaragavan, K. et al. (2010). Application of water quality index for groundwater quality assessment: thirumanimuttar sub-basin, Tamilnadu, India. *Environmental Monitoring and Assessment* 171 (1–4): 595–609.

Verma, S. and Kuila, A. (2019). Bioremediation of heavy metals by microbial process. *Environmental Technology and Innovation* 14: 100369.

Wang, S. and Zhao, X. (2009). On the potential of biological treatment for arsenic contaminated soils and groundwater. *Journal of Environmental Management* 90: 2367–2376.

Wen, X., Lu, J., Wu, J. et al. (2019). Influence of coastal groundwater salinization on the distribution and risks of heavy metals. *Science of the Total Environment* 652: 267–277.

Wongsasuluk, P., Chotpantarat, S., Siriwong, W., and Robson, M. (2018). Using hair and fingernails in binary logistic regression for bio-monitoring of heavy metals/metalloid in groundwater in intensively agricultural areas, Thailand. *Environmental Research* 162: 106–118.

World Health Organization (WHO) (2011) Guidelines for drinking water quality, vol. 4, Recommendations, 3rd edn, Geneva.

Wu, L., Liao, L., Lv, G. et al. (2013). Micro-electrolysis of Cr (VI) in the nanoscale zero-valent iron loaded activated carbon. *Journal of Hazardous Material* 254: 277–283.

Wu, C., Luo, Y., Deng, S. et al. (2014). Spatial characteristics of cadmium in topsoils in a typical e-waste recycling area in southeast China and its potential threat to shallow groundwater. *Science of the Total Environment* 472: 556–561.

Wuana, R.A. and Okieimen, F.E. (2011). Heavy metals in contaminated soils: a review of sources, chemistry, risks and best available strategies for remediation. *ISRN Ecology* 2011: 1–20.

Xie, Y., Chen, T., Lei, M. et al. (2011). Spatial distribution of soil heavy metal pollution estimated by different interpolation methods: accuracy and uncertainty analysis. *Chemosphere* 82 (3): 468–476.

Yan, G., Mao, L., Liu, S. et al. (2018). Enrichment and sources of trace metals in roadside soils in Shanghai, China: a case study of two urban/rural roads. *Science of the Total Environment* 631–632: 942–950.

Yang, G., Zhu, L., Santos, J.A.G. et al. (2017). Effect of phosphate minerals on phytoremediation of arsenic contaminated groundwater using an arsenic-hyperaccumulator. *Environmental Technology and Innovation* 8: 366–372.

Zhang, L.C. and Zhao, G.J. (1996). The species and geochemical characteristics of heavy metals in the sediments of Kangjiaxi River in the Shuikoushan Mine Area, China. *Applied Geochemistry* 11 (1–2): 217–222.

Zhang, W., Liu, H., Fan, X. et al. (2019). Removal of uranium from aqueous solution by a permeable reactive barrier loaded with hydroxyapatite-coated quartz sand: implication for groundwater remediation. *Geochemistry* https://doi.org/10.1016/j.chemer.2019.125545.

6

Nitrate Pollution in Groundwater and Their Possible Remediation Through Adsorption

Arun Lal Srivastav[1], *Naveen Patel*[2], *Uday Bhan Prajapati*[3], *and Vinod Kumar Chaudhary*[4]

[1]*Chitkara University School of Engineering and Technology, Chitkara University, Solan, Himachal Pradesh, India*
[2]*Department of Civil Engineering, Institute of Engineering and Technology, Dr. Ram Manohar Lohia Awadh University, Ayodhya, Uttar Pradesh, India*
[3]*Patanjali Herbal Research Department, Patanjali Research Institute, Haridwar, Uttarakhand, India*
[4]*Department of Environmental Sciences, Dr. Ram Manohar Lohia Awadh University, Ayodhya, Uttar Pradesh, India*

6.1 Introduction

Out of the five elements (*Mahabhutas*) of Indian philosophy, water is the fourth, followed in evolutionary order only by earth. While being thus considered one of the basic indivisible elements of creation, this alerts us to the fact that water is a more advanced evolutionary product than, for example, fire, and thus also more vulnerable to external influences. Lord Krishna declared in the *Bhagvad-gita* (7.8) *'I am the taste of water'*. In *Rig Veda*, the properties of good drinking water were described as follows: '*Sheetham* (cold to touch), *Sushihi* (clean), *Sivam* (should have nutritive value, requisite minerals, and trace elements), *Istham* (transparent), *Vimalam lahu Shadgunam* (its acid base balance should be within normal limits)' (Sadgir and Vamanrao 2003).

Water is one of the most precious resources among the resources available to humans. Water covers almost 70.9% of the Earth's surface, out of which oceans contain 96.5%, ice caps and glaciers 1.7%, groundwater 1.7%, and in air in the form of clouds, vapour, and precipitation nearly 0.001% (Gleick 1993). Out of all the water resources, only 2.5% of the water is fresh and the majority is in the form of groundwater or ice (98.8%). Water of safe drinking quality is crucial to humans and other forms of life. In the last decades, access to safe drinking water has improved in almost every part of the world, but still approximately 1 billion people lack water and over 2.5 billion lack proper access to adequate sanitation. However, it has been estimated in some of the research that more than half of the world population will be facing vulnerability related to water by 2025 (Kulshreshtha 1998). Maximum acceptable concentrations of inorganic and organic substances and microorganisms in water have been established internationally and also in many countries to ensure the safety of drinking water

Groundwater Geochemistry: Pollution and Remediation Methods, First Edition.
Edited by Sughosh Madhav and Pardeep Singh.

(WHO 2004). Around the world, shortage of water has become one of the major problems and the crisis of water at the global scale has attracted the attention of the world's researchers to find a sustainable solution to produce fresh drinking water (Mossad and Zou 2013; Patel et al. 2020a). Water contamination is one of the main causes of health problems in human beings. About 2.3 billion people are suffering from water-related diseases worldwide (United Nations Educational, Scientific and Cultural Organization [UNESCO] 2003). According to the reports of the WHO and UNICEF, in developing countries more than 2.2 million people die each year due to drinking of unclean water and inadequate sanitation (Azizullah et al. 2011). Generally, water quality parameters are categorized into chemical, physical, or biological. Drinking water quality must be very safe in order to protect the health of biological organisms. In order to protect human health, good quality drinking water should be made available for everyone as a fundamental necessity (Srivastav and Kaur 2020). However, in recent years, public concern about the quality of groundwater has increased mainly due to an increase in industrial and agricultural activities that resulted in the release of toxic pollutants such as synthetic organic chemicals, inorganic anions, and metal ions (Patel et al. 2019a,b; Srivastav et al. 2015; Velizarov et al. 2004). In India, more than 90% of the rural and suburban population are still dependent on water obtained from the ground for their daily activity. The CGWB (2010) of India reported that the majority of groundwater is generally safe for drinking purposes. However, there are some places where high levels of contaminants (arsenic, nitrate, fluoride, etc.) make it unfit for direct human consumption. Deterioration in the quality of drinking water may interrupt the daily life of humans (Jalba et al. 2010). Contaminated water consumption may cause serious diseases. Hence, the attention on risk management to provide safe drinking water has increased many folds (Tang et al. 2013).

In India, around 40% urban and 90% rural public depend on groundwater for their domestic needs (Ranjan et al. 2020; Srivastav and Ranjan 2020). Elevated levels of metallic chemicals, metalloid (arsenic), nitrate, and fluoride are found to be increased in groundwater, which causes the water to be unhealthy (Izah et al. 2016; Srivastav et al. 2014). Arsenic, fluoride, and iron contaminations of water have originated mostly from the rocks, whereas human activities (discharges from houses, agriculture, and industries) have added nutrients (nitrates, phosphates) and metallic contaminations (Singh et al. 2012, 2015). The modern lifestyle, industrial expansion, rapid urban extensions, agrochemical-based agriculture, and population explosions have adversely degraded the quality of groundwater (Patel et al. 2020b; Srivastav 2020; WHO 2008). Figure 6.1 shows the common types of anions found in water.

Various metals like cadmium and mercury along with other toxicants including nitrate, arsenic, and fluoride have been found to be increasing at a very fast rate in groundwater, making it unfit for drinking purposes and other domestic purposes. Arsenic, iron, and fluoride are the commonly observed contaminants and are found to be geogenic, whereas other pollutants including heavy metals, nitrates, and phosphates are anthropogenic in nature and have been generated by agricultural practices, domestic sewerage, industrial effluents, etc. Rapid modernization, urbanization, and industrialization have created enormous pressure on our natural resources including water resources and that had led to degradation of our groundwater quality at a much faster pace (Srivastav et al. 2013a). Nitrate, because of its high solubility, is considered to be one of the most prevalent groundwater contaminants

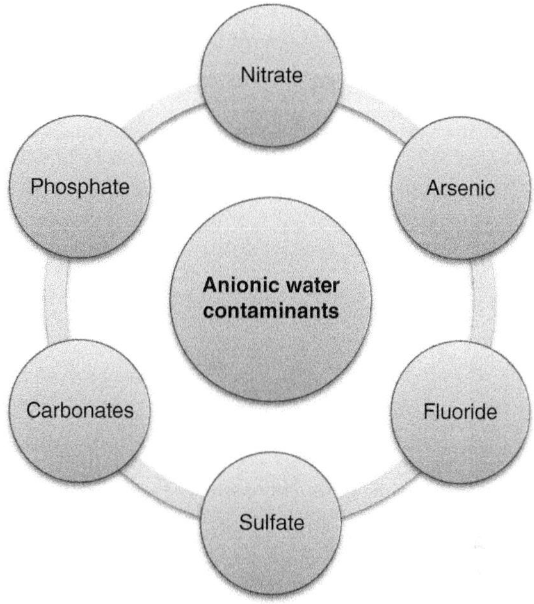

Figure 6.1 Common anionic water contaminants.

around the world, and hence imposes serious human health-related issues and eutrophication (Bhatnagar and Sillanpää 2011). Nitrogen, because of its seven valence states (+5 to −3), is considered to have a complex biogeochemical cycle with multiple biotic/abiotic transformations. There are several organic and inorganic nitrogen compounds that are also essential for survival of all forms of biological life (Vymazal 2007). Nitrite, ammonium, and nitrate are the most predominant forms of inorganic nitrogen found in wetlands (Bialowiec et al. 2012). Nitrate pollution is one of the primarily problems associated with groundwater and it has been found to impart serious biological effects. High concentrations can cause methemoglobinemia, a disease of infants, and have been cited as a risk factor in developing gastric and intestinal cancer, therefore this contaminant is of great concern. Groundwater contamination by nitrates is a worldwide problem mainly related to the use of fertilizers in intensive agriculture (Bhatnagar and Sillanpää 2011; Bouchard et al. 1992; Srivastav et al. 2013b). Various potential sources that are responsible for groundwater contamination with nitrate include: application of fertilizer for agricultural activities; discharge of wastewater from industries; septic tanks; and atmospheric deposition (Aelion and Conte 2004). Application of fertilizers that are inorganic in nature is considered to be one of the most important factors for groundwater pollution (Rao and Puttanna 2000). Nitrate, either via infiltration and runoff, enters into groundwater or surface water systems (Limbrick 2003). Generally, groundwater with high nitrate contamination is founded in the areas where agricultural manures and chemical fertilizers are used in agricultural fields (Vinten and Dunn 2001). Certain parameters like leakage of sewage, overutilization of animal waste and fertilizers, improper management of wastewater effluents, and wastewater without denitrification lead to considerable increase in the nitrate content of groundwater. Numerous aquifers get contaminated due to these activities, and this is quite common, especially in areas where groundwater is replenished from the surface directly. It is very difficult to remediate

Table 6.1 Nitrate contamination groundwater in India.

State	District affected by high nitrate contamination
Andhra Pradesh	Vishakhapatnam, East Godavari, Krishna, Prakasam, Nellore, Chittoor, Anantapur, Cuddapah, Kurnool, Mehboob-nagar, Rangareddy, Medak, Adilabad, Nalgonda, and Khammam
Jharkhand	Palamu, Sahebgunj
Bihar	Patna, Nawada, and Bhagalpur
Haryana	Ambala, Sonepat, Jind, Gurgaon, Faridabad, Hissar, Sirsa, Karnal, Kurukshetra, Rohtak, Bhiwani, and Mahendragarh
Himachal Pradesh	Kulu, Solan, and Una
Madhya Pradesh	Sehore
Maharashtra	Thane, Jaina, Beed, Nanded, Latur, Osmanabad, Solapur, Satara, Sangli, Kolhapur, Dhule, Jalgaon, Aurangabad, Ahmednagar, Pune, Buldana, Amravati, Akola, Nagpur, Wardha, Bhandara, Chandrapur, and Gadchiroli
Punjab	Patiala, Faridkot, Firozpur, Sangrur, and Bhatinda
Rajasthan	Jaipur, Churu, Ganganagar, Bikaner, Jalore, Barmer, Bundi, and Swai Madhopur
Tamilnadu	Coimbatore, Periyer, and Salem
Utter Pradesh	Orai, Jhansi, Lalitpur, Faizabad, Sultanpur, Maharajganj, Gorakhapur, and Deoria
West Bengal	Dinajpur, Malda, Birbhum, Nadia, Midnapur, Howrah, Murshidabad, and Purulia
New Delhi	City Shahdara and Mehrauli Blocks NCT (national capital territory)

Source: ENVIS, 2004

contaminated groundwater, hence our prime responsibility is to prevent the contamination of groundwater (Ceplecha et al. 2004). Contamination of the groundwater with nitrate has emerged as one of the serious matters of concern throughout the world. Parts of Europe, including Germany, Switzerland, and Great Britain, certain parts of the Israel, and United States have been facing issues related to groundwater (Elyanow and Persechino 2005). In Europe, nitrate concentration in groundwater exceeds the international recommendations for drinking water (50 $mgNO_{3-}/l$) in 22% of cultivated land (WHO 1993). Similar nitrate concentrations have been reported in Bulgaria, (Gatseva and Argirova 2008), Senegal (Sall and Vanclooster 2009), USA, China (Laegreid et al. 1999), Italy (Ghiglieri et al. 2009, and Spain (Mesa et al. 2002). Table 6.1 is the compilation of the districts in India affected with high nitrate concentration in groundwater.

A report related to quality of groundwater of 27 metropolitan cities has been released by the Central Pollution Control Board (CPCB). In this report, nitrate concentration above permissible limits have been reported in various parts of India, including: Faridabad, Delhi, Jaipur, Bhopal, Indore, Jabalpur, Nagpur, Nasik, Pune, Bangalore, Hyderabad, Vishakhapatnam, Patna, Kochi, Ahmedabad, Rajkot, Surat, Vadodara, Asansol, Dhanbad, Jamshedpur, Kolkata, Amritsar, Kanpur, Allahabad, Varanasi, and Mumbai (CPCB 2008). Once nitrate reaches the aquifers, they will remain polluted for decades, even when satisfactory measures are taken to reduce its leaching (WHO 2007).

6.2 Sources of Nitrate in Groundwater

Elevated nitrate levels in groundwater mainly result from human and animal wastes, and also because of excessive use of fertilizers. The risk of groundwater contamination has increased mainly due to excessive use of chemicals and fertilizers (Barrabes and Sa 2011; Bhatnagar and Sillanpää 2011; Wang et al. 2009). Groundwater contamination with nitrate is high in the areas of intensive agricultural activities and are usually found to be much higher than the permissible level in drinking water. There are various sources from which nitrate can come to groundwater including animal wastes, municipal wastes, soil organic matters, nitrogen fertilizers, septic tanks, landfills, and from the wastewater that had been discharged from industrial and domestic places (Barrabes and Sa 2011; Suthar et al. 2009; Trevisan et al. 2000; Wang et al. 2009). Nitrogen from human waste also appears to be an important source of nitrates, especially in rural areas lacking centralized water and sanitation systems. Figure 6.2 contains the sources of nitrate in water.

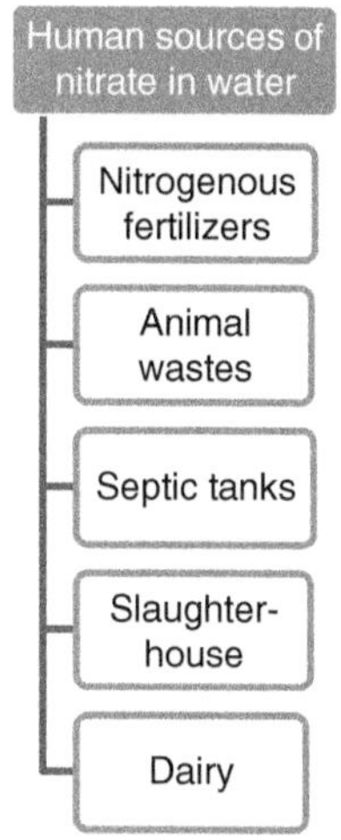

Figure 6.2 Sources of nitrate in water.

The other most common sources of nitrate are uncontrolled discharges of municipal wastes and industrial wastewaters, runoffs of septic tanks, processed foods, dairy, and meat products, and decomposition of decaying organic matter buried in the ground. Chan (2011) reported about vegetable borne nitrate and nitrite and the risk of methemoglobinemia in infants; those under three months old are particularly susceptible to risk. Very high concentrations (>2500 mg/kg) of nitrate in vegetables (especially leafy vegetables) have been reported in different countries (Du et al. 2010; Santamaria 2006).

6.3 Standards of Nitrate for Drinking Water

Setting standards for drinking water contaminants is a very difficult and imperfect process because it is not only based on scientific considerations but is also influenced by political, economic, and social issues. In fact, for scientists it is very difficult to predict the effects of the chemical ingestion in small amounts via water as data relating 'human health effects' to 'chemicals' in drinking water are very limited. Drinking water guideline values have been derived for a number of chemical constituents. A guideline value normally represents the concentration of a constituent that does not result in a significant risk to health over a lifetime of consumption (WHO 2006). Table 6.2 shows the concentration of nitrate recommended for drinking water.

Table 6.2 gives a summary of nitrate standards for drinking water as applied in various parts of the world. Carcinogenicity is the primary characteristic that is used for many contaminants, in order to determine the maximum contaminant levels (MCL). However, until now no conclusive epidemiological studies had been done in order to establish MCL for nitrate, and hence there is no link for nitrate to cancer in humans. The MCL for nitrate mandated by the USEPA is 10 mgN/l (Zhao et al. 2012).

Table 6.2 Maximum permissible nitrate concentration for drinking water.

Country/Organization	Level of allowable nitrate (mg/l)	References
India	45	IS: 10500 (2012)
WHO	50	WHO (2008)
USEPA	45	USEPA (2009)
EU	50	EU (2007)

6.4 High Nitrate Concentrations in Drinking Water and Their Impact on Human Health

Water containing nitrate in high concentrations can cause enormous health-related issues in humans. The impact of nitrate itself has been called primary toxicity. Intake of drinking water with high nitrate concentration leads to several problems, including: diarrhoea, vomiting, diabetes, abdominal pain, birth defects, spontaneous abortions, respiratory tract infections, increased infant mortality, hypertension, central nervous systems defects, and changes to the immune system (Samatya et al. 2006). Nitrate's secondary toxicity is caused mainly from the reduction of nitrate in the intestine into reactive nitrite ion. It has been implicated in methemoglobinemia, especially in infants (six months old). As ferrous ion in haemoglobin is oxidized by nitrite to ferric form, it leads to the formation of methemoglobin (Methb). The state of methemoglobinemia can be described by cyanosis, stupor, and cerebral anoxia. Skin colours of bluish or brownish grey, excessive crying in children, and irritability are the symptoms of methemoglobinemia (Samatya et al. 2006). In tertiary toxicity, the reaction between nitrite and secondary or tertiary amine in acidic medium takes place. Consequently, n-nitroso compounds are formed and some of them may cause carcinogenic, teratogenic, and mutagenic problems in human beings (Samatya et al. 2006). Nitrate concentration in high amounts causes serious health issues other than methemoglobinemia including non-Hodgkin's lymphoma and miscarriages. Acute poisoning may occur after the ingestion of plants or water contaminated with high nitrate concentration within 30 minutes to 4 hours of duration (Romano and Zeng 2009). Nitrate poisoning in animals may lead to death of cattle (Islam and Patel 2010).

6.5 Techniques for Nitrate Removal from Water

In order to abate nitrate from the aqueous medium, significant research has been carried out in diverse areas, as holistic development of any society can be achieved through conducting intensive research (Srivastav et al. 2019). Various types of treatment techniques are considered important in their adoption on the basis of scale of application, electricity requirement, and ease of operation. This paper attempts to summarize the recent advances of available techniques for nitrate removal from aqueous solutions.

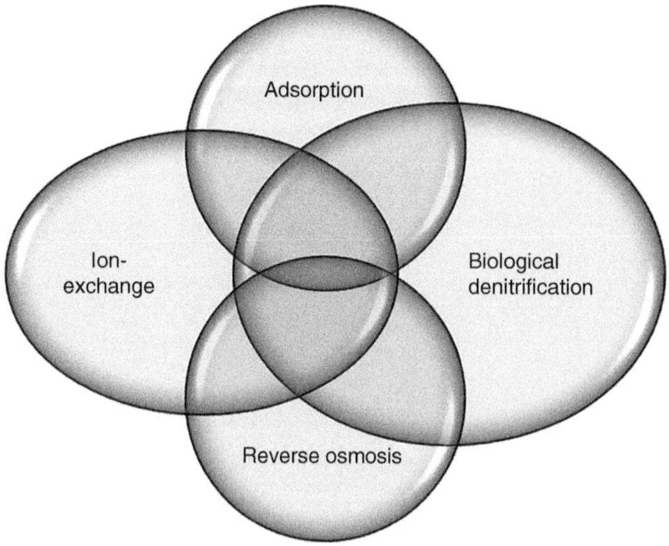

Figure 6.3 Universal techniques for nitrate removal from water.

Biological denitrification (Cameron and Schipper 2012; Zhao et al. 2012), catalytic reduction (Barrabes and Sa 2011; Hasnat et al. 2012), reverse osmosis (Schoeman and Steyn 2003), ion exchange (Milmile et al. 2011), and chemical reduction (Chen et al. 2005) are several methods that can be used for the removal of nitrate. Excellent reviews have been published for methods for the removal of nitrate from aqueous solutions (Bhatnagar and Sillanpaa 2010; Kapoor and Virarghavan 1997; Shrimali and Singh 2001). These processes can remove nitrate from water with varying degrees of efficiency, cost, and ease of operation. Processes including ion exchange, biological denitrification, and adsorption are found to more acceptable compared to other technology. For removal of nitrate from water, the ion-exchange process has been found to be more viable, whereas, biological denitrification has been found to be more appropriate in the case of nitrate removal from surface water (Kapoor and Virarghavan 1997). Duolite A 161, Duolite A 165, and Amberlite IRA 996 are the three strong base anion exchange resins that have been utilized by Peter et al. (1987) in order to abate the nitrate and sulphate present in excess concentration in water. Figure 6.4 shows various types of the low-cost adsorbents for the removal of nitrate from the water.

There are several types of adsorbents that have been reported in the literature, such as natural and synthesized materials. Some agricultural and industrial wastes were also used as adsorbents to purify the water against the contaminants (modified from Bhatnagar and Sillanpää 2011 and Singh et al. 2020). Hydrous metal oxides were also explored to remove water contaminants (including nitrate) from the water (Singh et al. 2012; Srivastav and Ranjan 2020; Srivastav et al. 2013a). Though, the studies have found that these anionic exchange resins were more selective to nitrate as compared to sulphate. Beltran de Heredia et al. (2006) used Amberlite IRN-78 ion-exchange resin for removal of nitrate from aqueous solutions. Another ion-exchange resins selective to nitrate has been reported by Samatya

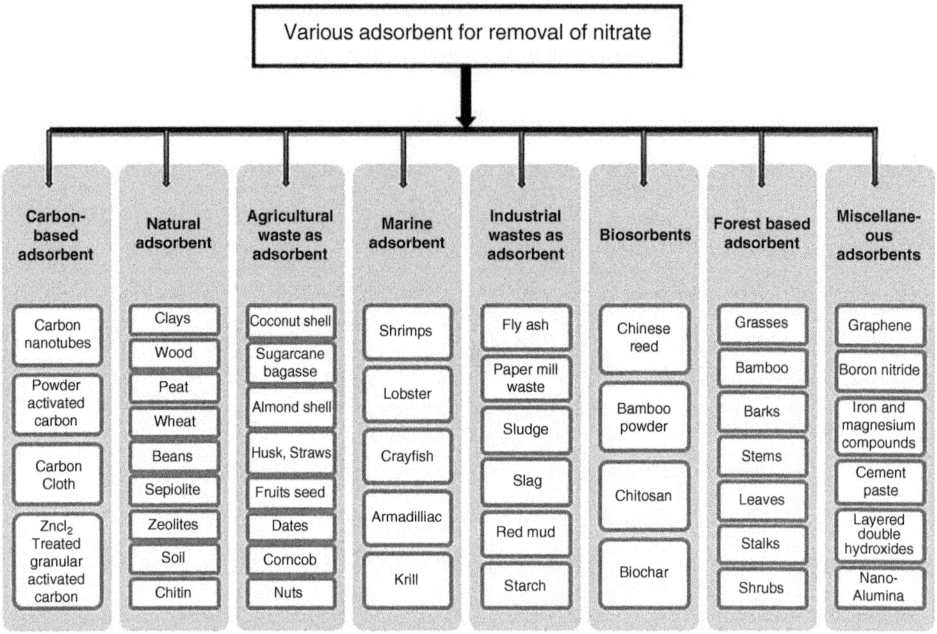

Figure 6.4 Low-cost adsorbents used for the removal of nitrate from the water. *Source:* Modified from Bhatnagar and Sillanpää (2011) and Singh et al. (2020).

et al. (2006) and has been named as Purolite A 520 E. Nitrate removal kinetics has been studied by Chabani et al. (2006) using ion-exchange resins, namely Duolite A 126, Purolite A 520 E, Amberlite IRA 900, and Amberlite IRA 996.

6.6 Challenges in Using Organic Adsorbent or Resin-Based Ion Exchangers

Stability of anion exchange resins is lower than cation exchange resins and during their application it undergoes degradation of the exchange group. Within a temperature range of 20–80 °C, strong basic groups of anion exchange resins are thermally unstable, and their capacity subsequently decreases. The anion exchange resins are also more susceptible to foulants (Ball and Harries 1988). According to Philipot and Patte (1982), there may be the release of undesirable substances occurring in treated water after the use of synthetic resins. Baciocchi et al. (2005) reported that although ion-exchange resins are less pH dependent, other common constituents of natural waters such as sulphates and nitrates reduce the efficiency of nitrate removal. Loss of capacity has been reported by Gonder et al. (2006) on using an anion exchanger that is fouled organically. The major problems in application of resin-based filters for water treatment include their thermal instability and bacteriological degradation.

6.7 Inorganic Materials Used in the Removal of Nitrate from Aqueous Solutions

A lower number of inorganic materials has been reported for the removal of nitrate from water. Abe and Ito (1965) worked on hydrated alumina for the removal of nitrate and found a slight preference of chloride over nitrate. The saturation capacity for anions increased with decreasing pH values. Zirconium oxide behaves both as cation and anion exchangers with point of zero charge (pH_{pzc}) at 6.5 values. It is found to be more selective for phosphate and chloride than nitrate ion (Vesely and Pekarek 1972). Inorganic ion exchangers are one of the probable materials for the removal of contaminants from aqueous solutions due to their specific selectivity towards the certain anions (Wassay et al. 1996). Bhatnagar et al. (2010) evaluated nano-alumina for the removal of nitrate from aqueous solutions. Batch adsorption experiments were performed as a function of contact time, initial nitrate concentration, temperature, pH, and effect of other interfering anions. Pseudo second-order kinetic model was found to fit well for nitrate adsorption kinetic data.

The maximum adsorption capacity of nano-alumina for removal of nitrate was found to be 4.0 mg/g at 25 ± 2 °C. Equilibrium was achieved at 4.1 pH value and nitrate sorption was well illustrated by the Langmuir isotherm model. The effect of competitive anions such as chloride, carbonate, fluoride, phosphate, and sulphate was observed on the adsorption on nitrate by performing nitrate adsorption under at 20 mg/l initial concentration. Presence of sulphate, carbonate, and chloride ions generally affects the adsorption capacity of nitrate. The removal rate of nitrate was founded to ~40–45% in the presence of chloride, whereas it was founded to be ~45–50% in case sulphate and carbonate.

It was observed that the outer-spherically sorbing anions, especially chloride and sulphate, can significantly interfere with the sorption of nitrate at elevated concentrations where the sorption competition may occur for the limited amount of sorption sites on nano-alumina. On the other hand, phosphate and fluoride (inner-spherically sorbing anions) showed little impact on nitrate sorption. Only few pores were present on the surface of nano-alumina observed by SEM (scanning electron microscopy) analysis and sorbent did not show any well-defined porous structure. 151.7 m^2/g and 1.09 cm^3/g are the BET (Brunauer–Emmett–Teller) surface area and total pore volume of nano-alumina that has been founded, respectively. Presence of γ-alumina phase in the sorbent was confirmed by XRD (x-ray diffraction) analysis (Bhatnagar et al. 2010). At 760/cm, a single peak of nano-alumina was observed above specified range of wave number. Whereas, two bands at 802 and 582/cm were also observed, which are the characteristics of Al–O vibration in Al_2O_3 (Lu et al. 2009). An absorption band at wave number 1620/cm was also observed which which could be due to presence of alumina (Tsuchida 1993). Fristche (1993) reported that yellow bismuth hydroxide precipitate can be used for the removal of nitrate. The mechanism of nitrate removal was exchange of hydroxyl ions with nitrate ions. Singh (1999) has observed the sorptive properties of various bismuth compounds for nitrate removal purpose and their feasibility in drinking water treatment. In order to maximize the capacity of removal of nitrate various preparatory conditions of hydrous bismuth oxides (HBOs) were studied. Singh and Ghosh (2000) concluded that only polymeric forms of HBO were found more efficient in the removal of nitrate from water. Recently, Singh et al. (2012) presented the method of preparation and HBO properties, which was for the removal of nitrate from aqueous solutions. Furthermore, Srivastav et al. (2014) and Singh et al. (2015) synthesized different types of hydrous bismuth-based media and successfully removed nitrate from the water. Thereafter, Ranjan et al. (2015, 2019) have varied the matrix of HBO in order to optimize the removal of fluoride as well as nitrate from the water. For the amendment, different types of cations were added in the matrix of HBO such as calcium, magnesium, ferric, etc. Because of the enhanced cationic characters of the HBOs, removal of anions (both nitrate and fluoride) was increased. Therefore, it was assumed that adding cations into the matrix of the adsorbent could be beneficial for the anion contaminant removal from the water.

6.8 Concluding Remarks

Nitrate is an important anionic problem for the world, including India. Excessive intake of nitrate through drinking water can cause several types of human health disorders. Because of this harm, several world authorities have recommended a maximum limit of nitrate (ranged from 45 to 50 mg/l) in drinking water. Moreover, nitrate can be removed using adsorption, ion exchange, and catalytic reduction. However, adsorption has been a preferred method of nitrate removal from water because of its easy operation and less costly venture. Mostly, organic nature adsorbents have been used to remove water contaminants. But due to fouling and growth of microbes on the bed, organic adsorbents would not be efficient after some extent. Therefore, the need of inorganic adsorbents has arisen. Inorganic sorptive materials/ion exchangers appear to be possible alternatives to organic resins. Adsorptive removal using inorganic materials/hydrous metal oxides for water treatment represent an innovative, attractive, and economic approach. A number of hydrous metal oxides have been used in sorptive removal of nitrate from aqueous solutions, and they have been found to have adsorptive properties. Most of the adsorbents are available in the form of fine powders, which cause difficulties at ground level applications. Therefore, finding an optimal carrier should be investigated that will enable the contaminants removal in the mode of column adsorption.

References

Abe, M. and Ito, I. (1965). Synthesis and properties of the phosphor antimonic (V) acid cation exchanger. *Nippon Kagaku Zasshi* 86: 817.

Aelion, C. and Conte, B. (2004). Susceptibility of residential wells to VOC and nitrate contamination. *Environ. Sci. Technol.* 38: 1648–1653.

Azizullah, A., Khattak, M.N.K., Richter, P., and Häder, D.P. (2011). Water pollution in Pakistan and its impact on public health – a review. *Environ. Int.* 37: 479–497.

Baciocchi, R., Chiavola, A., and Gavasci, R. (2005). Ion exchange equilibria of arsenic in the presence of high sulphate and nitrate concentrations. *Water Science and Technology: Water Supply* 5 (5): 67–74.

Barrabes, N. and Sa, J. (2011). Review: catalytic nitrate removal from water, past, present and future perspectives. *Appl. Catal., B* 104: 1–5.

Bhatnagar, A. and Sillanpää, M. (2010). Utilization of agro-industrial and municipal waste materials as potential adsorbents for water treatment-a review. *Chemical Engineering Journal* 157: 277–296.

Bhatnagar, A. and Sillanpää, M. (2011). A review of emerging adsorbents for nitrate removal from water. *Chem. Eng. J.* 168 (2): 493–504.

Bhatnagar, A., Kumar, E., and Sillanpa, M. (2010). Nitrate removal from water by nano-alumina: characterization and sorption studies. *Chem. Eng. J.* 163: 317–323.

Bhatnagar, A., Kumar, E., and Sillanpää, M. (2011). Fluoride removal from water by adsorption—a review. *Chem. Eng. J.* 171: 811–840.

Bialowiec, A., Davies, L., Albuquerque, A., and Randerson, P.F. (2012). The influence of plants on nitrogen removal from landfill leachate in discontinuous batch shallow constructed wetland with recirculating subsurface horizontal flow. *Ecol. Eng.* 40: 44–52.

Ball, M. and Harries, R.R. (1988). Ion exchange resin assessment. In: *Ion Exchange for Industry*. Chichester, West Sussex. England: Ellis Horwood Publishers.

Bouchard, D.C., Williams, M.K., and Surampalli, R.Y. (1992). Nitrate combination of groundwater sources and potential health effects. *JAMA* 7: 85–90.

Beltran de Heredia, J., Domınguez, J.R., Cano, Y., and Jimenez, I. (2006). Nitrate removal from groundwater using Amberlite IRN-78: Modelling the system. *Applied Surface Science* 252: 6031–6035.

Bureau of Indian Standards (BIS), Specifications for Drinking Water, IS: 10500: 2012. Bureau of Indian Standards, New Delhi, 1991/2012.

Cameron, S.G. and Schipper, L.A. (2012). Hydraulic properties, hydraulic efficiency and nitrate removal of organic carbon media for use in denitrification beds. *Ecol. Eng.* 41: 1–7.

Central Groundwater Board (CGWB). A report of Central Groundwater Board, Ministry of Water Resources; Government of India, 2010.

Central Pollution Control Board (CPCB). Status of Groundwater Quality in India–Part II. CPCB, Ministry of Environment and Forest, Government of India, New Delhi, 2008.

Ceplecha, Z.L., Waskom, R.M., Bauder, T.A. et al. (2004). Vulnerability assessments of Colorado groundwater to nitrate contamination. *Water Air Soil Pollut.* 159: 373–394.

Chabani, M., Amrane, A., and Bensmaili, A. (2006). Kinetic modelling of the adsorption of nitrates by ion exchange resin. *Chem. Eng. J.* 125: 111–117.

Chan, T.Y.K. (2011). Vegetable-borne nitrate and nitrite and the risk of methaemoglobinaemia. *Toxicol. Lett.* 200: 107–108.

Chen, Y.M., Li, C.W., and Chen, S.S. (2005). Fluidized zero valent iron bed reactor for nitrate removal. *Chemosphere* 59: 753–759.

Du, S.T., Jin, C.W., and Zhang, Y.S. (2010). Current situations and research progress of nitrate pollution in vegetables and their regulating strategies. *Sci. Agric. Sin.* 43: 3580–3589.

Elyanow, D. and Persechino, J., Advances in Nitrate Removal, GE Water and Process Technology, Technical Paper, 2005.

EU (2007). European Communities (Drinking Water) Regulations, http://www.environ.ie/en/Legislation/Environment/Water/FileDownLoad,1541,en.pdf, 2007

Fristche, U. (1993). Removal of nitrates and other anions from water by yellow bismuth hydroxide. *J. Environ. Sci. Health* 28 (9): 1903–1913.

Gonder, B.Z., Kaya, Y., Vergili, I., and Barlas, H.H. (2006). Regeneration of industrial district wastewater using a combination of Fenton process and ion exchange—a case study. *Resour., Conserv. Recycl.* 52 (2): 425–440.

Gatseva, P.D. and Argirova, M.D. (2008). High-nitrate levels in drinking water may be a risk factor for thyroid dysfunction in children and pregnant women living in rural Bulgarian areas. *Int. J. Hyg. Environ. Health* 211: 555–559.

Ghiglieri, G., Barbieri, G., Vernier, A. et al. (2009). Potential risks of nitrate pollution in aquifers from agricultural practices in the Nurra region, northwestern Sardinia, Italy. *J. Hydrol.* 379: 339–350.

Gleick, P.H. (1993). *Water in Crisis: A Guide to the World's Freshwater Resources*, vol. 13. Oxford University Press.

Hasnat, M.A., Ahamad, N., Nizam Uddin, S.M., and Mohamed, N. (2012). Silver modified platinum surface/H+ conducting Nafion membrane for cathodic reduction of nitrate ions. *Appl. Surf. Sci.* 258: 3309–3314.

Islam, M. and Patel, R. (2010). Synthesis and physicochemical characterization of Zn/Al chloride layered double hydroxide and evaluation of its nitrate removal efficiency. *Desalination* 256: 120–128.

Izah, S.C., Chakrabarty, N., and Srivastav, A.L. (2016). A review on heavy metal concentration in potable water sources in Nigeria: human health effects and mitigating measures. *Exposure Health* 8: 285–304.

Jalba, D.I., Cromar, N.J., Pollard, S.J.T. et al. (2010). Safe drinking water: critical components of effective inter-agency relationships. *Environ. Int.* 36 (1): 51–59.

Kapoor, A. and Virarghavan, T. (1997). Nitrate removal from drinking water-review. *J. Environ. Eng.* 123 (4): 371–380.

Kulshreshtha, S.N. (1998). A global outlook for water resources to the year 2025. *Water Resour. Manage.* 12 (3): 167–184.

Laegreid, M., Bockman, O.C., and Kaarstad, O. (1999). *Agriculture, Fertilizers and the Environment*, vol. 294. Norsk Hydro ASA, Porsgrunn, Norway: CABI Publishing.

Limbrick, K. (2003). Baseline nitrate concentration in groundwater of the chalk in South Dorset, UK. *Sci. Total Environ.* 314–316: 89–98.

Lu, C.L., Lv, J.G., Xu, L. et al. (2009). Crystalline nano tubes of Γ-Alooh and Γ-Al2O3: hydrothermal synthesis, formation mechanism and catalytic performance. *Nanotechnology* 20 (21): 1–9.

Mesa, J.M.C., Armendariz, C.R., and de la Torre, A.H. (2002). Nitrate intake from drinking water on Tenerife island (Spain). *Sci. Total Environ.* 302: 85–92.

Milmile, S.N., Pande, J.V., Karmakar, S. et al. (2011). Equilibrium isotherm and kinetic modeling of the adsorption of nitrates by anion exchange Indion NSSR resin. *Desalination* 276 (1–3): 38–44.

Mossad, M. and Zou, L. (2013). Study of fouling and scaling in capacitive deionisation by using dissolved organic and inorganic salts. *J. Hazard. Mater.* 244–245: 387–393.

Patel, N., Rai, D., Shahane, S., and Mishra, U. (2019a). Lipases: sources, production, purification, and applications. *Recent Pat. Biotechnol.* 13 (1): 45–56.

Patel, N., Shahane, S., Majumdar, R., and Mishra, U. (2019b). Mode of action, properties, production, and application of laccase: a review. *Recent Pat. Biotechnol.* 13 (1): 19–32.

Patel, N., Khan, M.D., Shahane, S. et al. (2020a). Emerging pollutants in aquatic environment: source, effect, and challenges in biomonitoring and bioremediation-a review. *Pollution* 6 (1): 99–113.

Patel, N., Pathak, P., Rai, D., and Chaudhary, V.K. (2020b). Biosensors used for monitoring of environmental contaminants. In: *Nanosensor Technologies for Environmental Monitoring* (eds. A. Inamuddin and M. Abdullah), 69–83. Springer.

Philipot, J.M. and Patte, A. (1982). Water denitrification by biological process. *Tech. Sci. Munic.* 774: 165–172.

Ranjan, M., Srivastav, A.L., and Shaktibala (2015). Effects of addition of cationic ligands in hydrous bismuth oxide on removal of fluoride from aqueous solutions. *Curr. Sci.* 108: 9–10.

Ranjan, M., Singh, P.K., Srivastav, A.L., and Kumar, V. (2019). Adsorptive properties of Cation added hydrous bismuth oxide on nitrate sorption. *J. Water Chem. Technol.* 41 (5): 283–291.

Ranjan, M., Singh, P.K., and Srivastav, A.L. (2020). A review of bismuth-based sorptive materials for the removal of major contaminants from drinking water. *Environ. Sci. Pollut. Res.* https://doi.org/10.1007/s11356-019-05359-9.

Rao, E.V.S.P. and Puttanna, K. (2000). Nitrates, agriculture and environment. *Curr. Sci.* 79: 1163–1168.

Romano, N. and Zeng, C. (2009). Evaluating the newly proposed protocol of incorporated potassium in nitrate toxicity experiments at different salinities: a case study with the tiger prawn, Penaeus monodon, juveniles. *Aquaculture* 289: 304–309.

Sadgir, P. and Vamanrao, A., Water in Vedic Literature. In Abstract Proceedings of the 3rd International Water History Association Conference, Alexandria, (http://www.iwha.net/a_abstract.htm), 2003.

Samatya, S., Kabay, N., Yuksel, U. et al. (2006). Removal of nitrate from aqueous solution by nitrate selective ion exchange resins. *React. Funct. Polym.* 66: 1206–1214.

Santamaria, P. (2006). Nitrate in vegetables: toxicity content, intake and EC regulation. *J. Sci. Food Agric.* 86: 10–17.

Schoeman, J.J. and Steyn, A. (2003). Nitrate removal with reverse osmosis in a rural area in South Africa. *Desalination* 155: 15–26.

Singh, P.K. (1999). Nitrate removal from water by bismuth based media. Ph.D. Thesis, Department of Civil Engineering. I.I.T. Kanpur

Singh, P.K. and Ghosh, D.K. (2000). Nitrate removal from water by bismuth based media. In: *Waste Recycling and Resource Management in the Developing World* (eds. B.B. Jana, R.D. Banerjee, B. Guterstam and J. Heeb), 456–459. Switzerland: University of Kalyani, India and Internat. Ecolog. Engg. Soc.

Singh, P.K., Srivastav, A.L., Ghosh, D.K., and Sharma, Y.C. (2012). Preparation and properties of hydrous bismuth oxides for nitrate removal from aqueous solutions. *Desalin. Water Treat.* 40 (1–3): 144–152.

Singh, P.K., Banerjee, S., Srivastav, A.L., and Sharma, Y.C. (2015). Kinetic and equilibrium modeling for removal of nitrate from aqueous solutions and drinking water by a potential adsorbent, hydrous bismuth oxide. *RSC Adv.* 5: 35365–35376.

Singh, S., Wasewar, K.L., and Kansal, S.K. (2020). Low-cost adsorbents for removal of inorganic impurities from wastewater. In: *Inorganic Pollutants in Water* (eds. P. Devi, P. Singh and S.K. Kansal), 173–203. Elsevier.

Shrimali, M. and Singh, K.P. (2001). New methods of nitrate removal from water. *Environ. Pollut.* 112: 351–359.

Sall, M. and Vanclooster, M. (2009). Assessing the well water pollution problem by nitrates in the small scale farming systems of the Niayes region, Senegal. *Agric Water Manag* 96: 1360–1368.

Srivastav, A.L. (2020). Chemical fertilizers and pesticides: role in groundwater contamination. In: *Agrochemicals Detection, Treatment and Remediation*, 143–159. Elsevier https://doi.org/10.1016/B978-0-08-102977-0.00019-6.

Srivastav, A.L. and Kaur, T. (2020). Factors affecting the formation of disinfection by-products in drinking water: human health risk. In: *Disinfection By-Products in Drinking Water* (ed. M.N.V. Prasad), 433–450. Elsevier https://doi.org/10.1016/B978-0-08-102977-0.00019-6.

Srivastav, A.L. and Ranjan, M. (2020). Inorganic water pollutants. In: *Inorganic Pollutants in Water* (eds. P. Devi, P. Singh and S.K. Kansal), 1–15. Elsevier https://doi.org/10.1016/B978-0-08-102977-0.00019-6.

Srivastav, A.L., Singh, P.K., Srivastava, V., and Sharma, Y.C. (2013a). Application of a new adsorbent for fluoride removal from aqueous solutions. *J. Hazard. Mater.* 263: 342–352.

Srivastav, A.L., Dubey, V.K., Singh, P.K., and Sharma, Y.C. (2013b). Safe ground water: I, contamination of ground water of Varanasi, Uttar Pradesh, India by nitrate, sulfates and alkalinity. *J. Indian Chem. Soc.* 90: 983–989.

Srivastav, A.L., Singh, P.K., Weng, C.H., and Sharma, Y.C. (2014). Novel adsorbent hydrous bismuth oxide for the removal of nitrate from aqueous solutions. *J. Hazard., Toxic Radioact. Waste* 04014028 (8): 1–8.

Srivastav, A.L., Singh, P.K., and Sharma, Y.C. (2015). Synthesis of a novel adsorbent, hydrous bismuth oxide (HBO2) for the removal of fluoride from aqueous solutions. *Desalin. Water Treat.* 55: 604–614.

Srivastav, A.L., Kaur, T., Rani, L., and Kumar, A. (2019). Scientific research production of India and China in environmental chemistry: a bibliometric assessment. *Int. J. Environ. Sci. Technol.* 2019 (16): 4989–4996. https://doi.org/10.1007/s13762-019-02306-6.

Suthar, S., Bishoni, P., Singh, S. et al. (2009). Nitrate contamination in groundwater of some rural areas of Rajasthan, India. *J. Hazard. Mater.* 171: 189–199.

Tang, Y., Wu, S., Miao, X. et al. (2013). Resilience to evolving drinking water contamination risks: a human error prevention perspective. *J. Clean. Prod.* 57: 228–237.

Trevisan, M., Padovani, L., and Capri, E. (2000). Nonpoint-source agricultural hazard index: a case study of the province of Cremona, Italy. *J. Environ. Manage.* 26 (5): 577–584.

Tsuchida, T. (1993). Preparation and reactivity of acicular α-Al2O3from synthetic diaspore, β-Al2O3·H2O. *Solid State Ion.* 63: 464–470.

UNESCO, Water for people water for life. The United Nations World Water Development Report, United Nations Educational, Scientific and Cultural Organization (UNESCO) and Berghahn Books, 2003

USEPA (2009). *Drinking water standards. Office of Drinking Water*. Washington, DC: US Environmental Protection Agency.

Van der Hoek, P., Van der Hoek, W.F., and Klapwijk, A. (1987). Nitrate removal from ground water - use of a nitrate selective resin and a low concentrated Regenerant. *Water Air Soil Pollution* 37: 41–53.

Vesely, V. and Pekarek, V. (1972). Synthetic inorganic ion exchangers: hydrous oxides acidic salts of multivalent metals. *Talanta* 19: 219–262.

Velizarov, S., Crespo, J.G., and Reis, M.A. (2004). Removal of inorganic anions from drinking water supplies by membrane bio/processes. *Rev. Environ. Sci. Bio/Technol.* 3: 361–380.

Vinten, A. and Dunn, S. (2001). Assessing the effects of land use on temporal change in well water quality in a designated nitrate vulnerable zone. *Sci. Total Environ.* 265: 253–268.

Vymazal, J. (2007). Removal of nutrients in various types of constructed wetlands. *Sci. Total Environ.* 380: 48–65.

Wang, Q.H., Feng, C.P., Zhao, Y.X., and Hao, C.B. (2009). Denitrification of nitrate contaminated groundwater with a fiber-based biofilm reactor. *Bioresour. Technol.* 100: 2223–2227.

Wassay, S.A., Haron, M., Uchiumi, A., and Tokunaga, S. (1996). *Water Res.* 30: 1143.

WHO (1993). *Guidelines for Drinking Water Quality. 1. Recommendations, 2nd Edition.* Geneva: World Health Organization.

WHO (2004). *Guidelines for Drinking-Water Quality*, 3e, vol. 375–378, 301–303. Geneva: World Health Organisation.

WHO (2006). *Guidelines for Drinking-Water Quality*, 4e. Geneva, ISBN: 9241546964: World Health Organization.

WHO (2007). *Nitrate and Nitrite in Drinking-Water. Background Document for Development of WHO Guidelines for Drinking-Water Quality*. Geneva: World Health Organization.

WHO (2008). *Guidelines for Drinking-Water Quality, 3rd ed. Recommendations. Incorporating 1st and 2nd Addenda*, vol. 1. Geneva: World Health Organization.

Zhao, Y., Zhang, B., Feng, C. et al. (2012). Behavior of autotrophic denitrification and heterotrophic denitrification in an intensified biofilm-electrode reactor for nitrate-contaminated drinking water treatment. *Bioresour. Technol.* 107: 159–165.

7

Organic Micropollutants in Groundwater

A Rising Concern for Indian Drinking Water Supplies

Manvendra Patel

School of Environmental Sciences, Jawaharlal Nehru University, New Delhi, India

7.1 Introduction

The twentieth century has seen the production and consumption of thousands of anthropogenic organic compounds by humans (Lapworth et al. 2012; Schirmer and Schirmer 2008; Schwarzenbach et al. 2006; Stuart and Lapworth 2013). These compounds have helped humans to improve living standards, and develop numerous products for medical, energy, technological, and agricultural purposes and day-to-day needs (Patel et al. 2019; Schirmer and Schirmer 2008). Injudicious use of these chemicals had led their presence in surface and groundwater sources around the world (Patel et al. 2019). Since the presence of organic compounds in aqueous sources are in ng/L to μg/L concentrations, these pollutants are termed organic micropollutants (Daughton 2004). Widespread presence of organic micropollutants in aquatic sources is a serious concern, especially for groundwater, as 2.5 billion people depend upon it globally (WWAP 2017). Concerns regarding the presence of organic micropollutants in both surface and groundwater are growing rapidly, due to their potential threats to environmental and human health (Daughton 2004; Daughton and Ternes 1999; Halling-Sørensen et al. 1998; Lapworth et al. 2012; Patel et al. 2019; Schwarzenbach et al. 2006). The presence of organic micropollutants in groundwater is still poorly studied (Pal et al. 2010, 2014; Stuart and Lapworth 2013).

Organic micropollutants include artificial sweeteners, pharmaceuticals, personal care products, per- and poly-fluoroalkyl substances, pesticides, phthalates, surfactants, hormones, disinfection/oxidation products, and industrial chemicals and products (Lapworth et al. 2012; Sackaria and Elango 2020; Stuart and Lapworth 2013). Metabolites, and transformation products of these mentioned compounds are also included in organic micropollutants (Stuart and Lapworth 2013). Together, all these groups comprise thousands of compounds. For example, pharmaceuticals alone have more than 5000 commonly used compounds (Patel et al. 2019). Improvements in analytical techniques led to the detection of previously undetected organic micropollutants in aqueous systems (Richardson and Ternes 2011).

Groundwater Geochemistry: Pollution and Remediation Methods, First Edition.
Edited by Sughosh Madhav and Pardeep Singh.

Insufficient knowledge is available about the effects of organic micropollutants on flora and fauna and even less about their long-term impacts on human health (Bartrons and Peñuelas 2017; Lapworth et al. 2012; Sehonova et al. 2018). Most of these compounds are still unregulated (Kolpin et al. 2002; Larsson 2014; Tijani et al. 2016). Except for a few compounds, drinking water and other specific guidelines are unavailable for mostly organic micropollutants. Due to insufficient knowledge, absence of guidelines, and the trace levels of their presence (ng/L to µg/L in general), the organic micropollutants are also called as 'Emerging Organic Contaminants' or 'Contaminants of Emerging Concern' (Daughton 2004; Daughton and Ternes 1999). To date, organic micropollutants have been extensively studied in wastewater and surface water. However, little knowledge is available regarding the occurrence of organic micropollutants in groundwater (Stuart and Lapworth 2013). Thus, this chapter summarizes the presence of organic micropollutants in groundwater, with special focus on the Indian scenario. Factors affecting the fate and transportation have also been explained.

7.2 Types of Organic Micropollutants

As mentioned, organic micropollutants include diverse groups of organic compounds. Some of these compounds, including pharmaceuticals and personal care products, are poorly characterized. Less is known about their fate, transportation, and persistence in environmental matrices, while much more knowledge is present about compounds such as pesticides (Lapworth et al. 2012; Stuart and Lapworth 2013). However, health risk assessments provided information regarding health risk potential (Schirmer and Schirmer 2008; Stuart and Lapworth 2013). Many of these compounds such as hormones and sterols can alter hormonal and reproductive balances in microorganisms, plants, and animals in the environment. Thus, these compounds are also called 'endocrine disruptors'. Micropollutants can suffer a variety of transformations (including hydrolysis, photolysis, and microbial degradations) in the environment. These products are called 'transformation products' and coexist with their parent compounds in the environment. Detailed information of organic micropollutants is provided below.

7.2.1 Pesticides

A wide range of chemicals such as fungicides, herbicides, insecticides, molluscides, nematicides, and rodenticides are included in pesticides. These compounds are produced for the protection of plants, crops, and crop foods from various diseases and crop-harming organisms (Ali et al. 2014; Kolpin et al. 1998; Schirmer and Schirmer 2008). Extensive usage for agricultural purposes has led pesticides into both terrestrial and aquatic environments (Ali et al. 2014). Both surface and groundwater sources have reported pesticides up to mg/L ranges (Ali et al. 2014). Numerous pesticides have been traced in groundwaters worldwide (Kolpin et al. 1998; Spliid and Køppen 1998; Walls et al. 1996). Compared to other micropollutants, pesticides in groundwater are much more studied and well-established (Stuart and Lapworth 2013). Transformation products of pesticides have long been discovered (Galassi et al. 1996; Kolpin et al. 1998, 2000, 2002, 2004). Pesticide degradation products have been reported at higher concentrations than their parent compounds in groundwater sources (Kolpin et al. 2004; Lapworth et al. 2006). Degradation products can be biologically active and many of them can be highly toxic in nature (Stuart and Lapworth 2013).

7.2.2 Pharmaceuticals

After being recognized in the aquatic systems in 1980s, the presence of pharmaceuticals has been reported in each and every aquatic matrix (Patel et al. 2019). A rapid rise in occurrence studies have provided numerous cases of pharmaceutical in groundwater sources (Phillips et al. 2015; Sui et al. 2015). Manufacturing industries, human excretions, unused pharmaceutical disposal, and veterinary and agricultural usages are the main pharmaceutical sources (Barnes et al. 2008; Patel et al. 2019). Pharmaceuticals are a large group of compounds and include antibiotics (ofloxacin, ampicillin, sulfamethoxazole), non-steroidal anti-inflammatory drugs (ibuprofen, diclofenac), analgesics and antipyretics (paracetamol), β-blockers (acebutolol), anti-depressants (amitriptyline), lipid-lowering agents (atorvastatin), antacids (ranitidine), antihistamines (cetirizine, salbutamol), anti-epileptics (carbamazepine) and iodinated X-ray contrast media (iopamidol, iopromide) (Patel et al. 2019). Pharmaceuticals such as amoxicillin, azithromycin, clarithromycin, ciprofloxacin, diclofenac, and erythromycin are included in the European watch list (Commission 2015, 2018) and erythromycin and quinolone are included in contamination candidate lists in the USA (USEPA 2016).

7.2.3 Personal Care Products

Personal care products are also a wide range of compounds comprising products used by humans in everyday life. Some of these products include colognes, lotions, shampoos, hair dyes, toothpastes, insect repellents, etc. These include parabens, (alkyl esters of p-hydroxybenzoic acid), UV filters/sunscreens (benzophenones, methoxycinnamates), insect repellents (N,N-diethyl-meta-toluamide), bactericides and antifungal agents (triclocarban, triclosan), polycyclic musks (galoxalide, tonalide), and lifestyle compounds such as artificial sweeteners (sucralose, aspartame), illicit drugs, and compounds such as caffeine and nicotine. Most of these products are released into wastewater directly after usage (Sackaria and Elango 2020). Wastewater treatment plants are the main source of personal care products in the environment. Previous studies suggested the ubiquitous presence of personal care products in aquatic sources including groundwater (Sackaria and Elango 2020). Many personal care products are also highly persistent in nature (Daughton and Ternes 1999).

7.2.4 Per- and Poly-fluroalkyl Substances

Per- and Poly-fluoroalkyl substances include a wide variety of industrial products. These materials are used in industries for adhesives, metal finishing and electroplating, electronics, fire-fighting foams, food packaging, insecticides, stain repellents, and textile coatings (Sackaria and Elango 2020). Both perfluorooctanesulfonate (PFOS) and perfluorooctanoic acid (PFOA) are placed in fourth contamination candidate lists in the

USA (USEPA 2016). Groundwater infiltration of PFAS-contaminated surface waters, including river water, wastewaters, and leaching from contaminated soils, are possible sources (Sharma et al. 2016).

7.2.5 Industrial Chemicals and Products

Large numbers of industrial chemicals and products were released by manufacturing industries and human usage into the environment. These chemicals include petroleum hydrocarbons, various organic solvents, polyaromatic compounds, plasticizers, bisphenols, phthalates, and others (Moran et al. 2007; Verliefde et al. 2007). Many of the industrial products are categorized as priority pollutants by various environmental agencies around the world. Drinking water guidelines are also available for many of these compounds.

7.2.6 Treatment By-products

Water and wastewater treatment processes such as chlorination, catalysis, photo-catalysis, ozonation, oxidation, and advanced oxidation processes can generate a large number of by-products (Patel et al. 2019). Presence of turbidity, organic matter, and other matrices can enhance the formation of these by-products. Some of these by-products, such as trihalomethanes and haloacetic acids, are well established and even have drinking water guidelines (Boorman 1999). The amount of research on emerging compounds such as N-nitrosodimethylamine (NDMA) formed during chlorination has also increased (Stuart and Lapworth 2013). NDMA is a potential carcinogen and is formed in large concentrations in wastewater chlorination (Mitch et al. 2003). Formation of toxic chlorinated by-products through the chlorination process led developed countries to phase out chlorine disinfection by ozone and chloramines. However, both ozone and chloramines can also form new toxic by-products (Richardson and Bowron 1985; Patel et al. 2019).

7.2.7 Other Micropollutants

Along with the above-mentioned organic micropollutants, food additives, flame and fire retardants, surfactants, hormones, and sterols are other commonly used micropollutants reported in groundwater. Food additives are a group of compounds used for food preservation, foam stabilization, pharmaceutical coatings, and other purposes. For example, triethyl citrate is used for foam stabilization and pharmaceutical coating, butylated hydroxyanisole and butylated hydroxytoluene as food preservatives; other food additives include camphor, citral, citronellol, phenylethyl alcohol, heliotropin, triacetin, and terpineol (Stuart and Lapworth 2013). Polybrominated diphenyl compound-based flame retardants and phosphate-based retardants such as tris-(2-chloroethyl)phosphate are commonly used in household, industrial, and consumer products as a non-flammable barrier (Rahman et al. 2001; Stuart and Lapworth 2013).

Surfactants are synthetic organic compounds used for detergents, shampoos, emulsifiers, hair conditioners, fabric softeners, paints, and foaming liquid preparations (Sackaria and Elango 2020). Anionic, cationic, amphoteric, and non-ionic surfactants exist. Surfactants are commonly used in households, industries, and hospitals. Thus, large amounts of these surfactants are released into the environment and can easily infiltrate groundwater too (Sackaria and Elango 2020).

7.3 Organic Micropollutants in Indian Groundwater

India is one of the largest groundwater users in the world, with ~80% of the rural and ~50% of the urban population depending upon groundwater for domestic purposes without major treatment (Chakraborti et al. 2011). Numerous studies reported and reviewed the presence of organic micropollutants in Indian surface waters (rivers, lakes, streams), wastewaters (influents and effluents), and groundwater sources (Ali et al. 2014; Balakrishna et al. 2017; Gani and Kazmi 2017; Philip et al. 2018). These studies have reported pesticides (Lapworth et al. 2018), pharmaceuticals (Jindal et al. 2015, Sharma et al. 2019), personal care products (Sharma et al. 2019), PFAS (Sharma et al. 2016), phthalates (Somashekar et al. 2015), hormones (Lorenzen et al. 2010), and artificial sweeteners (Sharma et al. 2019). Some of the important and recent studies reporting organic micropollutants in Indian groundwater sources are provided in Table 7.1.

India is one of the largest producers and consumers of pharmaceuticals (Van Boeckel et al. 2014). Hyderabad, Ahmedabad, Bangalore, Mumbai, and Dehradun are major pharmaceutical manufacturing hubs, along with numerous small and medium manufacturing locations across the country (Mathew and Unnikrishnan 2012). Of all the manufacturing hubs, pharmaceutical occurrence studies in groundwater were only conducted in Hyderabad (Fick et al. 2009; Rutgersson et al. 2014). Other manufacturing hubs and locations of India seriously lack groundwater monitoring studies of organic micropollutants, including pharmaceuticals. A groundwater assessment study performed in Hyderabad's wells showed the presence of 11 of the 12 tested pharmaceuticals (Fick et al. 2009). Cetirizine (550–28 000 ng/L), ciprofloxacin (44–14 000 ng/L), and terbinafine (90–1800 ng/L) were majorly detected pharmaceuticals in all the tested wells (Fick et al. 2009). Other detected pharmaceuticals were citalopram (76–1400 ng/L), enoxacin (80–1900 ng/L), enrofloxacin (23–67 ng/L), lomefloxacin (15–35 ng/L), metoprolol (90 ng/L), norfloxacin (21–31 ng/L), ofloxacin (26–160 ng/L), and trimethoprim (17–55 ng/L) in the well water sources from Hyderabad (Fick et al. 2009).

A study conducted in SAS nagar, Punjab, tested the presence of 40 selected pharmaceuticals in groundwater (Jindal et al. 2015). The study reported the presence of diclofenac (6040–48080 ng/L) and pitavastatin (Bdl-480 ng/L) concentrations in groundwater, respectively (Jindal et al. 2015). Pharmaceuticals including carbamazepine (27%) sulfanilamide (62%), and sulfamethoxazole (77%) were frequently present in Gangetic plains groundwater sources from Varanasi (Lapworth et al. 2018). The presence of other micropollutants including sucralose (an artificial sweetener), pesticides (chlorpyrifos, and phenoxyacetic acids as transformation product of numerous herbicides including 2,4-dichlorophenoxyacetic acid), and perfluoroalkyl substances (perfluorodecanoic acid, perfluorononanoic acid and perfluorooctanesulfonic acid) were also reported in groundwater sources from Varanasi (Lapworth et al. 2018). Study also confirmed that the organic micropollutant concentration in shallow groundwater sources is higher than those of deep groundwater sources (Lapworth et al. 2018).

Table 7.1 Organic micropollutants reported in Indian groundwater sources in the past 10 years.

Micropollutant	Location	Concentration (ng/L)	References
	Pharmaceuticals		
Carbamazepine	Ganga river basin	BDL (below detection limit)–27.2	Sharma et al. (2019)
Carbamazepine	Varanasi	<1000	Lapworth et al. (2018)
Ciprofloxacin	Ganga river basin	BDL–5	Sharma et al. (2019)
Ciprofloxacin	Well water, Hyderabad	770	Rutgersson et al. (2014)
Diclofenac	Ganga river basin	BDL–1.56	Sharma et al. (2019)
Diclofenac	SAS nagar, Punjab	6040–48080	Jindal et al. (2015)
Ibuprofen	Ganga river basin	BDL–49.4	Sharma et al. (2019)
Ketoprofen	Ganga river basin	BDL–23.4	
Pitavastatin	SAS nagar, Punjab	BDL–480	Jindal et al. (2015)
Sulfanilamide	Varanasi	>1000	Lapworth et al. (2018)
Sulfamethoxazole	Ganga river basin	BDL–4.13	Sharma et al. (2019)
Sulfamethoxazole	Varanasi	<1000	Lapworth et al. (2018)
	Personal care products and artificial sweeteners		
Caffeine	Ganga river basin	15–262	Sharma et al. (2019)
Cyclamate	Ganga river basin	BDL–0.26	
Diethyltolumide (DEET)	Ganga river basin	2.76–14.8	
Sucralose	Ganga river basin	0.5–25	
Sucralose	Varanasi	<1000	Lapworth et al. (2018)
Triclosan	Ganga river basin	BDL–10.2	Sharma et al. (2019)
	Pesticides		
Atrazine	Delhi	BDL–17000	Aslam et al., (2013)

(*Continued*)

Table 7.1 (Continued)

Micropollutant	Location	Concentration (ng/L)	References
Atrazine	Nakodar Tehsil, Jalandhar	0.0–90	Chaubey et al. (2020)
Butachlor	Hooghly basin, West Bengal	0–247	Mondal et al. (2018)
Chlorpyrifos	Prabhani, Maharashtra	41 200	Motekar Shrinivas (2014)
Chlorpyrifos	Varanasi	0.6–4	Lapworth et al. (2018)
Chlorpyrifos	Nakodar Tehsil, Jalandhar	0.0–0.30	Chaubey et al. (2020)
dichlorodiphenyltrichloroethane	Dibrugarh, Assam	6549	Mishra and Sharma (2011)
dichlorodiphenyltrichloroethane	Nagaon, Assam	6904	Mishra and Sharma (2011)
Dieldrin	Kasaragod, Kerala	BDL–130	Akhil and Sujatha (2012)
Endrin	Delhi	BDL–83	Mutiyar et al. (2011)
Endosulfan	Nanded, Maharashtra	1500–4000	Motekar (2011)
Endosulfan	Parbhani, Maharashtra	2200	Motekar (2011)
Endosulfan	Hisar, Haryana	BDL–206.3	Kaushik et al. (2012)
Endosulfan	Ambala, Haryana	BDL–164.2	Kaushik et al. (2012)
Endosulfan	Gurgaon, Haryana	BDL–913.8	Kaushik et al. (2012)
Heptachlor	Nanded, Maharashtra	10–30	Motekar (2011)
Hexachlorocyclohexane	Dibrugarh, Assam	6549	Mishra and Sharma (2011)
Hexachlorocyclohexane	Nagaon, Assam	6904	Mishra and Sharma (2011)
Melathions	Delhi	130–159	Thakur et al. (2015)
Melathions	Parbhani, Maharashtra	82 300	Motekar Shrinivas (2014)
Methyl parathion	Amravati, Maharashtra	BDL–90	Lari et al. (2014)
Monocrotophos	Nakodar Tehsil, Jalandhar	0.0–140	Chaubey et al. (2020)
Phorate	Nakodar Tehsil, Jalandhar	10–30	Chaubey et al. (2020)
Phospamidon	Parbhani, Maharashtra	4000	Motekar Shrinivas (2014)
Profenophos	Delhi	14 150–39 900	Thakur et al. (2015)

Simazine	Delhi	BDL–40 000	Aslam et al. (2013)
Hormones and estrogens			
Oestrogen (E2)	Delhi and Mathura	BDL–1.4	Lorenzen et al. (2010)
Glucocorticoid (DEX)	Delhi and Mathura	BDL–3	Lorenzen et al. (2010)
Phthalates and Plasticizers			
Dibutyl phthalate (DBP)	Mysore	44 160–207 781	Somashekar et al. (2015)
Per- and Poly-fluoroalkyl substances			
Pentafluorobenzoic acid	Ganga river basin	BDL–9.2	Sharma et al. (2016)
Perfluorodecanoic acid	Ganga river basin	BDL–0.1	
Perfluorodecanoic acid	Varanasi	<1–33	Lapworth et al. (2018)
Perfluoroheptanoic acid	Ganga river basin	0.5–3.5	Sharma et al. (2016)
Perfluorohexanoic acid	Ganga river basin	0.8–4.9	
Perfluoropentanoic acid	Ganga river basin	BDL–2.2	
Perfluorononanoic acid	Ganga river basin	BDL–0.2	
Perfluorononanoic acid	Varanasi	<1–33	Lapworth et al. (2018)
Perfluorooctanoic acid	Ganga river basin	BDL–0.8	Sharma et al. (2016)
Perfluorobutanesulfonic acid	Ganga river basin	BDL–4.9	
Perfluorodecane-sulfonic acid	Ganga river basin	BDL–0.06	
Perfluorohexanesulfonic acid	Ganga river basin	BDL–0.08	
Perfluorooctanesulfonic acid	Ganga river basin	BDL–1.1	
Perfluorooctanesulfonic acid	Varanasi	<1–33	

Source: Modified from Mishra et al.,(2011).

Another groundwater study conducted in Ganga river basin reported several pharmaceuticals, personal care products, and artificial sweeteners (Sharma et al. 2019). Tested pharmaceuticals, personal care products, and artificial sweeteners include acesulfame, acetaminophen, aspartame, atenolol, caffeine, carbamazepine, clofibric acid, cyclamate, DEET, diclofenac, hydrochlorothiazide, ibuprofen, ketoprofen, naproxen, saccharide, sucralose, sulfamethoxazole, triclocarban, and triclosan (Sharma et al. 2019). Except for atenolol and clofibric acid, all other pharmaceuticals and personal care products were reported in groundwater samples, with caffeine having the highest concentration (15–262 ng/L) as shown in Table 7.1. Carbamazepine (BDL-27.2 ng/L), ibuprofen (BDL-49.4 ng/L), and ketoprofen (BDL-23.4 ng/L) were other reported pharmaceuticals (Sharma et al. 2019). Cyclamate (BDL-0.26 ng/L) and sucralose (0.5–25 ng/L) were two artificial sweeteners detected in the groundwater sample (Sharma et al. 2019). Total sum of detected pharmaceuticals, personal care products, and artificial sweeteners ranges between 34 and 293 ng/L in groundwater samples from Ganga river basin (Sharma et al. 2019). The study showed that only the pharmaceutical, personal care products, and artificial sweeteners detected in surface water sources were also reported in groundwater samples. This shows a high correlation between the presence of organic micropollutants in surface water and groundwater sources (Sharma et al. 2019). Thus, organic micropollutants present in surface waters of a particular location become potential candidates for groundwater assessment studies too.

Groundwater sources of Ganga river basin also show the presence of perfluoroalkyl substances (perfluorodecanoic acid, perfluorononanoic acid and perfluorooctanesulfonic acid) in groundwater samples from Varanasi, India (Lapworth et al. 2018). PFAS concentration lies between <1–33 ng/L (Lapworth et al. 2018). Another study tested the presence of 21 per- and polyfluoroalkyl substances in groundwater samples from industrialized and urbanized regions of the Ganga river basin (Sharma et al. 2016). The presence of pentafluorobenzoic acid, perfluorodecanoic acid, perfluoroheptanoic acid, perfluorohexanoic acid, perfluoropentanoic acid, perfluorononanoic acid, perfluorooctanoic acid, perfluorobutanesulfonic acid, perfluorodecane-sulfonic acid, perfluorohexanesulfonic acid, perfluorooctanesulfonic acid were reported at concentrations up to 9.2 ng/L in groundwater samples (Sharma et al. 2016).

Of all the organic micropollutants, pesticides are the most studied and reported in the Indian groundwater sources (Sackaria and Elango 2020). Commonly reported pesticides in Indian groundwater sources are organochlorine pesticides, organophoshpate pesticides, and synthetic pyethroids (Sackaria and Elango 2020). Organochlorine pesticides are chlorinated hydrocarbon and commonly comprise aldrin, dichlorodiphenyltrichloroethane (DDT), dieldrin, endosulfan, hexachlorocyclohexane, and others (Sackaria and Elango 2020). Organophosphate pesticides are commonly used for insects and pest control in agricultural fields. Up to 10 different organophosphate pesticides were reported in Indian groundwater (Sackaria and Elango 2020). Commonly reported pesticides in Indian groundwater sources includes endosulfan, DDT, Aldrin, dieldrin, chlorpyrofis, and malathion (Sackaria and Elango 2020). Commonly reported pesticides in Indian groundwater sources are summarized in Table 7.1.

Some of the highest reported pesticide (hexachlorocyclohexane and dichlorodiphenyltrichloroethane) concentrations were reported from Dibrugarh and Nagaon in Assam as shown in Table 7.1 (Mishra and Sharma 2011). Mean hexachlorocyclohexane concentrations in Dibrugarh and Nagaon were 5168 and 5574 ng/L, respectively (Mishra and Sharma 2011). Mean dichlorodiphenyltrichloroethane concentrations were 6549 and 6904 ng/L for Dibrugarh and Nagaon, respectively (Mishra and Sharma 2011). In a recent study, pesticide (atrazine, chlorpyriphos, monocrotophos, phorate) concentrations were detected in groundwater samples from Nakodar Tehsil, Jalandhar (Chaubey et al. 2020). Observed values of the pesticides were compared with the estimates of Pesticide Impact Rating Index (PIRI) model (Chaubey et al. 2020). Observed values were found to be higher compared to the PIRI estimated values (Chaubey et al. 2020). Concentrations of atrazine, chlorpyrophos, monocrotophos, and phorate were 0.0–90, 0.0–30, 0.0–140 and 10–30 ng/L, respectively (Chaubey et al. 2020).

High concentrations of pesticides have been reported in groundwaters near contaminated rivers (Duttagupta et al. 2020). Of the 40 pesticides reported in Ganges river water, 39 pesticides were also reported from groundwater samples from the West Bengal basin (Duttagupta et al. 2020). Commonly detected pesticides (>50% frequency) were malathion, atrazine, lindane, methyl parathion, chlorpyrifos, diazinon, alachlor, and butachlor (Duttagupta et al. 2020). Surface water sources also showed higher concentrations than groundwater samples (Duttagupta et al. 2020). Similar results were also observed for organochlorine and organophosphate pesticides concentration in Amravati, Bhandara, and Yavatmal regions of Maharashtra (Lari et al. 2014). Commonly detected pesticides are α-Hexachlorocyclohexane (390 ng/L) and parathion-methyl (90 ng/L) in Amravati region, α-endosulphan (780 ng/L) in Yavatmal region and chlorpyrifos (250 ng/L) in Bhandara regions (Lari et al. 2014). Similar correlations of the presence of organic micropollutants (pharmaceuticals, personal care products, and artificial sweeteners) between surface waters and groundwaters were also observed (Sharma et al. 2019).

7.4 Organic Micropollutant Sources and Transportation Pathways

Both point and non-point sources of groundwater pollution exist in the environment (Jurado et al. 2012). Discrete locations of contamination with a single discharge source can be easily identified as a point source. Examples of point sources include industrial discharges, WWTPs, landfills, and accidental spills (Jurado et al. 2012). Pollution caused through a wide area source is a non-point source. For example, agricultural lands are the main non-point source of fertilizers, pesticides, and other agrochemical compounds (Chowdary et al. 2005). Urban areas are the potential sources of pesticides, pharmaceuticals, personal care products, illicit drugs, and industrial compounds through loss from WWTPs, water supplies, and urban run-offs (Jurado et al. 2012). Infiltration of river water to the aquifer are sources of all the micropollutants (Jurado et al. 2012). WWTPs are inefficient in organic micropollutant removal (Salgado et al. 2012). Thus, discharging treated WWTP effluents in surface water bodies can transfer micropollutants into groundwater through the hyporheic zone (Lewandowski et al. 2011). Managed aquifer recharge processes utilizing surface water or treated water sources can easily enrich groundwater with micropollutants (Sui et al. 2015). Landfill leachate contains a large number of contaminants such as pharmaceuticals, chlorinated solvents, PAH, phthalates, and caffeine. Reports of the presence of these compounds in groundwater below landfill sites show the infiltration of micropollutants in groundwater (Stuart et al. 2012).

Extremely high concentrations of organic micropollutants (caffeine and paraxanthine in 17000–23000 ng/L and 55000–65000 ng/L, respectively) in septic tanks is also a concern (Swartz et al. 2006). The release of domestic effluents through septic systems or directly into surface waters in rural and suburban regions can severely contaminate environmental matrices (including surface water bodies and soil) with organic micropollutants (Sui et al. 2015). Sewer leakages were also suggested as a potential micropollutant source into environmental matrices (Sui et al. 2015). Leaked sewers can spill large amounts of untreated wastewater with high micropollutant concentration into surface water bodies and soil directly (Kuroda et al. 2012). Underground sewage leakages in urban areas can be undetected for long time and release micropollutant-loaded wastewater into soil and groundwater (Sui et al. 2015).

A wide range of compounds including pharmaceuticals (antibiotics, nonsteroidal and steroidal anti-inflammatory drugs, ionophores), hormones and oestrogens (steroids, ergonovine, oxytocin, prostaglandins, progesterone, HCG, GnRH, and FSH), nutrient supplements, insecticides, and dewormers are used for enhanced animal, poultry, and fisheries growth and development (Bártíková et al. 2016; Patel et al. 2019). Animal antibiotic consumptions can exceed human consumptions (Sarmah et al. 2006). These compounds along with their metabolites can enter the environment (soil, agricultural fields and surface water) via inappropriate disposal of animal and poultry wastes and their applications as organic fertilizers (Bártíková et al. 2016). In fisheries, excessive applications of pharmaceuticals and other beneficial chemicals are used in order to achieve positive effects (Mo et al. 2017). Excess organic micropollutants reach soil and sediments of the water bodies; from there they can easily reach groundwater aquifers (Mo et al. 2017).

Source-pathway receptor models are applied for the assessment of micropollutant source and transportation pathways in the environment (Stuart et al. 2012). The model also estimates the possible human and environmental exposure routes of the micropollutants (Stuart et al. 2012). Aquatic transport is the main route of transportation for less volatile micropollutants such as pharmaceuticals (Patel et al. 2019), while aerial transportations can be major pathway of volatile micropollutants such as persistent organic pollutants and perfluorinated compounds (Figure 7.1) (Stuart et al. 2012). Different micropollutants can reach the environment through manufacturing industries, household, municipal, agricultural, and veterinary usage (Figure 7.1). Micropollutants from both terrestrial (landfill, soil, agricultural lands) and aquatic (surface water sources including rivers, streams, lakes, WWTPs and untreated effluents from industries, hospitals and municipal) sources can infiltrate groundwater (Figure 7.1). Through groundwater, organic micropollutants can directly reach humans as either treated or untreated drinking water (Figure 7.1).

7.5 Fate of Organic Micropollutants in Groundwater

Organic micropollutants suffer different fates and transportation behaviours after entering the environment (Postigo and Barceló 2015). The fate and transportation of organic micropollutants depends upon the physicochemical properties of the micropollutant

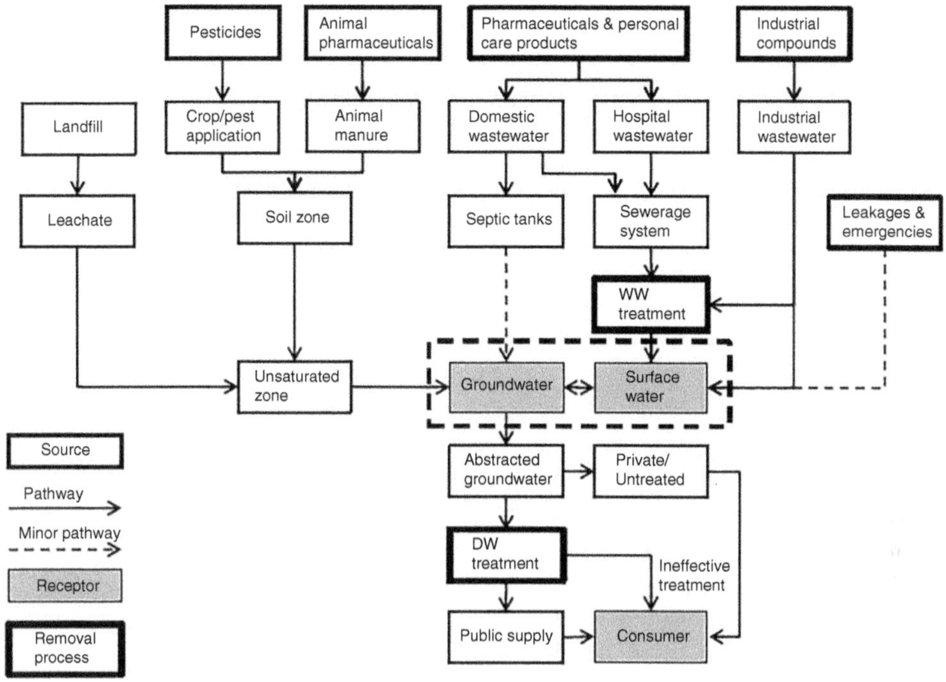

Figure 7.1 Organic micropollutant sources and transportation pathways in the aquatic environment. *Source:* Figure adopted with permission from [Lapworth et al. 2012] Copyright @ Elsevier.

studied as well as soil and aquifer characteristics. Soil properties such as lithology, hydraulic, textural, the presence and types of microbial communities, water residence time in soil, and the aquifer's redox state are important soil and aquifer properties affecting the fate and transformation of organic micropollutants (Burke et al. 2013; Chen et al. 2011; Lapworth et al. 2012). Organic micropollutant properties such as solubility in water, hydrophobicity, volatility, solid-water distribution coefficient (log K_d), dissociation constant (pK_a), octanol–water partition coefficient (log K_{ow}), and organic-carbon based sorption coefficient (log K_{oc}) also affect its fate and transportation behaviour in the environment (Mompelat et al. 2009; Pal et al. 2010, 2014). The sorption–desorption and degradation processes of organic micropollutants are redox sensitive. Thus, the presence of micropollutants in groundwater is highly affected by the aquifer's redox state (Postigo and Barceló 2015). For example, efficient removal of most pharmaceuticals is observed during oxic conditions in the aquifer compared to anoxic conditions (Burke et al. 2013). However, antibiotics such as clarithyromycin, clindamycin, and roxithormycin are exceptions and degrade more in anoxic conditions of the aquifer (Burke et al. 2013).

Other aquifer properties such as water table depth, substratum texture, and aquifer type also contribute to micropollutants' fate in the groundwater (Sackaria and Elango 2020). Shallow aquifers show relatively high micropollutant concentration compared to deep aquifers. This is due to the close proximity of the aquifer with contaminant sources (Lapworth et al. 2012). However, in deep aquifers fewer micropollutants are present, because of the degradations and reduced mobility with increase in depth (Lapworth et al. 2012). The substratum's geological character controls the micropollutant enrichment in groundwater (Sackaria and Elango 2020). Sandy, loamy, and silty textures with high porosity and hydraulic conductivity result in faster micropollutant movement into groundwater (Mutiyar et al. 2011). However, clay layers result in slow micropollutant movement because of poor permeability (Ghose and Saha 2009). For example, the presence of perfluorinated compounds in groundwater depends highly upon the sediment characteristics and the hydro-chemical properties of perfluorinated compounds (Weber et al. 2017).

7.6 Organic Micropollutant Analysis in Groundwater

Both gas chromatography (GC-MS, GC-GC-MS) and liquid chromatography (LC-MS, LC-MS-MS, HPLC-MS UPLC-MS, LC-qTOF-MS and RPLC) techniques have been successfully applied for the detection of organic micropollutants (Huntscha et al. 2012; Leendert et al. 2015; Schmidt 2018; Osenbrück et al. 2007). These techniques have successfully quantified organic micropollutants in drinking, ground, and surface waters. Proper micropollutant extraction using liquid–liquid extraction (LLE) and solid-phase extraction (SPE) are necessary before sample analysis. GC and LC coupled with MS and MS/MS are suitable for >1000 known, targeted, and suspected micropollutants (Moschet et al. 2014; Reemtsma et al. 2013; Robles-Molina et al. 2014; Wode et al. 2015). Still, many organic micropollutants are left unevaluated (Schmidt 2018). Therefore, application of LC and GC coupled with HRMS is suggested (Krauss et al. 2010; Schymanski et al. 2015; Schmidt 2018).

7.7 Conclusions and Future Strategies

Groundwater sources are widely contaminated with a large variety of organic micropollutants in ng/L to μg/L concentration ranges. These micropollutants include pesticides, pharmaceuticals, phthalates, perfluorinated compounds, surfactants, and others (Table 7.1). The highest reported pharmaceutical concentrations are diclofenac (48 080 ng/L in SAS Nagar Punjab) and sulfanilamide (>1000 ng/L in Varanasi), as shown in Table 7.1. For pesticides, the highest concentrations were reported for atrazine (up to 17 000 ng/L), simazine (up to 40 000 ng/L), and profenophos (up to 39 900 ng/L) in the groundwater sources of Delhi, and chlorpyriphos (41 200 ng/L) and malathions (82 300 ng/L) in Prabhani, Maharashtra (Table 7.1). Highly water soluble and polar compounds such as artificial sweeteners, some pesticides, and pharmaceuticals are the most abundant and ubiquitous micropollutants. Concentrations of other organic micropollutants are correlated with the production volume and usage rate. Owing to the availability of a large number of studies reporting the presence of organic micropollutants in Indian surface water, only very few groundwater studies are available. This points towards the severe lack of organic micropollutant studies in Indian groundwater sources. Correlations between the organic micropollutants detected in surface water samples with groundwater samples also point towards the possible presence of these contaminations in groundwater samples. Thus, surface water presence of organic micropollutants can form the baseline for the studies to be performed in the future.

References

Akhil, P. and Sujatha, C. (2012). Prevalence of organochlorine pesticide residues in groundwaters of Kasargod District, India. *Toxicological and Environmental Chemistry* 94: 1718–1725.

Ali, U., Syed, J.H., Malik, R.N. et al. (2014). Organochlorine pesticides (OCPs) in South Asian region: a review. *Science of the Total Environment* 476: 705–717.

Aslam, M., Alam, M., and Rais, S. (2013). Detection of atrazine and simazine in ground water of Delhi using high performance liquid chromatography with ultraviolet detector. *Current World Environment* 8: 323.

Balakrishna, K., Rath, A., Praveenkumarreddy, Y. et al. (2017). A review of the occurrence of pharmaceuticals and personal care products in Indian water bodies. *Ecotoxicology and Environmental Safety* 137: 113–120.

Barnes, K.K., Kolpin, D.W., Furlong, E.T. et al. (2008). A national reconnaissance of pharmaceuticals and in the United States—I Groundwater. *Science of the Total Environment* 402: 192–200.

Bártíková, H., Podlipná, R., and Skálová, L. (2016). Veterinary drugs in the environment and their toxicity to plants. *Chemosphere* 144: 2290–2301.

Bartrons, M. and Peñuelas, J. (2017). Pharmaceuticals and personal-care products in plants. *Trends in Plant Science* 22: 194–203.

Boorman, G.A. (1999). Drinking water disinfection byproducts: review and approach to toxicity evaluation. *Environmental Health Perspectives* 107: 207–217.

Burke, V., Duennbier, U., and Massmann, G. (2013). The effect of aeration on the removal of wastewater-derived pharmaceutical residues from groundwater–A laboratory study. *Water Science and Technology* 67: 658–666.

Chakraborti, D., Das, B., and Murrill, M.T. (2011). Examining India's groundwater quality management. *Environmental Science and Technology* 45: 27–33.

Chaubey, J., Singh, S., Singh, S., and Arora, H. (2020). Assessment of the impact of pesticide usage in groundwater aquifer of an agriculturally dominated area in North West India. *Water Practice and Technology* 15: 327–342.

Chen, H., Gao, B., Li, H., and Ma, L.Q. (2011). Effects of pH and ionic strength on sulfamethoxazole and ciprofloxacin transport in saturated porous media. *Journal of Contaminant Hydrology* 126: 29–36.

Chowdary, V., Rao, N., and Sarma, P. (2005). Decision support framework for assessment of non-point-source pollution of groundwater in large irrigation projects. *Agricultural Water Management* 75: 194–225.

Daughton, C.G. (2004). Non-regulated water contaminants: emerging research. *Environmental Impact Assessment Review* 24: 711–732.

Daughton, C.G. and Ternes, T.A. (1999). Pharmaceuticals and personal care products in the environment: agents of subtle change? *Environmental Health Perspectives* 107: 907–938.

Duttagupta, S., Mukherjee, A., Bhattacharya, A., and Bhattacharya, J. (2020). Wide exposure of persistent organic pollutants (PoPs) in natural waters and sediments of the densely populated Western Bengal basin, India. *Science of the Total Environment* 717: 137187.

European Commission 2015. European Commission Implementation Decision 2015/495/EC establishing a watch list of substances for Union-wide monitoring in the field of water policy pursuant to Directive 2008/105/EC of the European Parliament and of the Council. Off. J. Eur. Union.

European Commission 2018. European Commission Implementation Decision 2018/840 establishing a watch list of substances for Union-wide monitoring in the field ofwater policy pursuant to Directive 2008/105/EC of the European Parliament and of the Council and repealing Commission Implementing Decision (EU) 2015/495. Off.J. Eur. Union.

Fick, J., Söderström, H., Lindberg, R.H. et al. (2009). Contamination of surface, ground, and drinking water from pharmaceutical production. *Environmental Toxicology and Chemistry* 28: 2522–2527.

Gani, K.M. and Kazmi, A.A. (2017). Contamination of emerging contaminants in Indian aquatic sources: first overview of the situation. *Journal of Hazardous, Toxic, and Radioactive Waste* 21: 04016026.

Galassi, S., Provini, A., and Halfon, E. (1996). Risk assessment for pesticides and their metabolites in water. *International Journal of Environmental Analytical Chemistry* 65: 331–344.

Ghose, N.C. and Saha, D. (2009). Synthetic detergents (surfactants) and organochlorine pesticide signatures in surface water and groundwater of greater Kolkata, India. *Journal of Water Resource and Protection* 1.

Halling-Sørensen, B., Nielsen, S.N., Lanzky, P. et al. (1998). Occurrence, fate and effects of pharmaceutical substances in the environment-A review. *Chemosphere* 36: 357–393.

Huntscha, S., Singer, H.P., Mcardell, C.S. et al. (2012). Multiresidue analysis of 88 polar organic micropollutants in ground, surface and wastewater using online mixed-bed multilayer solid-phase extraction coupled to high performance liquid chromatography–tandem mass spectrometry. *Journal of Chromatography A* 1268: 74–83.

Jindal, K., Narayanam, M., and Singh, S. (2015). A systematic strategy for the identification and determination of pharmaceuticals in the environment using advanced LC-MS tools: application to groundwater samples. *Journal of Pharmaceutical and Biomedical Analysis* 108: 86–96.

Jurado, A., Vàzquez-Suñé, E., Carrera, J. et al. (2012). Emerging organic contaminants in groundwater in Spain: a review of sources, recent occurrence and fate in a European context. *Science of the Total Environment* 440: 82–94.

Kaushik, C., Sharma, H., and Kaushik, A. (2012). Organochlorine pesticide residues in drinking water in the rural areas of Haryana, India. *Environmental Monitoring and Assessment* 184: 103–112.

Kolpin, D.W., Barbash, J.E., and Gilliom, R.J. (1998). Occurrence of pesticides in shallow groundwater of the United States: initial results from the National Water-Quality Assessment Program. *Environmental Science and Technology* 32: 558–566.

Kolpin, D., Thurman, E., and Linhart, S. (2000). Finding minimal herbicide concentrations in ground water? Try looking for their degradates. *Science of the Total Environment* 248: 115–122.

Kolpin, D.W., Furlong, E.T., Meyer, M.T. et al. (2002). Pharmaceuticals, hormones, and other organic wastewater contaminants in US streams, 1999– 2000: a national reconnaissance. *Environmental Science and Technology* 36: 1202–1211.

Kolpin, D.W., Schnoebelen, D.J., and Thurman, E.M. (2004). Degradates provide insight to spatial and temporal trends of herbicides in ground water. *Groundwater* 42: 601–608.

Krauss, M., Singer, H., and Hollender, J. (2010). LC–high resolution MS in environmental analysis: from target screening to the identification of unknowns. *Analytical and Bioanalytical Chemistry* 397: 943–951.

Kuroda, K., Murakami, M., Oguma, K. et al. (2012). Assessment of groundwater pollution in Tokyo using PPCPs as sewage markers. *Environmental Science and Technology* 46: 1455–1464.

Lapworth, D.J., Gooddy, D., Stuart, M. et al. (2006). Pesticides in groundwater: some observations on temporal and spatial trends. *Water and Environment Journal* 20: 55–64.

Lapworth, D., Baran, N., Stuart, M., and Ward, R. (2012). Emerging organic contaminants in groundwater: a review of sources, fate and occurrence. *Environmental Pollution* 163: 287–303.

Lapworth, D., Das, P., Shaw, A. et al. (2018). Deep urban groundwater vulnerability in India revealed through the use of emerging organic contaminants and residence time tracers. *Environmental Pollution* 240: 938–949.

Lari, S.Z., Khan, N.A., Gandhi, K.N. et al. (2014). Comparison of pesticide residues in surface water and ground water of agriculture intensive areas. *Journal of Environmental Health Science and Engineering* 12: 11.

Larsson, D.J. (2014). Pollution from drug manufacturing: review and perspectives. *Philosophical Transactions of the Royal Society B: Biological Sciences* 369: 20130571.

Leendert, V., van Langenhove, H., and Demeestere, K. (2015). Trends in liquid chromatography coupled to high-resolution mass spectrometry for multi-residue analysis of organic micropollutants in aquatic environments. *TrAC Trends in Analytical Chemistry* 67: 192–208.

Lewandowski, J., Putschew, A., Schwesig, D. et al. (2011). Fate of organic micropollutants in the hyporheic zone of a eutrophic lowland stream: results of a preliminary field study. *Science of the Total Environment* 409: 1824–1835.

Lorenzen, G., Sprenger, C., Taute, T. et al. (2010). Assessment of the potential for bank filtration in a water-stressed megacity (Delhi, India). *Environmental Earth Sciences* 61: 1419–1434.

Mathew, G. and Unnikrishnan, M. (2012). The emerging environmental burden from pharmaceuticals. *Economic and Political Weekly*: 31–34.

Mishra, K. and Sharma, R.C. (2011). Contamination of aquatic system by chlorinated pesticides and their spatial distribution over North-East India. *Toxicology and Environmental Health Sciences* 3: 144.

Mitch, W.A., Sharp, J.O., Trussell, R.R. et al. (2003). N-nitrosodimethylamine (NDMA) as a drinking water contaminant: a review. *Environmental Engineering Science* 20: 389–404.

Mo, W.Y., Chen, Z., Leung, H.M., and Leung, A.O.W. (2017). Application of veterinary antibiotics in China's aquaculture industry and their potential human health risks. *Environmental Science and Pollution Research* 24: 8978–8989.

Motekar, S. 2011. Ground water contamination by organochlorine insecticide residues in the Godavari plain of Nanded district. Recent Research in Science and Technology.

Motekar Shrinivas, C. (2014). Liability for groundwater contamination by organochlorine and organophosphorus pesticide residues in Parbhani District. *Discovery, The International Daily Journal* 20: 65–69.

Mompelat, S., Le Bot, B., and Thomas, O. (2009). Occurrence and fate of pharmaceutical products and by-products, from resource to drinking water. *Environment International* 35: 803–814.

Mondal, R., Mukherjee, A., Biswas, S., and Kole, R.K. (2018). GC-MS/MS determination and ecological risk assessment of pesticides in aquatic system: a case study in Hooghly River basin in West Bengal, India. *Chemosphere* 206: 217–230.

Moran, M.J., Zogorski, J.S., and Squillace, P.J. (2007). Chlorinated solvents in groundwater of the United States. *Environmental Science and Technology* 41: 74–81.

Moschet, C., Wittmer, I., Simovic, J. et al. (2014). How a complete pesticide screening changes the assessment of surface water quality. *Environmental Science and Technology* 48: 5423–5432.

Mutiyar, P., Mittal, A., and Pekdeger, A. (2011). Status of organochlorine pesticides in the drinking water well-field located in the Delhi region of the flood plains of river Yamuna. *Drinking Water Engineering and Science* 4: 51–60.

Osenbrück, K., Gläser, H.-R., Knöller, K. et al. (2007). Sources and transport of selected organic micropollutants in urban groundwater underlying the city of Halle (Saale), Germany. *Water Research* 41: 3259–3270.

Pal, A., Gin, K.Y.-H., Lin, A.Y.-C., and Reinhard, M. (2010). Impacts of emerging organic contaminants on freshwater resources: review of recent occurrences, sources, fate and effects. *Science of the Total Environment* 408: 6062–6069.

Pal, A., He, Y., Jekel, M. et al. (2014). Emerging contaminants of public health significance as water quality indicator compounds in the urban water cycle. *Environment International* 71: 46–62.

Patel, M., Kumar, R., Kishor, K. et al. (2019). Pharmaceuticals of emerging concern in aquatic systems: chemistry, occurrence, effects, and removal methods. *Chemical Reviews* 119: 3510–3673.

Philip, J.M., Aravind, U.K., and Aravindakumar, C.T. (2018). Emerging contaminants in Indian environmental matrices–A review. *Chemosphere* 190: 307–326.

Phillips, P.J., Schubert, C., Argue, D. et al. (2015). Concentrations of hormones, pharmaceuticals and other micropollutants in groundwater affected by septic systems in New England and New York. *Science of the Total Environment* 512: 43–54.

Postigo, C. and Barceló, D. (2015). Synthetic organic compounds and their transformation products in groundwater: occurrence, fate and mitigation. *Science of the Total Environment* 503: 32–47.

Rahman, F., Langford, K.H., Scrimshaw, M.D., and Lester, J.N. (2001). Polybrominated diphenyl ether (PBDE) flame retardants. *Science of the Total Environment* 275: 1–17.

Reemtsma, T., Alder, L., and Banasiak, U. (2013). A multimethod for the determination of 150 pesticide metabolites in surface water and groundwater using direct injection liquid chromatography–mass spectrometry. *Journal of Chromatography A* 1271: 95–104.

Richardson, M.L. and Bowron, J.M. (1985). The fate of pharmaceutical chemicals in the aquatic environment. *Journal of Pharmacy and Pharmacology* 37: 1–12.

Richardson, S.D. and Ternes, T.A. (2011). Water analysis: emerging contaminants and current issues. *Analytical Chemistry* 83: 4614–4648.

Robles-Molina, J., Lara-Ortega, F.J., Gilbert-López, B. et al. (2014). Multi-residue method for the determination of over 400 priority and emerging pollutants in water and wastewater by solid-phase extraction and liquid chromatography-time-of-flight mass spectrometry. *Journal of Chromatography A* 1350: 30–43.

Rutgersson, C., Fick, J., Marathe, N. et al. (2014). Fluoroquinolones and qnr genes in sediment, water, soil, and human fecal flora in an environment polluted by manufacturing discharges. *Environmental Science and Technology* 48: 7825–7832.

Sackaria, M. and Elango, L. (2020). Organic micropollutants in groundwater of India—A review. *Water Environment Research* 92: 504–523.

Sarmah, A.K., Meyer, M.T., and Boxall, A.B. (2006). A global perspective on the use, sales, exposure pathways, occurrence, fate and effects of veterinary antibiotics (VAs) in the environment. *Chemosphere* 65: 725–759.

Schirmer, K. and Schirmer, M. (2008). Who is chasing whom? A call for a more integrated approach to reduce the load of micro-pollutants in the environment. *Water Science and Technology* 57: 145–150.

Schmidt, T.C. (2018). Recent trends in water analysis triggering future monitoring of organic micropollutants. *Analytical and Bioanalytical Chemistry* 410: 3933–3941.

Schwarzenbach, R.P., Escher, B.I., Fenner, K. et al. (2006). The challenge of micropollutants in aquatic systems. *Science* 313: 1072–1077.

Schymanski, E.L., Singer, H.P., Slobodnik, J. et al. (2015). Non-target screening with high-resolution mass spectrometry: critical review using a collaborative trial on water analysis. *Analytical and Bioanalytical Chemistry* 407: 6237–6255.

Sehonova, P., Svobodova, Z., Dolezelova, P. et al. (2018). Effects of waterborne antidepressants on non-target animals living in the aquatic environment: a review. *Science of the Total Environment* 631: 789–794.

Salgado, R., Marques, R., Noronha, J. et al. (2012). Assessing the removal of pharmaceuticals and personal care products in a full-scale activated sludge plant. *Environmental Science and Pollution Research* 19: 1818–1827.

Sharma, B.M., Bharat, G.K., Tayal, S. et al. (2016). Perfluoroalkyl substances (PFAS) in river and ground/drinking water of the Ganges River basin: emissions and implications for human exposure. *Environmental Pollution* 208: 704–713.

Sharma, B.M., Bečanová, J., Scheringer, M. et al. (2019). Health and ecological risk assessment of emerging contaminants (pharmaceuticals, personal care products, and artificial sweeteners) in surface and groundwater (drinking water) in the Ganges River Basin, India. *Science of the Total Environment* 646: 1459–1467.

Somashekar, K.M., Mahima, M., and Manjunath, K. (2015). Contamination of water sources in Mysore city by pesticide residues and plasticizer–A cause of health concern. *Aquatic Procedia* 4: 1181–1188.

Spliid, N.H. and Køppen, B. (1998). Occurrence of pesticides in Danish shallow ground water. *Chemosphere* 37: 1307–1316.

Stuart, M. and Lapworth, D. (2013). Emerging organic contaminants in groundwater. In: Smart Sensors for Real-Time Water Quality Monitoring (eds. S.C. Mukhopadhya and M.A. Springer), 259–284. Berlin: Springer-Verlag.

Stuart, M., Lapworth, D., Crane, E., and Hart, A. (2012). Review of risk from potential emerging contaminants in UK groundwater. *Science of the Total Environment* 416: 1–21.

Sui, Q., Cao, X., Lu, S. et al. (2015). Occurrence, sources and fate of pharmaceuticals and personal care products in the groundwater: a review. *Emerging Contaminants* 1: 14–24.

Swartz, C.H., Reddy, S., Benotti, M.J. et al. (2006). Steroid estrogens, nonylphenol ethoxylate metabolites, and other wastewater contaminants in groundwater affected by a residential septic system on Cape Cod, MA. *Environmental Science and Technology* 40: 4894–4902.

Thakur, S., Gulati, K., and Jindal, T. (2015). Groundwater contamination through pesticide usage in vegetable growing areas of Delhi. *International Journal of Multidisciplinary Research and Development* 2: 394–397.

Tijani, J.O., Fatoba, O.O., Babajide, O.O., and Petrik, L.F. (2016). Pharmaceuticals, endocrine disruptors, personal care products, nanomaterials and perfluorinated pollutants: a review. *Environmental Chemistry Letters* 14: 27–49.

USEPA. 2016. *Contaminant Candidate List 4-CCL 4* [Online]. Available: https://www.epa.gov/ccl/contaminant-candidate-list-4-ccl-4-0.

Van Boeckel, T.P., Gandra, S., Ashok, A. et al. (2014). Global antibiotic consumption 2000 to 2010: an analysis of national pharmaceutical sales data. *The Lancet Infectious Diseases* 14: 742–750.

Verliefde, A., Cornelissen, E., Amy, G. et al. (2007). Priority organic micropollutants in water sources in Flanders and the Netherlands and assessment of removal possibilities with nanofiltration. *Environmental Pollution* 146: 281–289.

Weber, A.K., Barber, L.B., Leblanc, D.R. et al. (2017). Geochemical and hydrologic factors controlling subsurface transport of poly-and perfluoroalkyl substances, Cape Cod, Massachusetts. *Environmental Science and Technology* 51: 4269–4279.

Walls, D., Smith, P., and Mansell, M. (1996). Pesticides in groundwater in Britain. *International Journal of Environmental Health Research* 6: 55–62.

Wode, F., van Baar, P., Dünnbier, U. et al. (2015). Search for over 2000 current and legacy micropollutants on a wastewater infiltration site with a UPLC-high resolution MS target screening method. *Water Research* 69: 274–283.

WWAP 2017. The United Nations World Water Development Report 2017: Wastewater, The Untapped Resource, WWAP (United Nations World Water Assessment Programme). UNESCO Paris.

8

Organic Pollutants in Groundwater Resource

Gurudatta Singh, Anubhuti Singh, Priyanka Singh, and Virendra Kumar Mishra

Institute of Environment and Sustainable Development, Banaras Hindu University, Varanasi, Uttar Pradesh, India

8.1 Introduction

Freshwater contamination because of human activities, as well as natural activities, have been a cause of concern. Numerous studies have acknowledged the prevalence of the freshwater pollution due to countless human activities, which include the inappropriate discharge of the industries as well as domestic wastewater, applications of the reprocessed wastewater in agricultural practices, and movement enabled through the subsurface geochemistry, as well as cyclical variations; these activities have frequently made freshwater susceptible to the variability of pollution by carbon-based pollutants (Kurwadkar 2019). The rapidly growing population and the increase of the agriculture, as well as industrialized activities, result in the increasing water claim as well as polluted water production. Subsequently, freshwater means were progressively visible to pollution through various bases like seepage via drainage systems as well as septic chambers, applications of manure in agrarian arenas, intentionally or unintended discarded disposals, exonerations of the polluted water wastes, and urban as well as rural rainstorm water overflow (Slack et al. 2005; Osenbrück et al. 2007; Barnes et al. 2008). In recent years, a huge amount of micro-organic pollutants had been found globally in shallow as well as freshwater (Barnes et al. 2008; Loos et al. 2009, 2010; Teijon et al. 2010; Valcárcel et al. 2011; Cabeza et al. 2012; Meffe et al. 2014).

Extraordinary growing populations have been combined with industrialized and agrarian activities, and can result in both amplified demands for freshwater as well as the cohort of polluted water. Conferring to ordinance 2006/118/EC, groundwater is the major source of freshwater in the European Union, and it is the most delicate. Freshwater might agonize pollutions through various sources, which include water seepage by the sewers as well as the septic tanks, leakage by the rivers, and application of manures, as well as agrochemicals, amongst others (Foster 2001; Vázquez-Suñé et al. 2010). As a result, a wide range of carbon-based contaminants could be found in the aquifers, posing risks to the freshwater qualities

Groundwater Geochemistry: Pollution and Remediation Methods, First Edition.
Edited by Sughosh Madhav and Pardeep Singh.

(Wolf et al. 2004). Therefore, the appropriate valuation of the freshwater qualities required the identification of such contaminants. To confrontat the abovementioned complications, it is essential to describe the freshwater quality principles. To date, regulation on freshwater pollution via carbon-based pollutants have simply affected insecticides. The environmental quality principles for both the individual constituents, which includes their significant metabolites as well as deprivation and response products (0.1 μg/L) and for the sum (0.5 μg/L), have been established (European Parliament and Council of the European Union 2006).

Carbon-based pollution comprises all the natural as well as artificial contaminants that can cause adversarial effects on the human's health or ecology environment. Sources of these organic contaminants can include pharmaceuticals, industries, agricultural activities, and daily household activities. The nature and sources of organic contaminants varies according to the processes and activities in the locality or region of the associated area of study. For example, in urban areas the nature and source of organic contaminants include production from sewage, wastewater treatment plants, and industrial activities. In a rural area, activities like agriculture and farming, wasteland applications, etc., take place. Transportation and fate of the subjected contaminant in any environment are generally controlled by the four general processes, namely advection, dispersion, interphase mass transfer, and transformation reactions. The fate of any definite pollutant in the environment is a function of the collective influence of these four general processes. Some of the organic pollutants are presented in Table 8.1.

Table 8.1 Some organic contaminants found in groundwater.

Serial No.	Contaminant	Potential health and other effects	Sources to groundwater
1	Pesticide	Causes poison, headache, wooziness, gastrointestinal symptoms, coldness, as well as sarcoma. Disrupts the nervous systems, thyroid glands, reproductive systems, as well as excretory system.	Moves into the environments through sources like herbicide, insecticide, fungicide, rodenticide as well as algicide.
2	Volatile organic compounds	May cause sarcoma as well as can disrupt the liver, anaemia, gastrointestinal disorders, skin problems, blurry visions, tiredness, weightiness, destruction of the nervous and respiratory systems.	Enters into the environs when used to manufacture plastic, colourant, latex, graces, diluents, crude oils, pesticides, toners, lacquers, splatters, sanitizers, petrol products, antibiotics, preservers, spot remover, paint remover, degreasers, and numerous materials.
3	Plasticizers, chlorinated solvents, benzo[a] pyrene, and dioxin	Causes sarcoma. Disrupts the nervous as well as reproductive systems, excretory systems, gastrointestinal, and liver.	Used as sealants, coatings, diluents, insecticides, plasticizers, constituents of petrol, disinfectants, as well as wood preservatives. Moves into the environs through indecorous discarded disposals, percolating overflow, permeable storage tanks, as well as manufacturing runoff.

8.2 Types of Organic Contaminants in Groundwater

8.2.1 Pesticides

Pesticides are compounds like carbamates or biological agents like viruses, bacterium, or fungus which dissuade, weaken, destroy, or otherwise depress pests. Pests may include insects, plants pathogens, unwanted plants, molluscs, birds, mammals, fishes, and nematodes, as well as microorganisms which destroy properties, cause annoyance, or spread diseases or their vectors (Suciu et al. 2016). Besides these assistances, pesticides too have their disadvantages, like possible toxicities in human beings as well as other organisms. The use of pesticides raises various concerns for the environment. Over 98% of squirted insecticide as well as 95% of herbicide reached places other than their targeted destinations; this includes nontargeted places like air and water as well as soils and pesticides are one of the causes of pollution in water (Kaur 2019). The involvement of pesticides into shallow water shrinks biodiversities, deteriorates habitats, and threatens organisms like dolphins, manatees and dugongs, and salmon (Flather et al. 1998). Because of this, many pesticides have been banned in many countries. One of the banned pesticides is dichlorodiphenyl-trichloroethane (DDT). Pests may develop resistance towards the pesticides, which leads to demanding new pesticides (Scott 1990). On the other hand, a large dosage of pesticides could be utilized to lessen the resistivity in the pests, though it can create the deteriorating of ambient problems of pollution. Insecticides can create severe as well as deferred well-being effects on the human beings who are exposed (Judson et al. 2009). The exposures to pesticides may cause variable adversarial effects on well-being, ranging from unpretentious irritation of the skin and eyes, to other adverse effects like distressing the nervous systems, or copying hormones, which causes generative complications, as well as causes sarcoma (Kapaj et al. 2006). Partial evidence is present for other harmful consequences from the exposures of pesticides, including neural and natal defects, as well as foetal death (Bassil et al. 2007). Types of pesticides and their uses in the aquatic environment are presented in Table 8.2.

8.2.2 Pharmaceuticals

Pharmacological items for consumption as well as their waste plays the most important part in degrading the surroundings. These medicines devise progressive and deleterious significance of various components in the environment; this includes biota in diverse means (Huerta et al. 2012). Numerous categories of pharmacological constituents have been identified by noteworthy deliberations by numerous progressive instrumental techniques in the shallow water, sub-surface water, underground water, domestic wastewater, and metropolitan wastewater, as well as manufacturing discharges. The usage of various pharmacological medicines on a global scale by humans has been recorded at approximately 100 000 tons/year (Kümmerer 2008). Our country has currently arisen as one of the top five pharmacological marketing regions in the world. Because of growing demands, frequently at an intercontinental level, pharmacological manufacturing is speedily rising. Numerous pharmacological industrial divisions were increasing in our country (Kumar et al. 2010) as was disposal of its discharges in the streams, either straight

Table 8.2 Types of pesticide and their uses in the environment.

Serial No.	Types of Pesticides	Uses	References
1.	Antimicrobials	Control germs and microbes	Hanson et al. (2019)
2.	Disinfectants	Control germs and microbes	Schaeck et al. (2016)
3.	Foggers	To kill insects	Santora (2018)
4.	Fungicides	Control fungal problems	Damicone (2014)
5.	Herbicides	Constrain the growth of undesirable vegetation	Gupta (2018)
6.	Insecticides	Control insects	Ware and Whitacre (2004)
7.	Insects growth Regulator	Interrupt the progression as well as reproduction in insects	Mondal and Parween (2000)
8.	Desiccants	To dry up living plant tissue	Sherlock et al. (2005)
9.	Defoliants	Cause plants to drop their leaves	Bi et al. (2005)
10.	Miticide	Control mites	Lindberg et al. (2000)
11.	Molluscicides	Control slugs, snails, and other molluscs	Iglesias et al. (2003)
12.	Mothballs	To eradicate textile pests by fumigation in closed flasks	Mayer et al. (2007)
13.	Pheromones	Disrupt the mating behaviour of insect	Welter et al. (2005)
14.	Repellents	To repel unwanted pests	Simpson and Simpson Hugh (1988)
15.	Rodenticide	Kills rodents	Perry et al. (1998)

or after incomplete treatments. Consequently, appropriate organization alongside complete as well as operative treatments of pharmacological manufacturing waste and of terminated medicines was mandatory to begin with up-to-date apparatuses as well as methods. The pharmacological medicines get excreted through urination as well as egestion by way of parental compounds as numerous as its metabolites (Dauti et al. 2014). The wastewater of the toilets get flushed by unpolluted water consequently; this is well-known as grey water. The grey water passes into the metropolitan sewers as well as gets into the connecting water streams and distresses the water superiority (Chander et al. 2016). There are no such confirmation that the pharmacological discharge into the water causes detriments in human beings, but some of the studies show and adversarial effect on freshwater life (Richardson and Ternes 2011). More than 4000 drugs prescribed and utilized by humans as well as animals eventually find their way into the environs. This may affect human beings as well as animals (Gupta 2011). These compounds make their way into systems in the terrestrial as well as riverine environment, which could affect the health as well as behaviours of different animals (Trombulak and Frissell 2000). Types of pharmaceutical compounds and their uses in the environment are presented in Table 8.3. Sources of pharmaceutical compounds in the groundwater are presented in Figure 8.1.

Table 8.3 Types of pharmaceuticals and their uses in the aquatic environment.

Serial No.	Pharmaceuticals	Uses	References
1	17 alpha-estradiol	Estrogenic hormone used in pharmaceuticals	Duesterberg et al. (2005)
2	Equilin	Estrogenic hormone used in pharmaceuticals	Hamid and Eskicioglu (2012)
3	Erythromycin	Pharmaceutical formulation as an antibiotic	Khashaba (2002)
4	Estradiol	Estrogenic hormone used in pharmaceuticals	Gibbs and Gabor (2003)
5	Norethindrone	Progresteronic hormone used in pharmaceuticals	Sanderson (2011)

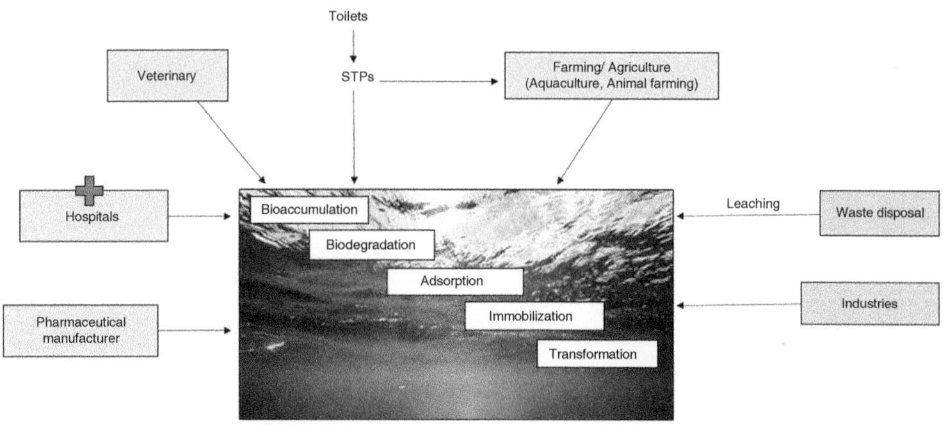

Figure 8.1 Sources of pharmaceuticals in groundwater.

8.2.3 Personal Care Compounds

Personal care products were pervasive in the riverine water environment globally, as well as being documented like pollutants of concern. Presently, pollutants of disquiet were categorized for their resolution, bioaccumulation, and toxicities (Brausch and Rand 2011). Personal care products were naturally energetic complexes that were predictable like pollutants in the environment on worldwide concerns because of their existence in environments worldwide (Monteiro and Boxall 2010). Personal care products may enter into the ecosystems via various paths, including both point as well as non-point sources. These compounds can usually be identified into the shallow waters with little deliberation (Daughton and Ternes 1999). This is probably the contributions of increasing water utilization at worldwide, the scarcities would be further aggravate the deliberations of personal care products (Petrović et al. 2011; Klaminder et al. 2015). Recently the investigation of lethal environmental effects of personal care products has revealed that the ecologically significant concentration of these components can revise environmental relations as well as relations

Table 8.4 Types of personal care compounds and their uses in the aquatic environment.

Serial No.	Personal care compounds	Activity	Uses	References
1.	Anthocyanins, Gallic caustic, epicatechin, conjugated flavonoids, proanthocyanidins, resveratrol, melanin	Antioxidant as well as drastic foraging stimulus of cell growth	Anti-aging Preservers	Soto et al. (2015)
2.	Ferulic caustic caffeic caustic, Gallic caustic, proanthocyanidins, resveratrol	Photoreceptor	Sun protectors Lip balm guards	Afrin et al. (2016)
3.	Tartaric caustic	Exfoliating Depigmentating pH regulator	Chemicals as well as physical exfoliating agents Anti-browning pH corrector	Soto et al. (2015)
4.	Oleic, linoleic, and linolenic acids Vitamins C and E	Antioxidants Nutritive	Antiaging Day and night creams	Hwang et al. (2014)

amongst the species. Personal care products could also encourage variations in the environment as well as functions, including biofilm primary productions, respirations as well as communal structures, bio-geochemical procedures, as well as growing invertebrates and also populous diminuendos (Richmond et al. 2017). Bacterial respirations could be repressed by personal care products. The usage of personal care products would probably continue to increase worldwide as the proportions of the global populace within contact to treatments upsurges, so the releasing of personal care products to the shallow water would probably increasing rather than subside in upcoming years (Schröder et al. 2016). Additionally, increasing water usage proficiency would probably contribute to increases of personal care product deliberations into the environment (Klaminder et al. 2015). Types of personal care compounds and their uses in the aquatic environment are presented in Table 8.4.

8.2.4 Volatile Organic Compounds (VOCs)

Volatile organic compounds (VOCs) are the chemicals or cleansers which are derivative of gasoline products. These are anthropogenic pollutants that arise from the manufacturing procedures (Churchill et al. 2001). Those pollutants could percolate in the groundwater through compound leaks or wastewater releases from manufacturing activity. VOCs were the categories of compounds that contain carbons, as well as vaporizes, certainly in the air at usual temperature. Volatile organic carbons were found in a variety of the marketable manufacturing, as well as domestic materials, including petroleum, chemicals, cleansers and degreasers, paints, toners, peroxides, and insecticides (Fujita and Campbell 2003). When VOCs were found in the environments, they were usually the outcome of human activities, like a leak or unsuitable removal wherever the chemicals had been allowed to

Table 8.5 Sources of volatile organic compounds in the groundwater.

Serial No.	Sources	Pollutants	References
1.	Chemical pulping process	• Terpenes, alcohols, phenols, methanol, acetones, chloroform • Reduced sulphur compounds • Organ-chlorine compounds	Niemelä and Alén (1999)
2.	Bleaching	• Acetone, methylene chloride, chloroform, choromethane, trichloroethane	Zhao et al. (2014)
3.	Wastewater Treatment Plant	• Terpenes, spirits, phenols, methanols, acetones, chloroforms	Hundža et al. (2019)
4.	Power Boilers	• SO2, NOx, fly ash, coarse particulates	Shemwell et al. (2002)
5.	Evaporator	• Alcohol, terpenes, phenols	Mastelić et al. (1992)
6.	Calcining	• Fine and coarse particulates	Ediz et al. (2010)

soak into the ground. When released into the environment, VOCs would pass at depth into the aquifers in the form of non-aqueous phase liquid, vaporizing at small concentration of VOCs by way of it being conveyed via the sub-surface (Srirattana et al. 2017). Examples of VOCs identified were dichloromethane, manufacturing solvents; trichloroethylene, utilized in the drainage systems cleansers; and tetrachloroethylene, utilized in the dry cleaning industries (Kargar et al. 2013). Coverage hazards are analysed on discrete ingredients of VOCs, from every perspective to activate various indications of effect on human well-being or infections. VOCs were accountable for a number of adversarial well-being effect specifically for nursing as well as pregnant mothers (Boots et al. 2012).

Once mixed into the groundwater these compounds become highly tenacious. These could be transferred via unsaturated regions in rejuvenate, into soil vapours, or as non-aqueous-phase liquids. Several VOCs were tarnished through bacteria into the aquifers, but others resist deprivation and could get transferred to the stretched expanses (Chapelle 1999). Source of VOCs in the groundwater are discussed in Table 8.5.

8.2.5 Hormones and Sterols

Since the beginning of worldwide industrial areas, steroidal oestrogen had become an emergent as well as severe concern. Globally, steroidal oestrogen includes estrone and estradiol as well as estriol, and poses severe extortions to the soil, plant life, and aquatic systems, as well as human beings. Certainly, oestrogen had received prominent attention in recent years, because of its quick accumulations within aquatic systems all over the world (Houtman 2010).

Steroidal hormones were naturally dynamic oestrogens, which were derivative of saturated fat as well as released via adrenal cortex, testes, ovary, and the placenta in human beings as well as other creatures. Steroidal components had also been found in vegetation (Ying et al. 2002; Hamid and Eskicioglu 2012). Steroidal hormones could be categorized as

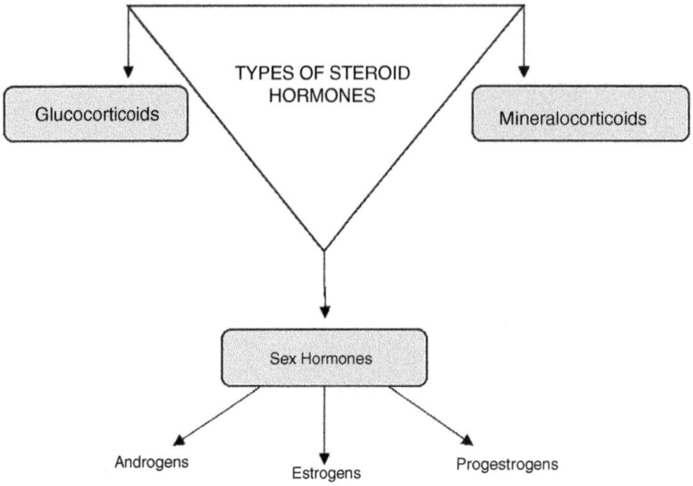

Figure 8.2 Types of steroid hormones in the aquatic environment.

natural or artificial oestrogens; as well as they may act as endocrine-disrupting chemicals. Conjugated hormones, which were also probably ecological threats, were made through the esterification of free steroids through glucuronide as well as sulphate group at the positions of C_3 or C_{17} (Hamid and Eskicioglu 2012). The physicochemical properties of steroidal hormone components are critical within orders to resolve their outcome into the aquatic system. The manifestation of naturally estrogenic hormones with little concentration into environs had been scrutinized by several researchers, as well as issues related to emerging contaminants (Richardson 2008). Globally, water has been contaminated by steroidal hormones through various unconfined methods, from sewage treatment plants as well as discharge through livestock's feed lots (Chen et al. 2010). Estrogenic hormones, as well as being utilized in CAFOs (concentrated animal feeding operations), had been identified into egestion and liquefied manure as well as solid waste collected through the cattle, inlet effluents, and in manure pragmatic unswervingly to farming lands (Adeel et al. 2017). Disputably, animal manure is the biggest source of estrogenic hormones in the natural environments. Undoubtedly, fowl, cows, and the manure of horses could contain the utmost amounts of estrogenic hormones (Andaluri et al. 2012). Numerous studies have recommended that increases in breast cancer in the manufacturing nations are connected to the exposures to endocrine-disrupting compounds, mainly estrogenic compounds. Estrogenic receptors ERα as well as ERβ were the two chief transducers of oestrogen actions and consequently significant objectives for these oestrogen like endocrine disrupters (Bolli and Marino 2011). Types of steroid hormones in the aquatic environment are presented in Figure 8.2.

8.2.6 Water Treatment By-Products

Some organic pollutants enter into drinking water after water treatment or dissemination through the absolute path. The World Health Organization's guiding principle, intended for consumable water quality (Water and WHO 2002), covers the substantial numbers of probable constituents through the water treatment or dissemination; nevertheless, only some of these constituents were of practical importance. This is significant so that the water source organizations can manage any of the compounds appropriately which were used by them (National Research Council 2004). In numerous circumstances, the greatest technique for controlling is through the managing practices, like optimizing the treatment processes, as well as regulating the constituents and compounds that come into contacting with the consumable water reasonably than by observing as well as analysing the chemicals (Monte et al. 2009). Decontamination by-products were designed once the chemicals were in the water source, which can react with the decontaminators (Richardson and Postigo 2011).

8.2.7 Chemicals Used in Treatments

8.2.7.1 Chlorine

Chlorine is the most extensively used in consumable water as a decontaminator (Gopal et al. 2007). The following reaction takes place very frequently after adding the chlorine to water:

$$Cl_2 + H_2O = HOCl + H^+ + Cl^-$$

The enormousness of the steadiness of hydrolysis constants are as the hydrolysis to the hypochlorous acid, HOCl, is effectively completed in the freshwater at $pH > 4$. Hypochlorous acid is a weak acid which dissociates incompletely in water as follows:

$$HOCl = H^+ + OCl^-$$

The term free chlorine denotes to the amount of the hypochlorous acid as well as hypochlorite ion. Subsequently HOCl is considered a more effective decontaminator than the OCl^-, as well as the free chlorine, even as hypochlorite, is more active than the combined chlorine (Wojtowicz 2000).

The utmost generally found chlorine decontamination by-products were the trihalomethanes (THM), halogenated acetic acids, halogenated acetonitriles, chloral hydrate, and the chlorinated phenols. Others includes chlorinated furanone MX, halopicrins, cyanogen halides, haloketones, and halo aldehydes (Gan et al. 2015). The halogenated decontamination by-products identified accounts for only about part of the total formed. Some of the chlorine compounds and their by-products are mentioned in Table 8.6.

8.2.8 Food Additives

Ingredients that are mixed into foodstuffs to preserve or recover their safety, freshness, taste, consistency, or appearance are known as food additives (Abdulmumeen et al. 2012). Numerous diverse foodstuff flavours have been technologically advanced over time to meet the needs of manufacturing the foodstuff, as producing the foodstuff on large scales was actually dissimilar to manufacturing them on a small scale at home. Flavourings were required to make sure that the processed foodstuff remained safe as well as in better conditions in its passage from the industrial unit or industrial kitchens, throughout the conveyance to the granaries as well as the shops, and lastly to the customers (King 2019). They were added deliberately to the foods to execute the assured technical determinations that customers repeatedly take for granted. The foodstuff flavourings were utilized for

Table 8.6 Some of the chlorine compounds and their by-products in the aquatic environment.

Serial No.	Chlorination of water supplies	By-products	Formation of by-products in water	References
1.	Free chlorine	HClO and ClO^-	Grounded on the flavour as well as odour verge of the free chlorine	Phattarapattamawong (1999)
2.	Trihalomethanes (THMs)	Chloroform, bromodichloromethane, dibromochloromethane, and bromoform	Made by the aqueous chlorination of humus constituents, of the solvable complexes concealed by the algae as well as by naturally happening nitrogenous composites	Abdel Halim (2014)
3.	Chloramine	Cyanogen chloride and Monochloramine		Na and Olson (2004)
4.	Chlorine dioxide	Chloride, chlorate, and chlorite	Reaction of sodium chlorite and chlorine	Qian et al. (2007)
5.	Ozone	CH_2O and other aldehydes, carboxylic acids, hydrogen peroxide, bromate, bromomethanes, brominated acetic acids, brominated acetonitriles, and ketones.		Marhaba and Bengraïne (2003)

several reasons, such as preservation, additives, and for colouring purposes. Antioxidants were likewise utilized which stop foodstuffs from reacting with oxygen, which can change the tastes or colours of the foods (Pereira de Abreu et al. 2012).

There are numerous possibly poisonous constituents in the environment that may contaminate foodstuffs consumed by customers. These include inorganic as well as organic constituents and might be devised after a wide range of sources. The food additives contain many of the toxic chemicals that can pollute freshwater by affecting its water quality. This can lead to affecting many aquatic organisms and their livelihood. Some of the food additive compounds and their effect on human health are accessible in Table 8.7.

8.3 Sources of Organic Contaminants in the Groundwater

Sources of organic pollutants can be classified in following ways:

1) Natural and anthropogenic sources
2) Point and Non-point sources

Table 8.7 Some of the food additive compounds with examples and their effect on human health.

Serial No.	Additives	Examples	Type of food	Effects on health	References
1.	Antioxidants	Ascorbic Acid	Fruits, Meat	Nausea, Vomiting, Heartburn, Headache	Stewart and Kim (2012)
2.	Colourings	Sunset yellow	Drinks, Sweets	Genetic damage	Kus and Eroglu (2015)
		Caramel	Drinks, Sweets, Soups	Hyperactivity, behavioural problems in children, cancer risk	Ramesh and Muthuraman (2018)
3.	Flavouring	Monosodium glutamate	Processed food	Asthma, headaches, and brain damage	Kazmi et al. (2017)
		Vanillin	Desert, Chocolate	Skin-related problems	Dervisoglu (2006)
4.	Preservatives	Sulphur dioxide	Juice	Asthma, allergy	DeChristopher et al. (2016)
		Sodium nitrate	Meat products	Damage to blood vessels	Larsen et al. (2010)

8.3.1 Natural and Anthropogenic Sources

Organic contamination includes all natural and synthetic contaminates that could cause adverse effects on human health or the environment. Sources of these organic contaminants can include pharmaceuticals, industries, agricultural activities, and daily household activities. On the basis of formation, the pollutants can be classified as natural or anthropogenic (produced by human activities).

8.3.1.1 Natural Sources

Several natural phenomena, physical processes, environmental activities, and biological processes are responsible for the formation of various kind of pollutants in the environment. These pollutants can include various kinds of natural organic matters produced, such as biological decomposition and transformation of dead plants and animal matter, and other environmental and physical activities like weathering of rocks and other atmospheric deposition processes viz., precipitation in the form of rain and snow, and movement of aerosols and fog from high altitude to lower altitude. Natural disasters, like floods and storms, can also be considered sources of organic contaminants of groundwater.

In groundwater, there are three main natural sources of organic matter: organic matter deposits such as buried peat, kerogen, and coal; soil and sediment organic matter; and organic matter present in waters infiltrating into the subsurface from rivers, lakes, and marine systems (Aiken and Kuniansky 2002). During the biological decomposition of the living matter several kinds of organic materials are produced. Collectively, these materials are called natural organic materials. These natural organic materials comprise thousands of organic compounds like humic and fulvic acids. These compounds upon leaching cause contamination in groundwater.

8.3.1.2 Anthropogenic Sources

These sources include the places from which pollutants are released due to human activities. These contaminants include synthetic organic compounds as well as VOCs. Synthetic organic compounds are widely used for various purposes, such as industrial manufacturing processes, human and animal healthcare, food production and processing activities, etc. The synthetic organic compound group includes herbicides, pesticides, and other chemicals that come from agriculture, urban stormwater runoff, or industrial activities. VOCs include chemical solvents or cleaners and their by-products. These compounds are derived from petroleum products and/or arise from industrial processes, such as carbon tetrachloride, tetrachloroethylene, trichloroethylene, etc.

The nature and sources of organic contaminants vary according to the processes and activities in the locality or region of the associated study area. For example, in an urban area the nature and source of organic contaminants include the production from sewage, wastewater treatment plants, and industrial activities. In a rural area, activities like agriculture and farming, wasteland applications, etc. take place. Due to these reasons the nature and source of contaminants produced by anthropogenic activities in rural area are different than the urban areas. However, natural phenomena are usually common in rural as well as urban areas, depicting the common nature and source of organic contaminants in both types of areas. A general view of organic contaminants of groundwater in natural and urban areas is given in Figure 8.3.

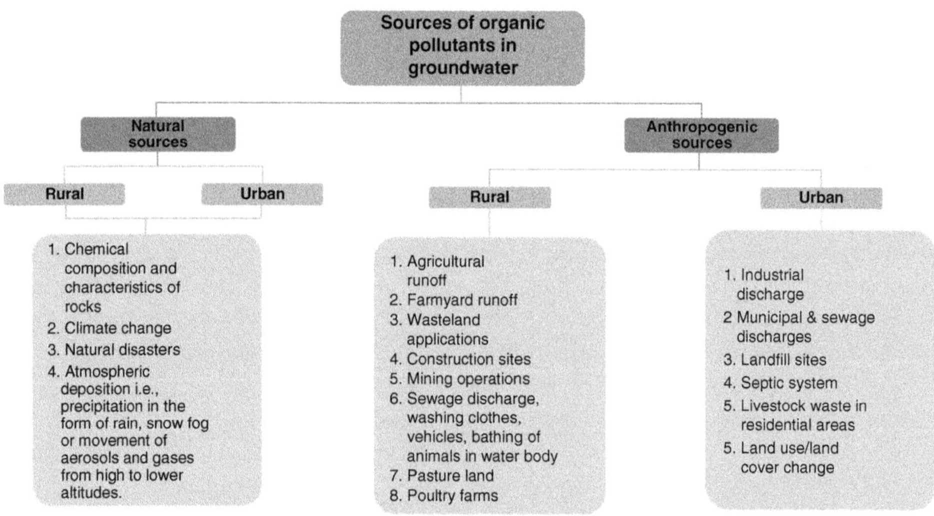

Figure 8.3 Source of organic pollutants in the groundwater.

8.3.2 Point and Non-Point Sources

The organic pollutants can be produced and mixed into the groundwater in several ways. Some of these may be easily identifiable as they are directly mixed into the water bodies, while others may be indirect or indistinguishable, i.e. diffused. On the basis of these facts the sources of contamination can be classified in point and non-point sources.

8.3.2.1 Point Sources

Point source pollution is defined as originating from discrete points, that is, a stationary location or fixed facility from which pollutants discharge or emit through a pipe, outfalls, and conveyance channels from either municipal wastewater treatment plants or individual waste treatment facilities (Lapworth et al. 2012).

The term 'point source' means any discernible, confined, and discrete conveyance, including but not limited to any pipe, ditch, channel, tunnel, conduit, well, discrete fissure, container, rolling stock, concentrated animal feeding operation, vessel, or other floating craft, from which pollutants are or may be discharged. This term does not include agricultural stormwater discharges and return flows from irrigated agriculture (United States Clean Water Act). Point sources are easier to find as the pollutants come from a definite and confined place. These sources include wastewater derived from domestic, industrial, or hospital premises and waste disposal sites, discharge pipes from oil refineries, and drainage ditches.

8.3.2.2 Non-Point Sources

Non-point source pollution is defined as originating from a non-discrete point, that is, a location which is not defined or not confined from which pollutants discharge. As the pollutants come from a non-confined place, i.e. from many places at once, non-point source pollution is harder to identify. Non-point source pollution takes place during precipitation (rainfall or snowmelt) that moves over the ground and gets mixed with natural and anthropogenic pollutants. As these pollutants move with the runoff, they leach to contaminate the groundwater.

Generally, agricultural land use is the most important source of non-point source pollution (Kronvang et al. 1995). During agricultural activities, several fertilizers and agrochemicals like pesticides and herbicides are applied to the croplands in order to increase the fertility. These fertilizers and agrochemicals leach into the groundwater and cause its organic contamination. Similarly, runoff from urban and agricultural areas and leakage from urban sewage systems are non-point sources of pollution (Trauth and Xanthopoulos 1997; Vázquez-Suñé et al. 2005). Other non-point sources include oil, grease, and toxic chemicals from urban runoff and energy production, sediments from improperly managed construction sites, crop and forest lands, and eroding streambanks. A brief overview of point and non-point sources of organic contaminates in groundwater is shown in Figure 8.4.

8.4 Fate of Organic Contaminants in the Groundwater

Transportation and fate of the subjected contaminant in any environment is generally controlled by the four general processes, namely advection, dispersion, interphase mass transfer, and transformation reactions. The fate of any definite pollutant in the

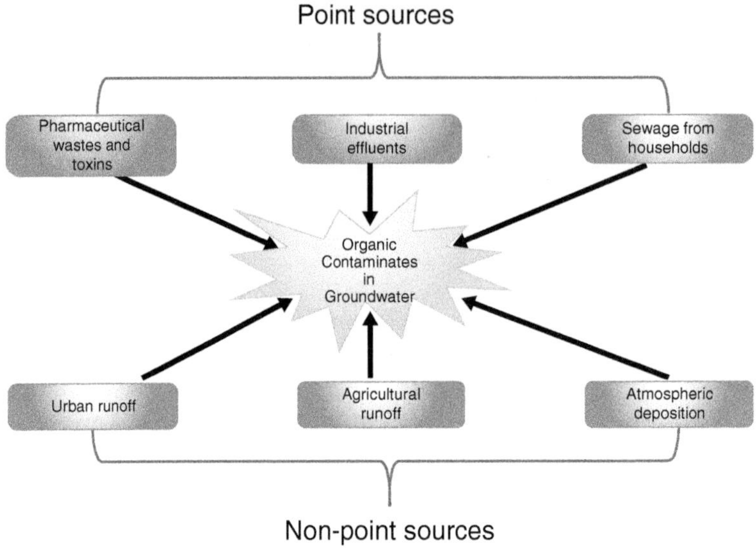

Figure 8.4 Point and non-point source of organic contaminants in the groundwater.

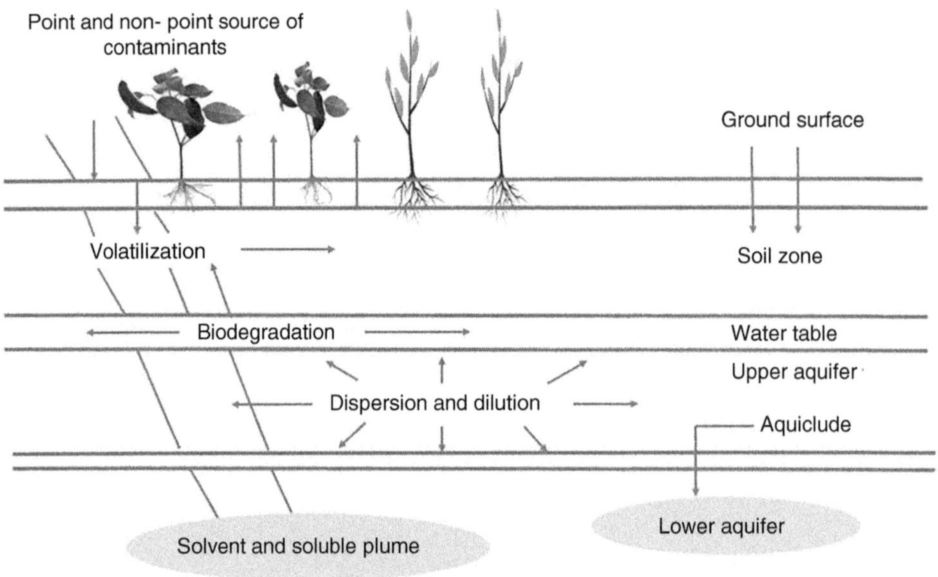

Figure 8.5 Fate of organic contaminants in the groundwater.

environment is a function of the collective influence of these four general processes (see Figure 8.5). The mutual influence of the four processes regulates the 'migration potential' and 'persistence' of a contaminant in the environment (Brusseau 2019). The spirit of any contaminant, hence its ability to infect the ways of its transportation in soil, water, or air, is fully controlled by these processes, or in other words the 'pollution potential'.

Various geochemical processes also influence the existence and the fate of organic Contaminants in groundwater of which adsorption, degradation, dilution, and transformation are the four main processes. Groundwater is not as much contaminated as other freshwater forms like rivers, ponds, streams, and lakes, and usually presents in lesser concentrations than influents and effluents from wastewater treatment plants, which are the chief cause of contamination to the groundwater environment. The basis of the presence of emerging organic contaminants into the groundwater resource are the process of sorption, generally to organic matter and clay minerals, and biological degradation or transformation (Lapworth et al. 2012).

For any contaminant to be dispersed in any environment mainly in the subsurface zone is influenced by its physio-chemical features like its solubilization potential in aqueous environment and other environmental characteristics (Wells 2007; Sedlak and Pinkston 2011). Undeniably, various other characteristic features of the contaminants like groundwater residence time, redox conditions, and total loading will be important in determining presence and persistence in the subsurface and groundwater.

The movement of organic compounds introduced to the soil surface will transfer through the soil zone to the unsaturated zone and finally into the saturated zone (Snyder et al. 2004; Zuehlke et al. 2004). The pathway of migration of the contaminants to the underground zone is controlled by the hydraulic/physical characteristics, sorption processes, and biodegradation controls which are elaborated in the next sections.

8.4.1 Physical Controls

The level of contamination is also influenced by the geological property of the zone of contamination. Groundwater contamination is affected by the degree of confinement of an aquifer, the hydraulic properties of the aquifer, and overlying soil (Worrall and Kolpin 2004).

8.4.2 Surface Sorption

The process of sorption can be defined as the contact of a contaminant with a solid, and it is divided into adsorption and absorption. The adsorption is the superficial contact between the solid surface and the contaminants on the other hand absorption refers to the deep penetration of the particles of contaminant within solids (Piwoni and Keeley 1990). There are numerous diverse aspects controlling the adhesion and penetration of contaminants into the solid surface of soil. These comprise chemical and physical features of the pollutant, composition of the exterior part of the solid, and the liquid media surrounding both. The level of contamination underneath the surface in both the saturated and unsaturated zones is lessened by the process of sorption into soil. Sorption of contaminants onto soil particles are well known for the chief role of soil carbon organic and clay for many molecules while sorption on geological constituents subterranean in the unsaturated zone and saturated zone has been not explored well (Casey et al. 2004; Wang et al. 2009; Olshansky et al. 2011).

8.4.3 Biodegradation

Literature suggests that it is also known that microorganisms degrade contaminants in the soil zone (Pedro and Illman 2006; Semple et al. 2007; Wick et al. 2007). However, in groundwater, microorganisms are not abundant and perhaps these are very less diversified; there are also variable redox reactions that may limit biological activity (Johnson et al. 1998). The rate of degradation of pollutants are quicker under aerobic conditions when related to anaerobic conditions (Barnes et al. 2004; Godfrey et al. 2007; Watanabe et al. 2008). Reducing environments have been shown to be vital in regulating the transport and fate of some organic complexes with high reduction potential (Williams et al. 1984; Holm et al. 1995).

8.5 Health Effects of Organic Contaminants

Numerous anthropogenic actions in our surrounding locality contaminate the environment where lethal toxic and health menacing contaminants are released into the soil, water, and air. Actions in filling stations and probable leaky subsurface storage tanks can frequently release compounds that also contaminate groundwater since the products available in these stations are composite blends of uncountable complexes, chiefly hydrocarbons. Examples of these contaminants are classified as VOCs, oxygenates, heavy metals, polycyclic aromatic hydrocarbon (PAH), phenols, chlorobenzenes, phthalates, and organochlorides.

The causes of these pollutants may be geogenic and anthropogenic in origin, but some common sources are leaching of rusted USTs (underground storage tanks), emulsion paints, laundry detergents in car wash and lube bays within stations, lubricants, petroleum, and engine fluids, and some described health impacts are brain damage, cancer, tumours, neonatal defects, anaemia, central and peripheral nervous system breakdown, damage to the kidney, liver, bones, autism, Crohn's disease, uremia, ulcerative colitis, celiac disease, skin, eye, and nose irritation, depression of the central nervous system in humans, headache, dizziness, narcosis, and death at high levels of exposure (Odipe et al. 2020)

8.5.1 Possible Health Impacts of some VOCs Found in the Groundwater Samples

- **1, 2 benzisothiazol-3-amine**
  Dermal exposure can cause skin sensitization and allergic contact dermatitis; also reported cases of asthma when inhaled (Kaur-Knudsen et al. 2012; Odipe et al. 2020).
- **Nonanal**
  Autism, Crohn's disease, uremia, ulcerative colitis, Celiac disease (Odipe et al. 2020; James et al. 2004; Francavilla et al. 2014)
- **Phthalic acid, isobutyl octyl ester**
  Hormone disruption that affects development and fertility (Koniecki et al. 2011)
- **Methyl palmitate**
  Allergic skin reaction, skin irritation (OSHA 2013)

Table 8.8 Organic contaminants and their effects on human beings.

Organic contaminants	Possible health effect	References
Benzene	Skin inflammation, eye irritation	European Chemicals Agency (ECHA) (2015b)
1,2 - Bis (trimethylsilyl) benzene	Causes annoyance in eyes	European Chemicals Agency (ECHA) (2015c)
trans-2-Decenoic acid	Sensation of skin burning	European Chemicals Agency (ECHA) (2015d)
n-Nonycyclohexane	Damage in CNS, headache	National Toxicology Program (2014) USEPA/IRIS (2015)
2-octenal, (E)	Damages eyes	
Heptanophenone	Uneasiness in respiration	
Napthalene, 2-methyl - Pento barbital	Vomiting, diarrhoea, anaemia, jaundice, euphoria, dermatitis	UNECE (2017)
Napthalene,1,6,7-trimethyl	Harmful for aquatic organisms	
Methoxyacetic acid, 4-tetradecyl ester	Irritates skin	

- **1, 6 - octadien-3-ol,3,7,-dimethyl**
  Skin, eye, and nose irritation (PubChem 2016)
- **m-fluoroanisol**
  Flammable liquid and vapour (OSHA 2013)
- **2-octenal, (E)**
  Eye damage, respiratory and skin irritation (OSHA 2013)
  See Table 8.8 for some organic contaminants and their possible effects on human beings.

8.6 Conclusion

One of the major problems that the twenty-first century is facing is environmental pollution, and research communities are paying close attention. The contamination of organic substances is posing threats to human health and harming our environment, so there is an immediate need to resolve this problem with efficient solutions. Cleaning up large-scale water pollution from mining activities and groundwater contamination from waste sites requires science-based decisions that take into account the specific hydrological conditions, the microbial and geochemical transformation pathways, and possible remediation technologies to choose the most effective strategies. Despite the anticipated advances in water treatment technologies, efforts to reduce introduction of problematic chemicals into the (aquatic) environment should be given highest priority. This requires the improvement of the scientific tools to identify those existing chemicals that need to be substituted and phased out, and political support will be needed to enforce such action. This chapter analyses the occurrence of groundwater contaminants, including natural and manmade.

Contaminants found in groundwater cover a broad range of physical, inorganic, and organic chemicals and radioactive parameters. The increasing trends of industrial effluents including municipal sewage are the main sources for contaminants, chiefly organic pollutants, in groundwater. The present study discusses the types and sources of contaminants, their movement, and remedial measures to stop further groundwater contamination in India. There is also a need to establish low-cost remediation methods to enable more reliable remediation to solve groundwater contamination problems.

References

Adeel, M., Song, X., Wang, Y. et al. (2017). Environmental impact of estrogens on human, animal and plant life: a critical review. *Environment International* 99: 107–119.

Abdel Halim, N.H., 2014. Fate of natural organic matter and formation of disinfection by-products in a conventional water treatment plant.

Abdulmumeen, H.A., Risikat, A.N., and Sururah, A.R. (2012). Food: its preservatives, additives and applications. *International Journal of Chemical and Biochemical Sciences* 1 (2012): 36–47.

Afrin, S., Giampieri, F., Gasparrini, M. et al. (2016). Chemopreventive and therapeutic effects of edible berries: a focus on colon cancer prevention and treatment. *Molecules* 21 (2): 169.

Aiken, G.R. and Kuniansky, E.L., 2002. US Geological Survey Artificial Recharge Workshop Proceedings: April 2–4, 2002, Sacramento, California. US Department of the Interior, US Geological Survey.

Andaluri, G., Suri, R.P., and Kumar, K. (2012). Occurrence of estrogen hormones in biosolids, animal manure and mushroom compost. *Environmental Monitoring and Assessment* 184 (2): 1197–1205.

Barnes, K.K., Christenson, S.C., Kolpin, D.W. et al. (2004). Pharmaceuticals and other organic waste water contaminants within a leachate plume downgradient of a municipal landfill. *Ground Water Monitoring and Remediation* 24: 119–126.

Barnes, K.K., Kolpin, D.W., Focazio, M.J., et al. 2008. Water-quality data for pharmaceuticals and other organic wastewater contaminants in ground water and in untreated drinking water sources in the United States, 2000–01. U. S. Geological Survey.

Bassil, K.L., Vakil, C., Sanborn, M. et al. (2007). Cancer health effects of pesticides: systematic review. *Canadian Family Physician* 53 (10): 1704–1711.

Bi, G., Scagel, C.F., Cheng, L., and Fuchigami, L.H. (2005). Effects of copper, zinc and urea on defoliation and nitrogen reserves in nursery plants of almond. *The Journal of Horticultural Science and Biotechnology* 80 (6): 746–750.

Bolli, A. and Marino, M. (2011). Current and future development of estrogen receptor ligands: applications in estrogen-related cancers. *Recent Patents on Endocrine, Metabolic & Immune Drug Discovery* 5 (3): 210–229.

Boots, A.W., van Berkel, J.J., Dallinga, J.W. et al. (2012). The versatile use of exhaled volatile organic compounds in human health and disease. *Journal of Breath Research* 6 (2): 027108.

Brausch, J.M. and Rand, G.M. (2011). A review of personal care products in the aquatic environment: environmental concentrations and toxicity. *Chemosphere* 82 (11): 1518–1532.

Brusseau, M.L. (2019). Physical processes affecting contaminant transport and fate. In: *Environmental and Pollution Science* (eds. M.L. Brusseau, I.L. Pepper and C. Gerba), 103–112. Cambridge, MA: Academic Press.

Cabeza, Y., Candela, L., Ronen, D., and Teijon, G. (2012). Monitoring the occurrence of emerging contaminants in treated wastewater and groundwater between 2008 and 2010. The BaixLlobregat (Barcelona, Spain). *Journal of Hazardous Materials* 239: 32–39.

Casey, F.X., Hakk, H., Šimůnek, J., and Larsen, G.L. (2004). Fate and transport of testosterone in agricultural soils. *Environmental Science & Technology* 38 (3): 790–798.

Chander, V., Sharma, B., Negi, V. et al. (2016). Pharmaceutical compounds in drinking water. *Journal of Xenobiotics* 6 (1).

Chapelle, F.H. (1999). Bioremediation of petroleum hydrocarbon-contaminated ground water: the perspectives of history and hydrology. *Groundwater* 37 (1): 122–132.

Chen, T.S., Chen, T.C., Yeh, K.J.C. et al. (2010). High estrogen concentrations in receiving river discharge from a concentrated livestock feedlot. *Science of the Total Environment* 408 (16): 3223–3230.

Churchill, J.E., Ashley, D.L., and Kaye, W.E. (2001). Recent chemical exposures and blood volatile organic compound levels in a large population-based sample. *Archives of Environmental Health: An International Journal* 56 (2): 157–166.

Damicone, J.P., 2014. Fungicide resistance management.

Daughton, C.G. and Ternes, T.A. (1999). Pharmaceuticals and personal care products in the environment: agents of subtle change? *Environmental Health Perspectives* 107 (suppl 6): 907–938.

Dauti, M., Alili-Idrizi, E., Ahmeti-Lika, S., and Malaj, L. (2014). Legal regulation and critical analysis for an effectively treatment of pharmaceutical waste. *International Journal of Social, Behavioral, Educational, Economic, Business and Industrial Engineering* 8 (11): 3660–3667.

DeChristopher, L.R., Uribarri, J., and Tucker, K.L. (2016). The link between soda intake and asthma: science points to the high-fructose corn syrup, not the preservatives: a commentary. *Nutrition & Diabetes* 6 (11): e234–e234.

Dervisoglu, M. (2006). Influence of hazelnut flour and skin addition on the physical, chemical and sensory properties of vanilla ice cream. *International Journal of Food Science & Technology* 41 (6): 657–661.

Duesterberg, B., Endrikat, J. and Schuermann, R., 2005. Extended use combination comprising estrogens and progestins. U.S. Patent Application 10/721,825.

European Chemicals Agency (ECHA) (2015b). European Chemicals Agency (ECHA) and data classification and on azidobenzene. Website: https://echa.europa.eu/informationon-chemicals/cl-inventory-database/-/discli/details/134767.

European Chemicals Agency (ECHA) (2015c). European Chemicals Agency (ECHA) and data classification and on 1,2 - Bis (trimethylsilyl) benzene. Website: https://echa.europa.eu/information-on-chemicals/cl-inventorydatabase/-/discli/details/191884.

European Chemicals Agency (ECHA) (2015d). European Chemicals Agency (ECHA) and data classification and on trans-2-Decenoic acid. Website: https://echa.europa.eu/information-on-chemicals/cl-inventorydatabase/-/discli/details/37727

European Commission (EC). Directive 2006/118/EC of the European Parliament and the Council of 12th of December 2006 on the protection of ground water against pollution and deterioration. Off J Europ Union 2006; L 372/19. 27/12/2006.

Ediz, N., Bentli, İ., and Tatar, İ. (2010). Improvement in filtration characteristics of diatomite by calcination. *International Journal of Mineral Processing* 94 (3–4): 129–134.

Foster, S.S. (2001). The interdependence of groundwater and urbanisation in rapidly developing cities. *Urban Water* 3 (3): 185–192.

Francavilla, R., Ercolini, D., Piccolo, M. et al. (2014). Salivary microbiota and metabolome associated with celiac disease. *Applied and Environmental Microbiology* 80 (11): 3416–3425.

Flather, C.H., Knowles, M.S., and Kendall, I.A. (1998). Threatened and endangered species geography. *BioScience* 48 (5): 365–376.

Fujita, E.M. and Campbell, D.E., 2003. Validation and Application Protocol for Source Apportionment of Photochemical Assessment Monitoring Stations (PAMS) Ambient Volatile Organic Compouund (VOC) Data.

Gopal, K., Tripathy, S.S., Bersillon, J.L., and Dubey, S.P. (2007). Chlorination byproducts, their toxicodynamics and removal from drinking water. *Journal of Hazardous Materials* 140 (1–2): 1–6.

Gan, W., Sharma, V.K., Zhang, X. et al. (2015). Investigation of disinfection byproducts formation in ferrate (VI) pre-oxidation of NOM and its model compounds followed by chlorination. *Journal of Hazardous Materials* 292: 197–204.

Gibbs, R.B. and Gabor, R. (2003). Estrogen and cognition: applying preclinical findings to clinical perspectives. *Journal of Neuroscience Research* 74 (5): 637–643.

Godfrey, E., Woessner, W.W., and Benotti, M.J. (2007). Pharmaceuticals in on-site sewage effluent and ground water, western Montana. *Groundwater* 45 (3): 263–271.

Gupta, R.C. (ed.) (2011). *Toxicology of Organophosphate and Carbamate Compounds*. Academic Press.

Gupta, P.K. (2018). Toxicity of herbicides. In: *Veterinary Toxicology* (ed. R.C. Gupta), 553–567. Cambridge, MA: Academic Press.

Hamid, H. and Eskicioglu, C. (2012). Fate of estrogenic hormones in wastewater and sludge treatment: a review of properties and analytical detection techniques in sludge matrix. *Water Research* 46 (18): 5813–5833.

Hanson, M.A., Dostálová, A., Ceroni, C. et al. (2019). Synergy and remarkable specificity of antimicrobial peptides in vivo; using a systematic knockout approach. *eLife* 8: e44341.

Holm, J.V., Rügge, K., Bjerg, P.L., and Christensen, T.H. (1995). Occurrence and distribution of pharmaceutical organic compounds in the groundwater downgradient of a landfill (Grindsted, Denmark). *Environmental Science & Technology* 29 (5): 1415–1420.

Houtman, C.J. (2010). Emerging contaminants in surface waters and their relevance for the production of drinking water in Europe. *Journal of Integrative Environmental Sciences* 7 (4): 271–295.

Huerta, B., Rodríguez-Mozaz, S., and Barceló, D. (2012). Pharmaceuticals in biota in the aquatic environment: analytical methods and environmental implications. *Analytical and Bioanalytical Chemistry* 404 (9): 2611–2624.

Hundža, M., Janiga, M., Solár, J., and Gondová, B. (2019). Water quality of the river Váh-Ružomberok (Slovakia), experience after 35-year water treatment. *Oecologia Montana* 28 (1): 1–6.

Hwang, E., Lee, D.G., Park, S.H. et al. (2014). Coriander leaf extract exerts antioxidant activity and protects against UVB-induced photoaging of skin by regulation of procollagen type I and MMP-1 expression. *Journal of Medicinal Food* 17 (9): 985–995.

Iglesias, J., Castillejo, J., and Castro, R. (2003). The effects of repeated applications of the molluscicidemetaldehyde and the biocontrol nematode Phasmarhabditishermaphrodita on molluscs, earthworms, nematodes, acarids and collembolans: a two year study in north-west Spain. *Pest Management Science: formerly Pesticide Science* 59 (11): 1217–1224.

James, S.J., Cutler, P., Melnyk, S. et al. (2004). Metabolic biomarkers of increased oxidative stress and impaired methylation capacity in children with autism. *The American Journal of Clinical Nutrition* 80 (6): 1611–1617.

Johnson, A.C., Hughes, C.D., Williams, R.J., and Chilton, P.J. (1998). Potential for aerobic isoproturon biodegradation and sorption in the unsaturated and saturated zones of a chalk aquifer. *Journal of Contaminant Hydrology* 30 (3–4): 281–297.

Judson, R., Richard, A., Dix, D.J. et al. (2009). The toxicity data landscape for environmental chemicals. *Environmental Health Perspectives* 117 (5): 685–695.

Kapaj, S., Peterson, H., Liber, K., and Bhattacharya, P. (2006). Human health effects from chronic arsenic poisoning–a review. *Journal of Environmental Science and Health, Part A* 41 (10): 2399–2428.

Kargar, M., Nadafi, K., Nabizadeh, R. et al. (2013). Survey of hazardous organic compounds in the groundwater, air and wastewater effluents near the Tehran automobile industry. *Bulletin of Environmental Contamination and Toxicology* 90 (2): 155–159.

Kaur, G. (2019). Herbicides and its role in induction of oxidative stress-a review. *International Journal of Environment, Agriculture and Biotechnology* 4 (4): 995–1004.

Kaur-Knudsen, D., Menné, T., and Christina Carlsen, B. (2012). Systemic allergic dermatitis following airborne exposure to 1, 2-benzisothiazolin-3-one. *Contact Dermatitis* 67 (5): 310–312.

Kazmi, Z., Fatima, I., Perveen, S., and Malik, S.S. (2017). Monosodium glutamate: review on clinical reports. *International Journal of Food Properties* 20 (sup2): 1807–1815.

King, P.L. (2019). *Lean for the Process Industries: Dealing with Complexity*. CRC Press.

Khashaba, P.Y. (2002). Spectrofluorimetric analysis of certain macrolide antibiotics in bulk and pharmaceutical formulations. *Journal of Pharmaceutical and Biomedical Analysis* 27 (6): 923–932.

Klaminder, J., Brodin, T., Sundelin, A. et al. (2015). Long-term persistence of an anxiolytic drug (oxazepam) in a large freshwater lake. *Environmental Science & Technology* 49 (17): 10406–10412.

Koniecki, D., Wang, R., Moody, R.P., and Zhu, J. (2011). Phthalates in cosmetic and personal care products: concentrations and possible dermal exposure. *Environmental Research* 111 (3): 329–336.

Kronvang, B., Grant, R., Larsen, S.E. et al. (1995). Non-point-source nutrient losses to the aquatic environment in Denmark: impact of agriculture. *Marine and Freshwater Research* 46 (1): 167–177.

Kumar, A., Bisht, B.S., Joshi, V.D. et al. (2010). Physical, chemical and bacteriological study of water from rivers of Uttarakhand. *Journal of Human Ecology* 32 (3): 169–173.

Kümmerer, K. (ed.) (2008). *Pharmaceuticals in the Environment: Sources, Fate, Effects and Risks*. Springer Science & Business Media.

Kurwadkar, S. (2019). Occurrence and distribution of organic and inorganic pollutants in groundwater. *Water Environment Research* 91 (10): 1001–1008.

Kus, E. and Eroglu, H.E. (2015). Genotoxic and cytotoxic effects of sunset yellow and brilliant blue, colorant food additives, on human blood lymphocytes. *Pakistan Journal of Pharmaceutical Sciences* 28 (1): 227–230.

Lapworth, D.J., Baran, N., Stuart, M.E., and Ward, R.S. (2012). Emerging organic contaminants in groundwater: a review of sources, fate and occurrence. *Environmental Pollution* 163: 287–303.

Larsen, F.J., Weitzberg, E., Lundberg, J.O., and Ekblom, B. (2010). Dietary nitrate reduces maximal oxygen consumption while maintaining work performance in maximal exercise. *Free Radical Biology and Medicine* 48 (2): 342–347.

Lindberg, C.M., Melathopoulos, A.P., and Winston, M.L. (2000). Laboratory evaluation of miticides to control Varroajacobsoni (Acari: Varroidae), a honey bee (Hymenoptera: Apidae) parasite. *Journal of Economic Entomology* 93 (2): 189–198.

Loos, R., Gawlik, B.M., Locoro, G. et al. (2009). EU-wide survey of polar organic persistent pollutants in European river waters. *Environmental Pollution* 157 (2): 561–568.

Loos, R., Locoro, G., Comero, S. et al. (2010). Pan-European survey on the occurrence of selected polar organic persistent pollutants in ground water. *Water Research* 44 (14): 4115–4126.

Marhaba, T.F. and Bengraïne, K. (2003). Review of strategies for minimizing bromate formation resulting from drinking water ozonation. *Clean Technologies and Environmental Policy* 5 (2): 101–112.

Mastelić, J., Grzunov, K., and Kravar, A. (1992). The chemical composition of Terpene alcohols and phenols from the essential oil and Terpene glycosides isolated from Thymus pulegioides growing wild in Dalmatia. *PlantaMedica* 58 (S 1): 678–680.

Mayer, J., Oxford & Hill Home Products, 2007. Packaging for chemical moth and insect repellent products. U.S. Patent Application 11/536,936.

Meffe, R., Kohfahl, C., Hamann, E. et al. (2014). Fate of Para-toluenesulfonamide (p-TSA) in groundwater under anoxic conditions: modelling results from a field site in Berlin (Germany). *Environmental Science and Pollution Research* 21 (1): 568–583.

Monte, M.C., Fuente, E., Blanco, A., and Negro, C. (2009). Waste management from pulp and paper production in the European Union. *Waste Management* 29 (1): 293–308.

Monteiro, S.C. and Boxall, A.B. (2010). Occurrence and fate of human pharmaceuticals in the environment. In: *Reviews of Environmental Contamination and Toxicology* (ed. D.M. Whitacre), 53–154. New York, NY: Springer.

Mondal, K.A.M.S.H. and Parween, S. (2000). Insect growth regulators and their potential in the management of stored-product insect pests. *Integrated Pest Management Reviews* 5 (4): 255–295.

Na, C. and Olson, T.M. (2004). Stability of cyanogen chloride in the presence of free chlorine and monochloramine. *Environmental Science & Technology* 38 (22): 6037–6043.

National Research Council (2004). *Water and Sustainable Development: Opportunities for the Chemical Sciences: A Workshop Report to the Chemical Sciences Roundtable.* National Academies Press.

National Toxicology Program 2014. Testing Status of Agents at NTP. Available from, as of Jul 31, 2014: http://ntp.niehs.nih.gov.

Niemelä, K. and Alén, R. (1999). Characterization of pulping liquors. In: *Analytical Methods in Wood Chemistry, Pulping, and Papermaking* (eds. E. Sjöström and R. Alén), 193–231. Berlin, Heidelberg: Springer.

Olshansky, Y., Polubesova, T., Vetter, W., and Chefetz, B. (2011). Sorption–desorption behavior of polybrominateddiphenyl ethers in soils. *Environmental Pollution* 159 (10): 2375–2379.

Odipe, O.E., Sawyerr, H.O., and Adewoye, S.O. (2020). Characterized organic pollutants and their health effects in sampled groundwater around Ilorin Metropolis. *International Journal of Environmental Protection and Policy* 8 (2): 36.

Osenbrück, K., Gläser, H.R., Knöller, K. et al. (2007). Sources and transport of selected organic micropollutants in urban groundwater underlying the city of Halle (Saale), Germany. *Water Research* 41 (15): 3259–3270.

OSHA, U., 2013. Globally harmonized system of classification and labelling of chemicals (GHS). United Nations Economic Commission for Europe (UNECE), 224.

Pedro, J.A. and Illman, W.A. (2006). *Bioremediation and Natural Attenuation: Process Fundamentals and Mathematical Models*. Wiley.

Pereira de Abreu, D.A., Cruz, J.M., and PaseiroLosada, P. (2012). Active and intelligent packaging for the food industry. *Food Reviews International* 28 (2): 146–187.

Perry, A.S., Yamamoto, I., Ishaaya, I., and Perry, R. (1998). Rodenticides. In: *Insecticides in Agriculture and Environment* (eds. A.A.S. Perry, I. Yamamoto, I. Ishaaya and R. Perry), 181–186. Berlin, Heidelberg: Springer.

Petrović, N.Z., Zahreddine, H., and Belić, M.R. (2011). Exact spatiotemporal wave and soliton solutions to the generalized (3+1)-dimensional nonlinear Schrödinger equation with linear potential. *PhysicaScripta* 83 (6): 065001.

Phattarapattamawong, S. (1999). Control of chlorinous odor in drinking. *Chemosphere* 39: 55–70.

Piwoni, M.D. and Keeley, J.W., 1990. Basic concepts of contaminant sorption at hazardous waste sites (ground-water issue) (No. PB-91-191007/XAB; EPA--540/4-90/053). Environmental Protection Agency, Ada, OK (United States). Robert S. Kerr Environmental Research Lab.

PubChem, (2016). National Center for Biotechnology Information. PubChem Database. Linalool, CID=6549, https://pubchem.ncbi.nlm.nih.gov/compound/Linalool (accessed on Nov. 12, 2019).

Qian, Y., Chen, Y., Jiang, Y., and Zhang, L. (2007). A clean production process of sodium chlorite from sodium chlorate. *Journal of Cleaner Production* 15 (10): 920–926.

Ramesh, M. and Muthuraman, A. (2018). Flavoring and coloring agents: health risks and potential problems. In: *Natural and Artificial Flavoring Agents and Food Dyes* (ed. A.M. Grumezescu), 1–28. Academic Press.

Richardson, S.D. (2008). Environmental mass spectrometry: emerging contaminants and current issues. *Analytical Chemistry* 80 (12): 4373–4402.

Richardson, S.D. and Postigo, C. (2011). Drinking water disinfection by-products. In: *Emerging Organic Contaminants and Human Health* (ed. A.M. Holban), 93–137. Berlin, Heidelberg: Springer.

Richardson, S.D. and Ternes, T.A. (2011). Water analysis: emerging contaminants and current issues. *Analytical Chemistry* 83 (12): 4614–4648.

Richmond, E.K., Grace, M.R., Kelly, J.J. et al. (2017). Pharmaceuticals and personal care products (PPCPs) are ecological disrupting compounds (EcoDC). *Elementa: Science of the Anthropocene* 5: 1–8.

Sanderson, H. (2011). Presence and risk assessment of pharmaceuticals in surface water and drinking water. *Water Science and Technology* 63 (10): 2143–2148.

Santora, M., 2018. Explosion in Apartment Highlights Risks of Using Chemical Foggers to Kill Insects. The New York Times. October, 19.

Schaeck, M., De Swaef, E., Van Den Broeck, W. et al. (2016). Germ-Free Sea bass Dicentrarchuslabrax larval model: a valuable tool in the study of host-microbe interactions. *Diseases of Aquatic Organisms* 117 (3): 177–185.

Schröder, T., Mouradian, S.L., Zheng, J. et al. (2016). Quantum nanophotonics in diamond. *JOSA B* 33 (4): B65–B83.

Slack, R.J., Gronow, J.R., and Voulvoulis, N. (2005). Household hazardous waste in municipal landfills: contaminants in leachate. *Science of the Total Environment* 337 (1–3): 119–137.

Scott, J.G. (1990). Investigating mechanisms of insecticide resistance: methods, strategies, and pitfalls. In: *Pesticide Resistance in Arthropods* (eds. R.T. Roush and B. Tabashnik), 39–57. Boston, MA: Springer.

Sedlak, D.L. and Pinkston, K.E. (2011). Factors affecting the concentrations of pharmaceuticals released to the aquatic environment. *Journal of Contemporary Water Research and Education* 120 (1): 7.

Semple, K.T., Doick, K.J., Wick, L.Y., and Harms, H. (2007). Microbial interactions with organic contaminants in soil: definitions, processes and measurement. *Environmental Pollution* 150 (1): 166–176.

Shemwell, B.E., Ergut, A., and Levendis, Y.A. (2002). Economics of an integrated approach to control SO2, NOx, HCl, and particulate emissions from power plants. *Journal of the Air & Waste Management Association* 52 (5): 521–534.

Sherlock, G., Block, W., and Benson, E.E. (2005). Thermal analysis of the plant encapsulation-dehydration cryopreservation protocol using silica gel as the desiccant. *CryoLetters* 26 (1): 45–54.

Simpson, H.M., Simpson H.M, 1988. Insect repellent cuffs for marine mooring lines. U.S. Patent 4,769,943.

Snyder, S.A., Leising, J., Westerhoff, P. et al. (2004). Biological and physical attenuation of endocrine disruptors and pharmaceuticals: implications for water reuse. *Groundwater Monitoring & Remediation* 24 (2): 108–118.

Soto, M.L., Falqué, E., and Domínguez, H. (2015). Relevance of natural phenolics from grape and derivative products in the formulation of cosmetics. *Cosmetics* 2 (3): 259–276.

Srirattana, S., Piaowan, K., Lowry, G.V., and Phenrat, T. (2017). Electromagnetic induction of foam-based nanoscalezerovalent iron (NZVI) particles to thermally enhance non-aqueous phase liquid (NAPL) volatilization in unsaturated porous media: proof of concept. *Chemosphere* 183: 323–331.

Suciu, G., Butca, C., Conu, R., et al. 2016, October. Rapid detection of pesticide residues based on telemetry platform. In 2016 12th IEEE International Symposium on Electronics and Telecommunications (ISETC) (pp. 95–98). IEEE.

Stewart, A.F. and Kim, E.D. (2012). Harmful effects of antioxidant therapy. In: *Male Infertility* (eds. S.J. Parekattil and A. Agarwal), 499–506. New York, NY: Springer.

Teijon, G., Candela, L., Tamoh, K. et al. (2010). Occurrence of emerging contaminants, priority substances (2008/105/CE) and heavy metals in treated wastewater and groundwater at Depurbaix facility (Barcelona, Spain). *Science of the Total Environment* 408 (17): 3584–3595.

Trauth, R. and Xanthopoulos, C. (1997). Non-point pollution of groundwater in urban areas. *Water Research* 31 (11): 2711–2718.

Trombulak, S.C. and Frissell, C.A. (2000). Review of ecological effects of roads on terrestrial and aquatic communities. *Conservation Biology* 14 (1): 18–30.

UNECE, 2017. About the GHS, Globally Harmonized System of Classification and Labelling of Chemicals (GHS). Available online: https://www.unece.org/trans/danger/publi/ghs/ghs_welcome_e.html.

USEPA/IRIS, 2015. Integrated Risk Information System. Available from, as of Jul 31, 2014: http://www.epa.gov/iris.

Vázquez-Suñé, E., Sánchez-Vila, X., and Carrera, J. (2005). Introductory review of specific factors influencing urban groundwater, an emerging branch of hydrogeology, with reference to Barcelona, Spain. *Hydrogeology Journal* 13 (3): 522–533.

Vázquez-Suñé, E., Carrera, J., Tubau, I., et al. 2010. An approach to identify urban groundwater recharge.

Valcárcel, Y., Alonso, S.G., Rodríguez-Gil, J.L. et al. (2011). Detection of pharmaceutically active compounds in the rivers and tap water of the Madrid Region (Spain) and potential ecotoxicological risk. *Chemosphere* 84 (10): 1336–1348.

Wang, C., Ding, Y., Teppen, B.J. et al. (2009). Role of interlayer hydration in lincomycin sorption by smectite clays. *Environmental Science & Technology* 43 (16): 6171–6176.

Ware, G.W. and Whitacre, D.M., 2004. An introduction to insecticides. The pesticide book, 6.

Watanabe, N., Harter, T.H., and Bergamaschi, B.A. (2008). Environmental occurrence and shallow ground water detection of the antibiotic monensin from dairy farms. *Journal of Environmental Quality* 37 (S5): S78–S85.

Water, S. and World Health Organization, 2002. Emerging issues in water and infectious disease. World Health Organization. 1–24.

Wick, L.Y., Remer, R., Würz, B. et al. (2007). Effect of fungal hyphae on the access of bacteria to phenanthrene in soil. *Environmental Science & Technology* 41 (2): 500–505.

Williams, G.M., Ross, C.A.M., Stuart, A. et al. (1984). Controls on contaminant migration at the Villa Farm Lagoons. *Quarterly Journal of Engineering Geology and Hydrogeology* 17 (1): 39–55.

Wells, M.J. (2007). Log DOW: key to understanding and regulating wastewater-derived contaminants. *Environmental Chemistry* 3 (6): 439–449.

Wojtowicz, J.A., 2000. Chlorine Oxygen Acids and Salts, Dichlorine Monoxide, Hypochlorous Acid, and Hypochlorites. Kirk-Othmer Encyclopedia of Chemical Technology.

Wolf, L., Held, I., Eiswirth, M., and Hötzl, H. (2004). Impact of leaky sewers on groundwater quality. *Actahydrochimicahydrobiologica* 32 (4–5): 361–373.

Worrall, F. and Kolpin, D.W. (2004). Aquifer vulnerability to pesticide pollution—combining soil, land-use and aquifer properties with molecular descriptors. *Journal of Hydrology* 293 (1–4): 191–204.

Welter, S., Pickel, C., Millar, J. et al. (2005). Pheromone mating disruption offers selective management options for key pests. *California Agriculture* 59 (1): 16–22.

Ying, Q.L., Nichols, J., Evans, E.P., and Smith, A.G. (2002). Changing potency by spontaneous fusion. *Nature* 416 (6880): 545–548.

Zhao, L.J., Zhang, Y.F., Hong, J.F., and Tu, W.W. (2014). Papermaking wastewater treatment-a brief review. In: *Advanced Materials Research*, vol. 926 (eds. P. Xu, H. Si, Y. Wang and P. Wang), 4276–4279. Trans Tech Publications Ltd.

Zuehlke, S., Duennbier, U., Heberer, T., and Fritz, B. (2004). Analysis of endocrine disrupting steroids: investigation of their release into the environment and their behavior during bank filtration. *Groundwater Monitoring & Remediation* 24 (2): 78–85.

9

Organic Pollutants of Global Concern in Groundwater Resources and Remediation Measures

Majid Peyravi and Parisa Nikpour

Department of Chemical Engineering, Babol Noshirvani University of Technology, Babol, Iran

Nomenclature

OPs	Organic pollutants	PBDEs	Polybrominated diphenyl ethers
PCPs	Personal care products	WTPs	Water treatment plants
WWTPs	Wastewater treatment plants	ATPs	Advanced treatment processes
ng/L	Nanogram per litre	PPCPs	Pharmaceutical and personal care products
TCS	Triclosan	AC	Activated carbon
AHTN	7-acetyl-1,1,3,4,4,6-hexamethyl-1,2,3,4-tetrahydronaphthalene	GAC	Granular activated carbon
HHCB	hexahydro-4,6,6,7,8,8-hexamethylcyclopenta-γ-2-benzo-pyran	PAC	Powdered activated carbon
UV	Ultraviolet	CNTs	Carbon nanotubes
DEET	N, N-diethyl-meta-toluamide	AOPs	Advanced Oxidation Processes
mg/L	Milligram per litre	V	Volt
PCE	Tetrachloroethylene	MF	Microfiltration
TCE	Trichloroethylene	UF	Ultrafiltration
1,1,1-TCA	1,1,1-trichloroethane	NF	Nanofiltration
1,1-DCA	1,1-dichloroethane	RO	Reverse osmosis
1,1-DCE	1,1-dichloroethylene	FO	Forward osmosis
PHs	Petroleum hydrocarbons	MD	Membrane distillation
PAHs	Polyaromatic hydrocarbons	MWCO	Molecular weight cutoff

Groundwater Geochemistry: Pollution and Remediation Methods, First Edition.
Edited by Sughosh Madhav and Pardeep Singh.

MTBE	Methyl tertiary-butyl ether	Da	Dalton
THMs	Trihalomethanes	nm	Nanometre
PFASs	Perfluoroalkyl and polyfluoroalkyl substances	MBR	Membrane bioreactor
PFAAs	Perfluoroalkyl acids	GO	Graphene oxide
FASAs	Perfluoroalkane sulfonamides	µm	Micrometre
FTOHs	Fluorotelomer alcohols	TFC	Thin film composite
PFOA	Perfluorooctanoic acid	cm^2	Square centimetres
PFOS	perfluorooctane sulfonic acid		

9.1 Introduction to Organic Pollutants in Groundwater Resources

Groundwater recourses are the most widely used water supplier in the world for drinking and utility. They are inevitably affected by some hazardous pollutants originating from significant population growth and industrial advances (Sililo 1999).

The organic compounds and their distribution or concentration were not considered vital, as long as they were detected in large scale in the groundwater recently. Then, the pollution of groundwater sources became a global concern and attracted attention toward the occurrence of the OPs, their environmental fate, and possible toxicity even at low amounts (Stuart and Lapworth 2013). The organic pollutants (OPs) are categorized as pharmaceuticals, surfactants, personal care products (PCPs), flame retardants, industrial compounds, water treatment by-products, lifestyle compounds, and pesticides (Kapelewska et al. 2018; Stuart and Lapworth 2013). Figure 9.1 represents a general schematic of OPs and their subclasses.

Significant concentrations of some endocrine-disrupting OPs (10^2–10^4 ng/L) were detected in groundwaters globally (Stuart and Lapworth 2013). Pharmaceuticals, lifestyle, and industrial compounds were the most detected pollutants in the groundwater with the maximum concentration of 40–10^4 ng/L. Carbamazepine, sulfamethoxazole, ibuprofen, caffeine, diclofenac, and bisphenol A were the most reported compounds globally (Lapworth et al. 2012). Pharmaceutical residues are the most frequently discovered OPs in all water matrices on every continent (Patel et al. 2019). The frequency and concentration of different pharmaceuticals depend on the location. For instance, painkillers had the highest concentration in Europe, but antibiotics were highest in Asia (Patel et al. 2019).

Although these compounds have specific applications, they severely affect human beings, and many regulations have been set to control their entrance into the environment. Landfill, pesticide use, animal waste, domestic/household, hospital, and industrial effluents are generally organic pollutant resources. Source: Kapelewska et al., (2018) and (Stuart and Lapworth 2013).

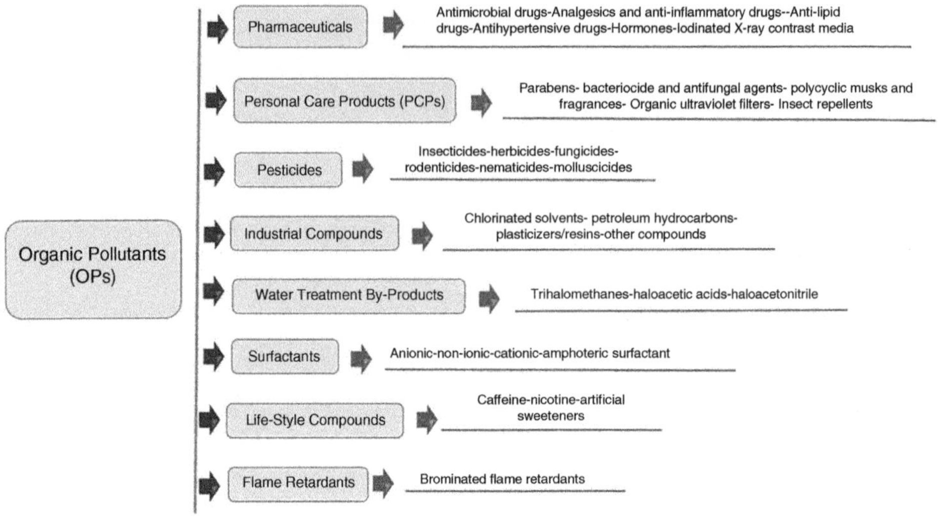

Figure 9.1 General view of OPs types and their compounds.

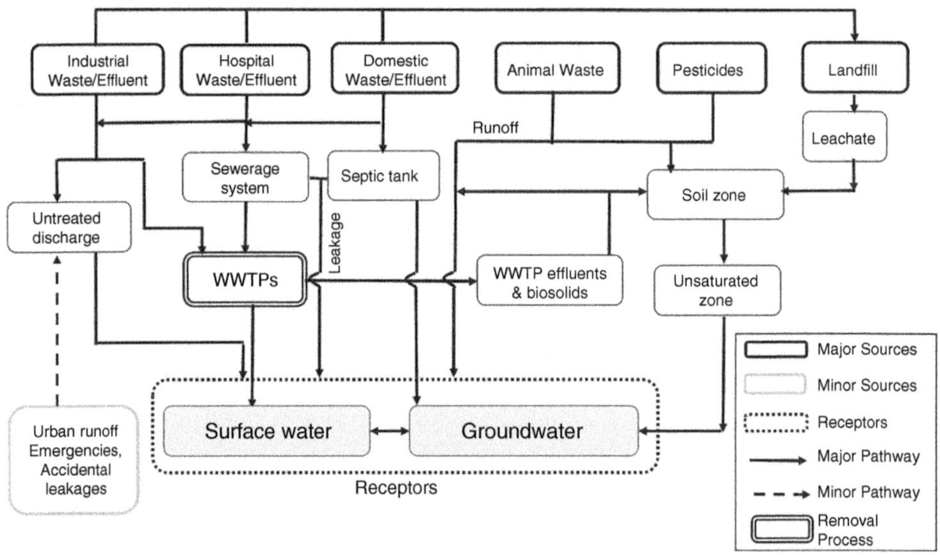

Figure 9.2 A schematic diagram of the probable sources, pathways, and receptors of OPs.

Figure 9.2 represents a schematic diagram of the possible sources, pathways, and receptors of OPs in the environment. Owing to the OPs resources, they can reach the groundwater by different paths. For instance, the available OPs in the landfill leachates the groundwater. The used pesticides and animal waste (may be used as nutrients for the soil) enter the land and groundwater. Household, hospital, and industrial effluents may reach the ground subsurface (directly) or the surface water and then groundwater after treatment in the wastewater treatment plants (WWTPs) (indirectly). Also, the biosolids or WWTPs effluents can enter the groundwater (Stuart and Lapworth 2013).

This chapter tries to provide sufficient knowledge about the groundwater OPs together with the necessary treatments. Firstly, it focuses on the types of compounds, their properties, probable origins, and hazards. Then it discusses the present remediation methods for organic compounds as well as the related advantages and disadvantages.

9.2 Organic Groundwater Pollutants: Types, Physical and Chemical Characteristics, and Associated Hazards

9.2.1 Pharmaceuticals

Pharmaceutical presence in groundwater sources has become a growing global concern (Kuczyńska 2019). Research performed in the developed countries confirmed the presence of more than 160 pharmaceuticals in the groundwater, surface water, and sewage (Taheran et al. 2016). A German research group reported 631 different pharmaceutical substances detected in 71 countries (Aus Der Beek et al. 2016). Domestic and hospital wastewater, manufacturer disposal, WWTPs, and landfill leachates are the usual resources of pharmaceutical pollutants (Yang et al. 2017; Ebele et al. 2020). The pharmaceuticals may transfer into the environment even after treatment in the WWTPs due to their partial elimination resulting from the low volatility, acutely low concentration, hydrophobicity, and complicated structures. Also, animal wastes are another pollution source, owing to the great dosage of drugs used for animals (Taheran et al. 2016). The discovered pharmaceuticals in the groundwater are categorized into several groups, based on their properties and targets. Brief information about each group, its related compounds, and the target is presented in Table 9.1 (Yang et al. 2017; Jjemba 2018).

Many pharmaceuticals can disappear fast in the environment, but their use in large quantities made them pseudo-persistent compounds. Despite the low concentration of pharmaceuticals in WWTPs, their residues can cause severe health problems (Yang et al. 2017). Even trace amounts of them in the water cycle and food chain, the usual routes for the pharmaceuticals entering the human body, can threaten human health (Organization 2012; Taheran et al. 2016). For instance, a gradual increase in antibiotic concentration in water may increase the persistence of *Escherichia coli*. This problem is intensified with population growth (Taheran et al. 2016; Yang et al. 2017).

9.2.2 Personal Care Products

PCPs are broadly applied in human life to improve its quality. These products generally consist of any care and washing products (such as soaps, detergents, perfumes, aftershaves, disinfectants, lotions), cosmetics, and even food (Yang et al. 2017; Jjemba 2018). They generally contain some chemical compounds which can be transmitted into the environment through human excretion or washing off after using (Stuart and Lapworth 2013; Serra-Roig et al. 2016). The PCPs can be released into the environment through WWTPs effluents. The high lipophilicity and low biodegradability of these compounds lead to their partial removal in the WWTPs (Cruz and Barceló 2015; Serra-Roig et al. 2016; Sharma et al. 2019). These compounds are placed in various groups.

Table 9.1 Pharmaceuticals: classes, related compounds, and targets.

Pharmaceutical types	Included Compounds	Target/Usage	References
Antimicrobial drugs: antibiotics, antifungal agents, antiprotozoal agents, and antiviral agents	Trimethoprim, ciprofloxacin, clofibric acid, lincomycin, sulfamethoxazole, and tetracycline	Protecting humans and animals from bacteria	(Yang et al. 2017; Jjemba 2018)
Analgesics and anti-inflammatory drugs	Codeine, aspirin, indomethacin, naproxen, ketoprofen, diclofenac, carbamazepine, paracetamol, acetaminophen (paracetamol), salicylic acid, salbutamol, and ibuprofen	Relieving pain	(Yang et al. 2017; Jjemba 2018)
Anti-lipid drugs or lipid regulators	Lescol XL, Zetia, Niaspan, Lovastatin, and Crestor	Reducing the concentration of cholesterol and triglycerides	(Yang et al. 2017; Jjemba 2018)
Antihypertensive drugs	Metoprolol tartrate, atenolol (Chlorthalidone), propranolol HCl (Inderal LA), Coreg (Carvedilol), Timolol maleate (Blocadren), Bisoprolol/HCTZ (Zebeta)	Using as β-blockers for cardiovascular diseases (angina and hypertension)	(Yang et al. 2017; Jjemba 2018)
Hormones	Natural estrogen (estrone, β-estradiol, and estriol) and synthetic 17α-ethinylestradiol	Regulation of physiology and behavior	(Kotowska et al. 2019, Yang et al. 2017).
Iodinated X-ray contrast media	Iopromide, iopamidol, and amidotrizoic acid	Applying for imaging of organs or blood vessels	(Nowak et al. 2020; Wang et al. 2019).

Source: Yang et al., (2017), John Wiley & Sons.

Parabens or preservatives, esters of para-hydroxybenzoic acid with an alkyl or benzyl group, play the preservative role in PCPs. Their antibacterial and fungicidal nature, low production cost, odorless nature, and stability in a broad pH spectrum make parabens good candidates for usage in pharmaceuticals, cosmetics, and food (Stuart and Lapworth 2013; Serra-Roig et al. 2016; Yang et al. 2017). Methylparaben, propylparaben, ethylparaben, and butylparaben are parabens with antimicrobial preservation potentials (Yang et al. 2017; Kapelewska et al. 2018). Endocrine disruption is reported as a paraben effect on human health (Serra-Roig et al. 2016).

Bacteriocide and Antifungal Agents

Triclosan (TCS), 5-chloro-2-(2,4-dichlorophenoxy) phenol, or 2,4,4u-trichloro-2u-hydroxy-diphenyl ether, is an aromatic hydrocarbon comprising of phenol and diphenyl ether. It is thermally stable, slowly biodegradable, lipophilic, and easily soluble in organic solvents, but not in water. It is broadly applied in PCPs such as deodorants, disinfectants, soap, shampoo, toothpaste, and mouthwash. Moreover, it is applied in sports clothing and plastics due to its good thermal stability (Hontela and Habibi 2013; Garcia-Morales et al. 2015; Yang et al. 2017; Kapelewska et al. 2018). TCS is degraded to several compounds, including methyltriclosan (biological methylation of TCS), dioxins, chlorophenols, and chloroform during the treatment process in WWTPs. TCS and its degradation products have a good persistence in the aqueous environment and could remain for a long time, even up to 40 years. They accumulate in the tissue of aquatics and disorder the endocrine system, possibly resulting in problems like cancer, reproduction disorders, and mutations in fish and mammals (Hontela and Habibi 2013; Garcia-Morales et al. 2015).

Polycyclic musks and fragrances are a major class of compound. They are semi-volatile compounds with poor aqueous solubility, high hydrophobicity/lipophilicity, and sorption tendency to the skin. They include nitro musks (musk ketone, musk xylene [common] and musk tibetene, musk moskene [not common]) and polycyclic musks (galaxolide and AHTN). 7-acetyl-1,1,3,4,4,6-hexamethyl-1,2,3,4-tetrahydronaphthalene (AHTN) and 1,3,4,6,7,8-hexahydro-4,6,6,7,8,8-hexamethylcyclopenta-γ-2-benzo-pyran (HHCB or galaxolide) are the most used musks in cosmetics, detergents, and PCPs (Yang et al. 2017; Jjemba 2018).

Organic ultraviolet (UV) filters or sunscreen agents have a wide application in PCPs, including lotions, beauty creams, shampoos, polymeric materials to protect from sunlight, and pharmaceuticals (Yang et al. 2017; Jjemba 2018). Most of them are highly insoluble in water and quite lipophilic. The benzophenones (like benzophenone, benzophenone-2, benzophenone-3, benzophenone-4), 3-(4-methyl benzylidene) camphor, and ethylhexyl methoxycinamate are common UV filters (Jjemba 2018; Kapelewska et al. 2018).

Insect repellents are products used to protect humans from pests. **N, N-diethyl-meta-toluamide (DEET)** is a kind of insect repellent known as a frequent chemical pollutant entering the environment through the wastewater effluent. DEET is sometimes called by different names such as detamid, diethyltoluamide, and N,N-diethyl-3-methylbenzamide. It was firstly applied in 1946 in the USA to repel mosquitos and became available for the public in 1957 (Merel and Snyder 2016; Lawrence et al. 2019). It is used to repel insects from the skin (in spray, liquid, or lotion forms), clothing, surfaces, and even for some animals (Merel and Snyder 2016; Yang et al. 2017; Kapelewska et al. 2018; Lawrence et al. 2019). It was first detected in Swedish municipal landfill leachate. Then, its occurrence was reported in the USA and it has gradually increased worldwide (Merel and Snyder 2016). DEET was observed in approximately all water matrices. However, it wouldn't have high toxic effects at reported concentrations ranging between 13 and over 1 000 000 ng/L (all water sources) exclusively for groundwater between 3481 (in Asia) and 13 500 ng/L (in America). Its maximum concentration in water is suggested 2.5 mg/L, more than the observed amounts (Merel and Snyder 2016). Also, the European Union announced that it is not a persistent, bioaccumulative, or toxic substance (Lawrence et al. 2019).

9.2.3 Pesticides

Pesticides are chemicals used to protect the crops from pathogens, fungi, insects, and weeds, enhancing the yields (Srivastav 2020). Also, they can protect humans against illnesses, such as dengue, encephalitis, filariasis, and malaria. Pesticides include insecticides, herbicides, fungicides, rodenticides, nematicides, and molluscicides. There are more than 500 types of pesticides used in agriculture (Srivastav 2020). Some pesticides include dichlorodiphenyltrichloroethane (DDT), mirex, endosulfan, lindane, aldrin, hexachlorocyclohexane, dieldrin, endrin, heptachlor, hexachlorobenzene, butachlor, atrazine, chlordane, methoxychlor, and toxophane. (Marican and Durán-Lara 2018). The application of pesticides intensively has increased in time, as they have become an inseparable tool in modern agriculture (Sjerps et al. 2019; Srivastav 2020). However, pesticides and their degradation products can deteriorate the quality of groundwater and threaten human health and ecosystems at a very high level. They can cause health problems such as headache, vertigo, gastrogenic diseases, infirmity, hyperirritability, itchiness of the skin, problems of the nervous system, paralysis, blurry vision or even blindness, flu, cancer, and death (Srivastav 2020).

9.2.4 Industrial Compounds

Industrial developments accompanied by complicated processes have resulted in the release of hazardous pollutants for human beings (Stuart and Lapworth 2013). Industrial compounds are categorized into four classes. The first class belongs to the chlorinated solvents (Stuart and Lapworth 2013). The groundwater pollution by chlorinated solvents originates from the great applications of tetrachloroethylene (PCE) and trichloroethylene (TCE) in dry-cleaning and metal processing industries (Hyldegaard et al. 2019; Kawabe and Komai 2019). The other solvents include 1,1,1-trichloroethane (1,1,1-TCA), 1,1-dichloroethane (1,1-DCA), 1,1-dichloroethylene (1,1-DCE), and vinyl chloride, with wide applications in chemical, electronic, and pharmaceutical industries (Yang et al. 2018; La Vigna et al. 2019). They are highly soluble in water and volatile with low sorption to sedimentation. However, they and even their degradation products are severely toxic with carcinogenic effects (Hyldegaard et al. 2019).

Petroleum hydrocarbons (PHs) are the other industrial pollutants entering the environment through exploration, production, and transport activities (Qian et al. 2018; Safdari et al. 2018). They mainly consist of aromatic and aliphatic compounds, which are severely poisonous and difficult to cope with (Medjor et al. 2018). The benzene, toluene, ethylbenzene, xylene isomers, phenols, and polyaromatic hydrocarbons (PAHs) are some PHs (Medjor et al. 2018; Poi et al. 2018). PAHs are hydrophobic compounds with low water solubility, high toxicity, and carcinogenic and mutagenic effects. Some PAHs can cause health problems such as eye and skin inflammation, attenuation of the neurological system, and even various cancers. The US Environmental Protection Agency declared 16 PAHs as precedence pollutants (Zhai et al. 2018). Methyl tertiary-butyl ether (MTBE) is an organic pollutant with high water solubility and biological and chemical stability. It is generally used as an additive to oxygenate gasoline to enhance its octane number and combustion efficiency (Tawabini and Makkawi 2018; Kucharzyk et al. 2019; Briganti et al. 2020). It causes some health problems like headache, vertigo, and nausea (Tawabini and Makkawi 2018).

The third class of industrial compounds is devoted to the plasticizers/resins such as bisphenols, adipates, and phthalates. Bisphenols (bisphenol A, bisphenol S, bisphenol AF, and bisphenol F) are generally applied in the plastics and food packaging to stabilize and prevent oxidation (Zhao et al. 2019). Endocrine disorders and negative effects on the hormonal system are the adverse effects of these compounds on human beings (Zhao et al. 2019). The final class is focused on the other compounds: some works consider them as emerging organic pollutants. 1, 4-Dioxane, a cyclic diether, has attracted great attention owing to its carcinogenic effects. Furthermore, its high water solubility, resistance to biodegradation, quick passage into the groundwater, and low sorption on materials make it very difficult to eliminate by conventional technologies (Karges et al. 2018). It had broad applications as a solvent and stabilizer in TCE, 1,1,1-TCA, and 1,1-DCE (Abe 1999; Anderson et al. 2012; Miao et al. 2018).

9.2.5 By-products in Water Treatment or Disinfection Process

Water treatment is usually done through various methods, in which by-products are produced accompanied by disinfection. These methods use some disinfectants, including chlorine, chlorine dioxide, chloramine, ozone, and UV light, that will react with natural organic/inorganic matters resulting in the production of some compounds named disinfection by-products (Ghoochani et al. 2017; Dubey et al. 2020). Trihalomethanes (THMs), haloacetic acids, and haloacetonitrile are such by-products that can cause cancer, liver and kidney disorders, cytotoxicity, and mutagenesis (Dubey et al. 2020). Chlorination is the most traditional method for disinfection because of chlorine's benefits, such as cost-effectiveness, simple operation, accessible technology, and high efficiency (Dobaradaran et al. 2020). However, it can generate THMs (Dobaradaran et al. 2020; Dubey et al. 2020) and HAAs (Ghoochani et al. 2017; Dobaradaran et al. 2020), necessitating the use of an alternative disinfectant like ozone or chlorine dioxide (primary disinfection) and chloramine (secondary disinfection) (Dobaradaran et al. 2020; Dubey et al. 2020). Chloramine for secondary disinfection produces N-nitrosamines, causing cancer itself.

9.2.6 Surfactants

Surfactants or surface-active agents are organic compounds comprising hydrocarbon chains (as the lipophilic moiety) and polar groups (as the hydrophilic or hydrophobic moiety) to lower the surface tension of liquids (Knepper and Berna 2003; Lechuga et al. 2016). They are broadly used in detergents and cleaners, paints, pesticide compositions, pharmaceuticals, wetting agents, and PCPs. They are categorized into four groups: anionic, non-ionic, cationic, and amphoteric (Knepper and Berna 2003).

Anionic surfactants are determined by a hydrophilic polar moiety with a negative charge. The surface-active properties and their effects on biomembranes result in extreme irritability. They can also play a preliminary role in skin roughness due to their adsorption on the skin surface (Pulce and Descotes 1996). Perfluoroalkyl and polyfluoroalkyl substances (PFASs) are anionic surfactants with an aliphatic carbon framework, and some fluorine atoms placed instead of hydrogens either thoroughly (per-) or partially (poly-) (Rahman et al. 2014). The high polarity and potent carbon–fluorine bonds have proposed them as highly stable substances in thermal and chemical conditions with extreme resistance to degradation. These anionic surfactants are used in food packaging, cookware coatings, cosmetics, and fire-fighting foams (Rahman et al. 2014; Munoz et al. 2017). Perfluoroalkyl acids (PFAAs), perfluoroalkane sulfonamides (FASAs), and fluorotelomer alcohols (FTOHs) are the three main groups of PFASs. PFAAs are observed in any type of water matrixes because of high solubility in water, low volatility, and simultaneous hydrophobic/hydrophilic moieties (Rahman et al. 2014). Hepatic and developmental toxicity, as well as bioaccumulation, are the reported health problems caused by some PFAAs (Munoz et al. 2017). Perfluorooctanoic acid (PFOA) and perfluoroooctane sulfonic acid (PFOS) are pollutants with repeated detection at the highest level in all of the PFAs (Ahrens and Bundschuh 2014; Rahman et al. 2014). The bioaccumulation of PFOA is lower than PFOS because of its various functional groups, together with shorter perfluorocarbon chain length (Ahrens and Bundschuh 2014).

Non-ionic surfactants are defined as the surfactants in which polar head groups are not electrically charged like alkyl, fluoroalkyl, or steroidal (Ahel et al. 1996; Sonia and Sharma 2014). For instance, alkylphenol polyethoxylates (APEOs) consist of an alkyl group (octylphenol or nonylphenol) and (mono, di or poly) ethoxylate. The biotransformation of APEOs results in the production of some compounds with higher lipophilicity. The resulted compounds are more toxic than the parent compounds and persistent in the aquatic environment (Ahel et al. 1996). They cause less local irritation and are less toxic than anionic surfactants (Pulce and Descotes 1996).

Cationic surfactants are quaternary ammonia compounds in which the surface-active parts have positive charges (Pulce and Descotes 1996). Cetrimonium chloride (hexadecyl trimethyl ammonium or cetyl trimethyl ammonium) is a cationic surfactant with widespread use in soap, detergent, emulsifiers, dispersants, and disinfectants. It has a transparent appearance and smells like ethanol. It is toxic and almost resistant to biodegradation. Also, it can produce poisoning products like hydrogen chloride and nitrogen oxides after thermal degradation (De Araújo and León 2018). Benzalkonium chloride is the other cationic surfactant that plays a preservation role in pharmaceuticals and personal care products (Isaac and Scheinman 2017). It can cause allergic dermatitis (Pulce and Descotes 1996; Isaac and Scheinman 2017).

9.2.7 Lifestyle Compounds

Lifestyle compounds like caffeine, nicotine, and artificial sweeteners can be considered groundwater pollutants that originate from sewage leakage (stuart and Lapworth 2013).

Caffeine is a natural alkaloid with $C_8H_{10}N_4O$ chemical molecular formula. It is a usual component of food, beverages, and pharmaceuticals, occurring naturally in coffee and cacao seeds (Buerge et al. 2003). Despite its efficient removal during the treatment process, its amount in water is much higher than that degraded. Caffeine residues are very stable in the water environment (Li et al. 2020).

Artificial sweeteners like acesulfame, saccharine, cyclamate, and sucralose are broadly used instead of sugar in beverages, food, and drugs. Due to the ineffectiveness of human metabolism on these compounds, they are excreted and enter the environment. Also, the acesulfame, cyclamate, and saccharin anionic nature at the typical pH of natural waters results in the mobility in the subsurface (Buerge et al. 2009). Acesulfame ($C_4H_5NO_4S$) is highly soluble and persistent during sewage treatment, but not absorbable. Also, laboratory tests confirmed its degradation by UV lights and sunlight that generate possibly persistent intermediate products (Robertson et al. 2016).

9.2.8 Flame Retardants

Flame-retardant substances are used to overcome fire hazards, especially in populated areas. Inorganic salts, phosphorous compounds, nitrogen-based compounds, and halogenated organic compounds (chlorinated paraffins and alicyclic compounds, and brominated aromatic compounds) are used for this purpose. Brominated flame retardants include polybrominated diphenyl ethers (PBDEs), 1,2,5,6,9,10-hexabromocyclododecane, tetra bromo bisphenol A, 1,2-bis-(2,4,6-tribromophenoxy) ethane, 2-ethyl-1-hexyl-2,3,4,5-tetrabromobenzoate, and bis (2-ethyl-1-hexyl) tetra bromophthalate (Fromme et al. 2016). PBDEs are chemically persistent, semi-volatile (low vapor pressure), and highly hydrophobic. Their propensity to accumulate in the lipids results in their bioaccumulation in fish and other species (Fromme et al. 2016; Annunciacão et al. 2017). They are incorporated into different products and polymer without chemical bonds, then leaching out easily and entering the environment (Annunciacão et al. 2017; Wu et al. 2019). They are also detected in the aquatic sediments and human blood and tissues (through inhalation, ingestion, and dermal exposure), affecting the immune and neurological system, as well as endocrine disorders due to their toxicity (Annunciacão et al. 2017; Wu et al. 2019).

9.3 OPs Remediation Methods

The frequencies and concentrations of pollutants detected in groundwaters were comparatively lower than surface waters. However, the polluted groundwater with serious harmful effects to human beings must be treated (Chen and Huang 2020).

There are diverse physical, chemical, biological, and thermal remediation methods to eliminate the OPs, classified into conventional processes used in WTPs, treatment processes used in WWTPs, and advanced treatment processes (ATPs) (Patel et al. 2019). Figure 9.3 shows a general view of the OPs remediation methods.

Wastewater treatment plants were not designed for the complete removal of all OPs. Generally, WWTPs can handle easily and moderately degradable organics in the mg/L range (Patel et al. 2019). The harmful pollutants (dissolved organic matter, larger [suspended] particles, and biodegradable solids) are eliminated through primary (sedimentation or coagulation) and secondary (mainly biological process) treatments (Yang et al. 2017; Kapelewska et al. 2018).

Bioremediations or biological treatments are traditional methods applied for pollutant degradation. Biodegradation can be performed by microbial cultures, including (bacteria, fungi, and alga), using biocatalysts, and as phytoremediation (Patel et al. 2019). Its performance highly depends on the nature of pollutants, reactor design, hydraulic retention time, the adsorption capacity of sludge, and sludge age (Yang et al. 2017). Biodegradation was used for groundwater polluted with chlorinated solvents (Němeček et al. 2018) and petroleum hydrocarbon (Rodríguez-Escales et al. 2013; Al-Hawash et al. 2018; Al-Hawash et al. 2019), industrial compounds (Fang et al. 2020), and THMs (Ansari et al. 2020). Ferrando and Matamoros (Ferrando and Matamoros 2020) applied immobilized microalgae-based systems for attenuation of pesticides (bromacil, atrazine, diuron, bentazone, and mecoprop) together with antibiotics and nitrates from groundwater both in the batch and

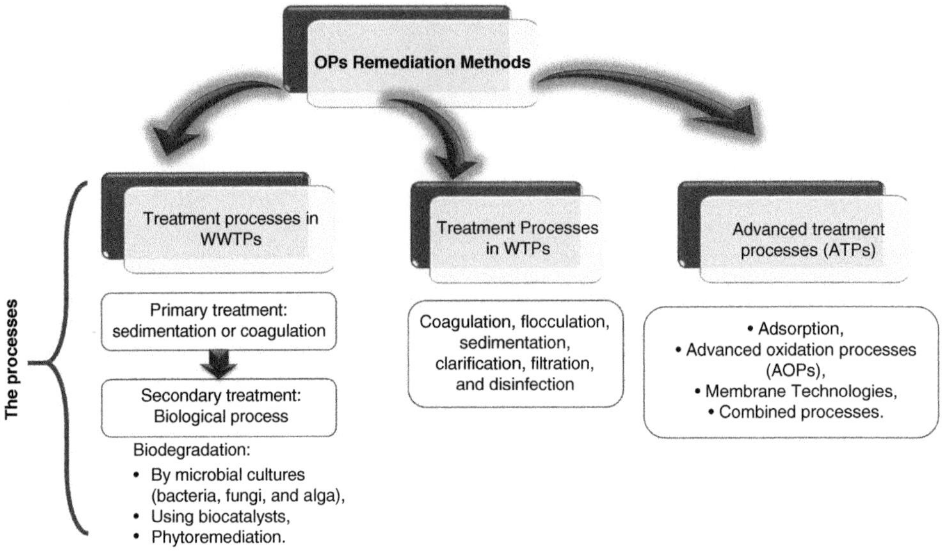

Figure 9.3 A general schematic of remediation methods used for OPs. *Source:* Modified from Patel et al., (2019).

continuous system. The immobilized microalgae increased the OPs average removal from 36 to 51% in a batch system. It could perform better in continuous mode with an average removal of 65% for OPs. Liang et al. (Liang et al. 2019) applied photocatalysis degradation with a novel composite (indirect Z-scheme system of $Ag@Ag_3PO_4/g\text{-}C_3N_4/rGO$) and visible light radiation to remove 2,2′,4,4′-tetrabrominateddiphenyl ether. The novel system could be a potential method for pollutant removal with 93.4% removal efficiency.

The WWTPs performance relies on pollutants' nature in terms of chemical and biological characteristics (hydrophilicity, solubility, volatility, biodegradability) and the adsorption capacity of the activated sludge (Yang et al. 2017, Kapelewska et al. 2018).

Water treatment plants (WTPs) are other alternatives that generally apply conventional processes, including coagulation, flocculation, sedimentation, clarification, filtration, and disinfection for water treatment. WTPs are unable to remove OPs completely (Patel et al. 2019).

For example, the applied processes in both WWTPs and WTPs cannot effectively eliminate pharmaceuticals and personal care products (PPCPs) due to the hydrophobicity of these compounds (Yang et al. 2017). Some pharmaceuticals such as trimethoprim, TCS, and metoprolol can lower removal efficiency with their inhibitory effects on the activated sludge bacterial community in the biological process (Yang et al. 2017). Also, conventional WTPs poorly eliminate PFAs because of their extremely low concentrations and high hydrophilicity (Rahman et al. 2014).

Then, the resistant and non-biodegradable compounds remain in the effluents and may enter the environment and pollute the water matrixes, including surface water and groundwater (Fanourakis et al. 2020). The ineffectiveness of theses conventional processes necessitates the use of ATPs. These processes include a diverse spectrum including adsorption by the various adsorbent, membrane technologies (forward osmosis, membrane distillation, reverse osmosis, nanofiltration, microfiltration, and ultrafiltration), advanced oxidation (ozonation, catalytic ozonation, UV, UV/H_2O_2, photocatalysis, electrochemical, and Fenton reaction), and combined processes (Patel et al. 2019).

Therefore, the final section defines the main ATPs used for treatment and presents the basic theory of each method, and their general advantages and drawbacks. Also, the reported probable and effective ATPs with a focus on membrane technology are tabulated.

9.4 Advanced Treatment Processes (ATPs): Theory, General Aspects, Advantages, Disadvantages, and Reported Researches

9.4.1 Adsorption

Adsorption is an ATP because of its operational simplicity, cost-effectiveness, suitable removal efficiency for low amounts of OPs, the feasibility of regeneration and reuse, reliability, and applicability to continuous and batch processes (Mansour et al. 2018). It is theoretically a surface phenomenon in which a target matter (organic/inorganic) adheres to the surface of an adsorbent (carbon-based material) by van der Waals and electrostatic interactions. (Fanourakis et al. 2020). The process is capable of removing even low amounts of synthetic and natural OPs from water. Its performance mainly depends on temperature, the pollutant concentration, particle size, exposure time, the characteristics of the adsorbate and adsorbent, and solution pH. Adsorption is a useful ATP for PPCP elimination from water because of the chemical structure and aromatic rings of PPCPs (Fanourakis et al. 2020).

Being inert, biocompatible, resistant to mechanical forces, and having high selectivity for the target pollutants, together with high adsorption capacity (depending on pore structure and surface charge), are properties essential for an efficient adsorbent (Fanourakis et al. 2020). Activated carbon (AC) is a conventional material for adsorptive treatments because of the high surface area resulting from its high microporosity in granular (granular activated carbon, GAC) and powder (powdered activated carbon, PAC) forms. Furthermore, graphene, graphene-based nanomaterials, carbon nanotubes (CNTs), and clay minerals are other alternatives suitable for OPs separation. The expensive regeneration of carbon-based adsorbents attracted attention toward inexpensive materials like polymeric materials (Fanourakis et al. 2020).

Biosorption is a kind of adsorption process with biomass or lignocellulosic materials (cellulose, hemicellulose, and lignin) as an adsorbent. Low expense and good adsorption capacity resulting from the various functional groups (hydroxyl, phenolic, carbonyl, and amines) of these materials attracted attention toward them (Sharma and Chaudhry 2020). Biochar is a carbon-based material produced through the pyrolysis of lignocellulosic materials (Weber and Quicker 2018; Tran et al. 2020). Generally, it has a large specific surface area, cation exchange capacity, highly porous structure, functional groups on its surface (-COOH-OH, and possible -NH_2), and the aromatic ring in its structure, helping to the effective adsorption of various aromatic pollutants (Weber and Quicker 2018; Wang and Wang 2019; Tran et al. 2020).

Metal–organic frameworks (MOFs) are a new generation of advanced adsorbents with high porosity, tunable geometric structures in terms of pore size and shape, and simple production. However, instability in the water and their hydrophilicity restricted their application for adoptions in watery solutions (Bhadra et al. 2017).

Table 9.2 reviews the adsorptive removal of different OPs. Despite the adsorption capability for separation of OPs, attaching of OPs to the adsorbent producing a new pollutant necessitates an additional separation step (especially expensive) to regenerate the adsorbent and restrict their application (Fanourakis et al. 2020). Generally, the adsorption process suffers from high production and regeneration expenses and probable reduction in efficiency by regeneration (Mansour et al. 2018).

9.4.2 Advanced Oxidation Processes (AOPs)

Oxidation is a chemical treatment method trying to remove the pollutant by an oxidizer. The process performance depends on the nature and present bonds in the oxidant. Effective oxidation needs an electropositive element or functional groups like double bonds, activated aromatic systems, and amino groups that have high electron density (Rahman et al. 2014). **Advanced oxidation processes** are in the aqueous phase with strongly reactive HO•, O_2•, and HO_2• (Patel et al. 2019). Reactive oxygen or free radicals are potent oxidants that make AOPs an effective instrument to degrade organic and inorganic pollutants. The hydroxyl radical is known for its high reactivity and oxidation capacity ($E^o = +2.80\,V$). In contrast, functional groups such as -Cl, -NO_2, and -COOH have lower reactivity due to electron-withdrawing nature (Wang and Wang 2016).

The AOPs mainly include ozonation, photolysis (UV), UV/ H_2O_2, O_3/UV, Fenton, and Fenton-like reactions, and electrochemical oxidation. **Ozonation** is a conventional oxidation process capable of the removal of widespread pollutants and disinfection. However, the low mineralization producing toxic by-products is the main drawback (Gozes et al. 2017). **Electrochemical oxidation** is an attractive process for trying to oxidize the OPs by reactive species produced via electricity, and no chemicals are used for oxidization, preventing the production of secondary waste. It is considered a technology for separation (electrodialysis and electrocoagulation) or degradation (anodic oxidation) (Kanakaraju et al. 2018). **Fenton reaction** is a kind of AOP presented in 1894. It has advantages like easy operation, low cost (due to the inexpensive reagent), the need for low chemical input, capability in the degradation of many PPCPs, and reduction of toxic effects to microorganisms (Vorontsov 2019; Sönmez et al. 2020). The Fenton process is defined as an oxidation treatment of OPs at the presence of the Fenton's reagent, a mixture of ferrous ion (Fe^{2+}; or different metal-based catalysts) and hydrogen peroxide (H_2O_2), producing hydroxyl radicals (Jain et al. 2018; Sönmez et al. 2020). The process performance relies on the concentration of Fe^{2+} and H_2O_2, the solution pH, and the OP properties. Fenton-like reactions generally include sono-Fenton, photo-Fenton, electro-Fenton, photo-electro Fenton, and ligand assisted Fenton (Jain et al. 2018). Fenton and photo-Fenton suffer from incomplete pollutant mineralization, having the best performance (highest rate) only at restricted pH, waste of H_2O_2, and low degradation rate for some chemicals (Vorontsov 2019). The

Table 9.2 The adsorptive removal of OPs by various adsorbents.

OP type	Compounds	Adsorbent	References
PPCPs	Diclofenac, naproxen, and carbamazepine	GAC	Varga et al. (2019)
	Carbamazepine, sulfamethoxazole, and paroxetine		Jaria et al. (2019)
	Carbamazepine and sildenafil citrate	PAC	Delgado et al. (2019)
	Ibuprofen and Triclosan		Kaur et al. (2018)
	Metronidazole	Polyaniline-polypyrrole copolymer	Aarab et al. (2020)
	Triclosan, oxybenzone, and *p*-chloro-*m*-xylenol	Polyaniline-derived porous carbons	Yoo et al. (2018)
	Paracetamol	Spherical and non-spherical biochar	Tran et al. (2020)
	Naproxen and clofibric acid	MOFs (chromium-benzenedicarboxylate and iron-benzenetricarboxylate)	Hasan et al. (2012)
	Naproxen, ibuprofen, and oxybenzone	MOF (chromium-benzenedicarboxylate)	Seo et al. (2016)
	Acidic ibuprofen; weakly acidic triclosan and oxybenzone; basic diclofenac sodium, and atenolol	Metal azolate framework-6-derived porous carbon	An et al. (2018)
	Diatrizoic acid, iopamidol, metformin, and carbamazepine	Natural clay	Mahouachi et al. (2020)
	Atrazine and imidacloprid	Biochar	Mandal et al. (2017)
	Pentachlorophenol	Modified chitosan	Shankar et al. (2020)
Pesticides	Malathion	Multi-walled carbon nanotubes	Dehghani et al. (2017)
	Diazinon	Multi-walled carbon nanotubes	Dehghani et al. (2019)
	TCE and trichloromethane	Biochar-supported nanoscale zero-valent iron	Qian et al. (2020)
	MTBE	ZSM-5 zeolite	Zhang et al. (2019)
Industrial Compounds	Bisphenol A	Magnesium ascorbyl phosphate graphene-based monolith	Fang et al. (2018)
		AC	Acosta et al. (2018)
		Magnetic CuZnFe2O4–biochar composite	Heo et al. (2019)

(*Continued*)

Table 9.2 (Continued)

OP type	Compounds	Adsorbent	References
Water Treatment By-products	THM	AC	Park et al. (2019)
		CNTs	Özdemir and Gungör (2018)
		Ion-exchange resins	Liu et al. (2018)
Surfactants	PFOA and PFOS	GAC	Park et al. (2020)
		PAC	Dudley (2012)
		Colloidal AC	Carey et al. (2019)
	Cationic surfactant (hexadecylpyridinium chloride) and anionic surfactant (sodium dodecylbenzene sulfonate)	Zero-valent iron nanoparticles	Abd El-Lateef et al. (2018)
Lifestyle Compounds	Caffeine	AC	(Anastopoulos et al. 2020; Beltrame et al. 2018)
		Graphene nanoplates	Anastopoulos et al. (2020)
		Carboxylic acid-functionalized UiO-66	Anastopoulos et al. (2020)
		MgAl-LDH/biochar composite	Dos Santos Lins et al. (2019)
	Nicotine	AC, clinoptilolite, GO, and mesoporous carbon	Anastopoulos et al. (2020)
	Sucralose (artificial sweetener)	AC	Soh et al. (2011)

determinative role of the metal's solubility in the homogenous catalyst (in which cerium and cobalt with high solubility are cytotoxic) and instability and recycle of heterogeneous catalysts are other problems (Wang and Wang 2016). **Sonolysis** is another advanced process that tries to degrade the pollutants by ultrasound irradiation (Kanakaraju et al. 2018). Trace PPCPs can be effectively removed by AOPs that change the polarity and functional groups of the PPCPs and make the water suitable for direct human contact (Gomes et al. 2017; Yang et al. 2017; Kanakaraju et al. 2018). In contrast, the oxidation is approximately an ineffective process for PFAS removal. The lack of any aromatic bonds or phenolic groups, the strong C—F bond, and -COOH and -SO_3H groups withdrawing electrons in the PFAs structure, probably prevent their oxidation (Rahman et al. 2014). Electrochemical oxidation was applied for pesticide (hexaclorocyclohexanes and chlorobenzenes) (Dominguez et al. 2018) and surfactant (cetrimonium chloride) (De Araújo and León 2018) removal. Also, industrial compounds like bisphenol A can be removed by various AOPs (Burgos-Castillo et al. 2018).

9.4.3 Membrane Separation

Although the chemical processes, including AOPs and photochemical degradation, suffer from degradation products, low efficiency, troublesome operation, high expense, and by-product production, physical separation through adsorption and membrane treatment generate no by-products. Despite physical adsorption, which demands an additional and expensive step to regenerate the adsorbent, the membrane technologies are more practical, effective, and economical for OP removal (Zhang et al. 2020).

Membrane technology gained remarkable popularity in OP removal because of its scalabilities, high quality of effluent, effectiveness, and modularity as well as capability to combine with other methods (Taheran et al. 2016). Microfiltration (MF), ultrafiltration (UF), nanofiltration (NF), reverse osmosis (RO), and forward osmosis (FO) systems are the current technologies applied for the physical removal of the OPs from water through size exclusion/screening, electrostatic repulsion, and hydrophobic adsorption (Taheran et al. 2016; Zhang et al. 2020).

Microfiltration and **ultrafiltration** are known as low-pressure processes separating pollutants through the sieving mechanism and retention of large compounds. UF and MF membranes have an ineffective performance in removal of OPs like PPCPs (Yang et al. 2017; Patel et al. 2019). Due to the MWCO of the UF membrane (10000–200 000 Da), which is higher than MWCO of the most PPCPs (200–400 Da), most PPCPs are separated with very low efficiency (Zhang et al. 2020). Some hydrophobic compounds can be partially separated by these membranes, indicating that hydrophobic adsorption plays the main role in separation instead of size exclusion (Patel et al. 2019; Zhang et al. 2020). MF membranes with larger pore sizes than PFASs (with an effective diameter of 1 nm) cannot separate them. (Rahman et al. 2014). MF and UF membranes can improve water quality used for groundwater discharge or reuse as tertiary treatments in WWTPs (Taheran et al. 2016). In contrast, high-pressure NF and RO systems showed high efficiency in separating PPCPs and charged PFASs (with a MWCO of 300 Da or greater) (Rahman et al. 2014; Taheran et al. 2016; Yang et al. 2017). Due to the MWCO of the RO membrane (<150 Da), the governing mechanism in separation is size exclusion. Moreover, electrostatic repulsion and hydrophobic adsorption are other mechanisms observed in the RO membrane (Zhang et al. 2020). Despite RO membranes capability in the removal of the most OPs from water, low permeation, high energy consumption, low membrane durability, membrane fouling, and high maintenance expenses have restricted their applications (Zhang et al. 2020).

NF membranes with a pore size of 0.5–2 nm and molecular weight cutoff (MWCO) of 200–1000 Da have separation performance between UF and RO membranes. NF membranes have high permeation flux and rejection of multivalent salts and organic molecules simultaneously (Zhang et al. 2020). NF systems need lower operating pressure than RO membranes decreasing the capital and operating costs (Taheran et al. 2016). These membranes represent high efficiency for PPCP removal because their MWCO overlaps the MWCO of the most PPCPs (200–400 Da). The removal mechanism of PPCPs is determined generally based on the membrane and the PPCP's nature. Size exclusion (the most important mechanism for polymeric membranes), hydrophobic adsorption (approximately permanent mechanism due to the hydrophobic nature of the membranes), and electrostatic exclusion (depending on the existence of charges on the membrane surface and PPCPs molecules) are three mechanisms involved in PPCP removal. Despite the hydrophobic adsorption, the size exclusion is a governing mechanism in neutral PPCPs removal, but the electrostatic exclusion in charged PPCPs. Due to the more extended MWCO range compared to UF membranes, NF membranes can remove more OPS, resulting in better removal efficiency and water quality. Moreover, these membranes have higher permeation flux and higher efficiency than the RO, under the same operating pressure (Zhang et al. 2020).

The high operating pressure used in the NF and RO processes increases the treatment cost and fouling/scaling, which has attracted attention toward **forward osmosis,** in which the osmotic gradient caused by a draw solution results in separation through a semipermeable membrane (Madsen et al. 2015).

Despite the membrane advantages, the pollutants are only separated by membranes without degradation, which results in the production of massive wastes requiring degradation or transformation (Taheran et al. 2016). Otherwise, the resulted concentration polarization will affect membrane performance. The **membrane bioreactor (MBR)** is an advanced technology that can overcome this problem. It takes the advantages of the membrane technologies inside the bioremediation system and produces effluent with high quality (Taheran et al. 2016).

9.4.4 Combined Processes

A combination of several processes can provide a good solution, especially for compounds with incomplete removal. For example, adsorption is frequently used as a pretreatment for AOPs processes (Patel et al. 2019). Also, a combination of membrane process with adsorption system is proposed to enhance separation performance (Taheran et al. 2016).

Another combined and innovative process is the membrane system accompanied by chemical transformation (catalytic oxidation, ozonation, photo-catalysis, photo-Fenton reactions, enzymatic digestion, and ultrasonic processes) degradation (Taheran et al. 2016). Table 9.3 reviews some recent research that used membrane technologies for OP removal.

9.5 Conclusion

The groundwater pollution by organic compounds has emerged as a global concern. OPs generally include PPCPs, pesticides, industrial compounds, water treatment by-products, surfactants, lifestyle compounds, and flame retardants. The significant concentrations of some OPs (>100 ng/L), which have adverse effects on human beings, necessitate effective removal processes. The specific properties of OPs like hydrophobicity and low biodegradability resulted in their partial removal in WWTPs. Therefore, some advanced treatment processes are proposed, including adsorption, AOPs, and membrane technologies.

AOPs and photochemical degradation suffer from degradation products, low efficiency, troublesome operation, high expense, and by-product formation. Furthermore, physical adsorption needs an additional and expensive step to regenerate the adsorbent. In contrast, membrane technologies are more practical, effective, and economical for OP removal, generating no by-product. UF and NF membranes have a poor performance in OP removal, mainly

Table 9.3 Some recent membrane technologies applied for OP removal.

Ops type	Compounds	Membrane Technology	Membrane: type/ modules	Key Results	Additional Information	References
PPCPs	Diclofenac sodium, metoprolol, carbamazepine, trimethoprim, chloramphenicol, and indomethacin	NF	Thin-film composite NF (DF30)	The highest rejection for negatively charged PPCPs: indomethacin and (95.4%) and diclofenac sodium (94.1%)	–	Xu et al. (2019)
	Atenolol, ibuprofen, and carbamazepine	NF	A dually charged NF membrane (consisting of a polyether sulfone UF membrane modified by charged polyelectrolytes (polydopamine and quaternate chitosan), MWCO: 935 Da	Good rejection for atenolol (76.22%), ibuprofen (89.85%), and carbamazepine (87.29%)	–	Ouyang et al. (2019)
	Tetracycline, mefenamic acid, atenolol, furosemide, ketoprofen, and diclofenac	MBR coupled with coagulation	Hollow fiber polyvinylidene fluoride	Enhancing removal efficiencies (17–23%) using poly aluminium chloride due to the increase of biodegradation and reduction of fouling	Using poly aluminium chloride and chitosan as coagulants to control membrane permeability and fouling	Park et al. (2018)
	Triclosan, caffeine, carbendazim, and acetaminophen	UF integrated with adsorption	An adsorptive membrane: a coating of an oxidized CNT on the inside of the UF membrane (polyvinyl chloride with 0.012 μm average pore size) in a U-shaped, hollow fiber module	Enhancing the adsorption capacity of the membranes for all compounds up to 34%.	Using poly aluminium chloride as a coagulant for coagulation pretreatment	Wang et al. (2018)
Water Treatment By-products	THMs	–	GO nanosheets membranes (functionalized by -F, -OH, and -H)	Rejection of THMs, but high permeation for water molecules	Investigation of THMs removal with a molecular dynamics simulation method	Ansari et al. (2020)

(*Continued*)

Table 9.3 (Continued)

Ops type	Compounds	Membrane Technology	Membrane: type/ modules	Key Results	Additional Information	References
Pesticides	Atrazine, 2, 6-dichloro benzamide, and desethyl-desisopropyl-atrazine	FO	A biomimetic aquaporin membrane provided by Aquaporin A/S	The higher pure water flux and rejection for all pesticides (over 97%) compared to standard cellulose acetate FO membrane	A TFC membrane with aquaporin proteins integrated into the active polyamide layer	Madsen et al. (2015)
	2-methyl-4-chlorophenoxy acetic acid and 2-methyl-4-chlorophenoxy propionic acid (two phenoxy acid herbicides) and 2,6-di-chlorobenzamide (transformation product)	RO/NF	Commercial flat sheet polyamide TFC	The increase in recovery, leading to the rise of rejection attributed to the pore blocking phenomenon. Decreasing the rejection of charged pesticides like phenoxy acids with feed concentration	The experiment was done on three groundwater sources in Denmark.	Fini et al. (2019)
	Phorate, parathion-methyl, dichlorvos, atrazine, and clofibric acid	MD (direct contact membrane distillation mode)	Commercial polytetrafluoroethylene flat sheet membrane (with 70–80% porosity, 0.2 μm average pore size, 179 μm thickness, and 137° contact angle)	High rejection (70–99%) for compounds with low hydrophobicity and vapor pressure. Lowered rejection (30–50%) for compounds with high vapor pressure hydrophobicity at 75% water recovery.	–	Plattner et al. (2018)
Industrial compounds	Bisphenol A	Filtration membrane at the pressure of 1–2 bar	Polysulfone and polysulfone/ GO nanocomposite membrane (containing 1 and 4% GO) in plate modulus	Enhancing the pure water flux of the nanocomposite membrane compared with the polysulfone membrane. The highest removal of bisphenol A by composite membrane (4% GO).	–	Nasseri et al. (2018)

Surfactant	PFOA and salts including NaCl, $CaCl_2$, and Na_2SO_4.	NF	A negatively charged NF membrane (a loose selective polyamide layer on a polyether sulfone support layer) with approximately 1.2 nm pore diameter	High rejection (~90%) as well as decreased scaling, and higher water recovery in comparison to the commercial NF membrane	–	Boo et al. (2018)
	PFOA	NF	Commercial NF membrane (aromatic polyamide advanced composite membrane in the spiral wound configuration) MWCO: 200 Da	Removal efficiency by 99.49–99.54%	–	Boonya-Atichart et al. (2016)
Lifestyle Compound	Caffeine	Submerged anaerobic MBR	Flat sheet chlorinated polyethylene membrane (0.2 μm nominal pore size)	87.5% average removal efficiency	–	Chen et al. (2018)
	Caffeine, diclofenac, and octylphenol	Integrated adsorption-NF system	NF270 manufactured by Film-Tech with the surface area of 46.2 cm^2	The better performance of the AC-NF set. Enhancing the performance by coagulation- AC pretreatment (removal efficiency of all three compounds >95%)	Applying a combined AC adsorption-NF membrane system (in two orientations AC-NF and NF-AC)	Huang et al. (2019)

Source: Modified from Taheran et al., (2016)

due to their higher pore size compared with OPs. Nanofiltration membranes are suitable alternatives for OPs removal. Their MWCO is between 200 and 1000 Da, which covers the MWCO of OPs, especially PPCPs. Then, they were applied for various types of OPs with good removal efficiencies, mainly over 90%.

The NF membrane is more suitable than the RO membrane for the removal of OPs because of high removal efficiency, lower energy consumption, and improved water permeability. But, membrane technologies can physically separate the pollutants without degradation and produce massive waste. A combination of chemical transformation or biodegradation with membrane processes can enhance membrane performance.

References

Aarab, N., Hsini, A., Essekri, A. et al. (2020). Removal of an emerging pharmaceutical pollutant (metronidazole) using PPY-PANi copolymer: kinetics, equilibrium and DFT identification of adsorption mechanism. *Groundwater for Sustainable Development* 11: 100416.

Abd El-Lateef, H.M., Ali, M.M.K., and Saleh, M.M. (2018). Adsorption and removal of cationic and anionic surfactants using zero-valent iron nanoparticles. *Journal of Molecular Liquids* 268: 497–505.

Abe, A. (1999). Distribution of 1, 4-dioxane in relation to possible sources in the water environment. *Science of the Total Environment* 227: 41–47.

Acosta, R., Nabarlatz, D., Sánchez-Sánchez, A. et al. (2018). Adsorption of Bisphenol A on KOH-activated Tyre pyrolysis char. *Journal of Environmental Chemical Engineering* 6: 823–833.

Ahel, M., Schaffner, C., and Giger, W. (1996). Behaviour of alkylphenol polyethoxylate surfactants in the aquatic environment—III. Occurrence and elimination of their persistent metabolites during infiltration of river water to groundwater. *Water Research* 30: 37–46.

Ahrens, L. and Bundschuh, M. (2014). Fate and effects of poly-and perfluoroalkyl substances in the aquatic environment: A review. *Environmental Toxicology and Chemistry* 33: 1921–1929.

Al-Hawash, A.B., Dragh, M.A., Li, S. et al. (2018). Principles of microbial degradation of petroleum hydrocarbons in the environment. *The Egyptian Journal of Aquatic Research* 44: 71–76.

Al-Hawash, A.B., Zhang, X., and Ma, F. (2019). Removal and biodegradation of different petroleum hydrocarbons using the filamentous fungus Aspergillus sp. RFC-1. *MicrobiologyOpen* 8: e00619.

An, H.J., Bhadra, B.N., Khan, N.A., and Jhung, S.H. (2018). Adsorptive removal of wide range of pharmaceutical and personal care products from water by using metal azolate framework-6-derived porous carbon. *Chemical Engineering Journal* 343: 447–454.

Anastopoulos, I., Pashalidis, I., Orfanos, A.G. et al. (2020). Removal of caffeine, nicotine and amoxicillin from (waste) waters by various adsorbents. A review. *Journal of Environmental Management* 261: 110236.

Annunciacão, D.L.R., Almeida, F.V., and Sodre, F.F. (2017). Method development and validation for the determination of polybrominated diphenyl ether congeners in Brazilian aquatic sediments. *Microchemical Journal* 133: 43–48.

Anderson, R.H., Anderson, J.K., and Bower, P.A. (2012). Co-occurrence of 1, 4-dioxane with trichloroethylene in chlorinated solvent groundwater plumes at US air force installations: fact or fiction. *Integrated Environmental Assessment and Management* 8: 731–737.

Ansari, P., Azamat, J., and Khataee, A. (2020). Computational study on the removal of trihalomethanes from water using functionalized graphene oxide membranes. *Chemical Physics* 531: 110589.

Aus Der Beek, T., Weber, F.A., Bergmann, A. et al. (2016). Pharmaceuticals in the environment – global occurrences and perspectives. *Environmental Toxicology and Chemistry* 35: 823–835.

Bhadra, B.N., Ahmed, I., Kim, S., and Jhung, S.H. (2017). Adsorptive removal of ibuprofen and diclofenac from water using metal-organic framework-derived porous carbon. *Chemical Engineering Journal* 314: 50–58.

Boo, C., Wang, Y., Zucker, I. et al. (2018). High performance nanofiltration membrane for effective removal of perfluoroalkyl substances at high water recovery. *Environmental Science and Technology* 52: 7279–7288.

Boonya-Atichart, A., Boontanon, S.K., and Boontanon, N. (2016). Removal of perfluorooctanoic acid (PFOA) in groundwater by nanofiltration membrane. *Water Science and Technology* 74: 2627–2633.

Briganti, A., Tuccimei, P., Voltaggio, M. et al. (2020). Assessing methyl tertiary butyl ether residual contamination in groundwater using radon. *Applied Geochemistry* 116: 104583.

Beltrame, K.K., Cazetta, A.L., De Souza, P.S. et al. (2018). Adsorption of caffeine on mesoporous activated carbon fibers prepared from pineapple plant leaves. *Ecotoxicology and Environmental Safety* 147: 64–71.

Buerge, I.J., Poiger, T., Muller, M.D., and Buser, H.-R. (2003). Caffeine, an anthropogenic marker for wastewater contamination of surface waters. *Environmental Science and Technology* 37: 691–700.

Buerge, I.J., Buser, H.-R., Kahle, M. et al. (2009). Ubiquitous occurrence of the artificial sweetener acesulfame in the aquatic environment: an ideal chemical marker of domestic wastewater in groundwater. *Environmental Science and Technology* 43: 4381–4385.

Burgos-Castillo, R.C., Sires, I., Sillanpää, M., and Brillas, E. (2018). Application of electrochemical advanced oxidation to bisphenol A degradation in water. Effect of sulfate and chloride ions. *Chemosphere* 194: 812–820.

Carey, G.R., Mcgregor, R., Pham, A.L.T. et al. (2019). Evaluating the longevity of a PFAS in situ colloidal activated carbon remedy. *Remediation Journal* 29: 17–31.

Chen, W.-H. and Huang, J.-R. (2020). *Adsorption of Organic Including Pharmaceutical and Inorganic Contaminants in Water toward Graphene-Based Materials. Contaminants of Emerging Concern in Water and Wastewater*. Elsevier.

Chen, R., Jiang, H., and Li, Y.-Y. (2018). Caffeine degradation by methanogenesis: efficiency in anaerobic membrane bioreactor and analysis of kinetic behavior. *Chemical Engineering Journal* 334: 444 452.

Cruz, S. and Barceló, D. (2015). *Personal Care Products in the Aquatic Environment*. Springer.

De Araújo, B.R. and León, J.J.L. (2018). Electrochemical treatment of cetrimonium chloride with boron-doped diamond anodes. A technical and economical approach. *Journal of Environmental Management* 214: 86–93.

Dehghani, M.H., Niasar, Z.S., Mehrnia, M.R. et al. (2017). Optimizing the removal of organophosphorus pesticide malathion from water using multi-walled carbon nanotubes. *Chemical Engineering Journal* 310: 22–32.

Dehghani, M.H., Kamalian, S., Shayeghi, M. et al. (2019). High-performance removal of diazinon pesticide from water using multi-walled carbon nanotubes. *Microchemical Journal* 145: 486–491.

Delgado, N., Capparelli, A., Navarro, A., and Marino, D. (2019). Pharmaceutical emerging pollutants removal from water using powdered activated carbon: study of kinetics and adsorption equilibrium. *Journal of Environmental Management* 236: 301–308.

Dobaradaran, S., Fard, E.S., Tekle-Röttering, A. et al. (2020). Age-sex specific and cause-specific health risk and burden of disease induced by exposure to trihalomethanes (THMs) and haloacetic acids (HAAs) from drinking water: an assessment in four urban communities of Bushehr Province, Iran, 2017. *Environmental Research* 182: 109062.

Dominguez, C.M., Oturan, N., Romero, A. et al. (2018). Removal of organochlorine pesticides from lindane production wastes by electrochemical oxidation. *Environmental Science and Pollution Research* 25: 34985–34994.

Dos Santos Lins, P.V., Henrique, D.C., Ide, A.H. et al. (2019). Evaluation of caffeine adsorption by MgAl-LDH/biochar composite. *Environmental Science and Pollution Research* 26: 31804–31811.

Dubey, S., Gusain, D., Sharma, Y.C., and Bux, F. (2020). *The Occurrence of Various Types of Disinfectant by-Products (Trihalomethanes, Haloacetic Acids, Haloacetonitrile) in Drinking Water. Disinfection By-products in Drinking Water*. Elsevier.

Dudley, L.-A. 2012. Removal of Perfluorinated Compounds by Powdered Activated Carbon, Superfine Powder Activated Carbon, and Anion Exchange Resin Unpublished thesis.

Ebele, A.J., Oluseyi, T., Drage, D.S. et al. (2020). Occurrence, seasonal variation and human exposure to pharmaceuticals and personal care products in surface water, groundwater and drinking water in Lagos state, Nigeria. *Emerging Contaminants* 6: 124–132.

Fang, Z., Hu, Y., Wu, X. et al. (2018). A novel magnesium ascorbyl phosphate graphene-based monolith and its superior adsorption capability for bisphenol A. *Chemical Engineering Journal* 334: 948–956.

Fang, Z., Gao, Y., Wu, X. et al. (2020). A critical review on remediation of bisphenol S (BPS) contaminated water: efficacy and mechanisms. *Critical Reviews in Environmental Science and Technology* 50: 476–522.

Fanourakis, S.K., Peña-Bahamonde, J., Bandara, P.C., and Rodrigues, D.F. (2020). Nano-based adsorbent and photocatalyst use for pharmaceutical contaminant removal during indirect potable water reuse. *npj Clean Water* 3: 1–15.

Ferrando, L. and Matamoros, V. (2020). Attenuation of nitrates, antibiotics and pesticides from groundwater using immobilised microalgae-based systems. *Science of the Total Environment* 703: 134740.

Fini, M.N., Madsen, H.T., and Muff, J. (2019). The effect of water matrix, feed concentration and recovery on the rejection of pesticides using NF/RO membranes in water treatment. *Separation and Purification Technology* 215: 521–527.

Fromme, H., Becher, G., Hilger, B., and Völkel, W. (2016). Brominated flame retardants–exposure and risk assessment for the general population. *International Journal of Hygiene and Environmental Health* 219: 1–23.

Garcia-Morales, R., Rodríguez-Delgado, M., Gomez-Mariscal, K. et al. (2015). Biotransformation of endocrine-disrupting compounds in groundwater: bisphenol A, nonylphenol, ethynylestradiol and triclosan by a laccase cocktail from Pycnoporus sanguineus CS43. *Water, Air, and Soil Pollution* 226: 251.

Ghoochani, M., Rastkari, N., Heibati, B. et al. (2017). Risk assessment of haloacetic acids in the water supply of Tehran, Iran. *Water Science and Technology: Water Supply* 17: 958–965.

Gomes, J., Costa, R., Quinta-Ferreira, R.M., and Martins, R.C. (2017). Application of ozonation for pharmaceuticals and personal care products removal from water. *Science of the Total Environment* 586: 265–283.

Hasan, Z., Jeon, J., and Jhung, S.H. (2012). Adsorptive removal of naproxen and clofibric acid from water using metal-organic frameworks. *Journal of Hazardous Materials* 209: 151–157.

Heo, J., Yoon, Y., Lee, G. et al. (2019). Enhanced adsorption of bisphenol A and sulfamethoxazole by a novel magnetic CuZnFe2O4–biochar composite. *Bioresource Technology* 281: 179–187.

Hontela, A. and Habibi, H.R. (2013). *Personal Care Products in the Aquatic Environment: A Case Study on the Effects of Triclosan in Fish. Fish Physiology*. Elsevier.

Huang, Z., Gong, B., Huang, C.-P. et al. (2019). Performance evaluation of integrated adsorption-nanofiltration system for emerging compounds removal: exemplified by caffeine, diclofenac and octylphenol. *Journal of Environmental Management* 231: 121–128.

Hyldegaard, B.H., Jakobsen, R., Weeth, E.B. et al. (2019). Challenges in electrochemical remediation of chlorinated solvents in natural groundwater aquifer settings. *Journal of Hazardous Materials* 368: 680–688.

Isaac, J. and Scheinman, P.L. (2017). Benzalkonium chloride: an irritant and sensitizer. *Dermatitis* 28: 346–352.

Jain, B., Singh, A.K., Kim, H. et al. (2018). Treatment of organic pollutants by homogeneous and heterogeneous Fenton reaction processes. *Environmental Chemistry Letters* 16: 947–967.

Jaria, G., Calisto, V., Silva, C.P. et al. (2019). Obtaining granular activated carbon from paper mill sludge–a challenge for application in the removal of pharmaceuticals from wastewater. *Science of the Total Environment* 653: 393–400.

Jjemba, P.K. (2018). *Pharma-Ecology: The Occurrence and Fate of Pharmaceuticals and Personal Care Products in the Environment*. Wiley.

Kanakaraju, D., Glass, B.D., and Oelgemöller, M. (2018). Advanced oxidation process-mediated removal of pharmaceuticals from water: a review. *Journal of Environmental Management* 219: 189–207.

Kapelewska, J., Kotowska, U., Karpińska, J. et al. (2018). Occurrence, removal, mass loading and environmental risk assessment of emerging organic contaminants in leachates, groundwaters and wastewaters. *Microchemical Journal* 137: 292–301.

Karges, U., Becker, J., and Puttmann, W. (2018). 1, 4-Dioxane pollution at contaminated groundwater sites in western Germany and its distribution within a TCE plume. *Science of the Total Environment* 619: 712–720.

Kaur, H., Bansiwal, A., Hippargi, G., and Pophali, G.R. (2018). Effect of hydrophobicity of pharmaceuticals and personal care products for adsorption on activated carbon: adsorption isotherms, kinetics and mechanism. *Environmental Science and Pollution Research* 25: 20473–20485.

Kawabe, Y. and Komai, T. (2019). A case study of natural attenuation of chlorinated solvents under unstable groundwater conditions in Takahata, Japan. *Bulletin of Environmental Contamination and Toxicology* 102: 280–286.

Knepper, T.P. and Berna, J.L. (2003). Surfactants: properties, production, and environmental aspects. *Comprehensive Analytical Chemistry* 40: 1–49.

Kotowska, U., Kapelewska, J., Kotowski, A., and Pietuszewska, E. (2019). Rapid and sensitive analysis of hormones and other emerging contaminants in groundwater using ultrasound-assisted emulsification microextraction with solidification of floating organic droplet followed by GC-MS detection. *Water* 11: 1638.

Kucharzyk, K.H., Rectanus, H.V., Bartling, C.M. et al. (2019). Use of omic tools to assess methyl tert-butyl ether (MTBE) degradation in groundwater. *Journal of Hazardous Materials* 378: 120618.

Kuczyńska, A. (2019). Presence of pharmaceutical compounds in groundwater with respect to land use in the vicinity of sampling sites. *Geologos* 25: 231–240.

La Vigna, F., Sbarbati, C., Bonfa, I. et al. (2019). First survey on the occurrence of chlorinated solvents in groundwater of eastern sector of Rome. *Rendiconti Lincei. Scienze Fisiche e Naturali* 30: 297–306.

Lapworth, D., Baran, N., Stuart, M., and Ward, R. (2012). Emerging organic contaminants in groundwater: a review of sources, fate and occurrence. *Environmental Pollution* 163: 287–303.

Lawrence, J.R., Waiser, M.J., Swerhone, G.D. et al. (2019). N, N-diethyl-m-Toluamide exposure at an environmentally relevant concentration influences river microbial community development. *Environmental Toxicology and Chemistry* 38: 2414–2425.

Lechuga, M., Fernández-Serrano, M., Jurado, E. et al. (2016). Acute toxicity of anionic and non-ionic surfactants to aquatic organisms. *Ecotoxicology and Environmental Safety* 125: 1–8.

Li, S., Wen, J., He, B. et al. (2020). Occurrence of caffeine in the freshwater environment: implications for ecopharmacovigilance. *Environmental Pollution* 263: 114371.

Liang, C., Zhang, L., Guo, H. et al. (2019). Photo-removal of 2, 2′ 4, 4′-tetrabromodiphenyl ether in liquid medium by reduced graphene oxide bridged artificial Z-scheme system of Ag@ Ag3PO4/g-C3N4. *Chemical Engineering Journal* 361: 373–386.

Liu, S., Huang, W., Shu, W. et al. (2018). Performance of L-493 macroprous resin for adsorption of trihalomethanes from water. *Water, Air, and Soil Pollution* 229: 242.

Madsen, H.T., Bajraktari, N., Helix-Nielsen, C. et al. (2015). Use of biomimetic forward osmosis membrane for trace organics removal. *Journal of Membrane Science* 476: 469–474.

Mahouachi, L., Rastogi, T., Palm, W.-U. et al. (2020). Natural clay as a sorbent to remove pharmaceutical micropollutants from wastewater. *Chemosphere* 258: 127213.

Mandal, A., Singh, N., and Purakayastha, T. (2017). Characterization of pesticide sorption behaviour of slow pyrolysis biochars as low cost adsorbent for atrazine and imidacloprid removal. *Science of the Total Environment* 577: 376–385.

Mansour, F., Al-Hindi, M., Yahfoufi, R. et al. (2018). The use of activated carbon for the removal of pharmaceuticals from aqueous solutions: a review. *Reviews in Environmental Science and Bio/Technology* 17: 109–145.

Marican, A. and Durán-Lara, E.F. (2018). A review on pesticide removal through different processes. *Environmental Science and Pollution Research* 25: 2051–2064.

Medjor, W., Namessan, O., and Medjor, E.A. (2018). Optimization, kinetics, physicochemical and ecotoxicity studies of Fenton oxidative remediation of hydrocarbons contaminated groundwater. *Egyptian Journal of Petroleum* 27: 227–233.

Merel, S. and Snyder, S.A. (2016). Critical assessment of the ubiquitous occurrence and fate of the insect repellent N, N-diethyl-m-toluamide in water. *Environment International* 96: 98–117.

Miao, Y., Johnson, N.W., Heck, K. et al. (2018). Microbial responses to combined oxidation and catalysis treatment of 1, 4-dioxane and co-contaminants in groundwater and soil. *Frontiers of Environmental Science and Engineering* 12: 2.

Munoz, G., Labadie, P., Botta, F. et al. (2017). Occurrence survey and spatial distribution of perfluoroalkyl and polyfluoroalkyl surfactants in groundwater, surface water, and sediments from tropical environments. *Science of the Total Environment* 607: 243–252.

Nasseri, S., Ebrahimi, S., Abtahi, M., and Saeedi, R. (2018). Synthesis and characterization of polysulfone/graphene oxide nano-composite membranes for removal of bisphenol A from water. *Journal of Environmental Management* 205: 174–182.

Němeček, J., Steinová, J., Špánek, R. et al. (2018). Thermally enhanced in situ bioremediation of groundwater contaminated with chlorinated solvents–A field test. *Science of the Total Environment* 622: 743–755.

Nowak, A., Pacek, G. and Mrozik, A. 2020. *Transformation and Ecotoxicological Effects of Iodinated X-Ray Contrast Media*.

Ouyang, Z., Huang, Z., Tang, X. et al. (2019). A dually charged nanofiltration membrane by pH-responsive polydopamine for pharmaceuticals and personal care products removal. *Separation and Purification Technology* 211: 90–97.

Özdemir, K. and Gungör, Ö. (2018). Development of statistical models for trihalomethane (THM) removal in drinking water sources using carbon nanotubes (CNTs). *Water SA* 44: 680–690.

Patel, M., Kumar, R., Kishor, K. et al. (2019). Pharmaceuticals of emerging concern in aquatic systems: chemistry, occurrence, effects, and removal methods. *Chemical Reviews* 119: 3510–3673.

Park, J., Yamashita, N., and Tanaka, H. (2018). Membrane fouling control and enhanced removal of pharmaceuticals and personal care products by coagulation-MBR. *Chemosphere* 197: 467–476.

Park, K.-Y., Yu, Y.-J., Yun, S.-J., and Kweon, J.-H. (2019). Natural organic matter removal from algal-rich water and disinfection by-products formation potential reduction by powdered activated carbon adsorption. *Journal of Environmental Management* 235: 310–318.

Park, M., Wu, S., Lopez, I.J. et al. (2020). Adsorption of perfluoroalkyl substances (PFAS) in groundwater by granular activated carbons: roles of hydrophobicity of PFAS and carbon characteristics. *Water Research* 170: 115364.

Poi, G., Shahsavari, E., Aburto-Medina, A. et al. (2018). Large scale treatment of total petroleum-hydrocarbon contaminated groundwater using bioaugmentation. *Journal of Environmental Management* 214: 157–163.

Plattner, J., Kazner, C., Naidu, G. et al. (2018). Removal of selected pesticides from groundwater by membrane distillation. *Environmental Science and Pollution Research* 25: 20336–20347.

Pulce, C. and Descotes, J. (1996). *Household Products. Human Toxicology*. Elsevier.

Qian, H., Zhang, Y., Wang, J. et al. (2018). Characteristics of petroleum-contaminated groundwater during natural attenuation: a case study in Northeast China. *Environmental Monitoring and Assessment* 190: 80.

Qian, L., Chen, Y., Ouyang, D. et al. (2020). Field demonstration of enhanced removal of chlorinated solvents in groundwater using biochar-supported nanoscale zero-valent iron. *Science of the Total Environment* 698: 134215.

Rahman, M.F., Peldszus, S., and Anderson, W.B. (2014). Behaviour and fate of perfluoroalkyl and polyfluoroalkyl substances (PFASs) in drinking water treatment: a review. *Water Research* 50: 318–340.

Robertson, W., Van Stempvoort, D., Roy, J. et al. (2016). Use of an artificial sweetener to identify sources of groundwater nitrate contamination. *Groundwater* 54: 579–587.

Rodríguez-Escales, P., Borras, E., Sarra, M., and Folch, A. (2013). Granulometry and surfactants, key factors in desorption and biodegradation (T. Versicolor) of PAHs in soil and groundwater. *Water, Air, and Soil Pollution* 224: 1422.

Safdari, M.-S., Kariminia, H.-R., Rahmati, M. et al. (2018). Development of bioreactors for comparative study of natural attenuation, biostimulation, and bioaugmentation of petroleum-hydrocarbon contaminated soil. *Journal of Hazardous Materials* 342: 270–278.

Seo, P.W., Bhadra, B.N., Ahmed, I. et al. (2016). Adsorptive removal of pharmaceuticals and personal care products from water with functionalized metal-organic frameworks: remarkable adsorbents with hydrogen-bonding abilities. *Scientific Reports* 6: 34462.

Serra-Roig, M.P., Jurado, A., Díaz-Cruz, M.S. et al. (2016). Occurrence, fate and risk assessment of personal care products in river–groundwater interface. *Science of the Total Environment* 568: 829–837.

Shankar, A., Kongot, M., Saini, V.K., and Kumar, A. (2020). Removal of pentachlorophenol pesticide from aqueous solutions using modified chitosan. *Arabian Journal of Chemistry* 13: 1821–1830.

Sharma, A. and Chaudhry, S.A. (2020). *Adsorption of Pharmaceutical Pollutants Using Lignocellulosic Materials. Green Materials for Wastewater Treatment*. Springer.

Sharma, B.M., Bečanová, J., Scheringer, M. et al. (2019). Health and ecological risk assessment of emerging contaminants (pharmaceuticals, personal care products, and artificial sweeteners) in surface and groundwater (drinking water) in the Ganges River Basin, India. *Science of the Total Environment* 646: 1459–1467.

Sililo, O.T. (1999). Groundwater contamination by organic chemicals in industrializing countries: the unseen threat. *IAHS Publication*: 23–28.

Sjerps, R.M., Kooij, P.J., Van Loon, A., and Van Wezel, A.P. (2019). Occurrence of pesticides in Dutch drinking water sources. *Chemosphere* 235: 510–518.

Soh, L., Connors, K.A., Brooks, B.W., and Zimmerman, J. (2011). Fate of sucralose through environmental and water treatment processes and impact on plant indicator species. *Environmental Science and Technology* 45: 1363–1369.

Sonia, T.A. and Sharma, C.P. (2014). 5-lipids and inorganic nanoparticles in oral insulin delivery. In: *Oral Delivery of Insulin*, 219–256. Cambridge: Woodhead Publishing.

Sönmez, G., Bahadir, T., and Işık, M. (2020). Removal of selected pharmaceuticals from tap water by the Fenton process. *International Journal of Environmental Analytical Chemistry* 100: 1–13.

Srivastav, A.L. (2020). *Chemical Fertilizers and Pesticides: Role in Groundwater Contamination. Agrochemicals Detection, Treatment and Remediation*. Elsevier.

Stuart, M. and Lapworth, D. (2013). *Emerging Organic Contaminants in Groundwater. Smart Sensors for Real-Time Water Quality Monitoring*. Springer.

Taheran, M., Brar, S.K., Verma, M. et al. (2016). Membrane processes for removal of pharmaceutically active compounds (PhACs) from water and wastewaters. *Science of the Total Environment* 547: 60–77.

Tawabini, B. and Makkawi, M. (2018). Remediation of MTBE-contaminated groundwater by integrated circulation wells and advanced oxidation technologies. *Water Science and Technology: Water Supply* 18: 399–407.

Tran, H.N., Tomul, F., Nguyen, H.T.H. et al. (2020). Innovative spherical biochar for pharmaceutical removal from water: insight into adsorption mechanism. *Journal of Hazardous Materials* 394: 122255.

Varga, M., Elabadsa, M., Tatár, E., and Mihucz, V.G. (2019). Removal of selected pharmaceuticals from aqueous matrices with activated carbon under batch conditions. *Microchemical Journal* 148: 661–672.

Vorontsov, A.V. (2019). Advancing Fenton and photo-Fenton water treatment through the catalyst design. *Journal of Hazardous Materials* 372: 103–112.

Wang, J. and Wang, S. (2016). Removal of pharmaceuticals and personal care products (PPCPs) from wastewater: a review. *Journal of Environmental Management* 182: 620–640.

Wang, J. and Wang, S. (2019). Preparation, modification and environmental application of biochar: a review. *Journal of Cleaner Production* 227: 1002–1022.

Wang, Y., Huang, H., and Wei, X. (2018). Influence of wastewater precoagulation on adsorptive filtration of pharmaceutical and personal care products by carbon nanotube membranes. *Chemical Engineering Journal* 333: 66–75.

Wang, X., Wang, Z., Tang, Y. et al. (2019). Oxidative degradation of iodinated X-ray contrast media (iomeprol and iohexol) with sulfate radical: an experimental and theoretical study. *Chemical Engineering Journal* 368: 999–1012.

Weber, K. and Quicker, P. (2018). Properties of biochar. *Fuel* 217: 240–261.

World Health Organization 2012. Pharmaceuticals in drinking-water.

Wu, Z., Han, W., Yang, X. et al. (2019). The occurrence of polybrominated diphenyl ether (PBDE) contamination in soil, water/sediment, and air. *Environmental Science and Pollution Research* 26: 1–23.

Xu, R., Zhang, P., Wang, Q. et al. (2019). Influences of multi influent matrices on the retention of PPCPs by nanofiltration membranes. *Separation and Purification Technology* 212: 299–306.

Yang, Y., Ok, Y.S., Kim, K.-H. et al. (2017). Occurrences and removal of pharmaceuticals and personal care products (PPCPs) in drinking water and water/sewage treatment plants: A review. *Science of the Total Environment* 596: 303–320.

Yang, J., Meng, L., and Guo, L. (2018). In situ remediation of chlorinated solvent-contaminated groundwater using ZVI/organic carbon amendment in China: field pilot test and full-scale application. *Environmental Science and Pollution Research* 25: 5051–5062.

Yoo, D.K., An, H.J., Khan, N.A. et al. (2018). Record-high adsorption capacities of polyaniline-derived porous carbons for the removal of personal care products from water. *Chemical Engineering Journal* 352: 71–78.

Zhai, Y., Yin, Z., Zhao, X. et al. (2018). Polycyclic aromatic hydrocarbons (PAHs) in the environment of Beijing, China: levels, distribution, trends and sources. *Human and Ecological Risk Assessment: An International Journal* 24: 137–157.

Zhang, Y., Jin, F., Shen, Z. et al. (2019). Adsorption of methyl tert-butyl ether (MTBE) onto ZSM-5 zeolite: fixed-bed column tests, breakthrough curve modelling and regeneration. *Chemosphere* 220: 422–431.

Zhang, Z., Zhang, W., and Lichtfouse, E. (2020). *Membranes for Environmental Applications.* Springer.

Zhao, X., Qiu, W., Zheng, Y. et al. (2019). Occurrence, distribution, bioaccumulation, and ecological risk of bisphenol analogues, parabens and their metabolites in the Pearl River Estuary, South China. *Ecotoxicology and Environmental Safety* 180: 43–52.

10

Impact of Industrial Effluents on Groundwater

Rishabh Jain[1,2], Anupma Thakur[1,2], Neera Garg[1,2], and Pooja Devi[1,2]

[1] *Academy of Scientific and Innovative Research (AcSIR), Ghaziabad, Uttar Pradesh, India*
[2] *Central Scientific Instruments Organisation, Chandigarh, India*

10.1 Introduction

Our planet Earth is often referred to as "the blue planet," since approximately 71% of its surface is covered with water. According to various global agencies including NASA (National Aeronautics and Space Administration), IWA (International Water Association), and UNESCO (United Nations Educational, Scientific and Cultural Organization), of all the water content present on Earth, only about 2.5% is freshwater content (Shiklomanov 2008). The direct utilization of the remaining portion is not feasible due to its salinity. While looking specifically at groundwater, this percentage is further reduced to 0.62% of the total water content on the planet (USBR 2019). The supplies of groundwater replenish through surface runoff, melting of glaciers or snow, and other forms of precipitation like rain (Heath 2004). Groundwater is the primary source of water consumption for half of the world as due to its quality, groundwater is considered fit for consumption for the living organisms. In particular, in countries like India, China, and the United States, it is utilized to a large extent to meet daily needs by humans (Margat and Van der Gun 2013). However, it is also the topmost extracted raw material from the earth crust to serve the three major sectors, i.e. agriculture, domestic, and industrial (Margat and Van der Gun 2013; Gleeson et al. 2016). For example, in agricultural countries like India, most irrigation activities use groundwater, to the extent of 40% (Siebert et al. 2010).

Groundwater has several advantages over surface water in terms of the absence of suspended matter, free from pathogenic bacteria, colourlessness, and constant temperature, although the chemical composition varies region-wise based upon the minerals present in the Earth crust in that region. The quality of groundwater is thus globally assessed in terms of water quality index (WQI). WQI, often determined by measuring the standard physicochemical parameters (e.g. pH, hardness, colour, biological oxygen demand [BOD], chemical oxygen demand [COD], turbidity, and total dissolved solids [TDS]), the

Groundwater Geochemistry: Pollution and Remediation Methods, First Edition.
Edited by Sughosh Madhav and Pardeep Singh.

concentration of dissolved ions (e.g. nitrate, sodium, potassium, aluminium, fluoride, chloride, and sulphides), heavy metals (arsenic, selenium, lead, mercury, and copper), and other minerals (Smith 1982). This WQI is mostly affected by the unregulated release of wastewater generated by these entities in water bodies without proper treatment, which is of prime environmental concern. In addition to the contamination, the significant increase in the utilization of groundwater put more stress on the groundwater table. This increased utilization is due to the escalated modernisation, technological advancement, industrialization, and urbanization.

In early times, wastewater management was considered a significant challenge and an environmental concern. For example, the sequence of events occurring during the transitional phase of the nineteenth and twentieth century led to the development of first centralized sewage systems in the United States and the United Kingdom. However, there was also exponential growth in the industrial sector, rise in population, and urbanization. This growth had adverse effects on the natural water resources, and hence the need to handle wastewater has arisen. During that critical period, the volume of generated wastewater was multiplying daily, and so was the pressure on government bodies and environmental agencies to handle the situation. At that time, there were various operational and technological challenges for the environment prevention agencies and governments for the management of industrial wastewater. During this period, especially in developing countries, around 90% of wastewater discharged directly into the environment without any pre-treatment. Thus, to deal with this crisis globally, an International Water Supply Association was founded just after the Second World War in 1947. Later, this foundation was accompanied by the International Association for Water Pollution Research in 1965, which then merged into a single body, which is known today as 'IWA'.

In the current scenario, roughly 80% of the untreated wastewater discharges into the environment, and around 1.7 billion individuals are dependent on groundwater and thus are at high health risk. Industrial and domestic effluents play a significant role in the distortion of groundwater quality and resources. The major industrial effluents which affect the groundwater quality are metals, metalloids, colouring dyes, surfactants, minerals, acids, and PPCPs. The primary industry sectors responsible for the outflow of these industrial effluents are paper mill industry, pharmaceutical industry, textile industry, brewery industry, soaps and detergent industry, soft-drink industry, tannery, and chemical industry. As a result, the quality and availability of clean potable water is declining daily. This contamination problem exists more in developing countries due to high input and maintenance cost of treatment plants, and lack of stringent regulatory practices of industrial/domestic discharge. In order to regulate the release of the chemical pollutants from different sources, various organizations like WHO (World Health Organization), UNESCO, USEPA (US Environmental Protection Agency), and BIS (Bureau of Indian Standards) are working on setting up their permissible limits (PL) in the groundwater. These safe limits ensure the usability of available groundwater in terms of quality and safety for end users.

According to UNESCO, more than one billion people live in countries with water scarcity and lack of access to water fit for consumption. This water shortage is also predicted to be the biggest concern in the future and for the world economy.

10.2 Industrial Effluents Affecting the Groundwater Quality

In today's era, many industrial effluents are making their ways into the groundwater and thus affecting its quality. This section has listed and discussed potential chemical threats, which are exponentially increasing their presence, and are of significant health concerns. These threats include heavy metals (e.g. arsenic, chromium, and lead), nitrates, synthetic dyes (e.g. dinitroaniline orange), pharmaceutical and personal care products (e.g. diclofenac, sulphonamides, and triclosan), and persistent organic pollutants (POPs) (e.g. polychlorinated biphenyls [PCBs] and furan).

10.2.1 Heavy Metals

Heavy metals are well-known for their high toxicity due to their non-biodegradable nature and bioaccumulation properties. The exponential increase in concentration of the heavy metal ions in groundwater has led to severe health effects on humans, including neurological, immunological, muscular, and reproductive disorders (World Health Organization 1996; Kaur et al. 2020). Thus, the Agency for Toxic Substances and Disease Registry and USEPA have jointly prepared the Priority List of Hazardous Substances. In the list, the elements such as mercury (Hg), arsenic (As), cadmium (Cd), selenium (Se), and lead (Pb) are indicated as the potential threats for the planet (Wasana et al. 2017). Naturally, these toxic elements exist in various regions, and often in unpredictable concentrations, depending upon the geological and climatic conditions. For example, arsenic ore exists as arsenate with iron and manganese (La Force et al. 2000). Though they have been widely used in various industrial sectors, as listed in Table 10.1, their associated adverse health effects on exposure beyond a permissible limit are of great concern. Water has been a significant source for heavy metals to enter into a biological system, as many industrial effluents make their way into groundwater, which is a primary source of consumption for drinking and other allied activities. Further, the high toxicity of heavy metals is attributed to their high density, sorbent nature, non-biodegradability, dissociation in water, complexation with inorganic and organic ligands (Mance 2012).

After considering the potential risks associated with heavy metal contamination, several international and national agencies accountable for the safety of the environment (including the WHO, USEPA, BIS, and the European Union) set the minimum PL for their consumption, shown in Table 10.1 (Martin and Griswold 2009). These limits are specifically for water matrix, to avoid the risks of getting affected by their elevated concentration through consumption. These elevated concentrations of the heavy metals in groundwater not only endanger human life but also affect crop production, aquatic life, and animals (Jaishankar et al. 2014). The wastewater discharges from industries, particularly from plating, mining, surface finishing and milling industries serve as the primary source of heavy metal contamination of groundwater.

10.2.2 Nitrates

The role of nitrates in affecting the quality of groundwater is very prominent, similar to heavy metals, and is a growing global concern. Nitrates are the part of the nitrogen cycle, consist of nitrogen and oxygen, and have negative footprints over the ecosystems (Shuval

Table 10.1 Heavy Metals: USEPA permissible limits (PL), applications, and health concerns (World Health Organization 2008; Martin and Griswold 2009).

S/N	Element (PL)	Occurrence/Applications	Health Concerns
1	Arsenic (0.010 ppm)	Dyes, paints, metals, semiconductors, drugs, soaps, fertilizers, and pesticides	Affects the production of red blood cells and white blood cells, darkening of the skin, the appearance of spots on the skin, lung cancer, skin cancer, liver disease, and at worst, death
2	Barium (2 ppm)	Fluorescent lamps, sparkplug electrodes, fireworks, bricks, and rubber	Fluctuation in blood pressure, weakness, diarrhoea, change in heartbeat pattern, paralysis, and at worst, it may even lead to death
3	Cadmium (0.005 ppm)	Fertilizers, electroplating, batteries, coatings, and plastics	Vomiting, diarrhoea, lung damage, breakable bones, kidney disease, and cancer
4	Chromium (0.1 ppm)	Alloy industry, protective coatings, cement, rubber, magnetic tapes, floor covering, and wood preservatives	Nose ulcers, breathing problems, cough, skin allergy, damage to nerve tissues, circulatory system, lungs, kidney, and may cause cancer as well
5	Copper (1.3 ppm)	Electrical wiring, fittings, valves, coins, kitchen utensils, currency coins, preservatives, fertilizers, refining, dye manufacturing, food additives, and building materials	Weakness, nausea, diarrhoea, hemolytic anaemia, jaundice, hypotension, cyanosis, damage liver, weakens the central nervous system, and in the worst case, may even lead to death
6	Lead (0.015 ppm)	Paints, batteries, X-ray shielding devices, ammunition, gasoline, pipe solder, toys, and cosmetics	Affects the nervous system, brain damage, kidney damage, inhibits sperm production in men, leads to miscarriage in pregnant women
7	Mercury (0.002 ppm)	Dental fillings, thermometers, light bulbs, switches, batteries, production of caustics soda, and chlorine gas	Kidney damage, liver damage, lungs damage, foetus damage, memory problems, change in vision, increase in heart rate or blood pressure, and death
8	Selenium (0.050 ppm)	Plastics, glass industry, paints, ink, enamels, rubber, and electronics industry	Selenosis, nausea, vomiting, diarrhoea, neurological disorders, nail brittleness, respiratory problems, and bronchitis
9	Silver (0.0001 ppm)	Solders, jewellery, electronic components, conductors, electrical contacts, and alloys	Argyria, breathing problems, stomach pains, skin allergy, throat and lung infection
10	Zinc (5 ppm)	Galvanization, corrosion-resistant alloys, rubber, and pesticides	Fever, vomiting, pulmonary distress, nausea, chills, diarrhoea, gastroenteritis, and manifests copper deficiencies

and Gruener 1972). Apart from crop agriculture, the wastewater discharges from industries, especially from paper mills, the explosive industry, and the agrichemical industry, contribute a significant portion of nitrates to the groundwater (Fadiran and Mamba 2005; Jatto et al. 2010; Alrumman et al. 2016). The permissible limit for the nitrate is 10 ppm set by European protection agency (EPA), to prevent groundwater contamination and probable environmental concerns (Kumar and Puri 2012). Further, nitrates can quickly sweep into the groundwater from the surface due to their excellent solubility, without getting affected by the leach fields. The primary source of its exposure to humans has been drinking water, crops, and non-vegetarian diets. In the human body upon consumption, first, it gets converted into nitrite by enzymatic reduction processes, carried out primarily by intestinal bacteria and oral commensal bacteria. Afterwards, in the stomach by non-enzymatic disproportionation under the acidic environment, nitrites may get reduced to nitric oxide. This reduction can occur in various other tissues of the human body by several biochemical reactions under different conditions and may lead to cardiovascular diseases. The excess consumption of nitrates leads to clinical health effects in the form of nitrate poisoning, which includes clinical symptoms such as vomiting, breathing difficulty, bluish skin, nausea, diarrhoea, weakness, and convulsions. The infants have a higher risk of getting affected in comparison to adults, due to lack of oxygen. Thus, many times, nitrate poisoning is also referred to as blue baby syndrome (Shuval and Gruener 2013). Several reports have also indicated the occurrence of methemoglobinemia and disruption of thyroid function in infants due to nitrate poisoning (Comly 1945; Lutyński et al. 1996). There are reports which indicate that nitrate affects not only the infants but also adults, especially pregnant women and people with pre-existing diseases. This poisoning increases the cancer risk in humans, especially for patients with bacterial overgrowth and achlorhydria (Linsha et al. 2018). In addition to that, excessive nitrate intake through drinking water during pregnancy may lead to premature babies, unwanted abortion, foetal deaths, congenital disabilities, retardation, and neonatal deaths (Ward et al. 2018).

Based on these facts and reports, many countries and environmental agencies opted for various strategies, legal policies, and awareness programmes. For example, the United States Department of Agriculture (USPA) has made a zero-tolerance policy towards adding nitrates in the diets of infants and children. Likewise, Germany and South Africa have strict PL of 4.4 ppm in comparison to US and EPA standards, which is 10 ppm (Kross 2002). Due to the health concerns raised by nitrates, various European countries banned the use of nitrates in baby bottle nipples. Even though these many measures are taken up by the international organization, exponentially increasing concentration of nitrates in the groundwater makes it a potential threat to the living ecosystem. In India, reports from the Central Groundwater Board indicate the increasing levels of nitrates in the groundwater above PL across 19 states (Suthar et al. 2009). The current practices of filtration, chlorination, boiling, and water softeners are not sufficient for the removal of nitrates from water. Alternative technologies based upon reverse osmosis, ion exchange, and electrodialysis are in practice. (Shrimali and Singh 2001).

10.2.3 Synthetic Dyes

Due to the importance of colour, the industrial production of synthetic dyes has gained importance in textile dyeing, printing inks, food colouring, and paint pigments (International Agency for Research on Cancer 2010; Liu et al. 2012). Of these, textile industries are a topmost user of dyes and have around 66% of the market share. These synthetic dyes provide not only adequate colour strength but are also cost-effective due to production enhancement. Despite the advantages, these dyes have environmental concerns when they enter into the groundwater, making the water unfit for consumption (Khattab et al. 2020). These synthetic dyes are generally categorized into four groups based on chemical and physical properties. The four major groups are azo dyes, phthalocyanine dyes, anthraquinone dyes, and formazan dyes (Cancer 2010). Among them, azo dyes, chiefly derived from benzidine (e.g. dinitroaniline orange), could be cancer-causing, toxic, and mutagenic to various living species, including humans. Due to these reasons, the European Union countries, including Germany (Narvekar and Srivastava 2002; Ahlström et al. 2005), have banned or restricted use of this category of dyes. However, in developing countries, most of the dyes are used by the industries without toxicity evaluation, and these are often released into surface waters or drainage system (Mathur and Bhatnagar 2007). From there, these dye effluents leak into the groundwater table and become part of the biological system.

In terms of health effects, numerous cases of skin irritation are reported by dermatologists due to contact of skin with synthetic dyes (Hassaan and El Nemr 2017). Further, these dyes can cause rhinitis, dermatitis, and asthma to the person working directly with them (Helaskoski et al. 2014). Sometimes, heavy metals associated with the dyes, especially in the case of phthalocyanine dyes, are found to cause toxicity. The wastewater from dye utilizing industries, especially textile industries, contains not only dyes but also acids, wetting agents, oxidizing agents, and detergents (Liu et al. 2012). Thus, the generated wastewater has high BOD, heavy metals, and is strongly coloured. Sometimes, these dyes react with disinfectants like chlorine and form undesirable toxic and often carcinogenic by-products (Kant 2011). When it comes in contact with groundwater, or surface water, it makes it unfit for consumption by changing its biological, chemical, and physical properties (Fernández et al. 2010).

10.2.4 Pharmaceutical and Personal Care Products (PPCPs)

PPCPs, including pharmaceutically active compounds, are often described as the unique category of groundwater contaminants which degrade the quality of groundwater. Mostly, these products are introduced into the environment by the consumers, but the industrial sector, including the pharmaceutical, cosmetics, dietary supplement, and fragrance industries contributes a significant portion of waste, which goes to the groundwater. Each industry has a different objective, e.g. in the case of pharmaceutical industries to develop drugs or medicines for the treatment and diagnosis of diseases. While achieving their respective objectives, they have negative footprints on the environment, especially water resources, including groundwater (Gottschall et al. 2012; Sui et al. 2015). These developed drugs or compounds can enter the groundwater at several phases of their lifespan, not only during the production phase but also the end-of-life phase (e.g. expired drugs) (Lapworth et al. 2012). There exist many PPCPs including analgesics (ibuprofen, dimethyl-aminophenazone, and diclofenac), antibiotics (sulfamethoxazole and ciprofloxacin) (McArthur and Tuckfield 2000; Fick et al. 2009), antiepileptics (carbamazepine, diazepam and primidone) (Heberer 2002; Sui et al. 2015), lipid-lowering agents (fenofibric acid, Gemfibrozil and etofibrate) (Cabeza et al. 2012; López-Serna et al. 2013), β-blockers (atenolol, betaxolol, and sotalol), antineoplastic drugs (cyclophosphamide, sulphonamides, and penicillin), oral contraceptives, insect repellents (n-diethyl-3-methyl-benzamide) (Schaider et al. 2014), sunscreen agents (octocrylene and ethyl-hexyl methoxycinnamate) (López-Serna et al. 2013), musks (galaxolide and tonalide), and other PPCPs, which have a massive impact on the quality of groundwater (Gottschall et al. 2012; Zemann et al. 2015). The significant pathways for these products to enter into groundwater are mainly in the form of residue during the manufacturing process, landfills, release from treatment plants, human and animal excretion (via urine or faeces), sewage discharges, antibacterial agent release (used for treatment in aquaculture or during intensive livestock treatments), and disposal of unused medicines (Knee et al. 2010; Sui et al. 2015). These products with the process of hydraulic exchange – vertical or lateral – can easily leak into the groundwater table from the polluted surface waters (Del Rosario et al. 2014).

The presence of toxic compounds in the PPCPs composition contaminates the groundwater which causes adverse health effects such as neurotoxicity, genotoxicity, hormonal changes, generation of reactive oxygen species, and growth inhibition (Lateef 2004). Other effects include antibiotic resistance (Martínez 2008), extinction of species (effect of diclofenac on the vulture species is well known to the world), and endocrine disruption. Their adverse effects are reported not only in humans but also to aquatic life, including fish, algae, bacteria, and plants. Further, they have the property to bioaccumulate and bio-magnify along the food chain, which increases the risk of exposure to humankind. Hence, their exposure in groundwater makes the water unfit for consumption not for the humans, but for other forms of life, specifically aquatic life which maintains the ecological balance. There exist several previous reports and studies related to their impact on the environment or groundwater resources. For example, higher concentration levels of erythromycin in groundwater can not only act as a growth inhibitor to *Microcystis flos-aquae* but also impose oxidative stress on the algae species (Ji et al. 2012; Machado and Soares 2019).

Similarly, antibiotics like ciprofloxacin are reported for their genotoxic effect as well as acute toxicity in the case of fishes (Martins et al. 2012). The exposure of diclofenac (an analgesic and anti-inflammatory drug) to the environment is referred to as 'Asian vulture crisis' as its high level resulted in an immense decline in the population of vulture species in Asia (Green et al. 2004). This incident put a ban on the use of diclofenac for veterinary treatment in India and Nepal, and this decision later normalized the situation (Prakash et al. 2012) (Lonappan et al. 2016). Likewise, the lipid regulator drugs like fenofibrate and clofibrate inhibit the basal ethoxyresorufin-o-deethylase (EROD) activity of rainbow trout hepatocytes (fish) (Laville et al. 2004). Another pollutant in this category is stimulants like caffeine, and its primary industrial contamination source includes industrial wastewater from tea, coffee, and soft drink industries. The other sources include landfills and sewage discharge (Seiler et al. 1999). There are reports that indicate its elevated concentrations in Singapore (Tran et al. 2014), Hawaii (Knee et al. 2010), California (Fram and Belitz 2011), and many other regions of the world. Perfluorosurfactant is another category of PPCPs persistently found in groundwater and used in an extensive range of consumer products including stain-resistant textiles, food packaging boxes, non-stick cookware, lubricants, and also utilized commercially in many industrial applications. Perfluorooctanoic acid, one among the perfluorosurfactants, was detected in elevated concentrations due to production facilities in US groundwater (Shin et al. 2011) and European groundwater (Loos et al. 2010).

There exist a number of other PPCPs contaminating the groundwater from industrial sources that are not discussed here, such as organophosphate flame retardants. From the above discussion, it is clear that there is a need to improve the treatment process of industrial discharges, especially from the pharmaceutical and personal care manufacturing industries, in order to protect the groundwater from PPCP contamination.

10.2.5 Persistent Organic Pollutants (POPs)

POPs include organic industrial chemicals such as PCBs present in transformer oils (Aganbi et al. 2019). POPs also constitute some pesticides like dichlorodiphenyltrichloroethane (DDT), chlordane, endrin, and hexachlorobenzene produced by agrichemical industries (Ang et al. 1989), and other industrial wastes, which include furans and dioxins (Fiedler 2003). They can bioaccumulate, move up the food chain, and travel vast distances when entering into the groundwater. The discharge from industries, especially oil manufacturing and agrichemical industries, contributes a significant portion of POPs into the groundwater. As these pollutants are not biodegradable, they can bioaccumulate into the fatty tissue of the living species over time as they have more excellent lipid fat solubility. Due to their semi-volatility, they can also spread globally via air currents and return to groundwater via rain or snow. This unique property is the reason for their presence, even in the arctic region. Based on these characteristics, they have the potential to affect living organisms, which include not only humans but also wildlife and aquatic life. Even in low concentrations, they can disrupt the neural activities and hormonal activities of the organisms.

PCBs are derived from biphenyl by replacing hydrogen atoms with chlorine atoms (Zhang et al. 2014). Due to their unique properties such as inertness, chemical resistance, resistance to photodegradation, and thermal stability (Duan et al. 2013), they find their applications in various industries such as oils in transformers and power capacitors for insulation, additives in paints, PVC (polyvinyl chloride) coatings, sealants, and adhesives (Batterman et al. 2009). However, owing to their toxic effects, their utilization is restricted or banned in some applications in several countries. However, the existing old transformers and power capacitors still contain a massive amount of these compounds. They can enter into the groundwater via waste discharges, oil spills and leakage, run-off, and deposition. Their penetration into the groundwater may result in homeostasis and liver damage upon consumption (Klaren et al. 2015).

Another important class of pollutants in the list of POPs are pesticides, which are a potential threat to the quality of groundwater. Sometimes, they are also referred to as plant protection products, since they are applied to increase crop yield by protection the plants from various disease, and their consumption is increasing in the field. Apart from agricultural lands, the significant potential sources from where pesticides enter into the groundwater are pesticide manufacturing plants or agrochemical industries, waste disposal sites, wastewater discharges, and sewage treatment plants. Most of the pesticides degrade when entering into the groundwater, forming new compounds which govern their toxicity. Although pesticides like DDT, chlordane, endrin, and hexachlorobenzene are persistent, they are not able degrade quickly, e.g. DDT reported having a half-life between 2 and 15 years. Similarly, hexachlorobenzene requires many years to breakdown, which leads to its accumulation, and its over-concentration poses a potential risk to human health, wildlife, and the environment. Because of these reasons, the usage of these pesticides is banned or restricted by various countries, including India.

10.3 Groundwater Contamination by Industrial Effluents: Case Study

In Section 10.2, we discussed various industrial effluents affecting the groundwater quality, namely heavy metals, nitrates, synthetic dyes, POPs, and PPCPs. In this section, we present the current scenario of ground contamination due to these industrial effluents with the help of recently reported case studies. The contamination reports of groundwater were released by several countries including India, China, Bangladesh, Pakistan, Japan, Bolivia, Vietnam, Mexico, Nicaragua, Ecuador, Chile, Argentina, Hungary, and the United States. Here, we are discussing two recent reports from China and India, which investigate the various parameters of groundwater quality and indicate the contamination from the industrial sources, respectively. First is the recent case study done by Zhai and his co-workers (2019) from China. In this, the groundwater along the eastern part of Songnen Plain, China (which covers around 12 districts of the region), is collected and assessed in terms of the impact on groundwater quality, shown in Table 10.2 (Zhai et al. 2019).

The selected region for this study has an independent groundwater system, which is impacted by the river Songhua and its tributaries. Further, the geolocation sampling sites for the collection of groundwater samples are shown in Figure 10.1a. From the collected 368 water samples, the authors have measured more than 20 parameters for the assessment of groundwater quality, i.e. TDS, pH, COD, potassium (K), manganese (Mn), calcium (Ca), sodium (Na), magnesium (Mg), bicarbonates, sulphates, chlorides, nitrates, nitrites, ammonia, fluoride, iodide, ron (Fe), copper (Cu), selenium (Se), zinc (Zn), lead (Pb), and arsenic (As). The spatial distribution of the measured data concerning each parameter is shown in Figure 10.1. Except for pH and TDS measurements, all other measurements were carried out by laboratory-based techniques. The statistical analysis of the independent groundwater system of the Songnen

Table 10.2 Analytics of measured parameters on the selected groundwater system.

S/N	Parameter	Minimum	Maximum	SD	Mean	E (%)	VC
1	TDS	335.36	1068.23	135.56	615.38	13.04	0.22
2	TH	165.53	596.57	79.89	318.57	17.94	0.25
3	pH	6.42	7.44	0.18	6.96	10.33	0.03
4	COD	0.42	2.85	0.5	1.5	0	0.34
5	Na	11.58	78.3	11.75	33.33	/	0.35
6	Mg	6.94	40.97	5.72	19.14	/	0.3
7	Ca	46.06	196.28	26.77	95.96	/	0.28
8	K	0.65	24.35	3.48	3.26	/	1.07
9	SO_4	1.75	118.68	19.35	44.8	0	0.43
10	HCO_3	99.21	604.6	97.17	263.96	/	0.37
11	Cl	5.52	130.37	25.21	48.16	0	0.52
12	NO_2-N	<0.001	1.87	0.1	0.04	0.27	2.38
13	NH_4-N	<0.025	0.98	0.23	0.26	16.03	0.9
14	NO_3-N	0.59	244.94	50.61	75.56	41.3	0.67
15	As	<0.001	0.03	0.01	0.01	20.11	0.86
16	Cu	<0.01	0.01	0	<0.01	0	0.5
17	Fe	0.08	13.19	2.78	2.84	61.69	0.98
18	Pb	<0.0001	0.03	0	<0.0001	24.46	1.5
19	Fe	0.08	13.19	2.78	2.84	61.69	0.98
20	Se	<0.002	0.01	0	<0.002	0	1
21	I	<0.001	0.02	0	<0.001	0	2
22	F	0.06	3.33	0.36	0.33	2.99	1.11
23	Mn	<0.001	2.06	0.4	0.73	28.26	0.55

Abbreviation: SD, standard deviation; E, excess rate; VC, variation coefficient.
The concentration of ions are in ppm (mg/L).
Source: Martin and Griswold (2009).

Plain is presented in Table 10.2. Although the spatial distribution indicates the level of pollutants varies with regard to the location, this is due to the presence of groundwater contamination sources near the particular sampling sites, as shown in Figure 10.1. From the study, the eight pollutants, which include arsenic, lead, iron, nitrite, nitrate, manganese, fluoride, and ammonia, are identified as eight primary effluents that are affecting the water quality and shown in the order of their concentration from high to low. Among them, the significant contribution is from heavy metals, nitrates, and POPs. The authors have also stated that this region is a substantial base for heavy industries. Due to this reason, industrial wastewater contributes a

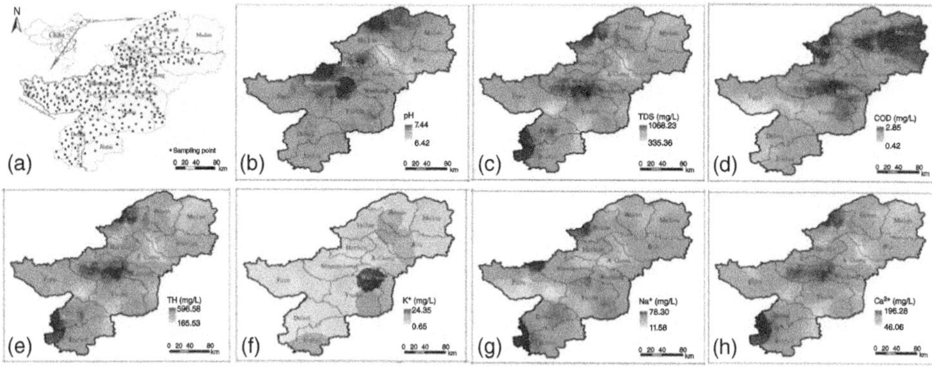

Figure 10.1 (a) Study area and sampling sites; (b) – (h) Spatial distribution of measured parameters in groundwater system of Songnen Plain. *Source:* Adapted with permission from reference (Zhai et al. 2019) © 2019, Elsevier.

lot in the groundwater contamination, and it is affecting the health of residents consuming the groundwater from the infected regions.

Likewise, for India, the Medak district of Telangana, India, was studied by Narsimha Adimalla and Ajay Kumar Taloor for the assessment of 13 different water quality parameters for groundwater quality evaluation. For this study, a total of 194 groundwater samples were collected each from different sampling sites in the region, as shown in Figure 10.2a (Adimalla and Taloor 2020). They have shown the spatial distribution of the parameter based upon their geographical location in their report (Adimalla and Taloor 2020).

The authors investigated 13 different parameters to obtain ground WQI of the region, which included various physicochemical parameters and dissolved ions. Among them, electrical conductivity (EC), pH, and TDS were measured in the field at the time of sampling and other parameters analysed using laboratory-based analytical methods. The overall results shown in Table 10.3 indicate groundwater contaminated with nitrate predominantly at many sites and about 57% of the sampling sites affected by the high concentration of fluoride ions. These contaminations make the groundwater unfit for drinking purposes at the effect sites. There exist a few sites with improper values of TH, TDS, Cl^-, EC, and Mg^{2+}. However, the concentration of ions such as Ca^{2+}, K^+, and Na^+ are within the maximum PL as per WHO guidelines.

Here, the author claims that among the 13 parameters screened and listed in Table 10.3, industrial wastes can be one of the primary factor responsible for the higher concentration of Mg^{2+} and nitrate in the region. These elevated levels, due to the rise in industrialization, can have adverse effects on the local population or ecosystem.

The presented case studies relate the role of industrial waste discharge in degrading water quality, thus necessitating a look into policy and technological measures for its control and implementation at the global level.

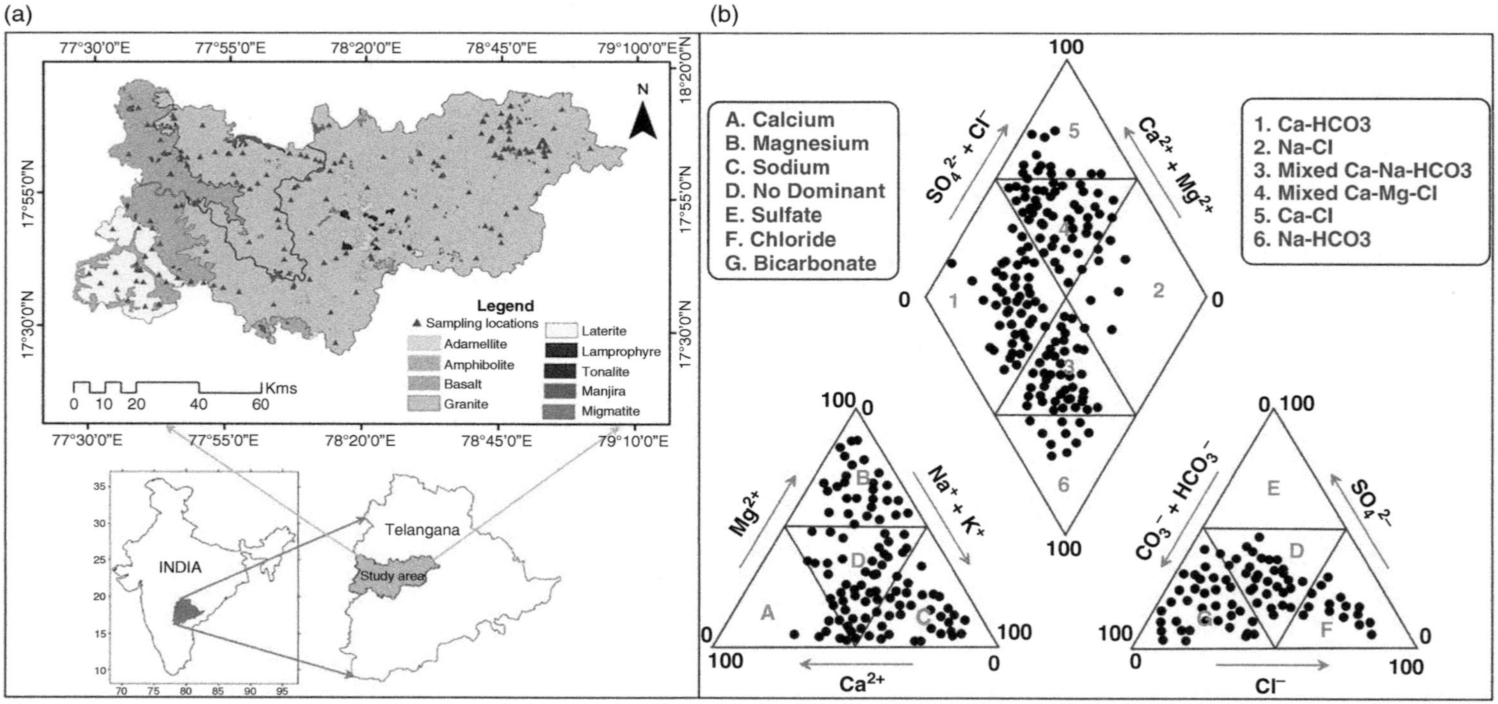

Figure 10.2 (a) Study area and sampling sites; (b) Piper trilinear distribution of measured parameters in groundwater system of Telegana, India. *Source:* Adapted with permission from reference (Adimalla and Taloor 2020) © 2020, Elsevier.

Table 10.3 Analytics of measured parameters on the groundwater system.

S/N	Parameter	Minimum	Maximum	SD	Mean	SEMPL (%)
1.	EC	168	3170	437	845	8
2.	pH	6.6	8.79	0.45	7.6	3
3.	TH	50	1550	142	241	3
4.	TDS	107	2028	279	541	42
5.	Cl^-	24	959	171	233	32
6.	SO_4^{2-}	21	328	54	161	22 (HDL)
7.	HCO_3^{2-}	18	527	87	219	13
8.	NO_3^-	4	440	67	73	50
9.	F^-	0.2	7.1	0.9	1.7	57
10.	Mg^{2+}	6	680	35	73	71
11.	Na^+	14	145	58	92	37
12.	Ca2+	10	164	43	68	/
13.	K+	1	24	3.5	3.3	5

Abbreviation: SD, standard deviation; HDL, Highest Desirable Limit; SEMPL, samples exceeding the maximum permissible limit); EC (electrical conductivity.
The unit of concentration of ions are in ppm (mg/L) and electrical conductivity (μS/cm).
Source: Adapted with permission for reference (Adimalla and Taloor 2020) © 2020, Elsevier.

10.4 Recommendation for Industrial Effluent Management: A Circular Economy Approach

As indicated earlier, groundwater is an essential resource for the sustainability of life on the planet and thus associated economic and industrial activities. The usage of groundwater is not limited to industries and agriculture but also has a significant share in various domestic activities. Despite these, groundwater contamination with various pollutants has increased exponentially. The industrial effluents have made a significant contribution to the degrading quality of groundwater. Thus, industrial effluent discharged from the industry should be pre-treated before its release into the environment. By applying effective pre-treatment processes, there is an opportunity to minimize or prevent the damage to the ecosystem and living organisms (humans, plants, wildlife, aquatic life, and microorganisms). Now more than ever, it has become essential to develop and deploy innovative solutions, technologies, and policies for sustainable development. Thus, there is a need to adapt and evolve in order to solve this global crisis. Further, the following recommendations should be taken into consideration to meet sustainable development goals:

1) Industrial wastewater is a rich source for multifaceted application and needs to be recycled under the concept of the circular economy. In order to recycle and reuse the water, there is a need to link the latest technological developments with various scientific breakthroughs in the field. An innovative mindset is required to ensure the implementation of

proper water management policies. These transformations and new creative ideas cannot come to existence on their own, and they require collaboration, consultation with experts, networking, and evidence.

2) There is a need for local regulatory bodies to monitor the source of these industrial effluents continuously and take preventive measures accordingly. A legal framework should be in place and enforced for the standardization of the treatment process and permitted level of pollutants in the wastewater. While framing these policies, it is essential to account for regional characteristics, which include soil composition, climatic factors, and many other aspects which affect groundwater quality. They must be strictly implemented to all industries without any exceptions, since the primary aim of these regulations is to have sustainable development while protecting natural water resources, including groundwater. This kind of development not only benefits humanity but all living beings.
3) Further, to achieve the sustainable development goals, there is a need to address the problem of the funding gap by identifying additional financial resources. These financial resources will help the scientific community to develop better screening and efficient removal technique for the pollutants, which are affecting the ecosystem.
4) Proper awareness, proper training, and education must be obligatory to human resources at all levels for efficient application of standards and policies laid down by different regulatory bodies.
5) Top world leaders should come forward and bring together researchers, scientists, technologists, and industries to solve this imminent threat to the environment.
6) Further, the existing standard guidelines for the treatment and disposal of industrial waste have been introduced by various agencies; based upon it, we have briefly described the generic strategy for the sustainability of water management. For effective treatment, wastewater had to go through the number of stages, i.e. preliminary stage, primary treatment stage, secondary treatment stage, and tertiary treatment stage. Although, in the case of industrial wastewater, it may need to go through extensive tertiary treatment depending on the chemical composition of the wastewater or industry domain. Often, industry skips this essential last step due to its high cost, large space allocation, skilled operators, chemicals, and equipment. This careless behaviour from industries often leads to contamination of groundwater, which results in more significant problems. In order to avoid such issues, regulatory bodies must come up with the standard guidelines for every type of industry depending on the location, which should be mandatory for industries to follow. In brief, a zero-waste discharge policy should be strictly implemented for large-, medium-, and small-scale industrial wastewater treatment plants to build up a circular economy.

10.5 Summary

Given the life on planet Earth's dependence on it, there is a need to protect groundwater. The groundwater needs protection from the unwanted discharge of industrial waste without proper treatment into the environment. These outflows contain many industrial effluents, which include heavy metals, pharmaceuticals and personal products, nitrates, synthetic dyes, and POPs, which are non-biodegradable. However, the scientific community has been reporting the presence of these pollutants for the past few years, but there is still room for improvement in order to detect more toxic compounds affecting the quality of groundwater. The effectiveness of onsite detection techniques is still at the nascent stage. These pollutants share unique common characteristics, i.e. bioaccumulation and biomagnification, which make them potential candidates for groundwater contamination. This contamination raises concerns not only for humans but for all living organisms. Once these pollutants enter the groundwater, their removal becomes extremely difficult or impossible, and this gives rise to the necessity of finding the contamination sources. In order to address these concerns, there is a need to develop an efficient regulatory framework for continuous monitoring of industrial effluents and their sources of contaminations. In addition to a better understanding of the potential threats, there is a necessity to find innovative ways to reuse wastewater, to avoid water scarcity in the near future.

References

Aganbi, E., Iwegbue, C.M., and Martincigh, B.S. (2019). Concentrations and risks of polychlorinated biphenyls (PCBs) in transformer oils and the environment of a power plant in the Niger Delta, Nigeria. *Toxicology reports* 6: 933–939.

Ahlström, L.-H., Eskilsson, C.S., and Björklund, E. (2005). Determination of banned azo dyes in consumer goods. *TrAC Trends in Analytical Chemistry* 24: 49–56.

Alrumman, S., Keshk, S., and El Kott, A. (2016). Water pollution: source and treatment. *American Journal of Environmental Engineering*: 88–98.

Adimalla, N. and Taloor, A.K. (2020). Hydrogeochemical investigation of groundwater quality in the hard rock terrain of South India using Geographic Information System (GIS) and groundwater quality index (GWQI) techniques. *Groundwater for Sustainable Development* 10: 100288.

Ang, C., Meleady, K., and Wallace, L. (1989). Pesticide residues in drinking water in the North Coast region of New South Wales, Australia, 1986–87. *Bulletin of Environmental Contamination and Toxicology* 42: 595–602.

Batterman, S., Chernyak, S., Gouden, Y. et al. (2009). PCBs in air, soil and milk in industrialized and urban areas of KwaZulu-Natal, South Africa. *Environmental pollution* 157: 654–663.

Cabeza, Y., Candela, L., Ronen, D., and Teijon, G. (2012). Monitoring the occurrence of emerging contaminants in treated wastewater and groundwater between 2008 and 2010. The Baix Llobregat (Barcelona, Spain). *Journal of hazardous materials* 239: 32–39.

Comly, H.H. (1945). Cyanosis in infants caused by nitrates in well water. *Journal of the American Medical Association* 129: 112–116.

Duan, X., Li, Y., Li, X. et al. (2013). Polychlorinated biphenyls in sediments of the Yellow Sea: distribution, source identification and flux estimation. *Marine pollution bulletin* 76: 283–290.

Fadiran, A. and Mamba, S. (2005). Analysis of nitrates and nitrites in some water and factory effluent samples from some cities in Swaziland. *Bulletin of the Chemical Society of Ethiopia* 19: 35–44.

Fernández, C., Larrechi, M.S., and Callao, M.P. (2010). An analytical overview of processes for removing organic dyes from wastewater effluents. *TrAC Trends in Analytical Chemistry* 29: 1202–1211.

Fick, J., Söderström, H., Lindberg, R.H. et al. (2009). Contamination of surface, ground, and drinking water from pharmaceutical production. *Environmental Toxicology and Chemistry* 28: 2522–2527.

Fiedler, H. (2003). Dioxins and furans (PCDD/PCDF). In: *Persistent Organic Pollutants* (ed. H. Fiedler), 123–201. Berlin, Heidelberg: Springer Berlin Heidelberg.

Fram, M.S. and Belitz, K. (2011). Occurrence and concentrations of pharmaceutical compounds in groundwater used for public drinking-water supply in California. *Science of the Total Environment* 409: 3409–3417.

Gleeson, T., Befus, K.M., Jasechko, S. et al. (2016). The global volume and distribution of modern groundwater. *Nature Geoscience* 9: 161–167.

Gottschall, N., Topp, E., Metcalfe, C. et al. (2012). Pharmaceutical and personal care products in groundwater, subsurface drainage, soil, and wheat grain, following a high single application of municipal biosolids to a field. *Chemosphere* 87: 194–203.

Green, R.E., Newton, I., Shultz, S. et al. (2004). Diclofenac poisoning as a cause of vulture population declines across the Indian subcontinent. *Journal of Applied ecology* 41: 793–800.

Helaskoski, E., Suojalehto, H., Virtanen, H. et al. (2014). Occupational asthma, rhinitis, and contact urticaria caused by oxidative hair dyes in hairdressers. *Annals of Allergy, Asthma and Immunology* 112: 46–52.

Hassaan, M.A. and El Nemr, A. (2017). Health and environmental impacts of dyes: mini review. *American Journal of Environmental Science and Engineering* 1: 64–67.

Heath, R.C. (2004). *Basic Ground-Water Hydrology*. Reston, VA: US Geological Survey.

Heberer, T. (2002). Tracking persistent pharmaceutical residues from municipal sewage to drinking water. *Journal of Hydrology* 266: 175–189.

International Agency for Research on Cancer (2010). *Some Aromatic Amines, Organic Dyes, and Related Exposures*. IARC Press, International Agency for Research on Cancer.

Jaishankar, M., Tseten, T., Anbalagan, N. et al. (2014). Toxicity, mechanism and health effects of some heavy metals. *Interdisciplinary toxicology* 7: 60–72.

Jatto, E., Asia, I., Egbon, E. et al. (2010). Treatment of waste water from food industry using snail shell. *Academia Arena* 2: 32–36.

Ji, K., Kim, S., Han, S. et al. (2012). Risk assessment of chlortetracycline, oxytetracycline, sulfamethazine, sulfathiazole, and erythromycin in aquatic environment: are the current environmental concentrations safe? *Ecotoxicology* 21: 2031–2050.

Kant, R. (2012). Textile Dyeing Industry an Environmental Hazard. *Natural Science* 4 (1) https://doi.org/10.4236/ns.2012.41004.

Kaur, M., Kumar, A., Mehra, R., and Kaur, I. (2020). Quantitative assessment of exposure of heavy metals in groundwater and soil on human health in Reasi district, Jammu and Kashmir. *Environmental Geochemistry and Health* 42: 77–94.

Khattab, T.A., Abdelrahman, M.S., and Rehan, M. (2020). Textile dyeing industry: environmental impacts and remediation. *Environmental Science and Pollution Research* 27: 3803–3818.

Knee, K.L., Gossett, R., Boehm, A.B., and Paytan, A. (2010). Caffeine and agricultural pesticide concentrations in surface water and groundwater on the north shore of Kauai (Hawaii, USA). *Marine Pollution Bulletin* 60: 1376–1382.

Klaren, W., Gadupudi, G., Wels, B. et al. (2015). Progression of micronutrient alteration and hepatotoxicity following acute PCB126 exposure. *Toxicology* 338: 1–7.

Kross, B. (2002). Nitrate toxicity and drinking water standards – a review. *Journal of Preventive Medicine* 10: 3–10.

Kumar, M. and Puri, A. (2012). A review of permissible limits of drinking water. *Indian Journal of Occupational and Environmental Medicine* 16: 40.

del Rosario, K.L., Mitra, S., Humphrey, C.P. Jr., and O'Driscoll, M.A. (2014). Detection of pharmaceuticals and other personal care products in groundwater beneath and adjacent to onsite wastewater treatment systems in a coastal plain shallow aquifer. *Science of the Total Environment* 487: 216–223.

La Force, M.J., Hansel, C.M., and Fendorf, S. (2000). Arsenic speciation, seasonal transformations, and co-distribution with iron in a mine waste-influenced palustrine emergent wetland. *Environmental Science and Technology* 34: 3937–3943.

Lateef, A. (2004). The microbiology of a pharmaceutical effluent and its public health implications. *World Journal of Microbiology and Biotechnology* 20: 167–171.

Lapworth, D., Baran, N., Stuart, M., and Ward, R. (2012). Emerging organic contaminants in groundwater: a review of sources, fate and occurrence. *Environmental Pollution* 163: 287–303.

Laville, N., Aït-Aïssa, S., Gomez, E. et al. (2004). Effects of human pharmaceuticals on cytotoxicity, EROD activity and ROS production in fish hepatocytes. *Toxicology* 196: 41–55.

López-Serna, R., Jurado, A., Vázquez-Suñé, E. et al. (2013). Occurrence of 95 pharmaceuticals and transformation products in urban groundwaters underlying the metropolis of Barcelona, Spain. *Environmental Pollution* 174: 305–315.

Liu, Y.-N., Tian, H., and Si, A.-H. (2012). Gliding arc discharge for decolorization and biodegradability of azo dyes and printing and dyeing wastewater. *Plasma Chemistry and Plasma Processing* 32: 597–607.

Lonappan, L., Brar, S.K., Das, R.K. et al. (2016). Diclofenac and its transformation products: environmental occurrence and toxicity-a review. *Environment International* 96: 127–138.

Loos, R., Locoro, G., Comero, S. et al. (2010). Pan-European survey on the occurrence of selected polar organic persistent pollutants in ground water. *Water Research* 44: 4115–4126.

Linsha, M., Liang, H., Xiaoyu, F., and Songlin, W. (2018). Nitrate and nitrite in health and disease. *Aging and Disease* 9: 938–945.

Lutyński, R., Steczek-Wojdyła, M., Wojdyła, Z., and Kroch, S. (1996). The concentrations of nitrates and nitrites in food products and environment and the occurrence of acute toxic methemoglobinemias. *Przeglad lekarski* 53: 351–355.

Machado, M.D. and Soares, E.V. (2019). Impact of erythromycin on a non-target organism: cellular effects on the freshwater microalga Pseudokirchneriella subcapitata. *Aquatic Toxicology* 208: 179–186.

Mance, G. (2012). *Pollution Threat of Heavy Metals in Aquatic Environments*. Springer Science and Business Media.

Margat, J. and Van der Gun, J. (2013). *Groundwater Around the World: A Geographic Synopsis*. CRC Press.

Martin, S. and Griswold, W. (2009). Human health effects of heavy metals. *Environmental Science and Technology Briefs for Citizens* 15: 1–6.

Martínez, J.L. (2008). Antibiotics and antibiotic resistance genes in natural environments. *Science* 321: 365–367.

Martins, N., Pereira, R., Abrantes, N. et al. (2012). Ecotoxicological effects of ciprofloxacin on freshwater species: data integration and derivation of toxicity thresholds for risk assessment. *Ecotoxicology* 21: 1167–1176.

Mathur, N. and Bhatnagar, P. (2007). Mutagenicity assessment of textile dyes from Sanganer (Rajasthan). *Journal of Environmental Biology* 28: 123–126.

Mcarthur, J.V. and Tuckfield, R.C. (2000). Spatial patterns in antibiotic resistance among stream bacteria: effects of industrial pollution. *Applied and Environmental Microbiology* 66: 3722–3726.

Narvekar, M. and Srivastava, A. (2002). Separation of banned amine isomers in relation to German Ban on azo dyes by derivatization on GC-MS. *Chromatographia* 55: 729–735.

Prakash, V., Bishwakarma, M.C., Chaudhary, A. et al. (2012). The population decline of Gyps vultures in India and Nepal has slowed since veterinary use of diclofenac was banned. *PLoS One* 7: e49118.

Schaider, L.A., Rudel, R.A., Ackerman, J.M. et al. (2014). Pharmaceuticals, perfluorosurfactants, and other organic wastewater compounds in public drinking water wells in a shallow sand and gravel aquifer. *Science of The Total Environment* 468–469: 384–393.

Seiler, R.L., Zaugg, S.D., Thomas, J.M., and Howcroft, D.L. (1999). Caffeine and pharmaceuticals as indicators of waste water contamination in wells. *Groundwater* 37: 405–410.

Shin, H.-M., Vieira, V.M., Ryan, P.B. et al. (2011). Environmental fate and transport modeling for perfluorooctanoic acid emitted from the Washington Works Facility in West Virginia. *Environmental Science and Technology* 45: 1435–1442.

Shiklomanov, I.A. 2008. *Water Distribution on Earth* [Online]. National Geographic. Available: https://www.nationalgeographic.org/media/water-distribution-earth [Accessed May 30 2020].

Shrimali, M. and Singh, K. (2001). New methods of nitrate removal from water. *Environmental Pollution* 112: 351–359.

Shuval, H.I. and Gruener, N. (1972). Epidemiological and toxicological aspects of nitrates and nitrites in the environment. *American Journal of Public Health* 62: 1045–1052.

Shuval, H.I. and Gruener, N. (2013). Infant methemoglobinemia and other health effects of nitrates in drinking water. In: *Proceedings of the Conference on Nitrogen as a Water Pollutant*, 183–193. Elsevier.

Siebert, S., Burke, J., Faures, J.-M. et al. (2010). Groundwater use for irrigation – a global inventory. *Hydrology and Earth System Sciences* 14: 1863–1880.

Smith, S. (1982). *Ground Water Hydrology for Water Well Contractors*. National Water Well Association.

Sui, Q., Cao, X., Lu, S. et al. (2015). Occurrence, sources and fate of pharmaceuticals and personal care products in the groundwater: a review. *Emerging Contaminants* 1: 14–24.

Suthar, S., Bishnoi, P., Singh, S. et al. (2009). Nitrate contamination in groundwater of some rural areas of Rajasthan, India. *Journal of Hazardous Materials* 171: 189–199.

Tran, N.H., Li, J., Hu, J., and Ong, S.L. (2014). Occurrence and suitability of pharmaceuticals and personal care products as molecular markers for raw wastewater contamination in surface water and groundwater. *Environmental Science and Pollution Research* 21: 4727–4740.

USBR. 2019. *Water Facts – Worldwide Water Supply* [Online]. Bureau of Reclamation. Available: https://www.usbr.gov/mp/arwec/water-facts-ww-water-sup.html [Accessed 2020].

Ward, M.H., Jones, R.R., Brender, J.D. et al. (2018). Drinking water nitrate and human health: an updated review. *International Journal of Environmental Research and Public Health* 15: 1557.

Wasana, H.M.S., Perera, G.D.R.K., Gunawardena, P.D.S. et al. (2017). WHO water quality standards Vs synergic effect(s) of fluoride, heavy metals and hardness in drinking water on kidney tissues. *Scientific Reports* 7: 42516.

World Health Organization (1996). *Guidelines for Drinking-Water Quality*. World Health Organization.

World Health Organization (2008). *Guidelines for Drinking-Water Quality*. World Health Organization.

Zemann, M., Wolf, L., Grimmeisen, F. et al. (2015). Tracking changing X-ray contrast media application to an urban-influenced karst aquifer in the Wadi Shueib, Jordan. *Environmental Pollution* 198: 133–143.

Zhai, Y.Z., Zheng, F.X., Zhao, X.B. et al. (2019). Identification of hydrochemical genesis and screening of typical groundwater pollutants impacting human health: a case study in Northeast China. *Environmental Pollution* 252: 1202–1215.

Zhang, Y.-F., Fu, S., Dong, Y. et al. (2014). Distribution of polychlorinated biphenyls in soil around three typical industrial sites in Beijing, China. *Bulletin of Environmental Contamination and Toxicology* 92: 466–471.

11

Impact on Groundwater Quality Resources Due to Industrial Effluent

Zeenat Arif, Naresh K. Sethy, P.K. Mishra, and B. Verma

Department of Chemical Engineering and Technology, Indian Institute of Technology (BHU), Varanasi, Uttar Pradesh, India

11.1 Introduction

Throughout the evolution of human civilizations, water has played an essential role. It is considered important for life to exist. Nature provides water, an important component for the existence of plants, animals, and human life. The cultural evolution started where there an adequate quantity and quality of freshwater. Freshwater is essential for the survival of living organisms and also for agriculture, industry, and domestic needs (Sharma et al. 2016). A reliable source of freshwater is groundwater, which is purer than surface water and is easily accessible (Saravi et al. 2013). Despite many priceless benefits, it is a poorly managed resource (Nwankwoala and Nwagbogwu 2012). Increasing industrialization and urbanization has become a matter of major concern because it leads to increase in abstraction of water from groundwater due to increased demand, and at the same time degrades water quality due to the presence of harmful pollutants discharged from industries (Singh et al. 2016; Ramesh et al. 2000).

Studies have shown that the health hazards caused by the presence of heavy metal (arsenic, cadmium, lead) in water leads to cardiovascular, neurological, and renal problems. The rapid industrialization and urbanization growth in the last two decades created unconstructive and adverse impacts on the environment. The major sources of groundwater pollution can be categorized as: (i) soil weathering and rock minerals; (ii) decomposition of organic materials; and (iii) discharge effluent from industries, sewage, and municipal waste. Industrial pollution is considered a major factor causing environment degradation (air, land, and water) (Ahamed and Loganathan 2017; Järup 2000). The discharge from both industries and household are rich in chemicals and biological matter that directly increase the demands on the oxygen present in water (Alexander et al. 2017; Ehya and Mosleh 2018; Mahmoudi et al. 2017). Thus, the polluted water has low levels of dissolved oxygen due to biological oxygen demand (BOD) and chemical oxygen demand (COD) (Afzal et al. 2018). Also, the industrial water waste contains chemicals and metals that are

Groundwater Geochemistry: Pollution and Remediation Methods, First Edition.
Edited by Sughosh Madhav and Pardeep Singh.

harmful to flora and fauna, human health, and the ecosystem because they degrade the quality of water by changing the properties of water such as pH, electrical conductivity (EC), total dissolved solids (TDS), etc. (Islam 2012). The contamination of groundwater due to discharged effluents of industrial, municipal, and agricultural waste containing pesticides, fertilizer residues, organic harmful dyes, and heavy metals is through the leaching process. Other sources of groundwater contamination include open dug wells that are considered to be worst in term of terms of faecal contamination and bacteriological analysis, which ultimately results in increased levels of microorganisms (Marcus and Maletic 2003).

Contaminated wastewater without any treatment when used for different activities creates adverse effects on human health. As well, if used for irrigation purposes, it affects the soil quality and crop of the agricultural system. So it becomes crucially important to access the quality of water before directly discharging into the environment (Bharti et al. 2013).

11.2 Quantity of Groundwater in India

In many states of India, the present groundwater status has reached a level of crisis. It becomes important for India to manage its groundwater resources so that growth and development of the country in the future should not get involved in any risk of serious implications. According to the Central Ground Water Board (CGWB) report on the total replenishable groundwater resources in 431.42 billion cubic metres (BCM) (Parmar 2017), the total replenishable groundwater for Ganga Basin is (39.6%); Godavari (9.42%), Brahmputra (6.15%), Indus (6.14%), and Krishna (6.12%) are the other four river basins which are above 5% of total replenishable groundwater resources in the country (Parmar 2017). As of April 2015, the statistics regarding water resources are shown in Table 11.1 (Water and Related Statistics, April 2015, Central Water Commission).

Considering the natural discharge as 35 BCM, the net actual groundwater availability for entire country is around 398 BCM. With the increasing population in India every year, the per capita water availability has shown a significant reduction of 15% in year 2011. Different scales within the districts are used to assess the resources of groundwater in the country, which include blocks/mandals/talukas/watersheds. A ratio of the annual groundwater extraction to the net annual groundwater availability is defined as groundwater level

Table 11.1 Statistics of groundwater resources.

Parameter	Unit (BCM/year)
Annual availability of water	1869
Usable water	1123
Surface water	690
Ground water	433

Source: Modified from Water and Related Statistics, April 2015, Central Water Commission.

Table 11.2 Groundwater level status in India from 1995 to 2011.

Groundwater level development	Explanation	% districts in 1995	% districts in 2004	% districts in 2009	% districts in 2011
0–70% – safe	Theses area have potential for the development of groundwater	92	73	72	71
70–90% – semi-critical	Areas in which cautious groundwater development is recommended	4	9	10	10
90–100% – critical	Area requires intensive monitoring and evaluation to maintain the groundwater level in safe criteria	1	4	4	4
>100% – over exploited	In future groundwater level development in these areas are connected to water conservation measures	3	14	14	15

Source: Based on Central Ground Water Board website.

development. This parameter helps in determining the quantity of groundwater available for use. A comparative study to assess the level of groundwater over past two decades is shown in Table 11.2 (Central Ground Water Board website).

The overexploitation of groundwater is having an adverse effect on stream–aquifer interaction. According to a CGWB 2014 report, approximately 16% of assessment units in India are overexploited. The dynamic groundwater resource of India is shown in Table 11.3 for year 2017.

In 2017–2018, India was seeing a 40% decline of groundwater level at an unsustainable rate. At this rate it was estimated that by the year 2030 nearly 60% of groundwater resources would be in a critical state.

11.3 Characterization of Groundwater Quality

There are different physicochemical parameters that need to be analysed, as given in standard manual of water and effluent water analysis to determine the quality of water. The details of different physicochemical parameters and the most common method used to determine these parameters is given in Table 11.4. Horton (1965) introduced the term water quality index (WQI), a tool to monitor water quality programmes in order to estimate water quality and to develop the correct measures to improve the quality of water. The WQI is calculated using Eqs. 11.1 and 11.2 (Jamshidzadeh and Barzi 2018).

$$\mathrm{WQI} = \Sigma Q_i W_i / SW_i \tag{11.1}$$

$$Q_i = 100\left[\left(V_n - V_i\right)/\left(V_s - V_i\right)\right] \tag{11.2}$$

Table 11.3 Glance at groundwater resource of India in year 2017.

Total annual groundwater recharge	**431.86 BCM**
Annual extractable groundwater resources	392.70 BCM
Annual groundwater extraction	48.69 BCM
Stage of groundwater extraction	63.33%
Categorization of assessment units (Block/Mandals/Firkas/Taluks)	
Total no. of assessed units	6881
Safe	4310
Semi-critical	972
Critical	313
Overexploited	1186
Saline	100

Table 11.4 Physicochemical parameters and their method of determination.

Parameter	Method
pH	Electrometric
Electrical conductivity	Electrometric
Total dissolved solids	Electrical conductance method
Total hardness	Titration by ethylene diamine tetra acetic acid (EDTA) Method
Calcium content	Titration by EDTA Method
Magnesium content	Titration by EDTA Method
Sodium Content	Flame photometry
Chloride	Titration by AgNO3 method (Mohr Method)
Alkalinity	Titration by H2SO4 method
Nitrate	Kjeldal method
Turbidity	Nephlometer

where Q_i – Quality rating scale Wi – Relative weight($1/S_i$), S_i- Std. permissible value, V_n- actual or test value, V_i- ideal value, V_s- Recommended standard value of the parameter. The standard and permissible limit of WQI as prescribed by the Bureau of Indian standard (BIS) and World Health Organization (WHO) is shown in Table 11.5 (Bunkar and Kumar 2019).

In a study by Ahamad et al. 2018, an effort was made to assess the groundwater quality in Varanasi by random collection from 10 different bore wells on the basis of WQI value. The study region on the basis of WQI value depicts that 93% of samples were categorized under excellent water, whereas the remaining 7% were in the good water category. In another study by Ahamad et al. 2019, the groundwater quality was assessed for New Delhi residents, which is adversely affected by Bhalswa Landfill. The WQI value was evaluated based on 17 physicochemical parameters. The value ranges from 61.92 to 276.52

Table 11.5 Water quality rating based on WQI.

WQI value	Rating of water quality
≤25	Excellent
>25–50	Good
>50–75	Poor
>75–100	Very poor
>100	Unsuitable

Source: Bunkar. N and Kumar. V, Water Quality Index for Assessment of Groundwater Quality Parameters in Udham Singh Nagar District of Uttrakhand.

(pre-monsoon) and 61.65 to 394.95 (post-monsoon). The results also concluded that from 30 samples, 3% were designated as good, 73% as poor water, and 23% very poor water during pre-monsoon whereas in post-monsoon it was estimated that around 7% were designated as good, 60% as poor water, and 33% very poor water.

11.4 Impacts of Industrial Activity on Groundwater in India

In industries, water has multiple applications such as cleaning, heating and cooling purposes, for generation of steam, as a solvent, as transporting media for dissolved substances, and as a constituent part of the industrial product. The abstraction of water by the industry is usually much greater than the actual consumption. In areas where there is scarcity of surface water resources, the industries meets their industrial requirements of water demand by using groundwater. The industries are creating environmental pressure because the actual amount abstracted by the industry is creating a negative impact on the quality of water discharged, thereby creating an imbalance between the volume of water withdrawn, consumed, and discharged. This is usually for one of the following reasons (WWAP 2006):

- Direct disposal of wastewater generated during the process without any treatment into ground via streams, rivers, and canals;
- Disposal of industrial wastewater to municipal sewer systems, which may or may not contain sewage treated effluent;
- Wastewater treatment on-site before disposal by either of the above.

One of the major challenges faced by urban authorities is the management of large volumes of wastewater. Sometimes there is the possibility to reclaim industrial wastewater so that it can be reused within the industry itself or by other users, but most of the effluent is directly returned to the water cycle without treatment. So the volume of these effluents and the hazardous contaminants in them, combined with their disposal mode, determine the pollution risk (Morris et al. 2003). After assessing the industrial impacts, issues related to groundwater quality are much more dominant than their quantity.

In the case of highly developed countries, the main concern is dealing with the industrial residual impacts, while for newly industrialized and developing countries, the major

Table 11.6 Common sources for groundwater contamination from industries.

Source	Contributory Factors
Underground and surface storage tanks	Undetected leakage due to poor maintenance/construction or inadequate bunding to retain major failures
Processes and effluent pipe work or other transfer system	Undetected leakage due to poor maintenance
Industrial sewers/collectors	Leakage due to poor maintenance
Soakaways, waste injection wells	Inappropriate disposal practice causing pollution
Bulk chemical storage areas	Leakage due to poor maintenance, poor handling, and storage procedures
Solid process-waste disposal sites	Leakage of leachate through poor construction
Accidental/catastrophic discharge	Plant fire, explosion, impact, and loss of material to ground
Liquid effluent and process lagoons	Leakage because of poor construction and maintenance

concern is related to the potential impact of rapidly growing, unplanned, poor monitoring of industrial activities. Industries generates a large amount of pollution load. In India, the advancement in science and technology resulted in industrial revolution, which in turn causes industrial pollution. Co-linearity in industrialization and population growth is the main cause of expansion of industrialization (Singh et al. 2018). During production processes, the water comes into contact with different harmful chemicals, heavy metals, organic and inorganic wastes, and even sludge (Rai 2008). These generated wastes are either dumped directly into rivers or other water bodies, which affects the status of our ecosystem. The different types of pollution load degrading the groundwater quality are listed in Table 11.6.

11.5 Industrial Structure and Industrial Pollution in India

The structural composition of industries determines the pollution in any country (Khan and Tarique 2015). The Ministry of Environment, Forest and Climate Change and Central Pollution Control Board (CPCB) have classified industries into three categories depending on the pollution index (PI). PI is defined as a number from 0 to 100; the increasing value of PI is considered as increasing degree of pollution load by the industry.

- Red category (more polluting industries) – Industries having PI score above 60. Example: textiles, tanneries, automobiles, pulp and paper, chemical, mining, nuclear power plant, explosives, slaughterhouses, thermal power plant, sugar, distillery units, etc.
- Orange category industries (somewhat polluting) – Industries having PI ranging between 41 and 59. Examples: dry cell battery, saree screenprinting, cotton spinning and weaving, glass manufacturing, printing press, etc.
- Green category industries (less polluting industries) – Industries having PI score ranging between 21 and 40. Examples: aluminium utensils, ayurvedic medicines, leather footwear, flour mills, optical lenses, biomass briquettes, PP film, carpentry and wooden furniture, etc.

- White category industries – Industries having PI score up to 20. Examples: air conditioners, fountain pen, medical oxygen, metal cap containers, repairing of electric motors, surgical and medical products repair.

11.6 Grossly Polluting Industries (GPIs) in India

Numerous industries with different types of products being manufactured are found in India. The nature of effluent generated through these does not remain the same throughout the year. It is dependent upon industry type within the cluster. Grossly polluting industries (GPIs) are defined as industries generating wastewater >100 KLD and/or hazardous chemical waste into water bodies.

The CPCB along with the State Pollution Control Board (SPCB) identified 1162 industrial units in 22 states/ union territories (UTs) as GPIs in the year 2011, out of which 992 industries were operational and 792 were complying with environmental standards. As per the most recent data from 2018, the CPCB and SPCB identified 2743 industries from 36 states/UTs that were classified as GPIs. Of these, 2497 units are operational, 2222 units are complying with the environmental standards, and 246 units are self-closed. State-wise status of GPIs for the year 2018 is given in Table 11.7. The increase of 136% in the number of industries is attributed to the fact that industries identified in 2011 covered only 22 states/UTs, while 2018 covers 36 states; this data indicates the increased industrialization/ development over the time in India. Also, compliance status has improved; it was 89% in 2018 while it was 79.8% in 2011 (www.indiaenvironmentportal.org.in/files/file/Pollution%20by%20Industries.pdf).

11.7 Various Kinds of Pollutants Released by Different Industries

Rivers are considered one of the chief contributors for human civilization. They play a significant role in recharging groundwater and help in controlling floods and supporting wildlife. A number of tanneries, textile, chemical plants, sugar and distilleries units, slaughterhouses, pulp and paper, and food and beverage units are situated along the banks of rivers, where they create water pollution due to the discharge of untreated wastewater and they deteriorate the quality of river water (Paul 2017). Some of the industries and their effluents are listed in Table 11.8.

11.7.1 Textile Industry

In India, this industry is considered an important income source as well as a cause of major pollution. It is the second-largest employer, providing employment to 45 million people in the nation, which will increase to 55 million by 2020 (Invest India 2019a). The textile industry is expanding with a varied range of sectors from traditional handloom, wool and silk, and handicrafts to the organized textile industry. In India's Gross Domestic Product (GDP), 2.3–13% was contributed by the national textiles industry as well as 12% of India's export

Table 11.7 State-wise status of GPIs.

State	No of Industries
Andaman & Nicobar	2
Andhra Pradesh	198
Arunachal Pradesh	97
Assam	0
Bihar	84
Chandigarh	0
Chattisgarh	2
Daman & Diu	0
Dadra & Nagar Haveli	0
Delhi	3
Goa	0
Gujarat	191
Haryana	660
Himachal Pradesh	0
Jammu & Kashmir	69
Jharkhand	45
Karnataka	4
Kerala	29
Lakshadweep	0
Madhya Pradesh	2
Maharashtra	4
Manipur	0
Mizoram	0
Meghalaya	0
Nagaland	0
Odisha	6
Puducherry	3
Punjab	5
Rajasthan	0
Sikkim	0
Tamil Nadu	0
Telangana	2
Tripura	0
Uttar Pradesh	1218
Uttarakhand	76
West Bengal	43

Table 11.8 List of industries distributed and their major discharges deteriorating the water quality.

Industries	Major component of industrial waste
Sugar	Sugar, molasses, alcohol, and heavy metals
Textile	Dyes and pigments, detergents, and surfactants
Tanneries	Acids, alkalis, chromium salts, tannins, solvents, dyes
Paper and pulp	Chlorinated complex products, dibenzo-p-dioxin, and benzofuran
Chemical fertilizer	Ammonia, urea, heavy metals
Thermal-power plant	Coal dust, CO_2, fly ash, chlorinated water, heavy metal residues
Petrochemical	Polycyclic and aromatic hydrocarbons, phenols, sulphides, naphthylenic acids, and other chemicals

Source: (Agale et al. 2013; Haneef and Akintug 2016; Islam et al. 2014; Pokhrel and Viraraghvan 2004; Yadav et al. 2007).

earnings. During its process an ample quantity of water is utilized by these units and generates water effluent containing dyes, heavy metals, organic and inorganic wastes, detergents, and surfactants, which deteriorates the quality of the receiving body and adversely affects the inhabitants (Verma et al. 2012). According to Manikandan et al. 2015, the textile effluents are turbid, coloured, loaded with organic and inorganic constituents, and low BOD/COD ratio, implying that a large proportion of non-biodegradable is present, which is unsuitable for direct discharge without any treatment. Kumar et al. 2020 in their report state that for an average-sized textile unit having a production capacity of 8000 kg of fabric/day, the amount of water consumed is nearly 1.6 million litres/day and it produces approximately 200–350 m^3 of wastewater/ton of finished products, thereby causing an average contamination of 100 kg COD/ton of fabric.

11.7.2 Leather Industry

India contributes 12.9% to the leather business from world's leather manufacture and produces around 3 billion square feet of leather yearly, from which around 9% represent the world's footwear manufacturers. It is known for its reliability in high trading earnings and is among 10 overseas exchange earners. India has access to 20 and 11% of the world's population of cows/buffalo and goats/sheep. The Leather business employs more than 4 million individuals (Invest India 2019b). The effluent generated from the tanneries carries chromium, salts, solids, sulphides, alkalinity, and lime, the main cause of water body contamination. The characteristics of tannery waste are strong colour, high chromium concentration, high BOD and COD, high TDS, and high total suspended solids (TSS).

11.7.3 Food-Related Industries

The estimated growth rate of the food industry in India was from $322 billion in 2016 to $543 billion in 2020. Food processing has a significant linking role for Indian farmers to consumers. The food processing industries consist of grain industries, edible oil industries, beverage, and dairy products manufacturing industries (Invest India 2020). The waste generated from food-processing industries includes poultry, dairy, meat waste, sugar processing, etc. These all are highly degradable and cause high oxygen (DO) depletion. The major industrial discharges are the organic contaminants rich in fats, proteins, and pathogens. The quality of effluents from food processing units have moderate to high BOD, high COD, TSS, TDS and great nutrient content such as nitrogen, phosphorous, heavy grease, and oil. The general treatment technology involves anaerobic management for, e.g. hybrid up-flow anaerobic sludge blanket reactor (HUASBR) followed by aerobic treatment.

11.7.4 Metal Industries

The Indian metal industry is classified into two primary divisions: iron-based (steel- carbon steel, ferrochrome steel, and hardened steel) and non-iron-based (copper, metal, tin, zinc, lead, manganese, and aluminium) metal ventures. The major processes of the metal industry are mining, ore refining, alloying, casting, and sheet/folding into foils. The common techniques used for manufacturing of metals include open hearth, oxygen incinerators, blast furnaces, and electric arc boilers (MOI 2015). A large amount of excess water is generated from steel industries from the positioning of coal, vent (fireplace), blast boiler washing, gases, etc. The characteristics of these water wastes are highly acidic and contain toxic elements like cyanogens, phenols, coke, metal, soluble bases, limestone, oils, etc.

11.7.5 Paper and Pulp Industry

According to an Indian Paper Mills Association evaluation, the paper and pulp industry contributes 4500 crore rupees to the exchequer and provides employment to more than 5 hundred thousand individuals over around 750 paper plants. The industry is categorized into four major segments: (i) printing and writing; (ii) packaging paper and board; (iii) specialty papers; and (iv) others and newsprint. The basic raw materials for the manufacturing processes are rice husk, wood, cellulose, fibres and waste-papers. A huge quantity of water is required for paper production; therefore it is considered a water exhaustive process and at the same time generates a large amount of wastewater contaminated with trichlorophenol, trichloroguicol, dichlorophenol, dichloroguicol, pentachlorophenol, dibenzo-p-dioxin and benzofuran (Kesalkar et al. 2012). These are responsible for high BOD, COD, SS, and when reaching the human body through the food chain they can cause reproductive problems, damage the immune system, and also cause cancer (Bhatngar 2015). The other adverse effect is that due to the dark colour of the waste paper it obstructs the photosynthetic process by decreasing the intensity sunlight, hence deteriorating the biota. Paper and pulp factories, notwithstanding being air polluters, produce a lot of inorganic contaminations, for example, sulphides, drying alcohols, and natural toxins including cellulose filaments, bark, wood sugars, and organic acids (PrintWeek India 2018).

11.7.6 Sugar Industry

India is producing 4.1 million tons of sugar from a total 380 factories with Uttar Pradesh having the second rank with 105 sugar industries producing approximately 1.1 million tons of sugar annually (Singh et al. 2018). Beside sugar as the main product, the mill generates different by-products also. The effluent generated from different steps of the process includes molasses, alcohol, other by-products, and liquid wastes, which cause pollution to the local rivers and lakes (Agale et al. 2013). The use of phosphoric acid and sulphur dioxide in the clarification step during production of sugar from cane juice is the main cause for algal bloom, which subsequently increases BOD, COD, SS, unpleasant odour and heavy metals (Moses et al. 2010).

11.7.7 Thermal Power Plant

India has more than 200 power plants, of which 116 are coal based, 65 are gas based and 19 are diesel based (Singh et al. 2018). These plants create thermal pollution because they release warm water, causing subsequent changes in DO level, disruption in ecological balance. Coal dust, fly ash, chlorinated water and presence of heavy metals like Hg, Bi, As, and Cr are the main constituents released from thermal power plants, thus creating water pollution (Sonwane 2015).

11.7.8 Petroleum and Oil Industry

India ranks fourth in the sale of liquefied petroleum gas (LPG). The anticipated oil demand in India is going to rise to 4% of the cumulative average growth rate (CAGR) during 2016–2030 while for the rest of the world except USA and China it was anticipated to be only 1% (Invest India 2019c). It has witnessed a steady increase in production and consumption of petroleum products. The production of petroleum products increased from 243.5 MMT (million metric tonnes) during 2016–2017 to 262.3 MMT in 2018–2019. Pollutants in this industry are released at each step, from manufacturing, to storing, and to transportation causing water pollution. The major pollutants include oil spillage, saline water, and phenolic mixtures. The release of phenolic wastes has a poisonous effect on plants, aquatic life, and humans. The presence of these compounds in water bodies leads to tumour formation, cancer, and mutation.

11.8 Current Status of Ground and Surface Water Pollution by Indian Industries

It is very clear that effluent discharges from industries have an adverse impact on ground and surface water quality. According to the study by Tariq et al. (2006), characteristics of effluents discharged from industries and their impact on quality of underground water were different according to the type of industry. For example, the pH of effluent in the aluminium industry was beyond the permissible limit when compared to other industries in the same way TSS of effluent of beverage industry was beyond the limit, while it was within the range for the remaining industries when parameters were compared with the NEQS (National Environmental Quality Standards). Also, it was observed that the BOD was exceeding the permissible limit. Different case studies were also done by the researchers in different regions of India for different types of industrial pollutants, and they have been summarized here.

Shankar et al. (2008) studied the influence of industrialization on groundwater quality in Peenya industrial area, Bangalore, India, and concluded that 77% of the samples were severely polluted and unfit for domestic use, which clearly interprets that groundwater is being polluted due to rapid industrialization.

Tariq et al. (2010) mentioned the impact of effluents from tanneries on groundwater and it was found that the groundwater in the locality of the tanneries was contaminated by chromium (Cr), cobalt (Co), cadmium (Cd), and Iron (Fe), and causes hazard to the surroundings. These metals and chemicals were released from different steps such as tanning/retaining, liming/delimbing, and finishing during leather manufacture. The greater chromium levels were mainly dangerous for human health and also make the water unsuitable for commercial, irrigation, and drinking purposes.

Singh et al. (2013) had a case study that was carried out and observed the effect of polluted surface water on groundwater at different sections of Budha Nullah. The untreated wastes of the industries are the main source for degrading water quality in Ludhiana. The investigation of Budha Nullah concludes that it had TDS of 1642 mg/L; high COD of 448 mg/L; BOD between 52 and 195 mg/L; most potable number (MPN) ranges from 240+ to 2400+/100 ml; Cr concentration was found to be 0.084 mg/L; Fe concentration was 0.913 mg/L; manganese (Mn) 0.095 mg/L; and nickel (Ni) 0.222 mg/L. According to the norms of CPCB, the water quality of Budha Nullah was categorized under E class, which implies not appropriate for drinking, bathing, irrigation, and industrial cooling.

Bhadra et al. (2013) mentioned the effect of industrial effluents on groundwater in Pali city, Rajasthan. The coloured industrial effluent was carried by Bandi River, which has alkaline nature and is rich in organics and other soluble salts. The water salinity increased with increased concentration of TDS, chloride, and sulphate. In the downstream, up to a 50 km distance the Bandi River had impact on the quality of groundwater in wells from Pali city to Nerda dam.

Chowdhury et al. (2015) studied the effluents of leather processing industries and their impact on water quality. It found that discharge effluent had high values of TSS, TDS, TS, BOD, COD, sulphate, chromium, arsenic (As), and cadmium (Cd) in the effluents sampled from three different selected leather industries. The values were far exceeding the permissible limits. It was also concluded that the physicochemical parameters of the effluents from each monitoring points were higher than the standard discarding limits.

Noukeu et al. (2016) performed the characterization of food processing industries' effluent. The investigation revealed that discharged effluents had enormously high levels of COD, BOD, TSS, phosphate (PO_4^{3-}), and nitrate (NO^{3-}) when compared with WHO standards. It was also recommended that biological treatment should be done since they had high values of organic materials.

Afzal et al. (2018) performed characterization in Hattar industrial estate, Haripur, of industrial effluents and groundwater and concluded that the effluents were contaminated with heavy metals, making them unsuitable for irrigation.

Kumar et al. (2020) mentioned the physicochemical analysis of few industrial effluents from Vapi industrial area, Gujarat. It was concluded that TDS, EC, chlorides, sulphates (SO_4^{2-}), COD, BOD, sodium (Na), and calcium (Ca) were above the permissible limit when compared to the WHO standards. Additionally, it was recommended that direct discharge of the effluent should not be carried out near water bodies without any treatment as they are unhealthy for irrigation purposes.

11.9 Groundwater Recharge Movement in India

Over last three to four decades, to overcome groundwater crisis and to capture the advantages of high level of runoff which was not done by natural recharges, recovery and restoring of groundwater through artificial recharge has become widespread in India. In fact, the expansion and growth in the artificial recharge has become to such an extent that now it is referred to as the 'Groundwater Recharge Movement'. Artificial groundwater recharge is defined as planning through work design to increase the recharging and percolation of surface water into groundwater aquifers in order to increase the quantity of groundwater for abstraction (Sakthivadivel 2007). Artificial groundwater recharge was used initially to increase the groundwater resources but now it has also been used for different purposes like surface runoff and floodwater conservation, water storage to reduce pumping cost, improving the water quality through SS filtration, and via dilution with naturally occurring groundwater.

Different methods and techniques have been employed to artificially recharge groundwater. These methods include (i) direct, (ii) indirect, and (iii) combination of direct and indirect methods. The direct method is the simplest and most widely used method. In this technique water is passed from the surface to the aquifer by simple infiltration. This infiltered water is percolated through the unsaturated zone to reach groundwater. Indirect recharge involves the installation of groundwater pumping facilities and infiltration galleries near the water bodies like streams or lakes in order to reduce the groundwater level and to induce infiltration from surface bodies. The third and last technique is recharging through integration of water resource development. The development scheme includes water recharge by means of reservoir/canal seepage, construction of check dam, etc. (CGWB (1995).

The cost and recovery of the artificial water recharge method is an important parameter to be analysed. The cost of recharge schemes is dependent on various factors including degree of water treatment, distance between source and recharge structure, cost of construction, and operation of the recharge structure. The cost of different artificial methods is shown in Table 11.9.

Table 11.9 Cost schemes of different artificial water recharge methods.

Methods	Capital cost/1000 m^3 of recharge structure ($)	Operational cost/1000 m^3/ year ($)
Injection well	551	21
Spreading channel	8	20
Recharge pit	515	2
Recharge pond	1	1
Percolation tank	6	1
Check dam	1	1

Source: (Sakthivadivel 2007).

11.10 Legislation, Policy Framework, and Challenges

The central government has formulated laws and model bills to improve the quality and quantity of water. The Model Bill for Ground Water Management was framed in 2011, in which the state has given the authority to enact their laws related to water management. In 2013, the National Water Framework Bill was formulated, which mainly focuses on demand management, usage efficiencies, and water pricing aspects. In response to the bill and model, the states have implemented ground water legislation. So far, rainwater harvesting has been made mandatory by 30 states and UTs (Ministry of Water Resource, Lok Sabha Unstarred Question No. 1758, July 30 2015 and Ministry of Water Resource, Lok Sabha Unstarred Question No. 716, December 3 2015). In 2015, the Ministry of Urban Development mandated rainwater harvesting in plots having size of 100 sq. m or more.

The Water Regulatory Authority (WRA) was also established in each state. Their main role is to fix and regulate the water tariff based on volumetric consumption and discharge. Different bodies having different functions were created at the central level related to groundwater. These include the Ministry of Water Resources, River Development and Ganga Rejuvenation, whose main role is the conservation and management of water in the country; the Ministry of Rural Development; and the Ministry of Environment, Forests and Climate Change are other framed bodies that are responsible for the prevention and control of pollution; this includes water pollution, and contamination of ground water. Apart from these, other major central institutions and their role related to groundwater are summarized in Table 11.10 (Ministry of Water Resources; Lok Sabha Question 2157, March 10 2015; PRS).

It has been stated by the government from time to time that is it is necessary to manage the groundwater as a community resource. The Easement Act, 1882 states that landowner has the right to abstract and dispose of water according to his own limit (Report of the High-Level Committee on Reorienting the Role and Restructuring of Food Corporation of

Table 11.10 Major central institutions and their role.

Institution	Function
Central Water Commission (CWC)	Initiating and coordinating schemes – related to conservation, quality monitoring, and utilization of water resources in collaboration with state governments.
Central Ground Water Board (CGWB)	Developing and disseminating technology for the sustainable use of groundwater. Implementation and monitoring policies for the sustainable management of groundwater resources. Estimation of groundwater resources.
Central Ground Water Authority (CGWA)	Regulate and management of groundwater resources.
Central Pollution Control Board (CPCB)	Implementation of the Water (Prevention and Control of Pollution) Act, 1974 to restore water quality

Source: Based on Lok Sabha Unstarred Question No. 2157, Ministry of Water Resource, River Development and Ganga Rejuvenation.

India, January 2015) and they are legally liable for any destruction caused to the water resource due to overextraction. This lack of regulation regarding overextraction of the resources has worsened the condition. The CGWB has the right to identify the overexploited area within the state and can only notify the owner, but they have no right to stop them from extraction of groundwater. Also, it becomes difficult for them to identify all such locations and penalize the owner because in India there are a huge number of users. Apart from this, the real-time monitoring of rivers, lakes, and groundwater bodies was not properly carried out by the Central Pollution Control Board and the CGWB to take proper and immediate action before the situation becomes worse.

11.11 Industrial Water Regulations

The problem associated with the management of industrial wastewater is obvious because of the poor coordination and regulation between regulatory bodies. Also, very few incentives were provided to industry owners if they are efficiently reusing water. Water tariffs are very low or sometimes ignored. This conflict between industry and local communities are increasing year by year over water allocation and water pollution and it will become unmanageable if it is not addressed now. The conflict is due to the exploitation of groundwater resources by the industries, diversion of water path to industries that was meant for irrigation, by the government are some of the major reasons over water use (Aggarwal and Kumar 2011). The principal agency, the Ministry of Water Resource (MoWR), is responsible for water but no power, rights, or authority are given to take action related to water pollution nor for the industrial use of water. In the same way, the Ministry of Industry (MoI), whose function is the planning and development of water resources for industrial purposes, has no control to regulate and monitor the water use by industries. On the other hand, the CPCB and SPCB regulate and monitor industrial water pollution and charge water cess based on the quantity of wastewater discharged, however they have no mandate to control the sourcing of water from different sources. Thus, these water control measures have not shown any significant improvement.

Although different wastewater treatments have been installed by the industries to meet the discharge norms. The common practices adopted by most of the industries is to measure the pollution concentration in a given quantity of water followed by dilution of the effluent by the freshwater to meet the required standard, because the water cost is very low in country. Secondly, separation of water that can be reused from different steps during production is not commonly done; as a result the clear water also contaminated when discharged with effluent. However, water pollution levels caused by industries can be reduced to a large extent if water pricing could be such that it encourages industries to conserve water.

11.12 Conclusion

The chapter deals with status and quality of the groundwater scenario in India and measures taken to reduce pollution that is degrading the water quality. The ever-increasing demand of the increasing human population leads to the parallel growth in industries, which causes serious problems if not regulated and monitored properly. The accidental discharges and inadequate management of untreated effluent are major concerns since they adversely affect the aquatic organisms, human health, water bodies. The discharge of untreated effluents by industries into the water bodies is a major source of groundwater pollution. It becomes necessary to curb the worst excesses committed by industries by means of legal restrictions on their production processes. Therefore in India, which is under a developing country, with the increasing alarming conditions of environmental pollution, its causes, and consequences, if not properly formulated and regulated, mere enactments and legislation are not enough to control. It can be easily said that there is direct correlation existing between water and human health, which will affect each other in the same manner and in same proportion. Thus it becomes significantly important that proper treatment and disposal of industrial wastewater should be carried out to mitigate its environmental impacts.

Efficient utilization of water in all industries should be optimized and an awareness for the scarcity of water resources should be fostered. It becomes crucially important to conserve water and eliminate pollution and minimize the losses. Promotion should be done through education, regulation, incentives, and disincentives to conservation water resources. Though different regulating bodies are regulating and controlling industries, the key problem lies in improper water resource management and poor coordination among different authorities, which should be avoided. The present condition of effluents released from industries has not been controlled, even after implementation of laws made by the government of India. The industries are recommended have a proper treatment system before being released into a water body. However, few of them have a proper treatment plant and dump effluents. Thus, this is a rising concern relating to the pollution of groundwater. The initiatives taken by government for installation of affordable pollution control equipment and providing incentives can promote industries to take up pollution control seriously.

References

Afzal, M.S., Ashraf, A., and Nabeel, M. (2018). Characterization of industrial effluents and groundwater of Hattar industrial estate, Haripur. *Advances in Agriculture and Environmental Science* 1 (2): 70–77.

Agale, M.C., Patel, N.G., and Patil, A.G. (2013). Impact of sugar industry effluents on the quality of groundwater from Dahiwad Village, district Dhule (M.S). *Archives of Applied Science Research* 5: 58–50.

Aggarwal, S.C. and Kumar, S. (2011). Industrial water Demand in India Challenges and Implications for Water pricing. In: *India Infrastructure Report*, 1e (ed. Infrastructure Development Finance Company), 274–284. Oxford University Press http://www.idfc.com/pdf/report/2011/Chp-18-Industrial-Water-Demand-in-India-Challenges.pdf.

Ahamad, A., Madhav, S., Singh, P. et al. (2018). Assessment of ground quality with special emphasis on nitrate contamination in parts of Varanasi City, Uttar Pradesh, India. *Applied Water Science* 8: 115.

Ahamad, A., Raju, N.J., Madhav, S. et al. (2019). Impact of non-engineered Bhalswa landfill on groundwater from quaternary alluvium in Yamuna flood plain and potential human health risk, New Delhi, India. *Quaternary International* 507: 352–369.

Ahamed, A.J. and Loganathan, K. (2017). Water quality concern in the Amaravathi River Basin of Karur district: a view at heavy metal concentration and their interrelationships using geostatistical and multivariate analysis. *Geology Ecology and Landscspe* 1 (1): 19–36.

Alexander, A.C., Ndambuki, J., Salim, R., and Manda, A. (2017). Assessment of spatial variation of groundwater quality in a mining basin. *Sustainability* 9 (5): 823.

Bhadra, B.K., Pathak, S., and Sharma, J.R. (2013). Impact of industrial effluents on groundwater around Pali city, Rajasthan using field and satellite data. *Journal of the Geological Society of India* 82 (6): 675–691.

Bharti, P.K., Kumar, P., and Singh, V. (2013). Impact of industrial effluents on ground water and soil quality in the vicinity of industrial area of Panipat city, India. *Journal of Applied and Natural Science* 5 (1): 132–136.

Bhatngar, A. (2015). Assessment of physico-chemical characteristics of paper industry effluents. *Rasayan Journal of Chemistry* 8: 143–145.

Bunkar, N. and Kumar, V. (2019). Water quality index for assessment of groundwater quality parameters in Udham Singh Nagar district of Uttrakhand. *International Journal of Current Microbiology and Applied Sciences* (Special Issue-8): 68–72.

CGWB (1995). *Report on Groundwater Resources in India*. New Delhi, India: Government of India.

Chowdhury, M., Mostafa, M.G., Biswas, T.K. et al. (2015). Characterization of the effluents from leather processing industries. *Environmental Processes* 2 (1): 173–187.

Ehya, F. and Mosleh, A. (2018). Hydrochemistry and quality assessment of groundwater in Basht plain, Kohgiluyeh-va-Boyer Ahmad Province, SW Iran. *Environmental Earth Sciences* 77 (5): 164.

Haneef, F. and Akintug, B. (2016). Quantitative assessment of heavy metals in coal-fired power plant waste water. *MATTER: International Journal of Science and Technology* 2: 135–149.

Horton, R.K. (1965). An index number system for rating water quality. *Journal of the Water Pollution Control Federation* 37 (3): 300–306.

Invest India (2019a). Snapshot: India – Knitting the future: India is among the world's largest producers of Textiles and Apparel. Retrieved on 10 May 2020 from www.investindia.gov.in/sector/textiles-apparel

Invest India (2019b). Snapshot: World's tannery. Retrieved on 10 May 2020 from www.investindia.gov.in/sector/leather.

Invest India (2019c). Snapshot: World's fastest-growing energy market: India is the 3rd largest energy and oil consumer in the world after China and the US. Retrieved on 5 January 2020 from www.investindia.gov.in/sector/oil-gas.

Invest India (2020). Snapshot: Food processing – A sunrise sector. Retrieved on 5 May 2020 from www.investindia.gov.in/sector/food-processing.

Islam, M. (2012). Effects of solid waste and industrial effluents on water quality of Turag River at Konabari industrial area, Gazipur, Bangladesh. *Journal of Environmental Science and Natural Resources* 5 (2): 213–218.

Islam, B.I., Musa, A.E., Ibrahim, E.H. et al. (2014). Evaluation and characterization of tannery wastewater. *Journal of Forest Products & Industries* 3: 141–150.

Jamshidzadeh, Z. and Barzi, M.T. (2018). Groundwater quality assessment using the potability water quality index (PWQI): a case in the Kashan plain Central Iran. *Environmental Earth Sciences* 77 (3): 59.

Järup, L. (2000). Cadmium overload and toxicity, Oxford journals. *Medicine & Health Nephrology Dialysis Transplantation* 17: 35–39.

Kesalkar, V.P., Khedikar, I.P., and Sudame, A.M. (2012). Physico-chemical characteristics of wastewater from paper industry. *International Journal of Engineering Resources* 2: 137–143.

Khan, M. and Tarique (2015). Industrial pollution in Indian industries: a post reform scenario. *Journal of Energy Research and Environmental Technology* 2: 182–187.

Kumar, V., Kumar, P., Singh, J., and Kumar, P. (2020). Current status of water pollution by integrated industrial hubs (IIHs) in India. In: *Environmental Degradation: Causes and Remediation Strategies*, vol. 1 (eds. V. Kumar, J. Singh and P. Kumar), 104–119. Agro Environ Media https://doi.org/10.26832/aesa-2020-edcrs-08.

Lok Sabha Unstarred Question No. 1758, Ministry of Water Resource, River Development and Ganga Rejuvenation, answered on July 30, 2015, http://164.100.47.192/Loksabha/Questions/QResult15.aspx?qref=20846&lsno=16.

Lok Sabha Unstarred Question No. 2157, Ministry of Water Resource, River Development and Ganga Rejuvenation, answered on March 10, 2015, http://164.100.47.132/LssNew/psearch/QResult16.aspx?qref=14305.

Lok Sabha Unstarred Question No. 716, Ministry of Water Resource, River Development and Ganga Rejuvenation, answered on December 3, 2015, http://164.100.47.192/Loksabha/Questions/QResult15.aspx?qref=24795&lsno=16.

Mahmoudi, N., Nakhaei, M., and Porhemmat, J. (2017). Assessment of hydrogeochemistry and contamination of Varamin deep aquifer, Tehran Province, Iran. *Environmental Earth Sciences* 76: 370.

Manikandan, P., Palanisamy, P.N., Baskar, R. et al. (2015). Physicochemical analysis of textile industrial effluents from Tripur city, TN, India. *International Journal of Advanced Research in Science and Engineering* 4: 93–104.

Marcus, A. and Maletic, J. I. Recovering Documentation-to-Source-Code Traceability Links using Latent Semantic Indexing. Proceedings 25th IEEE/ACM International Conference on Software Engineering (ICSE'03), Portland, May 3–10 (2003) 125–137.

MOI (2015). Maps of India: Metal Industry in India. Retrieved on 5 May 2020 from https://business.mapsofindia.com/india-industry/basic-metals.html.

Morris, B.L., Lawrence, A.R., Chilton, P.J. et al. 2003. Groundwater and its susceptibility to degradation: A global assessment of the problems and options for management. Early Warning and Assessment Report Series, RS, 03-3. United Nations Environment Programme, Nairobi, Kenya.

Moses, A.N., Destaings, N.N., Masindeand, N.E., and Miima, J. (2010). Effluent discharge by Mumias sugar company in Kenya: an empirical investigation of the pollution of River Nzoia Sacha. *Journal of Environmental Studies* 1: 1–30.

Noukeu, N.A., Gouado, I., Priso, R.J. et al. (2016). Characterization of effluent from food processing industries and stillage treatment trial with Eichhornia crassipes (Mart.) and Panicum maximum (Jacq.). *Water Resources and Industry* 16: 1–18.

Nwankwoala, H. and Nwagbogwu, C. (2012). Characteristics and quality assessment of groundwater in parts of Akure, South-Western Nigeria. *Journal of Environmental Science and Water Resources* 1 (4): 67–73.

Parmar, S.K. (2017). Present status of groundwater in India. *IOSR Journal of Mechanical and Civil Engineering* 14: 93–95.

Paul, D. (2017). Research on heavy metal pollution of river Ganga: a review. *Annals of Agrarian Science* 15: 278–286. https://doi.org/10.1016/j.aasci.2017.04.001.

Pokhrel and Viraraghvan, T. (2004). Treatment of pulp and paper mill wastewater- a review. *Science of the Total Environment* 333: 37–58.

PrintWeek India (2018). Indian paper industry moving out of the woods. Retrieved on 5 May 2020 from https://www.printweek.in/news/indian-paper-industry-moving-woods-29617.

Rai, P.K. (2008). Heavy metal pollution in aquatic ecosystems and its phytoremediation using wetland plants- an eco-sustainable approach. *International Journal of Phytoremediation* 10: 133–160.

Ramesh, R., Ramanathan, A.L., Ramesh, S. et al. (2000). Distribution of rare earth elements in the surficial sediments of the Himalaya river system. *Geochemical Journal* 34: 295–319.

Report of the High-Level Committee on Reorienting the Role and Restructuring of Food Corporation of India, January 2015, www.fci.gov.in/app/webroot/upload/News/Report%20of%20the%20High%20Level%20Committee%20on%20Reorienting%20the%20Role%20and%20Restructuring%20of%20FCI_English_1.pdf.

Sakthivadivel, R. (2007) The Groundwater Recharge Movement in India https://www.iwmi.cgiar.org/Publications/CABI_Publications/CA_CABI_Series/Ground_Water/protected/Giordano_1845931726-Chapter10.pdf.

Saravi, M., Keshavarzi, A., and Azareh, A. (2013). Modeling of SAR and sulfate concentration using artificial neural network approach. *International Journal of Agronomy and Plant Production* 4 (3): 499–506.

Singh, G., Singh, D.D., and Sharma, S.K. (2013). Effect of polluted surface water on groundwater: a case study of Budha Nullah. *Journal of Mechanical and Civil Engineering* 5: 1–8.

Singh, S., Raju, N.J., Gossel, W., and Wycisk, P. (2016). Assessment of pollution potential of leachate from the municipal solid waste disposal site and its impact on groundwater quality, Varanasi environs, India. *Arabian Journal of Geosciences* 9 (2): 1–12.

Singh, U., Singh, S., Tiwari, R.K., and Pandey, R.S. (2018). Water pollution due to discharge of industrial effluents with special reference to Uttar Pradesh, India – a review. *International Archive of Applied Sciences and Technology* 9: 111–121.

Shankar, B.S., Balasubramanya, N., and Reddy, M.M. (2008). Impact of industrialization on groundwater quality–a case study of Peenya industrial area, Bangalore, India. *Environmental Monitoring and Assessment* 142 (1–3): 263–268.

Sharma, S., Sharma, G., and Yadav, R.K. (2016). Impact of industrial development on groundwater quality- a case study of impact of effluent from Vishwakarma industrial area, Jaipur on ground water. *Bulletin of Environment, Pharmacology and Life Sciences* 5: 12–15.

Sonwane, S.M. (2015). Impact of effluent discharge from thermal power station on status of fish species of river Tapi at Bhusawal, district Jalgaon Maharashtra, IOSR. *Journal of Environmental Science and Toxicology in Food Technology* 9: 63–72.

Tariq, M., Ali, M., and Shah, Z. (2006). Characteristics of industrial effluents and their possible impacts on quality of underground water. *Soil Environment* 25 (1): 64–69.

Tariq, S.R., Shaheen, N., Khalique, A., and Shah, M.H. (2010). Distribution, correlation, and source apportionment of selected metals in tannery effluents, related soils, and groundwater – a case study from Multan, Pakistan. *Environmental Monitoring and Assessment* 166 (1–4): 303–312.

Verma, A.K., Dash, R.R., and Bhunia, P. (2012). A review on chemical coagulation/flocculation technologies for removal of colour from textile wastewaters. *Journal of Environmental Management* 93: 154–168.

Water and Related Statistics (April 2015). Central Water Commission. http://cwc.gov.in/sites/default/files/annual-report-cwc-2015-16.pdf

WWAP (2006). Water: a shared responsibility. The United Nations World Water Development Report 2. UNESCO, Paris and Berghahn Books, New York.

Yadav, A., Neraliya, S., and Gopesh, A. (2007). Acute toxicity levels and ethological responses of Channastriatus to fertilizer industrial wastewater. *Journal of Environmental Biology* 28: 159–162.

12

Effects of Acid Mine Drainage on Hydrochemical Properties of Groundwater and Possible Remediation

Anusha Vishwakarma[1], Sushil Kumar Shukla[2], Vinod Kumar Tripathi[3], Chandra Shekhar Dwivedi[4], Santosh Kumar Jha[5], and Ashutosh Tripathi[6]

[1]*Department of Environmental Science, Central University of Jharkhand, Ranchi, Jharkhand, India*
[2]*Department of Transport Science and Technology, Central University of Jharkhand, Ranchi, Jharkhand, India*
[3]*Department of Farm Engineering, Institute of Agricultural Sciences, Banaras Hindu University, Varanasi, Uttar Pradesh, India*
[4]*Department of Geoinformatics, Central University of Jharkhand, Ranchi, Jharkhand, India*
[5]*Department of Bioengineering, Birla Institute of Technology, Mesra, Ranchi, Jharkhand, India*
[6]*Amity Institute of Environmental Sciences, Amity University, Noida, Uttar Pradesh, India*

12.1 Introduction

Over several years, excavation of minerals has been considerably augmented to fulfil the growing demand for metals (Panda et al. 2016). On earth, water is the principal requirement for the survival of all life forms. In India, groundwater is the most reliable source of water for agricultural as well as for domestic purposes. More than 60 and 85% of the demand for water supplies for irrigation and drinking purposes, respectively, are reliant on groundwater. Because of overexploitation and pollution, groundwater assets are under threat. Water scarcity has put another pressure on water resources. The majority of Indian states like Uttar Pradesh, Tamil Nadu, Punjab, West Bengal, Telangana, and Haryana are facing groundwater contamination problems. There are various issues and pathways responsible for groundwater contamination, either natural (geogenic/pedogenic origin) or man-made (industrial, agricultural, and domestic). Out of all those contamination sources, mining is a major source of groundwater pollution. The mining of minerals is essential for the industrial development of the nation. India has ample mineral resources; in this way the mining industry is a noteworthy industry in India. Over several years, mining has been practised, as it is the major source of income worldwide (Oyewo et al. 2018). India is the third biggest coal-producing nation; that translates to a large number of mines all over the world (Sahu and Dey 2017).

Over the last 50 years, the problems associated with acid mine drainage (AMD) have become a major research issue. Mining activities cause several problems such as land degradation, deforestation, disposal of overburden, land subsidence, water pollution,

Groundwater Geochemistry: Pollution and Remediation Methods, First Edition.
Edited by Sughosh Madhav and Pardeep Singh.

AMD, air pollution, mine fires, etc. (Kamble and Kumbhar 2020). Groundwater quality deterioration is a major concern associated with underground and opencast coal mining in many coal fields of our country. Surface and underground mining practices underneath the piezometric level perpetually alter the hydraulic gradient, disturbing the quality of surface as well as groundwater (Singh et al. 2012). AMD is one of the most critical issues in the mining industry. It is an unavoidable by-product that generates naturally due to the extraction of metals and minerals. AMD is a non-point source water pollutant that results when sulphide-containing minerals, mainly pyrite, come in contact with air and water, oxidizing to produce ferrous sulphate and sulphuric acid. Naturally occurring sulphide-oxidizing (acidophilic) microorganisms and free oxygen available at the mine site accelerate the process of AMD formation. AMD generation is a long-lasting water quality issue in both working and abandoned mines. It's a major issue, particularly for those countries that are experiencing the issue of freshwater shortage, such as Australia and South Africa. According to the estimates in 1989, mine water had seriously damaged the watercourse and rivers, up to almost 19 300 km and 72 000 ha of lakes worldwide (CEQ 1989). It has many adverse impacts on the aquatic ecosystem of the receiving river, corrosion of mining machinery, and other environmental degradation. Due to its low pH value (approximately ≤ 2), elevated concentration of toxic metal ions (Pb, Cu, Zn, Mn, Al, Fe, etc.) (Tsukamoto and Miller 1999), and also the presence of anions like sulphates, chlorides, arsenates, nitrate, and high dissolved solid contents, AMD is considered a toxic pollutant (Dhir 2018).

It has a devastating effect on the biological (human life and aquatic life) and geological features (groundwater, surface water quality, and soil quality) of the adjacent environment. The unscientific management of AMD has led to water contamination and also raised the problem of groundwater pollution. AMD-contaminated surface and groundwater cause diseases upon ingestion that could be lethal to human beings (Andrea and John 2010). According to the USEPA survey conducted in the eastern US (1976) the general chemical composition of AMD generated from coal mining activities have pH as low as 2, sulphate concentration up to 4000 mg/L, and concentration of metals like Fe, Mn, and Al up to 440, 125 and, 270 mg/L, respectively (Sgambat et al. 1980). AMD occurs when sulphide minerals, chiefly pyrite, are present in coals. When it is exposed to air and water it is oxidized and hydrolysed in the presence of Thiobacillus bacteria to form sulphuric acid and ferrous sulphate, which further dissolve heavy metals (Akcil and Koldas 2006). AMD has a very multifaceted character and an extensively unstable composition from one mine to another. An AMD-contaminated water stream makes a generally unpleasant visual effect.

Due to the corrosive effect of AMD, it can have an impact on the economics of the mining operations. Because of its highly acidic nature, complex chemical composition, and presence of an elevated concentration of heavy metals, AMD causes serious effects on the aquatic ecosystem, influencing the valuable function of downstream waterways of the mining process.

India's commercial energy demand is highly dependent on coal. India is the world's third biggest coal producer country next only to China and the USA. The mining sector contributes around 4% in Indian gross domestic product (GDP) and offers work to more than one million people (Mehta 2002).

It is very difficult to predict the impact of AMD precisely because its composition varies mine to mine and seasonally, according to the area exposed to surface runoff during heavy rainfall, the effect of catchment discharge, and concentration of organic matters. Till now there have been no particular species present to identify the impact of AMD in the water body. It is hard to manage AMD generation and its treatment necessitates a high cost (Kefeni et al. 2017). In general, migration or preventing the formation of AMD from its source is viewed as the ideal alternative to limit the impact of AMD. A wide range of technologies are available for treatment of AMD that use physical, chemical (pH neutralization, solvent extraction, chemical precipitation, ion-exchange, micellar ultrafiltration, adsorption, reverse osmosis [RO], etc.) (Oyewo et al. 2018), and biological methods (aerobic and anaerobic constructed wetlands). For effective results, sometimes these technologies utilize more than one method of physical, chemical, and biological processes.

12.2 Processes for AMD Generation

As part of the rock weathering process, the chemistry of AMD generation is understandable within some environments, but as the variations in physicochemical characteristics and geochemistry vary with site location it rapidly turned out to be complex. Formation of AMD is a natural process, but it is accelerated by large-scale indiscriminate, unscientific mining and construction activities. The process of production of AMD is complex because it is affected by several chemical, biological and hydrological factors such as the morphology of pyrite, pH, presence or absence of bacteria (Evangelou and Zhang 1995), and rate of oxidation, which vary with environmental conditions.

12.2.1 Pyrite Oxidation

In the Earth's crust, pyrite is the most available sulphide mineral. Pyrite oxidation is the primary mechanism by which acid is released into mine drainage. Pyrite (FeS_2) is one of the most ubiquitous sulphides responsible for acid generation from metal and coal mines augmented with a high concentration of sulphate, Fe, Mn, Al, and also full of heavy metals such as Cd, Hg, Pb, etc. Pyrite oxidation succeeds when pyrite comes in exposure to oxygen and water; it oxidizes to form sulphate, which then turns into sulphuric acid (H_2SO_4) (Nordstrom and Southam 1997; Akcil and Koldas 2006). In general, materials that are rich in sulphide and poor in carbonates are good for producing acid drainage. The concentration of Fe^{3+}, pH level, temperature, O_2, E_h, and presence of microorganisms are some factors that affect the kinetics of pyrite oxidation (Blowes and Ptacek 2014). The breakdown mechanism of pyrite is expressed by the following oxidation and reduction reaction that further leads to AMD generation, as follows (Singer and Stumm 1970):

$$FeS_2 + {}^7/_2 O_2 + H_2O \rightarrow Fe^{2+} + 2SO_4{}^{2+} + 2H^+ \tag{12.1}$$

$$Fe^{2+} + {}^1/_4 O_2 + H^+ \rightarrow Fe^{3+} + {}^1/_2 H_2O \tag{12.2}$$

$$Fe^{3+} + 3H_2O \rightarrow Fe(OH)_2 + 3H^+ \tag{12.3}$$

$$FeS_2 + 14Fe^{3+} + H_2O \rightarrow 12Fe^{2+} + 2SO_4{}^{2+} + 16H^+ \tag{12.4}$$

Pyrite oxidizes in direct contact with water or oxygen to produce sulphate ions and acidity (Eq. 12.1). Replacement of oxygen by oxidation of ferrous iron into ferric ion is shown in Eq. 12.2. The formation of the yellow boy is one of the apparent effects of AMD. It is a yellow-orange colour precipitate of ferrous hydroxide ($Fe(OH)_3$). Hydrolysis and precipitation of ferric complexes cause the formation of a yellow boy, which can frequently be noticed at the bottom of a stream receiving acid mine water (Eq. 12.3). The introduction of more hydrogen ions leads to acidity of the receiving body (Eq. 12.4). Streams have its natural buffering capacity, which destroys completely at $pH \leq 4.2$, so while evaluating the AMD it is essential to measure the leftover concentration of hydrogen ion in solution (John and Goyal 2017). The kinetics of each reaction is different, hence no distinct rate law is available to explain the kinetics of sulphide oxidation for every different case (Evangelou and Zhang 1995). As per the study of Singer and Stumm 1970, it has been found that in the acidic pH region Fe^{3+} is the key pyrite oxidant, whereas in neutral-to-alkaline pH region, O_2 is anticipated as more active oxidant. The overall sequence of the reactions is acid-producing. In Eq. 12.5 complete oxidation of pyrite is presented, where ferric iron (produced from Eqs. 12.1 and 12.2) serves as the oxidizing agent (Pierre Louis et al. 2015; Kefeni et al. 2017). As compared to the reaction present in Eq. (12.1) this reaction (Eq. 12.5) is considered faster, recurring, and self-perpetuating, and continues until ferric iron gets depleted (Rose and Cravotta III 1998; Berghorn and Hunzeker 2001).

$$4FeS_2 + 15O_2 + 14H_2O \rightarrow 4Fe(OH) + 8SO_4^{2-} + 16H^+ \tag{12.5}$$

12.2.2 Sulphide Weathering

AMD is often referred to as acid rock drainage (ARD), produced by the weathering of sulphide carrying rocks and rich in acidity as well as metals too (Panda et al. 2016). In general, the weathering of sulphides in natural rocks releases sulphur which further turns to sulphuric acid, resulting in ARD or AMD. Generally, drainage from mine dump or waste is an outcome of sulphide weathering (chemical, biological, and hydrological) that causes the dissolution of heavy metals along with toxic elements like acids and cyanides, etc. (Panda et al. 2016). In the kinetic type weathering experiments (an electrochemical process between various sulphides that are in contact with each other), Kwong (1993) found that different composition of trace elements had importance in the stability of individual sulphides.

Although pyrite minerals are more prominent to acid drainage, there are also several sulphide-containing minerals that have the potential to contribute to mine drainage (Gray 1997; Moncur et al. 2009). Table 12.1 shows the list of sulphide minerals responsible for AMD generation.

12.2.3 Bacterial Activity

It has been reported that a number of acidophilic bacteria serve as biocatalysts that accelerate the rate of sulphide oxidation reaction. Microorganisms, namely *Acidithiobacillus ferrooxidans*, *Acidithiobacillus thiooxidans*, *Leptospirillum ferrooxidans*, *Sulfobacillus thermosulfidooxidans*, Sulfolobus, and Metallogenium are all related with arbitration of AMD generation from sulphide minerals at pH <4. Thiobacillus ferrooxidans bacteria is the most studied bacteria and it comprises an extremely vital group intended for pyrite

Table 12.1 Sulphide minerals responsible for AMD.

Mineral	Composition
Arsenopyrite	$FeS_2.FeAs$
Bornite	$CuFeS_4$
Chalcocite	Cu_2S
Chalcopyrite	$CuFeS_2$
Covellite	CuS
Galena	PbS
Millerite	NiS
Molybdenite	MoS_2
Pyrite	FeS_2
Pyrrhhdite	$Fe_{11}S_{12}$
Sphalerite	ZnS

Source: Sangita et al. (2010).

oxidation (Akcil and Koldas 2006; Panda et al. 2016). Its optimum growth has been observed between 25 and 35 °C. Among all these microbes' majority are Gram-negative, acidophilic mesophile, and chemolithoautotrophic who exhibit high metal tolerance (Baker and Banfield 2003). As compared to chemical action, only microbial action can produce 1000 times more acid drainage by dissolution of SO_4^- minerals under acidic conditions either directly or indirectly (Adams et al. 2005). Leptospirillum bacteria are iron oxidizers in acidic environment at temperature more than 20 °C (Johnson 1998). Sulfobacillus sp. and Sulfolobus sp. are thermophilic acidophiles that can oxidize both iron and sulphur. They can grow at temperature between 50 to 90 °C and pH range from 1–3. In general, in metallurgical processes sulphur-oxidizing bacteria are used to extract metals and minerals from ores.

There are two types of sulphide minerals: acid soluble and insoluble. Acid soluble minerals (chalcocite and sphalerite) are more vulnerable to form acid by means of bacteria, for example acidithiobacillus thiooxidans, which oxidize elemental sulphur and reduced inorganic sulphur compounds (RISCs) to sulphuric acid (Blowes and Ptacek 2014). On the other side, acid insoluble sulphide minerals (pyrite and tungstenite) remain steady in sulphuric acid diluted solution but are cause to undergo oxidative liquefaction. Two mechanisms, 'thiosulfate' and 'polysulfide', have been proposed by Schippers and Sand (1999) for the biological acid dissolution of sulphide minerals. Leptospirillum species bacteria is a more important sulphide oxidizer than O_2, that accelerates the sulphide minerals to break down and oxidize (Blowes and Ptacek 2014). It has been found that acidophilic bacteria obtain energy by oxidizing ferrous iron into ferric iron and accelerate breakdown of sulphide minerals and rate of oxidation (reaction 2) up to 100 times than abiotic oxidation (Singer and Stumm 1970). More recent results show that a complex microbial community is responsible for sulphide oxidation (Johnson and Hallberg 2003).

12.3 Effect of AMD on Groundwater Quality

Mining activity can diminish the quality of both surface as well as groundwater. AMD from abandoned coal mines affects the quality of both groundwater and surface water. But the vulnerability of AMD contamination into groundwater (particularly present in hard rock aquifers) is more profound compared to surface water (Kimmel 1983). Contamination of groundwater levels depend on the hydrogeological condition of the mine site, interaction between the soils, sediment, or rock through which the contaminated water flows into the receiving stream (Lottermoser 2007). A schematic overview of contamination of groundwater with different aspects of mining activity is shown in Figure 12.1. In case of underground mines, groundwater infiltrates into the mine and comes into the contact with pyritic coal and thus forms AMD. Opencast mining damages the geological structure of the Earth's crust so that the groundwater aquifer layers are filling out the coal mining pond and also causes the mechanical damages to landscape.

Leachates (AMD) originating from overburden dump and manufacturing waste from mines have the potential to penetrate through the ground surface to the subsurface layers and contaminated groundwater. As compared to underground mining, opencast or surface mining produces larger amounts of overburden (the waste rock that overlies the ore). Its highly acidic nature makes the mobility of heavy metals very easy. The presence of heavy metals in AMD poses toxicity and has potential to contaminate groundwater at very low concentrations. Coal ash leachates and AMD impacts groundwater quality by raising acidity, reducing oxygen concentration, and causing changes in the elemental composition (Carlson and Adriano 1993). During the rainy season, leaching of heavy metals (Cd, Cr, Cu, Pb, Hg, Zn, etc.) from mine spoils is easier, thereby polluting groundwater (Gupta and Nikhil 2016). Excess groundwater extraction may cause damage to nearby habitats and water bodies, and those many miles from the actual mine site. Groundwater frequently

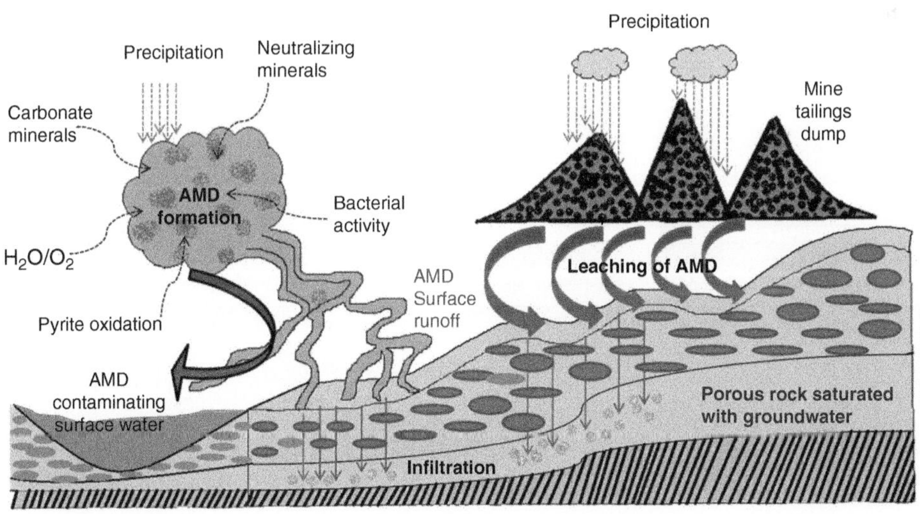

Figure 12.1 Schematic overview of contamination of groundwater near to mining sites.

flows through the cracks within the crushed sulphide minerals, which remain after an area has been mined. AMD deteriorates the buffering capacity of streams by neutralizing carbonate and bicarbonate ions into carbonic acid (Eqs. 12.6 and 12.7) (Kimmel 1983).

$$H^+ + CO_3^{2-} \leftrightarrow HCO_3^- \qquad (12.6)$$

Then

$$H^+ + HCO_3^- \leftrightarrow H_2CO_3 \qquad (12.7)$$

Carbonic acid (H_2CO_3) readily breaks down into H_2O and CO_2 (Eq. 12.8):

$$H_2CO_3 \rightarrow H_2O + CO_2 \qquad (12.8)$$

12.4 Treatment Technologies for AMD

Over the years, several widespread technologies have been studied for the prevention and remediation of AMD. Throughout the world every year a large amount of money is incurred to prevent AMD formation at its source. A proper chronological knowledge of sulphide oxidation is required for the careful disposal and management of AMD and also for careful handling of post mining overburdens and tailings. AMD treatment technologies mainly classified into two categories: (i) prevention or source control techniques and (ii) mitigational or remediation measures. Different possibilities to treat AMD and strategies to prevent contamination of water bodies are explained in Figure 12.2. In mining areas, treatment costs are too expensive to be employed at abandoned mine land (AML) sites (source of untreated AMD); simultaneously it is very essential to give proper concern to appropriate AMD prevention methods. A number of techniques are available to control AMD formation at source, but the use of corrective ones only leads to minimize the volume of mine drainage (Kontopoulos 1998; Johnson and Hallberg 2005a). Many processes are widely

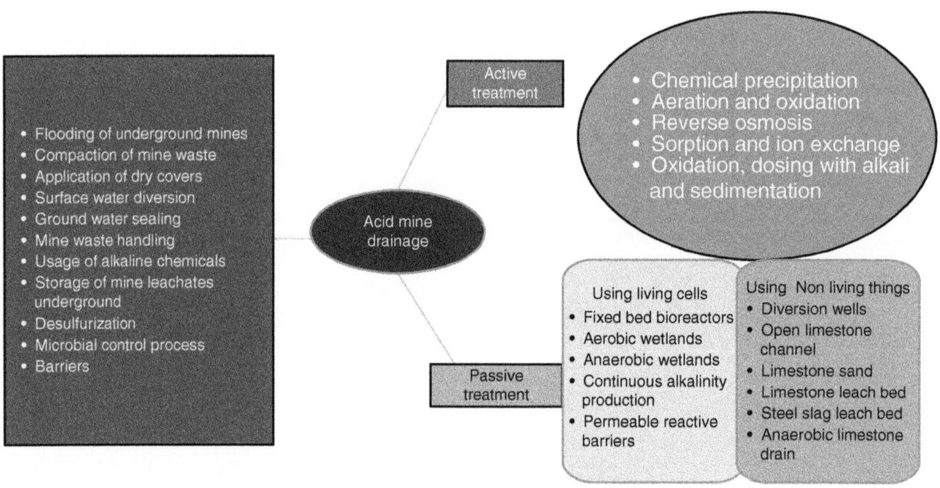

Figure 12.2 Flowchart for selecting a passive AMD treatment system.

used to treat AMD generally lead to formation of secondary waste, further increasing the cost of treatment (Simate and Ndlovu 2014). Conventional methods of AMD treatment use the combination of neutralization and precipitation followed by settling of precipitates in ponds (Kontopoulos 1998). AMD treatment technologies are chiefly divided into two major categories, namely active and passive treatment. Active treatment is also called chemical treatment, and it is more effective than passive treatment methods. An active treatment method involves the constant addition of chemicals to neutralize the pH and precipitate metals, whereas passive treatment methods include natural and biological ways to treat AMD (Moodley et al. 2018).

12.4.1 Active Treatment Technologies

The principle of active treatment methods for AMD is mainly based on wastewater treatment technologies (Younger et al. 2002). Different active treatment practices that are mostly used for AMD involves lime neutralization, chemical precipitation, oxidation, reduction, coagulation, solvent extraction, ion exchange, adsorption, and membrane processes like RO and filtration (Younger et al. 2002; Gaikwad and Gupta 2008). A number of factors affect the selection of an appropriate active treatment method, like concentration of Mn, Fe, TSS (total suspended solids), chemical composition of AMD, climate, terrain, available land area, etc. As per Means et al. (2003), after selecting a suitable method, a computer-based programme is used to design the specific components of the treatment system and to determine the cost of treatment. One of the main drawbacks of active treatment technology is its high operating costs and expensive maintenance. Most of the active treatment practices are based on chemical neutralization (pH adjustment) and hydroxide precipitation of metals (Jang and Kwon 2011).

12.4.1.1 Oxidation, Dosing with Alkali, and Sedimentation (ODAS)

There are a wide range of active treatment methods are available, but by far the most dominant one is ODAS (oxidation, dosing with alkali, and sedimentation), very common to industrial wastewater treatment system (USEPA 2004). But according to Younger et al. (2002) the most common order for AMD treatment is DAOS (dosing with alkali, oxidation, and sedimentation). Dosing with alkali is the first step, which is further followed by oxidation and sedimentation. Oxidation rates for dissolved metals in reduced form such as Fe^{2+} are strongly influenced by pH (Stumm and Morgan 1996). The first step may be pretreatment if AMD is fully loaded high suspended solids. The dosing with alkali step adds alkalinity to neutralize the solution and lower the concentrations of dissolved metals by forming metal hydroxides and oxyhydroxides. The aim of oxidation step is to oxidize reduced metals such as Fe^{2+} and Mn^{2+} into hydroxide form and remove oxide precipitates from AMD. The final step is sedimentation to remove the metal precipitates formed during the early stages of treatment (Trumm 2014).

12.4.1.2 Chemical Precipitation

Chemical precipitation is a conventional method that is widely used for treatment of AMD. It entails the addition of chemical reagents to alter the physical state of dissolved metals and suspended solids, followed by the separation of the precipitated solids by sedimentation or filtration from the cleaned water (Park et al. 2015). Amaral et al. (2016)

reported the use of this technology in sulphate-bearing AMD (coal and metal sulphides) and the separation was done via flotation with maximum removal percentage (80–82% of the feed content). Riodan et al. (1997) in their study combined chemical precipitation with biosorbent (residual brewery yeast as biosorbent media) and confirmed the capacity of this method, but he also mentioned that as this method generates a huge amount of sludge, which further needs treatment, it is not an environmental friendly method. Chemical precipitation efficiency depends on some factors such as the presence of ionic metals in solution, the precipitant used, and some other constituents that may restrain the precipitation reaction. Ferrous sulphate, ferric chloride, alum, lime, sodium hydroxide, and calcium hydroxide are some commonly used chemicals for raising pH. The hydroxide ions react with metals to form insoluble metal hydroxides (Eqs. (12.9) and (12.10))

$$Me^{2+} + 2OH^{-} \rightarrow Me(OH)_2(s) \tag{12.9}$$

$$Me^{3+} + 3OH^{-} \rightarrow Me(OH)_3(s) \tag{12.10}$$

12.4.1.3 Lime/Limestone Neutralization

The most economically preferred commercial method for treating AMD is lime. Neutralization and precipitation in a high-density sludge (HDS) happen under oxidative conditions. As a neutralizer limestone ($CaCO_3$) is a good alternative to lime for AMD treatment (Sun et al. 2000). Because of its low solubility, slow dissolution rate, and lower efficiency in mildly acidic mine water, use of limestone for neutralization of AMD has been limited (Cravotta and Trahan 1999). The steps involved in this process include: (i) initially, mine water is discharged to a rapid mix chamber with continued addition of hydrated lime (in dry and slurry form) to increase water pH up to 9. At this pH, most toxic metals become insoluble and precipitate. (ii) Then, it is passed through an aeration tank for oxidation and to further convert ferrous hydroxide precipitate into ferric hydroxide. (iii) The resulting water is then directed to a settling chamber, in that clean water overflow for discharge and heavy metal precipitates (sludge) are separated from the slurry for further treatment with a sludge-wasting side stream. According to the factors, volume, and composition of AMD, different variations of this process exist (Zinck and Griffith 2000).

The key drawback of this method is the requirement of high doses of alkaline reagents (lime or limestone) to raise pH for optimum removal of metals (McDonald and Grandt 1981). Different alkaline reagents are used to neutralize AMD are listed in Table 12.2. As AMD contains sulphate and calcium moreover, lime neutralization produces secondary wastes (Wang et al. 1996). The most commonly formed by-products of lime neutralization are gypsum ($CaSO_4$), calcium carbonate (additional by-product), and metal hydroxide sludges, which require high costs for disposal. A general equation for the process involved the productions of these by-products are:

$$Ca(OH)_2 + H_2SO_4 \rightarrow CaSO_4.2H_2O \tag{12.11}$$

$$H_2SO_4 + CaO \rightarrow CaSO_4 + H_2O \tag{12.12}$$

$$Ca(OH)_2 + CO_2 \rightarrow CaCO_3 + H_2O \tag{12.13}$$

Table 12.2 List of some neutralizing chemicals that are used to add alkalinity to AMD.

Neutralizing chemicals/ materials	Chemical Name	Characteristics	References
Hydrated lime	(Calcium Hydroxide) $Ca(OH)_2$	Hydrophobic, cost effective, low solubility at pH >7, requires simple dosing system, lime application varies with Mn concentration in water	Maree and du Plessis (1994);Skousen et al. (2000), (2018)
Soda ash	(Sodium carbonate) Na_2CO_3	Useful for low flow, to treat AMD with low acidity and metal content, higher reagent cost, poor sludge settling capacity	Skousen et al. (1996), (1999); Skousen (2002); Taylor et al. (2005)
Caustic soda	(Sodium hydroxide) NaOH	Efficient to treat AMD with high Mn level and low flow, high cost, dangerous to handle, poor sludge properties, freezing problems in cold weather	Skousen et al. (1996), (2000); Skousen (2014); Naidu et al. (2019)
Pebble quick lime	(Calcium oxide) CaO	Easily dissolved in water, increases pH, reduces heavy metal content, produces hydrated lime in reaction with water	Skousen et al. (2000); Othman et al. (2017); Skousen et al. (2018)
Ammonia	NH_3 or NH_4OH	Potentially able to treat AMD with high ferrous and Mn content, cheaper than caustic soda, difficult to handle	Faulkner (1990); Skousen et al. (1996), (1998), (2000)
Calcium peroxide	CaO_2	Slightly soluble in water, soluble in acids with formation of hydrogen peroxide, supplies alkaline condition to generate $Ca(OH)_2$	O'Neil (2013); Lu et al. (2017)
Fly ash	$CaCO_3$,$Ca(OH)_2$	Low cost, locally available, increases pH, causes precipitation of Fe oxyhydroxide, self-healing capacity	Pérez-López et al. (2007)
Kiln dust	CaO, $Ca(OH)_2$	Waste product of cement industry, strongly basic in nature, comprised of oxidized, anhydrous, micron-sized particles	Duchesne and Reardon (1998); Doye and Duchesne (2003); RoyChowdhury et al. (2015)
Magnesium hydroxide	$Mg(OH)_2$	Inexpensive, useful for high flow and highly acidic AMD	Skousen et al. (2000); Saha and Sinha (2018)
Magna lime	MgO	Cost effective, very reactive, dissolution rate is lower than quick lime, application requires proper mixing	Skousen et al. (2000); Skousen (2014a); Saha and Sinha (2018)
Phosphate rock	$X_5Y(PO_4)_3$	React with iron to form insoluble coatings, more expensive than other calcium-based chemicals	Evangelou (1995); Ziemkiewicz and Meek (1994)

12.4.1.4 Oxidation and Aeration

Aeration is the process of introducing oxygen into mine water, which combines with metals and forms metal oxides, which gets precipitated. A number of aerators exist for this process, but in active treatment plants the most commonly used is mechanical surface aerator. This aerator is used to treat highly acidic and moderate water (Saha and Sinha 2018) but it is not very efficient as it increases the addition of alkaline reagents (mainly, lime) to accelerate iron oxidation (Ackman and Kleinmann 1991). It has been studied by Budeit (2007) that accelerated oxidation is a good cost-effective option for treatment of alkaline mine water fully loaded with iron. It can reduce land area 10 times and capital cost up to half. Maelstrom Oxidizer is a commercially available, gravity-flow, and very effective system of active treatment method which can reduce chemical costs up to 46% at the site (Budeit 2007). It works by transferring oxygen into water by injecting high volumes of low-pressure air by means of a blower that achieves oxygen saturation within minutes. A simple, low-maintenance system has been invented by the Bureau of Mines to mix and aerate mine water in a pipeline. It consists of a jet pump to introduce air by Venturi action, and a static mixer, which induces turbulent flow. Simultaneous aeration and neutralization (sodium hydroxide (NaOH)) results iron oxidation within very short duration. Operation and maintenance costs are also extremely low (Ackman and Kleinmann 1991). Another aeration treatment system is in-line aeration (ILS); it is an easy and reasonably priced move towards aeration of mine water. It has very high oxidation rate through the mixing and aeration action of the jet pump and static mixer.

12.4.1.5 Sorption and Ion Exchange

Removal of heavy metals is a serious concern in the treatment of wastewater. Amongst all available technological options for the effective removal of heavy metals, ion exchange and adsorption methods are the most popular and important (Omer et al. 2003). A number of natural materials and low-cost waste products from agricultural or industrial activities have the potential to act as adsorbents (Bailey et al. 1999), such as dead biomass, red mud, blast furnace slag, fly ash, clay, tree bark, tea leaves, and natural zeolite (Ahmet et al. 2007; Motsi et al. 2009). Ion exchange is a stoichiometric and reversible chemical reaction in which ions from water solution are exchanged for a similarly charged ion attached to an immobile solid particle (i.e. ion-exchange resin) (Akpor and Muchie 2010). This process is very similar to biosorption. The majority of ion-exchange resins are too costly so to discard resins fully loaded with metals after a single use is not a cost-effective solution. The resins must therefore be regenerated, and regeneration again yields a solution of heavy metals which are hard to dispose of (Gaikwad 2010). Because of its metal recovery ability, ion exchange is a very promising method for active on-site treatment of AMD. Both sorption and ion exchange technologies are well suited for remediation of abandoned, remote mine sites or for sites located in cold areas. Various ion-exchange resins are available to achieve pollutant-free water, but a variety of silica-based and polymeric resins are broadly used for metal recovery or removal (Taylor et al. 2005). Because of convenience, ease of operation, and simplicity of design, the adsorption process is considered better (Faust and Aly 1987). Adsorption is the process of accumulation of a substance at the interface between two phases such as solid and liquid. At the surface of sorbents, there are unbalanced forces (viz. van der Waals forces) of attraction which are responsible for adsorption (Bhatnagar and

Sillanpää 2010). The use of sorption received consideration for treatment of mine water rich in base metals (such as Pb, Cu, Zn) (Barton 1978) and is also effective in removing sulphate (Cao et al. 2011).

12.4.1.6 Reverse Osmosis (RO)

RO is a semi-permeable membrane technique that acts as a molecular filter to remove all dissolved minerals above 99% (Akpor and Muchie 2010). RO is a process that allows the passage of water through the membrane and retains the dissolved and other suspended matter. It is a very efficient process for the removal of multivalent and monovalent ions from solution (Nidal et al. 2012). The resulting concentrated by-product solutions make eventual recovery of metals more feasible. Treatment of AMD by RO has been in use for decades; this has very high potential to eliminate contaminants (for example viruses, bacteria, and different metals even present in small concentrations) up to 99.8% (Anil et al. 2013). RO treated AMD contains high-quality water that can be suitable to use for potable or industrial use, even it can treat AMD up to the level of drinking water requirement (Atikinson 1997) whereas the leftover concentrated briny solution has Fe, sulphate, and metals (Skousen et al. 1998). In comparison with nanofiltration, ultra pressure reverse osmosis (ULPRO) has more removal efficiency that can remove heavy metal and total conductivity up to 97 and 96%, respectively (Zhong et al. 2007). Though, the major drawback of this technique is less water recovery efficiency (13.8–15.3%) as well as its high operational and maintenance cost (Pisa and Gulikova 2005). It requires high pressure to increase water recovery efficiency which further enhances the treatment cost.

12.4.2 Passive Treatment Technologies

Treatment of AMD by passive methods was initially started in the eastern USA's Appalachian coal field (Skousen et al. 2017). In recent years, various passive methods with different removal efficiency have been established that vary as per the composition and flow of water. It includes number of different techniques, which is described here. In comparison with active treatment systems, it is more convenient to use passive methods as they do not require continuous addition of chemicals, have minimal energy consumption, and also low maintenance and operation costs (Berghorn and Hunzeker 2001). The theory of this system is based on the application of naturally occurring physical, geochemical, and biological processes, and also on the flow of mine drainage through a filter material to remove pollutants from mine waters (Skousen et al. 2000). This system is appropriate to treat mine drainage with low flow rate and acidity are below 800 mg/L^{-1} as $CaCO_3$ (Taylor et al. 2005). Passive technologies perform in a controlled environment, not in receiving water body. Natural wetlands are a perfect example of passive treatment, in which saturated soils along with supporting vegetation provide favourable conditions in their rhizosphere to eliminate pollutants from water. Typha, Juncus, Phragmites, and Scirpus species are some examples of specific plants which are used in constructed wetlands that often consist of gravel, soil, and organic matter providing support to microbial attachment (Tsukamoto et al. 2004). A passive treatment system has good efficiency to remove sulphate, metal content, reduce acidity, and controlling salinity (Berghorn and Hunzeker 2001;

Taylor et al. 2005). Compared with chemical treatment, passive systems require longer retention times and greater space, provide less certain treatment efficiency, and are subject to failure in the long term.

12.4.2.1 Biological System

A biological treatment method is based on the capacity of plants to uptake pollutants as well as on microorganisms to produce alkalinity via reductive processes and hold metallic action (Johnson and Hallberg 2005a). There are also some methods that use microbes for treatment of AMD such as methanogenesis, denitrification, sulphate reduction, ammonification, and Fe and Mn reduction. The advantage of using biological treatment systems includes its low construction costs, easy maintenance, and good treatment efficiency (Dhir 2018). The biological treatment method is further classified into two categories: 'active systems' and 'passive systems.' The passive biological approaches employ the combination of aerobic and anaerobic wetlands have been applied for large-scale treatment of AMD. On the other hand, the active biological system uses offline sulphidogenic bioreactors.

12.4.2.1.1 Aerobic and Anaerobic Constructed Wetlands (AeWs and AnWs) Aerobic wetlands are shallow ponds (approximately 1–3 ft) that work similarly to natural wetlands, which facilitate wastewaters to flow over the substrate surface. It may possibly consist of more than one cell connected with one another that allows water to pass through it gradually under gravity and developing surface horizontal flow (Favas et al. 2016). The filling materials (rocks, soil, gravel, etc.) at the lower subsurface of ponds provide enough retention time and aeration for metal oxidation and precipitation. They can remove iron at the rate of 10–20 $g/m^2\ d^{-1}$ (Hedin et al. 1994). Aerobic wetlands promote metal oxidation and hydrolysis and oxidation of Fe^{+2} to Fe^{+3}, thereby causing precipitation of Fe^{+3} hydroxides and physical retention of Fe, Al, and Mn oxyhydroxides. This process creates a suitable environment for the growth of plants (e.g. Typha, Equisetum, Juncus), microorganisms (algae, protozoa, and bacteria), and mosses (e.g. Sphagnum), (Kadlec and Knight 1996; Favas et al. 2016). Aquatic plants are characteristically used to trap suspended particles and provide oxygen by the roots and rhizomes to support the growth of bacterial colonies. These bacteria serve as catalysts in the oxidation of contaminants and purify water (LópezPamo et al. 2002). The effectiveness of aerobic wetlands depends on metal concentrations, DO content, pH and acidity of AMD, presence of active microbial biomass, and detention time of the water in the wetland (Skousen et al. 2000). Significant alkalinity is required for the efficient performance of wetlands (URS repot 2003).

Anaerobic wetlands are lined or unlined shallow water ponds (1–3 m) comprising a substrate of organic matter (such as compost and sawdust) and limestone gravel. They are principally based on chemical and microbial reactions to neutralize acidity and promote metal precipitation. Anaerobic wetland ponds consist of two different substrates: (i) alkaline material and (ii) organic compounds. The upper substrate (0.1–0.5 m thick) contains a layer of organic matter that promotes excavation of DO, sulphate decline, and alteration of Fe^{3+} into Fe^{2+} (Skousen et al. 1998), and the lower subsurface (0.5–1 m thick) comprises alkaline material (limestone) to neutralize mine drainage as well as reduce the retention period required for treatment of AMD. It percolates through the permeable organic substrate that requires high oxygen demand resulting in an anaerobic environment (Skousen et al. 2000). Due to activity of sulphate-reducing bacteria (SRB) activity, anaerobic wetlands generate alkalinity, which use organic matter and sulphate as sources of carbon and electron acceptors, respectively (Taylor et al. 2005). Electron donors accelerate the reductive reaction and continue to provide an organic environment from compost. Wieder (1992) described that as per season and age of wetland, the efficiency and mechanism for treatment of AMD changes. Organic sorption materials (such as peat, compost, and sawdust) can initially remove 50–80% of metals from AMD (Brodie et al. 1988). Anaerobic wetlands are the most preferable method for treatment of AMD rich in DO, ferric iron, or aluminium. In the aerobic surface layers this type of wetland encourages metal oxidation and hydrolysis (Dhir 2018).

12.4.2.1.2 Vertical Flow Wetlands (VFWs)/Successive Alkalinity Producing System (SAPS) A new anaerobic passive treatment method that has been developed by Kepler and McCleary (1994) for treatment of AMD is vertical flow wetland; it is also called successive alkalinity producing system (SAPS) (Skousen et al. 2000). Vertical flow wetland is basically a combination of anaerobic wetland and anoxic limestone drains (ALDs) system. It involves the application of organic matter as well as a layer of limestone for producing alkalinity and removal of contaminants. In VFSs or SAPS, a pond is set up at the top of the system and the hydraulic head drives water from pond to flows in a downward direction passing through a bed of organic matter and limestone before getting drained out of the system. The principal mechanisms of this method involve: (i) as mine drainage passes through the layer of organic matter it removes oxygen, reduces concentration of Fe and Al through filtration, and reduces sulphate content also; (ii) the reduced water with less metal content then flows downward through limestone, which adds alkalinity to water (Skousen 2014). Subsequently with the help of drainage pipes the water is passed to a settling pond to be re-oxidized and settle out the precipitated iron hydroxides from solution (URS report 2003). In comparison with anaerobic horizontal flow wetlands, VFSs gives more residence time for contact of water with organic matter and limestone. As said by Demchak et al. (2001) in typical SAPS, mine water is ponded up to the depth of 50–150 cm with average thickness of layer of organic compost is 15–50 cm and layer of limestone is 50–200 cm. Compared with conventional anaerobic wetlands this technique can remove acidity more than 10 times per unit area (50 $gd^{-1}\ m^{-2}$) (Dietz and Stidinger 1996; Watzlaf 1997).

12.4.2.1.3 Bioreactors (SRB) Bioreactors system is verified as one of the most sustainable, eco-friendly, and cost-effective approach towards AMD treatment (Favas et al. 2016). Bioreactors work in stationary, continuous, and semi-continuous flow modes (Dhir 2018) and causes the biomineralization of selective transition metals (Fe and Zn) present in AMD (Favas et al. 2016). Because of the crucial role of sulphate-reducing bacteria in anaerobic bioreactors they are also known as sulphate-reducing bioreactors (URS report 2003). They are one of the widely used passive treatment techniques. Sulphate-reducing bacteria are a group of chemo-organotrophic anaerobic bacteria, largely belonging to the genera of Desulfomicrobium, Desulfobacter, Desulfovibrio, and Desulfotomaculum (RoyChowdhury et al. 2015). A detail characteristic of different SRBs used to treat AMD is given in Table 12.4. A number of microbial processes cause excavation of pollutants from AMD includes methanogenesis, denitrification, sulphate reduction, and Fe and Mn reduction processes.

Adsorption and precipitation are mainly responsible for metal removal in the form of carbonates, hydroxides, or sulphides. Anaerobic sulphate-reducing bioreactors are compiled with mixed layers of limestone and organic material, that allows AMD to pass through it vertically and then discharged to the drainage system (RoyChowdhury et al. 2015). Limestone produces alkalinity and also provide surface for biofilm development to speed up sulphate removal. By releasing HCO_3^-, SRB produce alkalinity (Eq. 12.14), precipitate metals, and reduce sulphate using electron donors (organic acid, alcohol, etc.) from AMD waters (Neculita et al. 2007).

$$2CH_2O(aq) + SO_4^{2-} + H^+ \rightarrow H_2S + 2HCO_3^- \tag{12.14}$$

SRB produce sulphides to precipitate as metal sulphides such as Cd, Cu, Fe, Pb, Hg, Ni, and Zn, whilst producing alkalinity. Additionally, it involves the chemical precipitation of metal (Me) and removes to some extent by co-precipitation with other metal sulphides (Eq. 12.15) (Figueroa 2005)

$$H_2S + Me^{+2} \rightarrow MeS + 2H^+ \tag{12.15}$$

Over the decades, different types of bioreactor have been studied for sulphate and heavy metal removal from AMD in active and passive treatment techniques (Table 12.3).

Acidithiobacillus ferrooxidans, an obligate acidophile, oxidizes a variety of RISCs and possesses the ability to oxidize ferrous ion (Dhir 2018). Heterotrophic bacteria such as Pseudomonas, Clostridium, and Desulfovibrio reduce Mn and Fe by using them as final electron acceptors under anaerobic conditions (Chen et al. 2000).

12.4.2.1.4 Permeable Reactive Barriers (PRB) Permeable reactive barrier (PRB) is the most recommended approach to prevent the contamination of groundwater with mine drainage (Blowes and Ptacek 2014). It is an in-situ remediation technique restraining barrier to interrupt AMD discharge into groundwater (Taylor et al. 2005). The construction of this system involves excavation of a pit at right angles in the pathway of flow of mine water where it contaminates groundwater. It is an effective passive method to treat groundwater rich in organic and inorganic contaminants such as PCE (perchloroethylene), TCE (trichloroethylene), VOCs (volatile organic compounds), radionuclides, and metals (DOE 2018). The barriers or the excavated parts use in this technique are filled with a number of reactive materials (such as limestone, zerovalent ions, organic matters, etc.) (Favas et al. 2016) with capacity to absorb, precipitate, or degrade the contaminants directly or indirectly via microbial activity. The opening of a trench is made transverse to the flow (Benner et al. 2002). The dissolution of limestone and reductive bacterial action produce alkalinity raises pH of solution. It diminishes sulphates and precipitate metals as sulphides, hydroxides, and carbonates. The general composition of filling material for barrier is a combination of degraded manure and straw (25%), composted green waste (25%), and limestone (50%) (Waybrant et al. 2002; Dhir 2018). Sasaki et al. (2008) in their laboratory study investigated that PRB column along with GBFS (granulated blast furnace slag) is effectual in immobilization of arsenate in groundwater affected by AMD.

Table 12.3 Names of different types of bioreactors.

Types	Mechanism of treatment	Advantages	Disadvantages	References
Anaerobic Wetlands/ Compost Bioreactors	Use layer of compost and limestone, filter suspended and colloidal materials, adsorption of heavy metals, eliminate DO, reduce Fe^{3+} to Fe^{2+}, sulphate-reducing bacteria causes biodegradation, treat net alkaline mine water	Cost-effective, high acidity removal, do not require continuous maintenance, remove sulphate	Longer retention period. Large surface area, efficiency decreases with time	Johnson and Hallberg (2005b), Dhir (2018), RoyChowdhury et al. (2015)
Fixed/Packed Bed Bioreactor	Consists of acidophilic, iron- and sulphur-oxidizing acidophiles, removes iron by oxidation and precipitation, bacteria immobilize as biofilm	Low cost, space saving and energy efficient, low sludge waste	Bacterial cell wash-out at high flow rate, detachment of cell from support media, less efficient to remove suspended solid	Johnson and Hallberg (2005b), Johnson and Younger (2005), Mujuru and Mutanga (2017)
Permeable Reactive Barriers (PRBs)	Microbiological processes generate alkalinity, remove metals as sulphides, hydroxides, and carbonates, operate on the same basic principles as compost bioreactors	Cost-effective, passive technique, no need of external energy for water movement, no need of construction of above-ground facilities, ability to treat multiple contaminants	Remediation takes long time, limited information regarding long-term performance of PRB, loss of reactant potency and permeability, restricted to shallow plumes	Johnson and Hallberg (2005a), Yeh et al. (2010), Snow (1999)
Continuous stirred tank reactors	Produce biohydrogen, rapid dissolution of reagents, uniform and continuous mixing of solution enhance the reaction rate of ferrous iron oxidation	Easy design, uncomplicated functioning, efficient consistent stirring, and proper maintenance of temperature and pH, simple measurement the rate of reaction directly	Incapable to support large number of fermenting microbes, potential harm to sensitive microbes, poor retention of biomass, high washout of biomass	Mazuelos et al. (1999), Yasin et al. (2019), Saratale et al. (2019)
Iron-Oxidizing Bioreactors	Packed bed bioreactor, contains iron-oxidizing prokaryotes that oxidizes Fe^{3+}into Fe^{2+}, *Acidithiobacillus ferrooxidans* oxidize inorganic sulphur compounds			Long et al. (2003), Johnson and Hallberg (2005a), Dhir (2018)

(Continued)

Table 12.3 (Continued)

Types	Mechanism of treatment	Advantages	Disadvantages	References
Upflow Anaerobic Sludge Blanket Reactor (UASB)	Suspended-growth reactor, attain high microbial population by promoting granulation	Higher treatment efficiency, low energy required, low operational and maintenance cost, low sludge production	Need post treatment to remove pathogens, require long time for microbial growth	Daud et al. (2018), Khanal et al. (2017), Gunasekaran et al. (2019)
Sulphidogenic Bioreactors	Generate alkalinity, remove metals as insoluble sulphides, hydrolysis, and precipitation of metals	Biosulphidogenesis reduce acidity, reducing engineering complexity, H_2S easily transferred to offline reactor vessel than HS, allow in-situ precipitation of transition metals	Efficiency is limited, depends on the function of anaerobic micro flora	Johnson and Hallberg (2005a), Johnson (2000), Dhir (2018)
Offline Sulphidogenic Bioreactors	Optimize production of hydrogen sulphide, include two techniques: BioteQ (based on sulphur reduction) and Thiopaq (based on sulphate reduction)	Low construction cost, effective to remove sulphate	Limited efficiency, working efficiency depends on anaerobic microflora, sensitive to modest acidity	Panda et al. (2016), Johnson and Hallberg (2005a)
Sulphate-Reducing Bioreactors	Use chemoorganotrophic anaerobic sulphate reducing bacteria, precipitation of metals as metal sulphides adsorption of metals occurs on the substrate material, eliminate metals through biosorption	Useful to treat AMD with high metal content, reduce acidity, produce good quality water	Not performed long term	Choudhary and Sheoran (2012), Neculita and Zagury (2008), Neculita et al. (2007), Johnson and Hallberg (2005b)
Rotating Biological Contactors	Biosorption of metals by immobilized microbial biofilms, support medium, remove metals as sulphides, hydroxides, and carbonates and raise pH,	Very effective to remove Fe, low energy requirement, no need of skilled person, low operational cost	Lower treatment efficiency, low pathogen removal	Dhir (2018), Ghodeif (2013)
Membrane bioreactors	Use ultrafiltration and microfiltration membranes for sedimentation and filtration, combine membrane technology with biological reactors	Generate high quality effluent, Flexible operating conditions, less area requirement,	High capital costs for membranes, fouling of membrane, need of high operating pressures	Ahuja et al. (2014), Kaur et al. (2019)

Table 12.4 Biological characteristics of SRB.

Taxonomic groups	Genera/Genus	Optimal growth temperature	Characteristics	References
Gram-negative mesophilic SRB	Desulfobacterium,Desulfovibrio, Desulfomicrobium, Desulfobacter, Desulfomonile, Desulfococcus, Desulfosarcina, etc.	20 to 40 °C	Diverse group with variety of shapes and physiological traits	Castro et al. (2000), Doshi (2006)
Gram-positive spore forming SRB	Desulfomaculum	20 to 40 °C (while some survive at higher temperature)	Form heat-resistant endospores	Castro et al. (2000), Doshi (2006)
Bacterial thermophilic SRB	Thermodesulfobacterium, Thermodesulfovibrio	65 to 70 °C	Inhabit high- temperature environments such as geothermal vents	Castro et al. (2000), Doshi (2006)
Archaeal thermophilic SRB	Archaeoglobus	>80 °C	Found only in marine hydrothermal vents	Castro et al. (2000); Doshi (2006)

12.4.2.2 Geochemical/Abiotic System

Geochemical system is a direct treatment of AMD using alkalinity producing substances and based on the retention period of alkaline agents with AMD. A detailed and thorough review has been discussed by Skousen et al. (2017) on different geochemical treatment technologies, specifying the design factors and efficiency of organic matters. The abiotic system also includes the application of both types of treatment methods, namely 'active method' (aeration and use of lime, limestone) as well as 'passive method' (such as anoxic lime stone drains) (Johnson and Hallberg 2005a).

12.4.2.2.1 Anoxic Limestone Drains (ALDs) An ALD system is basically a drain that is filled with crushed limestones, sealed by a geotextile or plastic bottom liner and then covered by soil or clay to avoid entry of oxygen and escape CO_2 (Skousen et al. 2017). In general, ALD is built near to a tailing pile to catch mine drainage that has not yet been exposed to oxygen. To uphold the anoxic conditions inside the drain, the drain is made up to 1–2 m depth (LópezPamo et al. 2002). Due to the anoxic condition of the system, it inhibits the precipitation of oxides and hydroxides. After filling the drain with limestone, mine drainage is constrained to pass through a substrate of limestone gravel that is impermeable to both air and water. The movement of mine water throughout the drain causes the dissolution of limestone, which generates alkalinity (HCO^{3-} + OH^{-}) and raises pH of water (Favas et al. 2016). This system can produce maximum alkalinity up to 300 mg L^{-1} as $CaCO_3$. The effectiveness of this system is good for a short period but as time increases it slowly precipitates hydroxide, starts losing its permeability, and hence causes the failure of the system (Dhir 2018). As per the study of Johnson and Hallberg (2005a), the diameter of drain usually ranges from narrow (0.6–1.0 m) to wide (10–20 m), and length around 30 m (RoyChowdhury et al. 2015). A well-designed ALD has the potential to increase alkalinity about 200 mg L^{-1} as $CaCO_3$ equivalent.

12.4.2.2.2 Open Limestone Channels (OLCs) Open limestone channels (OLCs) are open channels lined with coarse limestone. It is one the easiest passive treatment methods for treatment of AMD. Mine water is diverted through the channel, and the dissolution of limestone introduces alkalinity to AMD and increases pH. According to the length and slope of the channel, its performance varies. A slope of more than 20% shows good efficiency to remove precipitates from suspension and the maximum removal of pollutants from AMD (with acidity 500–2600 mg L^{-1} as $CaCO_3$) has been recorded at slopes between 45–60% (Dhir 2018). The guidelines for ALD design have been made by the US Bureau of Mines (Hedin et al. 1994). OLC is a good alternative of ALD to treat oxygenated discharges that contain high amounts of Fe^{+3}or Al. Though, design of an OLC must require high flow velocity, turbulence, vertical gradient, and large area for effective performance that keeps precipitates in suspension thus minimizing the armouring of limestone (Berghorn and Hunzeker 2001). To control the penetration of limestone into the groundwater table, it is important to use impermeable liners below the limestone layer (Berghorn and Hunzeker 2001).

12.4.2.2.3 Limestone Leach Beds (LLBs) Limestone leach beds (LLBs) are normally small shallow sinks. They contain a layer (up to 1 m) of limestone gravel that is filled with AMD to remove contamination via neutralization and metal precipitation. For treatment, AMD must be diverted through a drain and then rapidly flush out the system to flow the treated AMD in a settling pond (Hedin et al. 2013). Black et al. (1999) reported that a 30-minute retention period with limestone is adequate to remove 50% of acidity from slightly acidic mine water and also sufficient to remove majority of proton acidity in pH 3.0, metal-free water. It is a pre-treatment method for treatment of AMD with high content of Fe and Al, low pH (<3.0), and DO ($<1\,mg\,L^{-1}$) by either upward or downward flow practice. Ziemkiewicz et al. (2002) proclaimed that construction of LLBs at the upstream end of OLCs is more beneficial, increases working lifetime, and also reduces the length of the stream to give better quality treated water. Thorne and Pitzer (2003) illustrated that LLBs are efficient to treat acidic mine water with low metal concentration. LLBs can be constructs at the flow of mine water discharge as well as in underground mines for large-scale practice.

12.4.2.2.4 Steel Slag Leach Beds (SLBs) Slag leach bed (SLB) is a popular and potentially effective method to treat AMD-affected watercourse (Goetz and Riefler 2014). Steel slag is a waste by-product of steel manufacturing and is also obtained from metal recovery processes. Ziemkiewicz (1998) described that SLBs is a cost-effective method as it utilizes steel slag (a by-product) to add alkalinity and precipitate metals into AMD. It has been demonstrated that steel slag can generate alkalinity more than $2000\,mg\,L^{-1}$ into distilled water (Skousen et al. 2017). SLBs (e.g. check dams) work similarly as OLD and they are also channels or trenches, but the only difference is that they use slag as filling materials instead of limestone. It can be constructed or directly enforced along the existing waterways. A variety of slag materials are obtainable as by-products and its composition varies with different steel manufacturing methods. Blast furnace slag is one of the most widely acceptable materials for SLB (Sasaki et al. 2008). Blast furnace slags are rich calcium silicate glasses. These glasses remain unstable in water and raise pH of water during decomposition. The advantage of this system is its low operation and maintenance cost. The effectiveness of this system is badly affected by the presence and dissolution of trace metal (viz. Al, Mg, Fe, Ti, Mn, Si, etc.) from slag material (Skousen et al. 2017). SLBs are most suitable for treatment of AMD with the properties average acidity range < 1000 (mg $CaCO_3$/L), Avg. Flow Rate < 20(L/s), Oxygen Concentration – Ambient, Typical pH range > 1.5, and maximum pH achievable> 10 (Taylor et al. 2005).

12.4.2.2.5 Diversion Wells It is another simple method to add alkalinity to AMD-affected waterways. They were first used in Norway and Sweden for treatment of rainfall-induced stream acidity (Arnold 1991). Diversion wells consist of a cylindrical tank (made up of metal or concrete) filled with crushed limestone aggregate. Contaminated mine water is diverted through pipes to the bed of limestone with good velocity that can agitate and fluidize the bed. For effectual treatment in every few days the filling limestones must be replenished (Watzlaf et al. 2004). The hydraulic force causes the turbulent mixing of limestone aggregates into fine particles to prevent surfaces from armouring. AMD moves upward and floods over the 'well', and after that is diverted back into the stream (Berghorn and Hunzeker 2018). During this it also floods out the suspended metal flocs formed by neutralization and hydrolysis reactions from the top of the tank by the water current (Skousen et al. 2017). Limestones keep on dissolving and adding alkalinity to the system; therefore a long length of stream is required to improve downstream water quality (Watzlaf et al. 2004).

12.4.2.2.6 Limestone Sand The limestone sand system employed the use of sand-sized limestone within streambeds of AMD-contaminated watersheds (Zurbuch 1996). In this technique, limestone sand is suspended by the stream flow and adds alkalinity to the base flow conditions as it is transported all through the watershed (Watzlaf et al. 2004). The length of the stream matters to progress the water quality downstream. It is important to replenish limestone loads at some time intervals. The limestone particles can become armoured, but the stirring and cleaning in the waterbed offers clean surfaces for reactions (Metesh et al. 1998). If a sufficient amount of limestone is available for continuous addition, then we can say that this method is more cost reliable to treat AMD than other conventional passive treatment methods (Skousen et al. 2017).

12.4.2.3 Other Types of Methods

Table 12.5 lists some other AMD treatment methods, along with their advantages and limitations.

12.5 Conclusion

Worldwide only 0.28% of groundwater is available. This small amount of water resource is under threat of contamination by AMD water. In coal mining areas, the major reason for the pollution of groundwater is infiltration of AMD into water table. Surface mining activity changes the hydrogeological condition of the region that consequently forms depression cones and lowers the groundwater table. Disposal of mine tailing waste (mainly coal waste and ashes) in open pits significantly impacts the groundwater quality. This chapter clearly indicates that AMD causes several changes in the physicochemical parameters of groundwater. AMD is highly acidic in nature and creates a conducive environment for higher mobility of heavy metals. The important technology with reasonably good potential for treatment of AMD has been discussed. The use of PRBs shows the most promising approach in controlling the contamination of groundwater with AMD. Use of dry clay covers and cover with sludge to cap mine waste is the most desirable way to restrict exposure of mine tailings to oxygen and water that further inhibit the formation of AMD. Ion exchange and RO are suitable options for treatment of mine water up to a reasonable quality. The treated mine water can be used in several activities like in industrial and agricultural activities, as coolant and heating agents, for electricity generation using fuel cell technology, and solar pond. Further, the effective technologies have the potential for treatment of poor quality water containing a wide range of pollutants if needed. It is the need of the hour to cope with the situation of variation in the climate and achieve the sustainable development goal.

Table 12.5 List of some other AMD treatment methods and practices with their advantages and limitations.

Techniques	Description	Advantages	Limitations	References
DAF (dissolved air flotation)	Very effective to separate solid–liquid, involve coagulation and flocculation, through air bubbles transfer flocs to water surface, inject air into the solution to form air bubbles	High separation kinetics, effectively remove TSS, BOD. Produce good quality water, high rate operation	High energy consumption,	Ross et al. (2003), Rodrigues and Rubio (2007). Ratnayaka et al. (2009), Rao et al. (2014)
Hydroxide precipitation	Precipitate metals such as metal hydroxide use lime, caustic lime as precipitant, increase pH	Cost-effective, easy to handle, efficient to remove specific components	Produces large amount of sludge and gypsum, costly disposal require for secondary waste, need large quantity of reagents	Feng et al. (2000), Robb and Robinson (1995), Ziemkiewicz et al. (2003)
Electrocoagulation process	Works on electrochemistry, entails the formation of coagulants in situ by electrodissolution using sacrificial anodes, mainly Al and Fe upon application of an electric current between electrodes,	Low dosing of reagents, low energy consumption, can be run on renewable energy sources, easy automation, more efficient than conventional chemical coagulation, combine oxidation, coagulation, and precipitation	Operational issues with the EC process is electrode passivation, require maintenance, lack of systematic reactor design	Mamelkina et al. (2017) Oncel et al. (2013)
Phytoremediation	Use metal hyperaccumulator plants, involve phytoextraction and phytostabilization, biodegradation of pollutants, immobilization of heavy metals	Inexpensive, sustainable, environment-friendly, employed on large scale	Removal efficiency is slower, complete removal cannot be possible if solution has high concentration of heavy metals, chances of consumption of plants by animals	Das (2018), Hamed et al. (2006), RoyChowdhury et al. (2015)
Nanoremediation	Use nano-sized particles, zero-valent iron (ZVI), effectively neutralize acid, absorb metals on the surface of nano iron particles	Very quick and efficient	Nano-size and high mobility of ZVI may cause ecotoxicity, risk of bio-accumulating on living beings, expensive, not possible to employ on large scale	Das (2018)

(*Continued*)

Table 12.5 (Continued)

Techniques	Description	Advantages	Limitations	References
Electrochemical covers	Prevent AMD formation by steel cathode and Mg anode prevent O_2 migration,	Effective, produce good quality water	High capital cost of anode and cathode, regular replacement of anodes	Taylor et al. (2005)
Mn removal beds (MRBs)	Mn removal from AMD, shallow channels filled with fist-sized limestone rock, use aerobic heterotrophic bacteria that form Mn precipitates using organic matter as energy source	Absorb further amount of several trace metals	Occasional maintenance required for smooth working, limestone channels get blocked with silt and other materials, dissolved ferrous Fe create obstacle to complete Mn precipitation	Skousen et al. (2017), Dhir (2018)
RAPS	Produce alkalinity, contain layer of limestone and compost, flushed periodically to remove precipitated metals	Increase pH, promote reducing condition, offer sites for metal adsorption	High capital cost, pore clogging by gypsum and metal precipitates, require high maintenance, permeability lower with time	Watzlaf et al. (2004), Taylor et al. (2005)
Gas redox and displacement system (GaRDS)	Passive technique work on oxygen displacement, provide underground atmosphere to keep metal sulphide stable, retard sulphide oxidation	Low cost, rapid to install, effective to prevent contamination of groundwater with AMD	Not applicable to all sites, long term performance is doubtful	Watzlaf et al. (2004), Taylor et al. (2005), Taylor (2000)
Crystallization	Pre-treat AMD to adjust pH and reduce metal content, high temperature and pressure endorse crystallization	Reduce sulphate concentration below 200 mg/L, high efficiency, energy saving	Specifically work to remove one component, removal mechanism is more complex	Taylor et al. (2005), Lu et al. (2017)
Membrane technology	Require initially filtrate mine water for treatment and pH adjustment	Complete removal of contaminants	Costly, brine formation, membrane fouling	Buzzi et al. (2013), Martí-Calatayud et al. (2013), Tischler (2013)

References

Ackman, T.E. and Kleinmann, R.L.P. (1991). An in-line system for treatment of mine water. *International Mine Waste Management News* 1 (3): 1–4.

Ahmet, G., Ertan, A., and Ismail, T. (2007). Lead removal from aqueous solution by natural and pretreated clinoptilolite: adsorption equilibrium and kinetics. *Journal of Hazardous Materials* 146: 362–371.

Ahuja, S., Larsen, M.C., Eimers, J.L. et al. (eds.) (2014). *Comprehensive Water Quality and Purification*. Amsterdam: Elsevier.

Akcil, A. and Koldas, S. (2006). Acid mine drainage (AMD): causes, treatment and case studies. *Journal of Cleaner Production* 14: 1139–1145. https://doi.org/10.1016/j.jclepro.2004.09.006.

Akpor, O.B. and Muchie, M. (2010). Remediation of heavy metals in drinking water and wastewater treatment systems: processes and applications. *International Journal of Physical Sciences* 5: 1807–1817.

Amaral, F.J., Azevedo, A., Etchepare, R., and Rubio, J. (2016). Removal of sulfate ions by dissolved air flotation (DAF) following precipitation and flocculation. *International Journal of Mineral Processing* 149: 1–8.

Adams, D.J., Ogden, U.T., Pennington, P. et al. (2005). The role of microorganisms in acid rock drainage. In SME Annual Meeting February.

Arnold, D.E. (1991) Diversion wells-a low-cost approach to treatment of acid mine drainage. In: Proceedings of the 12th West Virginia surface mine drainage task force symposium, Morgantown, WV

Atikinson, S. (1997). Reverse osmosis makes mine-water drinkable. *Membrane Technology* 1997 (4): 7–10.

Bailey, S.E., Trudy, J., Olin, T.J. et al. (1999). A review of potentially low-cost sorbents for heavy metals. *Water Research* 33 (11): 2469–2479.

Baker, B.J. and Banfield, J.F. (2003). Microbial communities in acid mine drainage. *FEMS Microbiology Ecology* 44: 139–152. https://doi.org/10.1016/S0168-6496(03)00028-X.

Barton, P. (1978). The acid mine drainage. In: *Sulfur in the Environment. Part II Ecological Impacts* (ed. J.O. Nriagu), 313–358. New York: Wiley.

Black, C., Ziemkiewicz, P., and Skousen, J. (1999) Construction of a limestone leach bed and preliminary water quality results in Beaver Creek. In: Proceedings of the 20th West Virginia surface mine drainage task force symposium, Morgantown, WV

Benner, S.-G., Blowes, D.-W., Ptacek, C.-J., and Mayer, K.-U. (2002). Rates of sulfate reduction and metal sulfide precipitation in a permeable reactive barrier. *Applied Geochemistry* 17: 301–320.

Berghorn, G.H. and Hunzeker, G.R. (2001). Passive treatment alternatives for remediating abandoned-mine drainage. *Remediation Journal: The Journal of Environmental Cleanup Costs, Technologies & Techniques* 11 (3): 111–127.

Berghorn, G.H. and Hunzeker, G.R. (2018). Passive Treatment Alternatives for Remediating Abandoned- Mine Drainage. *Remediation Journal* 11 (3): 111–127. doi: 10.1002/rem.1007.

Bhatnagar, A. and Sillanpää, M. (2010). Utilization of agro-industrial and municipal waste materials as potential adsorbents for water treatment-a review. *Chemical Engineering Journal* 157: 277–296. https://doi.org/10.1016/j.cej.2010.01.007.

Blowes, D.W. and Ptacek, C.J. (2014). *The Geochemistry of Acid Mine Drainage*, 11e. Elsevier Ltd.
Brodie, G.A., Hammer, D.A., and Tomjanovich, D.A. (1988). An evaluation of substrale types in constructed wetlands acid drainage trealment systems. In: *Mine Drainage and Surface Mine Reclamation*, vol. 1). Info. Circular 9183, 389–398. Pittsburgh, PA: U.S. Bureau of Mines.
Budeit, D. (2007). Role of accelerated oxidation for removal of metals from mine drainage. *Proceedings of the American Society of Mining and Reclamation* 1: 121–131. https://doi.org/10.21000/JASMR07010121.
Buzzi, D.C., Viegas, L.S., Rodrigues, S. et al. (2013). Water recovery from acid mine drainage by electrodialysis. *Minerals Engineering* 4: 82–89.
Cao, W., Dang, Z., Zhou, X.-Q. et al. (2011). Removal of sulphate from aqueous solution using modified rice straw: preparation, characterization and adsorption performance. *Carbohydraate Polymers* 85 (3): 571–577. https://doi.org/10.1016/j.carbpol.2011.03.016.
Carlson, C.L. and Adriano, D.C. (1993). Environmental impacts of coal combustion residues. *Journal of Environmental Quality* 22 (2): 227–247.
Castro, H.F., Williams, N.H., and Ogram, A. (2000). Phylogeny of sulfate-reducing bacteria1. *FEMS Microbiology Ecology* 31: 1–9. https://doi.org/10.1111/j.1574-6941.2000.tb00665.x.
Chen, B.-Y., Utgikar, V.-P., Harmon, S.-M. et al. (2000). Studies on biosorption of zinc(II) and copper(II) on Desulfovibrio desulfuricans. *International Biodeterioration & Biodegradation* 46: 11–18.
Choudhary, R.-P. and Sheoran, A.-S. (2012). Performance of single substrate in sulphate reducing bioreactor for the treatment of acid mine drainage. *Minerals Engineering* 39: 29–35.
Council on Environmental Quality (CEQ), and Interagency Advisory Committee on Environmental Trends. 1989. Environmental Trends. Council on Environmental
Cravotta, C.A. and Trahan, M.K. (1999). Limestone drains to increase pH and remove dissolved metals from acidic mine drainage. *Applied Geochemistry* 14: 581–606. https://doi.org/10.1016/S0883-2927(98)00066-3.
Das, P.K. (2018). Phytoremediation and Nanoremediation: emerging techniques for treatment of acid mine drainage water. *Defence Life Science Journal* 3: 190. https://doi.org/10.14429/dlsj.3.11346.
Daud, M.K., Rizvi, H., Akram, M.F. et al. (2018). Review of upflow anaerobic sludge blanket reactor technology: effect of different parameters and developments for domestic wastewater treatment. *Journal of Chemistry*.
Demchak, J., Morrow, T., and Skousen, J. (2001). Treatment of acid mine drainage by four vertical flow wetlands in Pennsylvania. *Geochemistry: Exploration, Environment, Analysis* 1: 71–80.
Dhir, B. (2018). Biotechnological tools for remediation of acid mine drainage (removal of metals from wastewater and leachate). In: *Bio-Geotechnologies for Mine Site Rehabilitation*, 67–82. Elsevier.
Dietz, J.M. and Stidinger, D. M. (1996). Acidic mine drainage abatement in an anaerobic sub-surface flow wetland environment - case history of the treatment system at Corsica, PA. p. 531–540. In Proceedings of the American Society for Surface Mining and Reclamation Annual Meeting. (Knoxville, TN, May 15–20, 1996)
Doe, U. (2018). NETL rare earth recovery projects meeting and exceeding expectations. https://www.energy.gov/fe/articles/netl-rare-earth-recovery-projects-meeting-and-exceeding-expectations.

Doshi, S.M. (2006) Bioremediation of Acid Mine Drainage Using Sulfate-Reducing Bacteria US Environmental Protection Agency; Office of Solid Waste and Emergency Response and Office of Superfund Remediation and Technology Innovation, 65.

Doye, I. and Duchesne, J. (2003). Neutralization of acid mine drainage with alkaline industrial residues: laboratory investigation using bath-leaching test. *Applied Geochemistry* 18: 1197–1213.

Duchesne, J. and Reardon, E.J. (1998). Determining controls on element concentrations in cement kiln dust leachate. *Waste Management* 18: 339–350.

Evangelou, V.P. (1995). Potential microencapsulation of pyrite by artificial inducement of ferric phosphate coatings. *Journal of Environmental Quality* 24: 535–542.

Evangelou, V.P. and Zhang, Y.L. (1995). Critical reviews in environmental science and technology a review: pyrite oxidation mechanisms and acid mine drainage prevention a review: pyrite oxidation mechanisms and acid mine drainage prevention. *Critical Reviews in Environmental Science and Technology* 25: 37–41. https://doi.org/10.1080/10643389509388477.

Faulkner, B. (1990). *Handbook for the Use of Ammonia in Treating Mine Water*. Charleston, WV: West Virginia Mining and Reclamation Assoc.

Faust, S.D. and Aly, O.M. (1987). *Adsorption Process for Water Treatment*. Stoneham: Butterworths Publishers.

Favas, P.J.C., Sarkar, S.K., Rakshit, D. et al. (2016). *Acid Mine Drainages from Abandoned Mines: Hydrochemistry, Environmental Impact, Resource Recovery, and Prevention of Pollution*. Elsevier Inc.

Feng, D., Aldrich, C., and Tan, H. (2000). Treatment of acid mine water by use of heavy metal precipitation and ion exchange. *Minerals Engineering* 13 (6): 623–642.

Figueroa, L. 2005. Microbial ecology of anaerobic biosystems treating mining influenced waters. Presented at the Mine Water Treatment Technology Conference, Pittsburgh, PA, August 15–18, 2005.

Gaikwad, R.W. (2010). Review and research needs of active treatment of acid mine drainage by ion exchange. *Electronic Journal of Environmental, Agricultural & Food Chemistry* 98: 1343–1350.

Gaikwad, R.W. and Gupta, D.V. (2008). Review on removal of heavy metals from acid mine drainage. *Applied Ecology and Environmental Research* 6: 79–96. https://doi.org/10.15666/aeer/0603.

Geiger, A. and Cooper, J. (2010). Overview of airborne metals regulations exposure limits health effects, and contemporary research. *Cooper Environmental Services* 4 (3): 3–4.

Ghodeif, K. (2013). Baseline Assessment Study for Wastewater Treatment Plant for Al Gozayyera Village, West Kantara City, Ismailia Governorate, Egypt. Network of Demonstration Activities for Sustainable Integrated Wastewater Treatment and Reuse in the Mediterranean: Cairo, Egypt.

Goetz, E.R. and Riefler, R.G. (2014). Performance of steel slag leach beds in acid mine drainage treatment. *Chemical Engineering Journal* 240: 579–588. https://doi.org/10.1016/j.cej.2013.10.080.

Gray, N.F. (1997). Environmental impact and remediation of acid mine drainage: a management problem. *Environmental Geology* 30: 62–71. https://doi.org/10.1007/s002540050133.

Gunasekaran, M., Merrylin, J., Usman, T.M. et al. (2019). Biohydrogen production from industrial wastewater. In: *Biofuels: Alternative Feedstocks and Conversion Processes for the*

Production of Liquid and Gaseous Biofuels (eds. A. Pandey, C. Larroche, C.-G. Dussap, et al.), 733–760. Amsterdam: Academic Press.

Gupta, S.K. and Nikhil, K. (2016). Ground water contamination in coal mining areas: a critical review. *International Journal of Engineering and Applied Sciences* 3 (2): 69–74.

Hamed, M.M., Badawi, J.C., El-Housseini, M., and Abdel-Shafi, E. (2006). Phytoremediation of heavy metal contaminated soil using Sorghum virgatum. *Journal of Engineering and Applied Sciences* 53: 731–748. https://doi.org/10.1016/j.ccr.2005.01.001.

Hedin, R.S., Nairn, R.W., and Kleinmann, R.L.P. (1994). "Passive treatment of Coal Mine Drainage," US Bureau of Mines Info. Circ. 1994.

Hedin, R.S., Weaver, T., Wolfe, N., Watzlaf, G. (2013) Effective passive treatment of coal mine drainage. In: Proceedings of the 35th National association of abandoned mine land programs conference, Daniels, WV, USA

Jang, M. and Kwon, H. (2011). Pilot-scale tests to optimize the treatment of net-alkaline mine drainage. *Environmental Geochemistry and Health* 33: 91–101. https://doi.org/10.1007/s10653-010-9353-3.

John, R. and Goyal, M. (2017). Acid Mine Drainage: Sources, Impacts, and Prevention, epub.

Johnson, D.B. (1998). Biodiversity and ecology of acidophilic microorganisms. *FEMS Microbiology Ecology* 27: 307–317.

Johnson, D.B. (2000). Biological removal of sulfurous compounds from inorganic wastewaters. In: *Environmental Technologies to Treat Sulfur Pollution: Principles and Engineering* (eds. P. Lens and L. Hulshoff Pol), 175–206. London7 International Association on Water Quality.

Johnson, D.B. and Hallberg, K.B. (2003). The microbiology of acidic mine waters. *Research in Microbiology* 154: 466–473. https://doi.org/10.1016/S0923-2508(03)00114-1.

Johnson, D.-B. and Hallberg, K.-B. (2005a). Acid mine drainage remediation options: a review. *Science of the Total Environment* 338: 3–14.

Johnson, D.-B. and Hallberg, K.-B. (2005b). Biogeochemistry of the compost bioreactor components of a composite acid mine drainage passive remediation system. *Science of the Total Environment* 338: 81–93.

Johnson, K.-L. and Younger, P.-L. (2005). Rapid manganese removal from mine waters using an aerated packed-bed bioreactor. *Journal of Environmental Quality* 34: 987–993.

Kadlec, R. and Knight, R. (1996). *Treatment Wetlands*, 436. Boca Raton, FL. USA: CRC/Lewis Publishers.

Kamble, P.H. and Kumbhar, J.S. (2020). An Overview of Impact of Coal Mining On Water Resources. *International Journal of Scientific Research in Multidisciplinary Studies* 5: 115–121.

Kaur, H., Hippargi, G., Pophali, G.R., and Bansiwal, A.K. (2019). Treatment methods for removal of pharmaceuticals and personal care products from domestic wastewater. In: *Pharmaceuticals and Personal Care Products: Waste Management and Treatment Technology* (eds. M.N.V. Prasad, M. Vithanage and A. Kapley), 129–150. Butterworth-Heinemann.

Kefeni, K.K., Msagati, T.A.M., and Mamba, B.B. (2017). Acid mine drainage: prevention, treatment options, and resource recovery: a review. *Journal of Cleaner Production* 151: 475–493. https://doi.org/10.1016/j.jclepro.2017.03.082.

Kepler, D.A. and McCleary, E.C. (1994). Successive alkalinity-producing systems (SAPS) for the treatment of acidic mine drainage. p. 195–204. In: Proceedings, International Land Reclamation and Mine Drainage Conference, April 24–29, 1994, USDI, Bureau of Mines SP 06A-94. Pittsburgh, PA.

Khanal, S.K., Giri, B., Nitayavardhana, S., and Gadhamshetty, V. (2017). Anaerobic bioreactors/digesters: design and development. In: *Current Developments in Biotechnology and Bioengineering* (eds. D.-J. Lee, V. Jegatheesan, H.H. Ngo, et al.), 261–279. Amsterdam: Elsevier.

Kimmel, W.G. (1983). The impact of acid mine drainage on the stream ecosystem. In: *Pennsylvania Coal: Resources, Technology and Utilization* (eds. S.K. Majumdar and W.W. Miller), 424–437. Pennsylvania Academy of Science.

Kontopoulos, A. (1998). Acid mine drainage control. In: *Effluent Treatment in the Mining Industry* (eds. S.H. Castro, F. Vergara and M.A. Sánchez), 57–118. University of Concepción.

Kumar Pabby, A., Rizvi, S.A.A., and Sastre, A.M. (2013). *Handbook of Membrane Separations: Chemical, Pharmaceutical, Food and Biotechnological Applications*. London, New York: CRC Press, Taylor & Francis Group.

Kwong, Y.T.J. (1993). Prediction and Prevention of Acid Rock Drainage from a Geological and Mineralogical Perspective; MEND Project 1.32.1; Mine Environment Neutral Drainage (MEND): Ottawa, Canada.

López Pamo, E., Aduvire, O., and Barettino, D. (2002). Tratamientos pasivos de drena jesácidos de mina: estado actual y perspectivas de futuro. *Boletín geológico y minero* 113 (1): 3–21.

Long, Z.E., Huang, Y.H., Cai, Z.L. et al. (2003). Biooxidation of ferrous iron by immobilized Acidithiobacillus ferrooxidans in poly(vinyl alcohol) cryogel carriers. *Biotechnology Letters* 25: 245–249.

Lottermoser, B.G. (2007). *Mine Wastes: Characterization, Treatment and Environmental Impacts*, 2e, 304. Berlin: Springer.

Lu, H., Wang, J., Wang, T. et al. (2017). Crystallization techniques in wastewater treatment: an overview of applications. *Chemosphere* 173: 474–484.

Mamelkina, M.A., Cotillas, S., Lacasa, E. et al. (2017). Removal of sulfate from mining waters by electrocoagulation. *Separation and Purification Technology* 182: 87–93. https://doi.org/10.1016/j.seppur.2017.03.044.

Maree, J.P. and du Plessis, P. (1994). Neutralisation of acidic effluents with limestone. *Water Science Technology* 29 (9): 285–296.

Martí-Calatayud, M.C., Buzzi, D.C., García-Gabaldón, M. et al. (2013). Sulfuric acid recovery from acidminedrainagebymeansofelectrodialysis. *Desalination* 343: 120–127.

Mazuelos, A., Romero, R., Palencia, I. et al. (1999). Continuous ferrous iron biooxidation in flooded packed bed reactors. *Minerals Engineering* 12 (5): 559–564.

McDonald, D.G. and Grandt, A.F. (1981). Limestone- lime treatment of acid mine drainage-full scale. EPA Project Summary, EPA-600/S7–81-033.

Means, B., McKenzie, B., and Hilton, T. 2003. A computer-based model for estimating mine drainage treatment costs. Proceedings of the 24th Annual West Virginia Surface Mine Drainage Task Force Symposium, Morgantown, WV.

Mehta, P.S. (2002). "The Indian Mining sector: Effects on the Environment and FDI inflows." In Conference on foreign direct investment and the environment, pp. 7–8.

Metesh, J.J., Jarrell, T., et al (1998) Treating Drainage From Abandoned Mines in Remote Areas. 2600:1–18

Moncur, M.C., Jambor, J.L., Ptacek, C.J., and Blowes, D.W. (2009). Applied geochemistry mine drainage from the weathering of sulfide minerals and magnetite. *Applied Geochemistry* 24: 2362–2373. https://doi.org/10.1016/j.apgeochem.2009.09.013.

Moodley, I., Sheridan, C.M., Kappelmeyer, U., and Akcil, A. (2018). Environmentally sustainable acid mine drainage remediation: research developments with a focus on waste/by-products. *Minerals Engineering* 126: 207–220. https://doi.org/10.1016/j.mineng.2017.08.008.

Motsi, T., Rowson, N.A., and Simmons, M.J.H. (2009). Adsorption of heavy metals from acid mine drainage by natural zeolite. *International Journal of Mineral Processing* 92: 42–48. https://doi.org/10.1016/j.minpro.2009.02.005.

Mujuru, M. and Mutanga, S.S. (2017). *Management and Mitigation of Acid Mine Drainage in South Africa: Input for Mineral Beneficiation in Africa*. Africa Institute of South Africa.

Naidu, G., Ryu, S., Thiruvenkatachari, R. et al. (2019). A critical review on remediation, reuse, and resource recovery from acid mine drainage. *Environmental Pollution* 247: 1110–1124. https://doi.org/10.1016/j.envpol.2019.01.085.

Neculita, C.M. and Zagury, G.J. (2008). Biological treatment of highly contaminated acid mine drainage in batch reactors: Long-term treatment and reactive mixture characterization. *Journal of Hazardous Materials* 157: 358–366. https://doi.org/10.1016/j.jhazmat.2008.01.002.

Neculita, C.M., Zagury, G.J., and Bussière, B. (2007). Passive treatment of acid mine drainage in bioreactors using sulfate-reducing bacteria: Critical review and research needs. *Journal of Environmental Quality* 36 (1): 1–16.

Nidal, M., Mohamed, K., and Chris, J.W. (2012). *Membrane Modification: Technology and Applications*, vol. 19, 477–449. CRC Press.

Nordstrom, D.K. and Southam, G. (1997). Geomicrobiology of sulfide mineral oxidation. In: *Geomicrobiology: Interactions between Microbes and Minerals* Reviews in Mineralogy, vol. 35 (eds. J.F. Banfield and K.H. Nealson), 361–390. Washington, DC: Mineralogical Society of America.

Omer, Y., Yalcin, A., and Fuat, G. (2003). Removal of copper, nickel, cobalt and manganese from aqueous solution by kaolinite. *Water Res.* 37: 948–952.

Oncel, M.S., Muhcu, A., Demirbas, E., and Kobya, M. (2013). Journal of environmental chemical engineering a comparative study of chemical precipitation and electrocoagulation for treatment of coal acid drainage wastewater. *Biochemical Pharmacology* 1: 989–995. https://doi.org/10.1016/j.jece.2013.08.008.

O'Neil, M.J. (2013). *The Merck Index –an Encyclopedia of Chemicals, Drugs, and Biologicals*, 294. Cambridge, UK: Royal Society of Chemistry.

Othman, A., Sulaiman, A., and Sulaiman, S.K. (2017). The use of quicklime in acid mine drainage treatment. *Chemical Engineering Transactions* 56: 1585–1590.

Oyewo, O.A., Agboola, O., Onyango, M.S. et al. (2018). Current methods for the remediation of acid mine drainage including continuous removal of metals from wastewater and mine dump. In: *Bio-Geotechnologies for Mine Site Rehabilitation* (eds. M.N.V. Prasad, P.J. de Campos Favas and S.K. Maiti), 103–114. Elsevier.

Panda, S., Mishra, S., and Akcil, A. (2016). Bioremediation of acidic mine effluents and the role of sulfidogenicbiosystems: a mini-review. *Euro-Mediterranean Journal for Environmental Integration* 1: 1–9. https://doi.org/10.1007/s41207-016-0008-3.

Park, S.M., Shin, S.Y., Yang, J.S. et al. (2015). Selective recovery of dissolved metals from mine drainage using electrochemical reactions. *Electrochimica Acta* 181: 248–254.

Pérez-López, R., Cama, J., Nieto, J.M., and Ayora, C. (2007). The iron-coating role on the oxidation kinetics of a pyritic sludge doped with fly ash. *Geochimica et Cosmochimica Acta* 71: 1921–1934.

Pierre Louis, A.-M., Yu, H., Shumlas, S.L. et al. (2015). Effect of phospholipid on pyrite oxidation and microbial communities under simulated acid mine drainage (AMD) conditions. *Environmental Science and Technology* 49: 7701–7708. https://doi.org/10.1021/es505374g.

Pisa, V. and Gulikova, E. (2005). Desalting of acid mine drainage by reverse osmosis method – field tests 9th International Mine Water Congress Oviedo, Spain, International Mine Water Association IMWA.

Rao, M.N., Sultana, R., and Kota, S.H. (2014). *Solid and Hazardous Waste Management*. BS Publications.

Ratnayaka, D.D., Brandt, M.J., and Johnson, M. (2009). *Water Supply*. Butterworth-Heinemann.

Riodan, C., Bustard, M., Putt, R., and Mchale, A.P. (1997). Removal of uranium from solution using residual brewery yeast: com- binedbiosorption and precipitation. *Biotechnology Letters* 19: 385–387.

Robb, G.A. and Robinson, J.D. (1995). Acid drainage from mines. *The Geographical Journal* 161 (1): 47–54.

Rodrigues, R.T. and Rubio, J. (2007). DAF - dissolved air flotation: potential applications in the mining and mineral processing industry. *International Journal of Mineral Processing* 82: 1–13.

Rose, A.W. and Cravotta, C.A. III (1998). Geochemistry of coal mine drainage. *Coal Mine Drainage Prediction Pollution and Prevention in Pennsylvania* 1: 1–22.

Ross, C.C., Valentine, G.E., Smith, B.M., and Pierce, J.P. (2003). Recent advances and applications of dissolved air flotation for industrial pretreatment. Presented at: The Industrial Water/ Wastewater Program North Carolina AWWA/WEA Conference, Greensboro-North Carolina.

RoyChowdhury, A., Sarkar, D., and Datta, R. (2015). Remediation of acid mine drainage-impacted water. *Current Pollution Reports* 1: 131–141. https://doi.org/10.1007/s40726-015-0011-3.

Saha, S. and Sinha, A. (2018). A review on treatment of acid mine drainage with waste materials: a novel approach. *Global Nest Journal* 20: 512–528.

Sahu, L. and Dey, S. (2017). Enrichment of carbon recovery of high ash coal fines using air fluidized vibratory deck separator. *International Journal of Coal Science Technology* 4: 262–273. https://doi.org/10.1007/s40789-017-0172-3.

Sangita, K., Udayabhanu, G., and Prasad, B. (2010). Studies on environmental impact of acid mine drainage generation and its treatment: an appraisal. *Indian Journal of Environmental Protection* 30 (11): 953e967.

Saratale, G.D., Saratale, R.G., Banu, J.R., and Chang, J.S. (2019). Biohydrogen production from renewable biomass resources. In: *Biohydrogen* (eds. A. Pandey, S.V. Mohan, J.-S. Chang, et al.), 247–277. Amsterdam: Elsevier.

Sasaki, K., Nukina, S., Wilopo, W., and Hirajima, T. (2008). Removal of arsenate in acid mine drainage by a permeable reactive barrier bearing granulated blast furnace slag: column study. *Materials Transactions* 49 (4): 835e844.

Schippers, A. and Sand, W. (1999). Bacterial leaching of metal sulfides proceeds by two indirect mechanisms via thiosulfate or via polysulfides and sulfur. *Applied Environmental Microbiology* 65: 319–321.

Sgambat, J.P., LaBella, E.A., and Roebuck, S. (1980). *Effects of Underground Coal Mining on Ground Water in the Eastern United States*, vol. 1. Industrial Environmental Research Laboratory, Office of Research and Development, US Environmental Protection Agency.

Singer, P.C. and Stumm, W. (1970). Acid mine drainage: the rate-determining step. *Science* 167: 1121–1123.

Singh, A.K., Mahato, M.K., Neogi, B. et al. (2012). Environmental geochemistry and quality assessment of mine water of Jharia coalfield, India. *Environmental Earth Sciences* 65: 49–65. https://doi.org/10.1007/s12665-011-1064-2.

Simate, G.S. and Ndlovu, S. (2014). Acid mine drainage: challenges and opportunities. *Journal of Environmental and Chemical Engineering* 2: 1785–1803.

Skousen, J. (2002), A brief overview of control and treatment technologies for acid mine drainage, In: Proceedings of National Mtg of the American Society of Mining and Reclamation, pp 879–899.

Skousen, J. (2014). Overview of acid mine drainage treatment with chemicals. Chapter 29. In: *Acid Mine Drainage, Rock Drainage, and Acid Sulfate Soils: Causes, Assessment, Prediction, Prevention, and Remediation.* https://doi.org/10.1002/9781118749197. ch29.

Skousen, J., Hilton, T., and Faulkner, B. (1996). Overview of acid mine drainage treatment with chemicals. *Green Lands* 26: 36–45.

Skousen, J.-G., Rose, A.-W., Geidel, G. et al. (1998). Handbook of technologies for avoidance and remediation of acid mine drainage, Morgantown, WV, National Mine Land Reclamation Center, p. 131

Skousen, J., Rose, A., Geidel, G. et al. (1999). Handbook of Technologies for Avoidance and Remediation of Acid Mine Drainage. National Mine Land Reclamation Center, West Virginia University, Morgantown, WV. https://www.osmre.gov/resources/library/ghm/hbtechavoid.pdf

Skousen, J.G., Sexstone, A., and Ziemkiewicz, P.F. (2000). Acid mine drainage control and treatment. *Reclamation of Drastically Disturbed Lands*: 1–42.

Skousen, J.G., Demchak, J., and Morrow, T. (2014). Treatment of acid mine drainage by four vertical flow wetlands in Pennsylvania. *Geochemistry: Exploration, Environment, Analysis* https://doi.org/10.1144/geochem.1.1.71.

Skousen, J., Zipper, C.E., Rose, A. et al. (2017). Review of passive systems for acid mine drainage treatment. *Mine Water Environment* 36: 133–153. https://doi.org/10.1007/s10230-016-0417-1.

Skousen, G., Ziemkiewicz, P.F., and Mcdonald, L.M. (2018). Acid mine drainage formation, control and treatment: Approaches and strategies. *The Extractive Industries and Society* 6 (1): 241–249. doi: 10.1016/j.exis.2018.09.008.

Snow, D.H. (1999) Overview of Permeable Reactive Barriers. Unpublished paper.

Stumm, W. and Morgan, J.J. (1996). *Aquatic Chemistry: Chemical Equilibria and Rates in Natural Waters.* New York: Wiley.

Sun, Q., L.M. McDonald Jr., J.G. Skousen, Effects of armoring on limestone neutralization of AMD, in: 2000 West Virginia Surface Mine Drainage Task Force Symposium, Morgantown, WV, 2000, pp. 1–10.

Taylor, J. and Waring, C. (2000). The passive prevention of ARD in underground mines by displacement of air with a reducing gas mixture: GaRDS. *Mine Water and the Environment* 20: 2–7.

Taylor, J., Pape, S., and Murphy, N. A summary of passive and active treatment technologies for acid and metalliferous drainage (AMD). Fifth Australian workshop on acid drainage. Vol. 2931. 2005.

Thorne, B., Pitzer, K. (2003) Response of fish populations in the Little Sandy watershed. In: Proceedings of the 24th West Virginia surface mine drainage task force symposium, Morgantown, WV

Tischler, J.S., Klein, R., Mühling, M., and Schlömann, M. (2013). Bioremediation of Mine Water. *Advances in Biochemical Engineering/Biotechnology* 141: 109–172. doi: 10.1007/10.

Trumm, D. (2014). Selection of active and passive treatment systems for AMD - Flow charts for New Zealand conditions. *New Zealand Journal of Geology and Geophysics* 53 (2–3): 195–210. https://doi.org/10.1080/00288306.2010.500715.

Tsukamoto, T.-K. and Miller, G.-C. (1999). Methanol as a carbon source for microbiological treatment of acid mine drainage. *Water Research* 33: 1365–1370.

Tsukamoto, T.-K., Killion, H.-A., and Miller, G.-C. (2004). Column experiments for microbiological treatment of acid mine drain- age: low-temperature, low-pH and matrix investigations. *Water Research* 38: 1405–1418.

United States Environmental Protection Agency (USEPA) (2004). Primer for municipal wastewater treatment systems. Office of wastewater management. EPA 832-R-04-001.

URS Report, (2003) Passive and semi-active treatment of acid rock drainage from metal mines-state of the practice, Prepared for U.S. Army Corps of Engineers, Concord, Massachusetts, by URS Corporation, Portland, ME.

Wang, W., Zhenghe, X., and Finch, J.A. (1996). Fundamental study of an ambient temperature ferrite process in the treatment of acid mine drainage. *Environmental Science and Technology* 30 (8): 2604–2608.

Watzlaf, G.R. (1997) Passive Treatment of Acid Mine Drainage in Down-Flow Limestone Systems. Proceedings America Society of Mining and Reclamation, 611–622. doi: 10.21000/JASMR97010611

Watzlaf, G.R., Schroeder, K.T., Kleinmann, R.L. et al. (2004). The passive treatment of coal mine drainage. United States Department of Energy National Energy Technology Laboratory Internal Publication, 1–72.

Waybrant, K.-R., Ptacek, C.-J., and Blowes, D.-W. (2002). Treatment of mine drainage using permeable reactive barriers: column experiments. *Environmental Science and Technology* 36: 1349–1356.

Wieder, R.K. 1992. The Kentucky wetlands project: A field study to evaluate man-made wetlands for acid coal mine drainage treatment. Final Report to the U.S. Office of Surface Mining, Villanova University, Villanova, PA.

Yasin, M., Cha, M., Chang, I.S. et al. (2019). Syngas fermentation into biofuels and biochemicals. In: *Biofuels: Alternative Feedstocks and Conversion Processes for the Production of Liquid and Gaseous Biofuels* (eds. A. Pandey, C. Larroche, C.-G. Dussap, et al.), 301–327. Amsterdam: Academic Press.

Yeh, C.-H., Lin, C.-W., and Wu, C.-H. (2010). A permeable reactive barrier for the bioremediation of BTEX-contaminated groundwater: Microbial community distribution and removal efficiencies. *Journal of hazardous materials* 178 (1–3): 74–80. Retrieved from http://www.ncbi.nlm.nih.gov/pubmed/20122795.

Younger, P.L., Banwart, S.A., Hedin, R.S. (2002) Mine Water(2001) OECD Environmental Outlook. 1–14. doi: 10.1787/9789264188563-en

Zinck, J.M. and Griffith, W.F. 2000. An assessment of HDS-type lime treatment processes – efficiency and environmental impact. In: ICARD 2000. Proceedings from the Fifth International Conference on Acid Rock Drainage. Society for Mining, Metallurgy, and Exploration, Inc. Vol II, 1027–1034

Zhong, C.M., Zhen-Liang, X., Fang, X.H., and Cheng, L. (2007). Treatment of acid mine drainage (AMD) by ultra-low-pressure reverse osmosis and nanofiltration. *Environmental Engineering Science* 24: 1297–1306.

Ziemkiewicz, P.F. (1998) Steel slag: applications for AMD control. In: Proceedings of the 1998 conference on hazardous waste research, Snowbird, UT, pp 44–62

Ziemkiewicz, P., Meek, F. (1994). Long term behavior of acid forming rock: results of 11-year field studies. Proceedings, International Land Reclamation and Mine Drainage Conference 49–56.

Ziemkiewicz, P.F., Skousen, J., Simmons, J.S. (2002) Long-term performance of passive acid mine drainage treatment systems. In: Proceedings of the 23rd WV surface mine drainage task force symposium, Morgantown, WV.

Ziemkiewicz, P.F., Skousen, J.G., and Simmons, J. (2003). Long-term performance of passive acid mine drainage treatment systems. *Mine Water Environment* 22: 118–129.

Zurbuch, P.E. (1996) Early results from calcium carbonate neutralization of two West Virginia rivers acidified by mine drainage. In: Proceedings of the 17th WV surface mine drainage task force symposium. Morgantown, WV.

13

Impact of Electronic Waste Pollutants on Underground Water

Juhi Khan[1], Amrish Kumar[2], Ajay Giri[3], Dan Bahadur Pal[4], Anamika Tripathi[2], and Deen Dayal Giri[5]

[1]*Department of Botany, IFTM University, Moradabad, Uttar Pradesh, India*
[2]*Pollution Ecology Research Laboratory, Department of Botany, Hindu College, Moradabad, Uttar Pradesh, India*
[3]*Department of Botany, Banaras Hindu University, Varanasi, Uttar Pradesh, India*
[4]*Department of Chemical Engineering, Birla Institute of Technology, Mesra, Ranchi, Jharkhand, India*
[5]*Department of Botany, Maharaj Singh College Saharanpur, Saharanpur, Uttar Pradesh, India*

13.1 Introduction

Electronic wastes (e-waste) are emerging pollutants in present-day society due to excessive use and unmanaged disposal of the electronic equipment. The major electronic equipment of today includes computers and their accessories, cellular phones, televisions, LCD/Plasma TVs, Ipods, washing machines, printers, mobiles and refrigerators, etc. (Lahiry 2019). Thus, in a broad sense e-waste are discarded, obsolete, electrical devices (Anusree and Balasubramanian 2019). These wastes contain several toxic substances harmful to the terrestrial and aquatic ecosystems. Due to use, e-waste pollution problems are present in almost every country. The pollutants of e-waste pose major problems due to toxic substances, if they are not disposed according to the protocols (Kiddee et al. 2013). Asian countries like India and China, having low labour cost and less strict rules and regulations, makes them suitable for dumping sites for e-waste for retrieving metals and valuables by developed countries (Gaidajis et al. 2010). About 50–80% of e-waste of developed countries is dumped in Asia and Africa. There was about 62% increase in total e-waste generation in a span of six years in India (Ahmed and Panwar 2015). India's annual e-waste generation was ~2 million tonnes, and it ranks fifth among major e-waste producing countries (U.S.A. > China > Japan > Germany) as per the global e-waste monitor in 2017. In 2016–2017, only 0.036 million tonnes of e-waste was treated in India (Baldé et al. 2017). The government of India has shown concern about the minimization of waste for environmental sustainability. The optimal and efficient use of natural resources has been emphasized to recycle the disposed e-waste. All concerned are supposed to care about the environment while ensuring the economic growth and quality of life (Rajya Sabha

Groundwater Geochemistry: Pollution and Remediation Methods, First Edition.
Edited by Sughosh Madhav and Pardeep Singh.

Secretariat 2011). As per a United Nations report on e-waste in World Economic Forum (2019), the waste reaching streams was about 48.5 million tonnes in 2018 (Lahiry 2019). The huge amount of e- waste generation, unregulated accumulation, and illegal recycling practice can create environmental issues posing threats to human health (Luo et al. 2011). In the future, development of economy and society will further increase these problems in the world.

The e-waste can be broadly categorized into two groups: hazardous and non-hazardous. E-waste mainly contains ferrous and non-ferrous metals, plastics, glass, wood, printed circuit boards, concrete, ceramics, rubber, and other items. Major constituents of e-waste are iron and steel (50%), plastics (21%), non-ferrous metals (13%) and other constituents (~ 16%). In addition to iron, other commonly used metals in e-waste are copper, aluminium, and valuable gold, silver, palladium, platinum etc. The e-waste becomes hazardous mainly due to the presence of elements like Pb, Hg, As, Cd, Se, Cr(VI), and flame retardants in high quantity. There may be the presence of over one thousand different substances in e-waste that are toxic and could cause serious environmental pollution upon disposal (BAN and SVTC 2002).

13.2 Recycling of e-waste

The e-waste is recycled formally as well as informally. In the informal e-waste recycling, the obsolete electronic equipment is broken by hammering. The dismantled gadgets and motherboards are used to extract precious Cu, Ag, and Au metal from the circuits. During extraction of these metals from motherboards, workers are exposed to different metals and deadly toxic fumes. The circuit boards are cooked in open flames or in shallow pans to extract copper, which exposes workers to fumes containing lead. The extraction of gold by acid bath of circuit board chips releases even more harmful gases into the vicinity that are very hazardous to the worker's health and environment (He et al. 2017).

13.3 Pollutants in e-waste

Electronic equipment is composed of organic and inorganic materials. Both are released in environment during processing of electronic wastes, and both organic and inorganic heavy metals are released into the environment (Singh et al. 2015). After entry into the environment they cause serious problems in air, soil, and water which makes them unsuitable for their inhabitants directly or indirectly. The generalized flow of pollutants from the e-waste to groundwater is presented in the flowchart given in Figure 13.1.

13.3.1 Inorganic Heavy Metals in e-waste

Pollutants or toxins in e-waste are typically concentrated in circuit boards, batteries, plastics, and LCDs (liquid crystal displays) (Rajya Sabha Secretariat 2011). Table 13.1 show the major heavy metal pollutants in waste electrical and electronic equipment (Alexander and Bernd 2008).

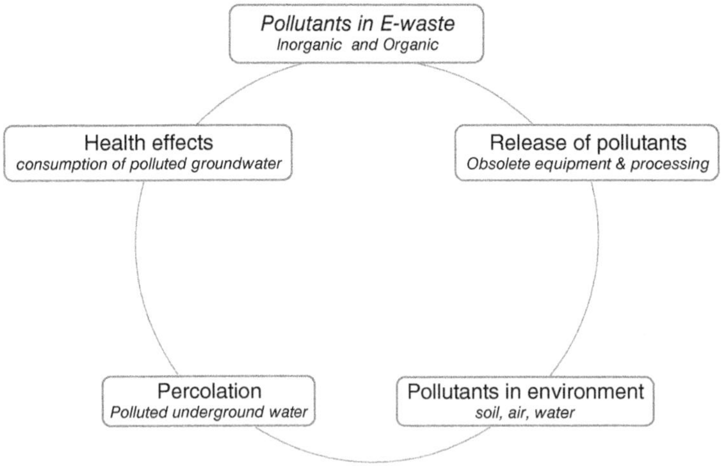

Figure 13.1 Generalized cycle of e-waste pollutants.

Table 13.1 Major inorganic pollutants present in e-waste.

	Inorganic Pollutant	**Electronic equipment/component**
1.	Arsenic	LEDs, diodes, semiconductors, solar cells, microwaves
2.	Barium	Electron tubes, additives of lubricant, plastic, and rubber filler
3.	Cadmium	Alloys, batteries, cathode ray tubes, circuit boards, batteries, monitor, pigments, solder
4.	Copper	Circuits, cables, ribbons, coils, pigments
5.	Lead	circuit boards, LEDs, rechargeable batteries, transistors, solar, lithium batteries, PVC stabilizers, lasers, thermoelectric elements
6.	Lithium	Cell phone batteries, other rechargeable batteries such as those used in photography and videography.
7.	Mercury	Components in copper machines and steam irons, batteries in clocks and pocket calculators, switches, LCDs
8.	Selenium	Photoelectric cells, pigments, photocopiers, fax machines

This equipment, if mishandled after the end of their lives, contaminate the soils with heavy metals. These polluted soils will take one or two centuries to remediate. This long life of metal pollution remediation is paid by many generations in terms of their health. The problem is further intensified and amplified by complex contamination by many heavy metals (Qin et al. 2008). The main sources of excessive heavy metals in the soil are due to settling of atmospheric dust and depositions, irrigation of agricultural fields with metal-contaminated sewage, improper heaping of the industrial solid waste, flowing of effluents and tailing of mining, and application of pesticides and fertilizers containing metal. The entry of various heavy metals in the atmosphere takes place through gas and dust produced by burning of coal for energy production and transport of mineral for metallurgy, as well as by production of construction materials.

13.3.2 Organic Pollutants in e-waste

The harmful materials present in e-waste include metals and compounds formed during their processing such as combustion, partial combustion, and restoration or renovation products. Products of incomplete combustion include all organic compounds emitted during partial burning and reformation products. The later are organic compounds formed due to the interaction of different ingredients during combustion. The organic compounds which are recalcitrant to degradation and remain in the environment for long time periods are called persistent organic pollutants (POPs). There are several such compounds that are found in e-waste. Brominated flame retardants fall under such category. The common brominated flame retardants are polybrominated diphenyl ethers (PBDEs), polychlorinated biphenyls (PCBs), polycyclic aromatic hydrocarbons (PAHs), polychlorinated dibenzo-p-dioxins (PCDDs), and furans. The POPs significantly affect both the environment and human health in long-term exposure (Brigden et al. 2008).

13.3.3 Entry of e-waste Organic Pollutants into Groundwater

The water which is present below the surface of earth in rock, sand, gravel, and soil is called groundwater. The groundwater level is maintained by the percolation of water from rainfall. Groundwater and surface water are the easily available drinking water sources for human and their dependent animals. Most rural homes are dependent on groundwater drinking water, so supply of adequate, clean groundwater is their basic necessity.

Groundwater gets contaminated in different ways:

- Natural contamination of groundwater: dissolving rocks and minerals, contamination of shallow groundwater by bacteria or pathogens or organic matter
- Agricultural activity-related pollutants: percolation of contaminated water from animal feedlots, chemicals used in agriculture such as plant nutrients, herbicides, pesticides
- Industrial and municipal waste contaminants
- Household septic system contaminants

In addition to the above-mentioned sources of contamination of groundwater, electronic waste and their processing products significantly affect groundwater quality, especially the organic compounds. Improperly disposed e-waste ultimately leads to toxin entry in the groundwater. The inorganic heavy metals of soil get dissolved in water, and percolated to groundwater. The groundwater may touch larger water bodies like lakes or rivers or end up in the ocean. Once the groundwater is contaminated, it is very difficult to protect crops, grazing animals, and human beings from exposure with the pollutant. This groundwater entry of toxins can be distributed to various ecosystems and food chains in different ways to affect human beings. So, e-waste pollutant entry to groundwater is a potential threat to the quality of groundwater. Some organic pollutants most commonly originating from e-waste are described in the next sections.

13.3.3.1 Polybrominated Diphenyl Ethers (PBDEs)

The highly toxic PBDEs is one of the POPs emitted at informal e-waste recycling sites that can bioaccumulate in food chains. PBDE can be transported long-range in the environment. They have been reported from different environmental samples of soil, air, and water as well as detected in animals and humans from different countries across the world (Chen et al. 2012).

The persistent PBDES are likely to enter the underground water and pollute it. So, development of an adequate groundwater protection plan is a necessity in different countries. The PBDE are supposed to enter groundwater through the fractured bedrock with thin overburden cover. However, little is known about transport processes from the ground surface to depth. A study conducted to investigate the potential of groundwater contamination by PBDE flame retardants in Canada indicated that natural fractured bedrock aquifer contains PBDEs. These compounds previously not measured in groundwater were detected in greater concentration in the aquifer compared to surface water bodies. They mainly reached the aquifer through septic tanks, manure, and the atmosphere. The study indicated that further survey regarding PBDE concentrations in groundwater is required in such areas to find out the potential sources of contamination (Levison et al. 2012). PDBEs affect activity of thyroid, and can disrupt endocrine and reproductive health (Ohajinwa et al. (2019).

13.3.3.2 Polychlorinated Biphenyls (PCBs)

The another important POP present in e-waste that deleteriously influences human health after entering in the food chain through the environment are organo-halogen compounds like PCBs (Borthakur 2016). PCBs are manmade organics composed of H, C, and Cl atoms in their structure. This light yellow-coloured compound is a constituent of electronic equipment like capacitor, plastics, paint, transformer, dye, and rubbers. The environmental persistent PCBs are emitted and enter in the environment especially during the crude process of informal e-waste recycling as well as during uncontrolled burning of municipal solid waste. Dismantling of PCBs containing waste also releases them into the atmosphere (Xing et al. 2008). These compounds remain for long periods and are recycled between air, water, and soil. The aerial particulate PCBs get deposited on land in dry conditions and washed with water during rainy to surface water bodies. They remain suspended in water and then settle down on the bottom of water bodies with time. They may be transported long distance in air as evident from their presence in snow and sea water. They can reach to underground water with gravitational water. These harmful compounds, when present in the potable ground water, can potentially pose a variety of adverse health effects including cancer and other serious problems immune system, endocrine system, nervous system, and reproductive system. The organo-halogen compounds interfere with TH signalling, leading to altered gene expression of their concerned gene in humans (Zheng et al. 2017).

13.3.3.3 Polycyclic Aromatic Hydrocarbons (PAHs)

PAHs are C and H-containing compound with multiple aromatic rings (five or more) having low water solubility and low vitality. These PAHs naturally occur in petroleum products such as crude oil, gasoline, and coal. They are released in the environment during burning of petroleum products, oil, gas, garbage, coal, and tobacco. They are also notably released in wildfire. Due to their carcinogenic and teratogenic effects these PAHs are well-known global pollutants. Moreover, during the e-waste dismantling process the incomplete combustion of printed circuit board, wire insulation, LCD, TV, PVC (polyvinyl chloride) material, and computer with cathode ray tube at a relatively low temperature leads to the release of PAHs. PAHs are also formed during a spill or incomplete burning of oil and coal, garbage or other organic substances, agricultural runoff etc. High levels of PAHs were determined in atmospheric particulates collected from e-waste dismantling areas. The potential sources of PAHs are indiscriminate dumping and illegal open burning of e-waste (Adeyi and Oyeleke 2017). The water insolubility of PAHs limits their mobility in the environment. The concentration of PAHs in river and estuarine sediments vary depending upon proximity to municipality and industrial discharge points. PAHs are present at high concentrations in the groundwater and soils of many hydrocarbon industries area. Products containing PAHs compound are harmfully affect quality of soil, sub-soil, and groundwater, as well as human health by degrading ecosystem and vegetation (Manmi et al. 2019).

13.3.3.4 Polychlorinated Dibenzo-p-dioxins (PCDDs)

The PCDDs are a toxic group of about 75 different dioxin compounds. The main sources of dioxins in the environment are e-waste burning and recycling processes (Dai et al. 2020). Zennegg et al. (2014) reported that the release of these compounds unintentionally enter into the environment during manual disassembly of equipment, circuit boards, and shredding, roasting, and acid-stripping of metals, as well as open burning of e-waste. The lipophilic nature of PCDD makes them suitable for bioaccumulation in the food chain and they remain resistant to biological and chemical degradation (Sadler and Connell 2012). As per a WHO report, the half-life of PCDD in the body ranges between 7 and 11 years (W.H.O. 2016). Today about 50% of the world's population is dependent on groundwater for drinking and other purposes. But groundwater resources are increasingly threatened by such chemical contamination.

Lopez and colleagues investigated groundwater samples in France and they reported the presence of dioxins in more than 10% of investigated samples (Lopez et al. 2015). Many studies proved high concentration of PCDD throughout a core of collected groundwater (Grant et al. 2011; MacKay and Gschwend 2001). Water samples collected from three remote coastal and six offshore sites in the Baltic Sea analysed for PCDD content found 2.6 kg (21 g of TEQ) in the mixed surface layer of the whole open Baltic (Broman et al. 1991). PCDD levels in offshore, near shore, and tributary sediments of Lakes Superior were between 5 to 18 000 $pg\,g^{-1}$ dw in Lake Superior and 3 to 6100 $pg\,g^{-1}$ dw in Lake Huron (Shen et al. 2009). The PCDD concentration in the sediments of Haihe River, the most polluted river in seven largest basins in China, was 1.3–26 pg I-TEQ g^{-1} dry weight (Liu et al. 2007).

Humans exposed to dioxins may develop skin hyperpigmentation and lesions. Dioxins affect lipid metabolism and may lead to development of fatty liver disease. They alter function of liver enzymes, e.g. increased P_{450} dependent monooxygenase activity, serum glutamic pyruvic transaminase (SGPT), and serum glutamic oxaloacetic transaminase (SGOT). In addition, they may alter shape and size of different organs such as spleen, testes, or ovary by hyperplasia or atrophy. Bone marrow depletion, systemic haemorrhage, general weakness and drastic weight loss, suppressed immune system, and endocrine and nervous system abnormalities have also been reported from this well-known carcinogen. In addition, PCDD also damages reproductive system, kidney, and stomach. Occurrences of soft-tissue sarcoma and lymphoma have been reported due to regular or accidental exposure with dioxin. The dioxins have been found in blood serum, adipose tissue, and even in breast milk of the general population of advanced countries (Mukerjee 1998). One to four-hundred-year-old Alaskan Eskimo tissues frozen in the Arctic showed the presence of small amount of dioxins (Schecter 1994). PCDD is a potent teratogenic and feto-toxicant of animals and it has been shown to promote carcinogenesis in rat liver. Severe chloracne are the common signs of PCDD exposure in humans and other animals such as rhesus monkeys, rabbits, cattle, and hairless mice (Vos et al. 1991).

13.3.3.5 Furans

Furans, a highly flammable and colourless volatile liquid, has a boiling point near room temperature. About 135 different persistent dibenzofurans (PCDFs) are detected in air, water, and soil. They are produced by natural sources as well as by anthropogenic activities like combustion activities in power generation, e- waste incineration, and metal smelting (Rodriguez et al. 2008). The presence of furans has been detected in freshwater, wastewater, groundwater, and drinking water sources through wells, surface water, and in swimming areas (CEPA 1990). The PCDFs are stable tricyclic, planar, aromatic compounds which are nonpolar, poorly water soluble, and lipophilic (Alcock and Jones 1996). Hazardous and carcinogenic furans are found in very small quantities in the ecosystem but due to their persistence and long half-life of 10–20 years, they exert a drastic effect (Urban et al. 2014). Water samples from the Dead Sea and Salt lakes in Western Australia has been shown to contain several alkylated or halogenated furans in small quantities (Krause et al. 2014). The groundwater samples of monitoring wells and multilevel samplers of Ottawa, Canada, are reported to contain trace quantities of furans and very persistent freon (F113) in high concentration (Lesage et al. 1990).

In Japan and Taiwan, women with furan exposure of milligrammes in contaminated rice oil developed reproductive anomalies and infant mortality though chloracne, the most persistent effect of contact. The international agencies like the USEPA (United States Environmental Protection Agency) and WHO (Word Health Organization) list it as human carcinogen and a possible endocrine disruptor (OEHHA 2005). In addition, furans can cause diabetes, neurotoxicity, immunotoxicity, skin toxicity, teratogenicity, endocrine effects, and a predisposition to cancer. Human immune response is heavily damaged by these compounds, which makes them susceptible to infectious diseases (Kerkvliet 1995) and led to decreased population of neurons in the brain, slow down of the neural transmission network, as well as affecting hormones of gonad and thyroid gland (Kakeyama and Tohyama 2003). Severe exposure to furans leads to loss of sebaceous glands and atrophy of hair follicles (Hebert et al. 1990), keratinisation of dermal epithelium (Panteleyev et al. 1997), as well as prevalence and complications from endometriosis (Rier and Foster 2003). Furans act as endocrine disrupters and decrease the production of thyroid hormones (Vulsma 2000) and morphological damage of the liver (Smith et al. 1998).

Similar diseases are reported in wild and domestic animals like horses, cows, and rabbits (Stohs 1990): severe endometriosis in female monkeys (Rier et al. 1993); at low dose exposure development of cyst in rats and mice (Heilier et al. 2008); and low reproductive capability of rodent female and disrupted sperm production in rodent male (Petroff et al. 2001) are the other effects encountered. In the test animals ectopic testes, vaginal pouches, and nipple retention were reported (Gray Jr and Kelce 1996). Wildlife are differentially affected by furan exposure. It was found to be responsible for small baculum in young male otters and small penises in alligators, whereas it caused cryptorchidism in the Florida panther, sex reversal in fish, and altered social behaviour in birds (Gray Jr 1998).

Table 13.2 BTz and MBZ present in the groundwater and other water sources.

Compound detected	Water Source	Cartridge Used	Elution solvent	Recovery %	Limit of quantification	References
BTZ, MBZ	Effluent Influent Groundwater Tap water	Oasis HLB 500	3×2 mL DCM/ CH_3OH (1:1)	79–108	9.8–36.7 ngL^{-1}	Liu et al. (2011)
BTZ, MBZ	Groundwater	Oasis HLB 200	7 mL CH_3OH / ACO (6 : 4)	95–113	50 ngL^{-1}	Reemtsma et al. (2010)
MBZ, DMBZ	River Groundwater	Merck EN 200	10 mL DCM	ND	3.3–6.7 ngL^{-1}	Koroša et al. (2016)
BTZ, MBZ	Effluent River Groundwater,	Oasis HLB 10	0.1% HCOOH in CH_3OH	94–124	0.2–4.2 ngL^{-1}	Yang et al. (2018)

13.3.3.6 Benzotriazole (BTz) and Methyl-benzotriazole (MBZ)

Benzotriazole (BTz) and their methyl derivative methyl-benzotriazole (MBZ) are multipurpose materials with a range of industrial, commercial, and household applications. They are mixed in films, paint coatings, and paints for better stability in ultraviolet light. They are used in electrochromic, organic solar cells and organic light-emitting diodes due to their solubility, optical, and electronic properties. They provide many desirable properties to polymers and increases usability in different devices (Herrero et al. 2013). They were initially patented as metal corrosion inhibitors in aqueous heat exchange. They are presently used in antifreeze liquids as well as industrial applications in cooling, braking, cutting and metal-working fluids, aircraft de-icing fluids, and household detergents. BTZs and MBZs are released from a variety of sources, including electronics, and enter the environment. Their high aqueous mobility makes their presence ubiquitous in different water bodies like groundwater, rivers, lakes, seas, treated and untreated wastewater, and drinking water. Its concentration was 2.0×10^{-8} gL^{-1} in water sample of Rhine (Netherlands) and it was detected in three rivers in Trinidad. Some reports of their presence in the water are listed in the Table 13.2.

13.4 Groundwater Quality

Groundwater, the world's most valuable natural resource, is under great threat due to its contamination by various sources. Protection of groundwater from contamination is one of the challenging issues at present. Groundwater resources are widely utilized for domestic, irrigation, and industrial purposes. Clean usable water without any contamination is needed for everyone. Different factors are used to estimate the quality and suitability of water for drinking. The physical and chemical factors used to decide the suitability include range of pH, electrical conductivity, total hardness, total dissolved solids, and presence and level of different elements like Ca, Mg, Na, K, bicarbonate, sulphate, and chloride

(Prasanth et al. 2012). The concentration of these ions and salts should be well within the permissible limit defined by the World Health Organization or specified by country such as the Bureau of Indian Standards, India (WHO 2011).

The pH value of potable water is supposed to be slightly basic. As per the guidelines of the WHO and BIS on drinking water quality, the pH value should be between 6.5 and 8.5. Electrical conductivity states the presence of ions and salts in groundwater. Higher electrical conductivity means groundwater is rich in salts. Its permissible limit for drinking water is 1500 micro-siemens per centimetre. Hardness of water is decided by presence of bicarbonate, chloride, and sulphates of calcium and magnesium, and indicates the capacity to react with soap. The desirable limit of hardness is 200 mgL^{-1} and permissible limit is 600 mgL^{-1} in India. There are several inorganic minerals, metals, salts, ions as well as organic compounds that are dissolved in groundwater. The total mass of dissolved components in water denote total dissolved solids (TDS). Groundwater having high TDS is not suitable for drinking purposes and low TDS water is safe to drink. As per WHO and BIS, desirable limit ranges between 500-600 mgL^{-1} and permissible limit ranges between 1000-2000 mgL^{-1}.

13.5 Global Scenario of e-waste Generation

Use of electronic equipment is higher in developed countries compared to developing countries, so per capita e-waste generation is also higher in these countries. In the global e-waste statistics of 2016, total e-waste generation in the US was 11.3 Mt, followed by Brazil and Mexico (Baldé et al. 2017). The Basel convention (1989) guide in reference to the 'trans-boundary movement of hazardous waste and their disposal' was adopted in 1992 and currently more than 160 parties. However, some countries used to export their hazardous e-waste to many developing countries by ratifying the convention. These countries are also not free from the danger of groundwater contamination with pollutants. There are about 1200 identified polluted sites needing contaminated water treatment (Mulligan et al. 2001) and ~ 63% of them are polluted by heavy metals like Cd, Hg, Cu, Pb, Ni, and Zn, the most hazardous pollutants according to the USEPA. The total production of e-waste in 2016 by Canada was about 724 kt (Baldé et al. 2017), and ~140 kt yr^{-1} is disposed in landfills (Kumar and Holuszko 2016). The country lacks national legislation on e-waste management and non-profit organizations used to collect and recycle e-waste. China, the world's largest manufacturer and consumer of electronic products, have the fastest e-waste annual growth rate, ranging between 13 and 15% (Ignatuschtschenko 2017).

The major concern about groundwater contamination is leachate generated from the huge amount of e-waste that percolates to the groundwater. Han and co-workers assessed the groundwater quality near landfills and identified 96 different groundwater pollutants that were composed of 12 different salts, 5 metal ions, 15 heavy metals, 2 microbiological pollutants, and a large number of 62 xenobiotic organic compounds (Han et al. 2016). A mix of these pollutants leaks from the landfill to contaminate the groundwater and degrade water quality. Groundwater quality near landfills was distance dependent. Water quality was severely degraded in 200 m, however the groundwater contaminants were detectible within a range of 1000 m of a landfill. The evidence clearly indicates landfill sites should be quite distant from the major population or cities.

The growth and development of information communication technology hazardous e-waste have been dumped in Africa, ignoring Basel Convention restricting the movements of hazardous waste from one country to other. Imported junk and not repairable electronics were ~75% dumped in landfills without any controls (Schmidt 2006). Such dumping leads to the degradation of groundwater quality, as in the case of Thohoyandou landfill site (Ghana, South Africa); the groundwater in nearby areas has high level of mercury, affecting both the environment and human health (Nevondo et al. 2019). This Agbogbloshie slum of Ghana is a very large dumpsite of e-waste in this country (Stein 2013). The samples of groundwater from another landfill site, Kumasi, in Ghana, showed presence of high levels of Pb, Fe, Cd, and Cr beyond the acceptable limits of the WHO. The water quality index of the majority of water samples (> 85%) was of poor quality, indicating the need to enhance the understanding of physicochemical parameters and management of groundwater resources in the Oti community (Boateng et al. 2019). Only a few African countries have their own formal government policy for e-waste management.

Considering the threat of groundwater quality in different parts of the country, the government of India is looking for effective remediation processes. Recently, the Central Pollution Control Board and Bhopal Municipal Corporation signed a memorandum of understanding to set the country's first e-waste clinic at Bhopal. In this initiative, e-waste will be collected door to door and deposited in the clinic for their segregation, processing, and disposing of waste (TOI 2019). The success of the trial can lead to an increased number of such clinics in India to protect groundwater contamination from e-waste.

13.6 Evaluation of Genotoxic Effect of Groundwater

13.6.1 Ames Test

In this test, *Salmonella typhimurium* is used as the test organism. The bacterial cells are grown overnight to get a bacterial suspension of about 10^8 cells per mL. An equal amount of diluted test sample (0.1 mL) and bacterial suspensions (0.1 mL) mixed in 2.5 mL molten top agar (0.6%), having NaCl (0.5%), is poured over surface of Vogel–Bonner minimal agar plate. After solidification of top agar, plates are incubated in the dark (37 °C, 48 hours). The number of revertants and surviving colonies are counted in the plates against a background lawn. For the positive control 4-Nitroquinoline 1-oxide is used, whereas sterilized distilled water serves as a negative control. The ratio of mutant colonies on the positive control and negative control, i.e. mutation rate; when itexceeds 2, it is said that the test compound has mutagenic activity.

13.6.2 SOS/umu-Test

The SOS/umu-test could be performed in test tubes without S9 metabolic activation. The *S. typhimurium* grown overnight in TGA medium supplemented with ampicillin (20 μg mL^{-1}) is diluted to achieve OD_{595} about 0.28. For positive control, various source water concentrations (equivalent to 0.25, 0.5, 1, and 2 L of original water), DMSO 100 μL, 600 μL bacterial culture ($OD_{595=}$ 0.28), and 4-NQO (0.5 μg·mL^{-1}) are used as a positive control. DMSO is used in negative control. After exposure of 3 hours, 2.7 mL buffer (pH 7.0) and 10 μL chloroform are added to the mixture and placed on incubator shaker (150 rpm, 10 minutes), followed by addition of 500 μL ONPG solutions in PBS. The genotoxic activities are expressed in term of β-galactosidase units and in enzyme induction ratios of control. Water is considered non-genotoxic, if the IR2 and dose-response relation are observed.

The study of groundwater samples collected from Guangdong province show presence of organochlorine pesticides and other polyaromatic hydrocarbons in the samples. The water sample showed potential genotoxicity and mutagenicity effect like frameshift mutation in *Salmonella typhimurium* strain TA98 in the Ames test and SOS/umu test (Guan et al. 2017). The environment relevant concentrations PBDE in range of 0.01 and 1 μM played a role in the dysregulation of oxidative stress, DNA damage, and gene expression in bronchial epithelial cells. The effect of PBDE 47 were more prominent compared to the other PBDE (Montalbano et al. 2020). In Brazil, well water samples collected near landfill showed presence of organophosphorus flame retardants such as OPFRs (organophosphate ester flame retardants) and PBDEs as well as some new brominated flame retardants that were capable to alter gene function (Cristale et al. 2019).

13.7 Effect on Underground Life Forms

Most of the biological diversity is localized near the Earth's surface but many organisms are adapted to the photic underground habitats' nearly constant temperature, dark, and humidity with limited sources of nutrients (Schmidt and Hahn 2012). The narrowly distributed species with peculiar traits live in subsurface air and water-filled underground spaces (Humphreys 2006). A limited amount of information is available regarding these organisms due to their access. Life forms are mainly archaea, bacteria fungi, protozoan, and micro and macro invertebrates, whereas vertebrates are very uncommon (Griebler and Lueders 2009). Despite recognition of the ecosystem and life forms present, we are concerned about quality of groundwater neglecting underground endemic biota maintaining its quality (Reboleira et al. 2013) and playing important role in ecosystem services and lotic ecosystems like rivers and springs dependent on groundwater (Griebler et al. 2014). Such biota are extremely sensitive to the organic pollutants of the groundwater.

13.8 Groundwater Remediation

Groundwater is a precious resource to all inhabitants on the planet. Protection of groundwater from contamination is an issue of global importance and one of the most important challenges of the century. There are three main stages in its remediation: identification of source of pollution, control of pollutants diffusion, and remediation of the polluted site (Kui and Xiangping 2019). Remediation practice can be performed in-situ as well as ex-situ. In the remediation process, pollutants are either removed or transformed into harmless ones. The in-situ remediation can be done by physical and chemical methods (Kisku et al. 2015). The methods include nanotreatment, pump and treat, soil vapour extraction treatment, thermal treatment, and permeable reactive barriers (PRBs), as well as bioremediation.

13.8.1 Pump and Treat

In this conventional method, groundwater from the subsurface is extracted through wells and trenches to the surface where it is treated ex situ before discharge to the groundwater. The aboveground treatment includes carbon adsorption, air stripping, biological reactors, or precipitation (EPA 2011).

13.8.2 Nanotreatment

This eco-friendly technology calls for effective remediation of polluted sites by applying reactive nanomaterials for the detoxification and transformation of pollutants. Nanomaterials detoxify the pollutant by chemical reduction or catalysis or both in remediation of pollutants (Karn et al. 2009). The distinctive features of nanomaterials, such as small size, high surface area, and novel surface coatings make them versatile and they have a wide application compared to large-sized particles (Smarta et al. 2006). The materials could be nano-scale iron or carbon nanotubes, however there are risks associated with the use of nanomaterials in remediation.

13.8.3 Reactive Barrier

The low-cost and easy construction of reactive barriers lead to wide acceptability of this technology. The barrier could be of three classes: horizontal barrier, vertical barrier, and cover barrier (Kui and Xiangping 2019). The contaminated groundwater flowing through the reactive barrier is converted into insoluble precipitates. Vertical barriers reduce percolation of wastewater to the deep aquifers. Slurry walls are subsurface barriers that impede or stop movement of groundwater. Horizontal barriers are very useful in controlling downward movement of metal contaminants. Landfill sites are lined by synthetic material or clay. The liner materials are strong, permeable, and flexible, and possess high tensile strength that prevents the percolation of heavy metals from the leachate of landfill into the deep aquifers.

In addition to the physical techniques, there are chemical techniques to combat the issue of groundwater pollution. The chemical techniques have been discussed in different chapters in this book. Chemical techniques can be implemented ex-situ or in-situ. When deciding to perform the process of in-situ chemical treatment, caution should be taken in selecting the chemical agents to prevent unintentional additional contamination of landfill area under treatment. The common chemical treatments are oxidation, reduction, and neutralization, diminishing the mobility and toxicity of metal pollutants by converting them into inactive forms.

13.9 Conclusion

Considering the literature, we can conclude safely that the e-waste processing releases pollutants in the environment either intensely or un-intensely. The organic and inorganic pollutants released are potentially toxic to different life forms on the Earth's surface water and groundwater-dependent plant, animals, and human beings. The long-term exposure of organic chemicals present in e-waste were able to affect the gene expression in different test organisms and they were even so potent as to change the normal cells to uncontrolled dividing cancerous cells. It is of urgent need to protect the groundwater from various type of pollutants by strict regulation of e-waste release.

References

Janz, A. and Bilitewski, B. (2008). Hazardous substances in waste electrical and electronic equipment. In: *E-Waste: Implications, Regulations and Management in India and Current Global Best Practices* (ed. R. Johri), 93–120. New Delhi: TERI.

Ahmed, S. and Panwar, R.M. (2015). Hazardous constituents of e-waste and predictions for India. *Proceedings Institution of Civil Engineering Waste Resource Management* 169: 83–91.

Alcock, R.E. and Jones, K.C. (1996). Dioxins in the environment: a review of trend data. *Environmental Science & Technology* 30 (11): 3133–3143.

Anusree, P.S. and Balasubramanian, P. (2019). Awareness and Disposal Practices of E-waste with Reference to Household Users in Kochi City. *International Journal of Recent Technology and Engineering (IJRTE)* 8 (1S4): 293–299.

Adeyi, A.A. and Oyeleke, P. (2017). Heavy metals and polycyclic aromatic hydrocarbons in soil from e-waste dumpsites in Lagos and Ibadan, Nigeria. *Journal of Health & Pollution* 7: 71–84.

Baldé, C.P., Forti, V., Gray, V. et al. (2017). *The Global E-Waste Monitor*. Bonn, Germany: United Nations University.

BAN, SVTC (2002). The Basel Action Network and Silicon Valley Toxics Coalition February 25, 2002). Exporting Harm: The High-Tech Thrashing of Asia.

Boateng, T.K., Opoku, F., and Akoto, O. (2019). Quality of leachate from the Oti landfill site and its effects on groundwater: a case history. *Applied Water Science* 9: 33.

Borthakur, A. (2016). Health and environmental hazards of electronic waste in India. *Journal of Environmental Health* 78 (8): 18–23.

Brigden, K., Labunska, I., Santillo, D., and Johnston, P. (2008). Chemical Contamination at E-Waste Recycling and Disposal Sites in ACCRA and Korforidua, Ghana. Greenpeace Research Laboratories; Amsterdam, Netherlands: p. 24. Technical note.

Broman, D., Naf, C., Rolff, C., and Zebuhr, Y. (1991). Occurrence and dynamics of polychlorinated dibenzo-p -dioxins and dlbenzofurans and polycyclic aromatic hydrocarbons in the mixed surface layer of remote coastal and offshore waters of the Baltic. *Environmental Science & Technology* 25: 1850–1864.

CEPA, Canadian Environmental Protection Act. (1990). Polychlorinated dibenzodioxins and polychlorinated dibenzofurans. Ontario. Canadian Environmental Protection Act.

Chen, Y., Li, J., Liu, L., and Zhao, N. (2012). Polybrominated diphenyl ethers fate in China: a review with an emphasis on environmental contamination levels, human exposure and regulation. *Journal of Environmental Management* 113: 22–30.

Cristale, J., Belé, T.G.A., Lacorte, L., and Marchi, M.R.R. (2019). Occurrence of flame retardants in landfills: a case study in Brazil. *Environmental Research* 168: 420–427.

Dai, Q., Xu, X., Eskenazi, B. et al. (2020). Severe dioxin-like compound (DLC) contamination in e-waste recycling areas: An under-recognized threat to local health. *Environment International* 139: 105731.

EPA, (2011) Cost analysis for selected groundwater clean-up project: Pump and treat system and permeable reactive barriers. Environmental Protection Agency.

Gaidajis, G., Angelakoglou, K., and Aktsoglou, D. (2010). E-waste: environmental problems and current management. *The Journal of Engineering Science and Technology Review* 3 (1): 193–199.

Grant, S., Mortimer, M., Stevenson, G. et al. (2011). Facilitated transport of dioxins in soil following unintentional release of pesticide-surfactant formulations. *Environmental Science & Technology* 45: 406–411.

Gray, L.E. Jr. (1998). Xenoendocrine disrupters: laboratory studies on male reproductive effects. *Toxicology Letters* 102: 331–335.

Gray, L.E. Jr. and Kelce, W.R. (1996). Latent effects of pesticides and toxic substances on sexual differentiation of rodents. *Toxicology and Industrial Health* 12: 515–531.

Griebler, C. and Lueders, T. (2009). Microbial biodiversity in groundwater ecosystems. *Freshwater Biology* 54: 649–677.

Griebler, C., Malard, F., and Lefébure, T. (2014). Current developments in groundwater ecology—from biodiversity to ecosystem function and services. *Current Opinion in Biotechnology* 27: 159–167.

Guan, J., Yamazaki, Y., Chand, D. et al. (2017). Novel mechanisms of ALK activation revealed by analysis of the Y1278S neuroblastoma mutation. *Cancers* 9: 149.

Han, Z., Ma, H., Shi, G. et al. (2016). A review of groundwater contamination near municipal solid waste landfill sites in China. *Science of the Total Environment* 569–570: 1255–1264.

He, K., Sun, Z., Hu, Y. et al. (2017). Comparison of soil heavy metal pollution caused by e-waste recycling activities and traditional industrial operations. *Environmental Science and Pollution Research* 24: 9387–9398.

Hebert, C., Harris, M., Elwell, M., and Birnbaum, L. (1990). Relative toxicity and tumor-promoting ability of 2, 3, 7, 8-tetrachlorodibenzo-p-dioxin (TCDD), 2, 3, 4, 7, 8-pentachlorodibenzofuran (PCDF), and 1, 2, 3, 4, 7, 8-hexachlorodibenzofuran (HCDF) in hairless mice. *Toxicology and Applied Pharmacology* 102 (2): 362–377.

Heilier, J.F., Donnez, J., and Lison, D. (2008). Organochlorines and endometriosis: a mini-review. *Chemosphere* 71 (2): 203–210.

Herrero, P., Borrull, F., Pocurull, E., and Marcé, R.M. (2013). Efficient tandem solid-phase extraction and liquid chromatography-triple quadrupole mass spectrometry method to determine polar benzotriazole, benzothiazole and benzenesulfonamide contaminants in environmental water samples. *Journal of Chromatography. A*: 22–32.

Humphreys, W.F. (2006). Aquifers: the ultimate groundwater-dependent ecosystems. *Australian Journal of Botany* 54: 115.

Ignatuschtschenko, E. (2017). E-waste management in China: bridging the formal and informal sectors. *Journal of Chinese Governance* 2 (4): 385–410.

Kakeyama, M. and Tohyama, C. (2003). Developmental neurotoxicity of dioxin and its related compounds. *Industrial Health* 41 (3): 215–230.

Karn, B., Kuiken, T., and Otto, M. (2009). Nanotechnology and in situ remediation: a review of the benefits and potential risks. *Environmental Health Perspectives* 117: 1813–1831.

Kerkvliet, N.I. (1995). Immunological effects of chlorinated dibenzo-p-dioxins. *Environmental Health Perspectives* 103 (9): 47.

Kiddee, P., Naidu, R., and Wong, M.H. (2013). Electronic waste management approaches: an overview. *Waste Management* 33: 1237–1250.

Kisku, D., Abhisekh, H.S., Singh, S., and Singh, T.B. (2015). *International Journal of Advanced Research* 3 (9): 1095–1104.

Koroša, A., Auersperger, P., and Mali, N. (2016). Determination of micro-organic contaminants in ground water (Maribor, Slovenia). *Science of the Total Environment* 571: 1419–1431.

Krause, T., Tubbesing, C., Benzing, K., and Schöler, H.F. (2014). Model reactions and natural occurrence of furans from hypersaline environments. *Biogeosciences* 11: 2871–2882.

Kui, H. and Xiangping, L. (2019). Application status of vertical barrier technology in Site; contamination remediation. *IOP Conference Series: Earth Environmental Science* 242 (05): 2029.

Kumar, A. and Holuszko, M. (2016). Electronic waste and existing processing routes: a Canadian perspective. *Resources* 5 (4): 35.

Lahiry, S. (2019). Recycling of e-waste in India and its potential. down to http://earth.org.in. Down to Earth.

Lesage, S., Jackson, R.E., Priddie, M.W., and Riemann, P.G. (1990). Occurrence and fate of organic solvent residues in anoxic groundwater at the gloucester landfill, Canada. *Environmental Science & Technology* 24: 559–566.

Levison, J., Novakowski, K., Reiner, E.J., and Kolic, T. (2012). Potential of groundwater contamination by polybrominated diphenyl ethers (PBDEs) in a sensitive bedrock aquifer (Canada). *Hydrogeology Journal* 20 (2).

Lopez, B., Ollivier, P., Togola, A. et al. (2015). Screening of French groundwater for regulated and emerging contaminants. *Science of the Total Environment* 518–519: 562–573.

Liu, H., Zhang, Q., Wang, Y. et al. (2007). Occurrence of polychlorinated dibenzo-p- dioxins, dibenzofurans and biphenyls pollution in sediments from the Haihe River and Dagu Drainage River in Tianjin City, China. *Chemosphere* 68: 1772–1778.

Liu, Y.S., Ying, G.G., Shareef, A., and Kookana, R.S. (2011). Simultaneous determination of benzotriazoles and ultraviolet filters in ground water, effluent and biosolid samples using gas chromatography-tandem mass spectrometry. *Journal of Chromatography. A* 1218 (31): 5328–5335.

Luo, C., Liu, C., Wong, Y. et al. (2011). Heavy metal contamination in soil and vege tables near e-waste processing sites, South China. *Journal of Hazardous Materials* 186: 481–490.

MacKay, A. and Gschwend, P.M. (2001). Enhanced concentrations of PAHs in groundwater at a coal tar site. *Environmental Science & Technology* 35: 1320–1328.

Manmi, D.A.M.A.A., Abdullah, T.O., Al-Jaf, P.M., and Al-Ansari, N. (2019). Soil and groundwater pollution assessment and delineation of intensity risk map in Sulaymaniyah City, NE of Iraq. *Water* 11: 2158.

Montalbano, A.M., Albano, G.D., Anzalone, G. et al. (2020). Cytotoxic and genotoxic effects of the flame retardants (PBDE-47, PBDE-99 and PBDE-209) in human bronchial epithelial cells. *Chemosphere* 245: 125600.

Mukerjee, D. (1998). Health impact of polychlorinated dibenzo-p-dioxins: a critical review. *Journal of the Air & Waste Management Association* 48 (2): 157–165.

Mulligan, C.N., Yong, R.N., and Gibbs, B.F. (2001). *Engineering Geology* 60: 193–207.

Nevondo, V., Malehase, T., Daso, A.P., and Okonkwo, O.J. (2019). *Water SA* 45 (2): 225–231.
OEHHA California (2005). *Public Health Goal for TCDD in Drinking Water- Draft*. California: Office of Environmental Health Hazard Assessment.
Ohajinwa, C.M., van Bodegom, P.M., Osibanjo, O. et al. (2019). Health risks of polybrominated diphenyl ethers (PBDEs) and metals at informal electronic waste recycling sites. *International Journal of Environmental Research and Public Health* 16: 906. https://doi.org/10.3390/ijerph16060906.
Panteleyev, A.A., Thiel, R., Wanner, R. et al. (1997). 2, 3, 7, 8-tetrachlorodibenzo-p-dioxin (TCDD) affects keratin 1 and keratin 17 gene expression and differentially induces keratinization in hairless mouse skin. *Journal of Investigative Dermatology* 108 (3): 330–335.
Petroff, B.K., Roby, K.F., Gao, X. et al. (2001). A review of mechanisms controlling ovulation with implications for the anovulatory effects of polychlorinated dibenzo-p-dioxins in rodents. *Toxicology and Industrial Health* 158 (3): 91–107.
Prasanth, S.V.S., Magesh, N.S., Jitheshlal, K.V. et al. (2012). *Applied Water Science* 2 (3): 165–175.
Qin, Y.S. et al. (2008). Study on the influences of combined pollution of heavy metals Cu and Pb on soil respiration. *Journal of Anhui Agricultural Sciences* 36 (3): 1117–1128.
Rajya Sabha Secretariat. (2011). E-waste in India -Research Unit.
Reboleira, A.S.P.S., Abrantes, N., Oromí, P., and Gonçalves, F. (2013). Acute toxicity of coppersulfate and potassium dichromate on stygobiont Proasellus: general aspects ofgroundwater ecotoxicology and future perspectives. *Water, Air, and Soil Pollution* 224: 1550.
Reemtsma, T., Miehe, U., Duennbier, U., and Jekel, M. (2010). Polar pollutants in municipal wastewater and the water cycle: occurrence and removal of benzotriazoles. *Water Research* 44 (2): 596–604.
Rier, S. and Foster, W.G. (2003). Environmental dioxins and endometriosis. *Seminars in Reproductive Medicine* 21 (2): 145–154.
Rier, S.E., Martin, D.C., Bowman, R.E. et al. (1993). Endometriosis in rhesus monkeys (Macaca mulatta) following chronic exposure to 2, 3, 7, 8-tetrachlorodibenzo-pdioxin. *Toxicological Sciences* 21 (4): 433–441.
Rodriguez, C., Cook, A., Devine, B. et al. (2008). Dioxins, furans and PCBs in recycled water for indirect potable reuse. *International Journal of Environmental Research and Public Health* 5 (5): 356–367.
Stein, C. (2013). Inside Ghana's Electronic Wasteland. Available online: http://www.aljazeera.com/indepth/features/2013/10/inside-ghana-electronic-wasteland 2013 103012852580288.htmL.
Sadler, R. and Connell, D. (2012). Global distillation in an era of climate change. In: *Organic Pollutants Ten Years after the Stockholm Convention- Environmental and Analytical Update* (eds. T. Puzyn and A. Mostrag-Szlichtyng), 191–216. Croatia: InTech.
Schecter, A. (1994). *Dioxins and Health*, 1994. New York: Plenum Press.
Schmidt, C.W. (2006). *Environ Health Perspect* 114 (4): A232–A235.
Schmidt, S.I. and Hahn, H.J. (2012). What is groundwater and what does this mean to fauna? – An opinion. *Limnologica* 42 (1): 1–6. https://doi.org/10.1016/j.limno.2011.08.002.
Singh, A., Gangwar, C., Kumar, A., and Tripathi, A. (2015). Insinuation for distribution of heavy metals in soil samples derived from the E-waste burning areas of Moradabad, India. *IJPAS* 03: 2394–5710.

Shen, L., Gewurtz, S.B., Reiner, E.J. et al. (2009). Occurrence and sources of polychlorinated dibenzo-p-dioxins, dibenzofurans and dioxin-like polychlorinated biphenyls in surficial sediments of Lakes Superior and Huron. *Environmental Pollution* 157: 1210–1218.

Smarta, S.K., Cassady, A.I., Lua, G.Q., and Martina, D.J. (2006). The biocompatibility of carbon nanotubes. *Carbon* 44 (6): 1034–1047.

Smith, A.G., Clothier, B., Robinson, S. et al. (1998). Interaction between iron metabolism and 2, 3, 7, 8-tetrachlorodibenzo-p-dioxin in mice with variants of the Ahr gene: a hepatic oxidative mechanism. *Molecular Pharmacology* 53 (1): 52–61.

Stohs, S.J. (1990). Oxidative stress induced by 2, 3, 7, 8-tetrachlorodibenzo-p-dioxin (TCDD). *Free Radical Biology and Medicine* 9 (1): 79–90.

T.O.I. Times of India: India's first e-waste clinic in Bhopal soon 2019. https://timesofindia.indiatimes.com/city/bhopal/countrys-first-e-waste-clinic-in bhopalsoon/article show/71342753.cms.

Urban, J.D., Wikoff, D.S., Bunch, A.T. et al. (2014). A review of background dioxin concentrations in urban/suburban and rural soils across the United States: implications for site assessments and the establishment of soil cleanup levels. *Science of the Total Environment* 466: 586–597.

Vos, J.G., van Loveren, H., and Schuurman, H.-J. (1991). Immunotoxicity of dioxin: immune functionand host resistance in laboratory animals and humans. In: *Biological Basis for Risk Assessment of Dioxins and Related Compounds* (eds. M.A. Gallo, R.J. Scheuplein and K.A. van der Heijden), 79–93. New York: Cold Spring Harbor Laboratory Press.

Vulsma, T. (2000). Impact of exposure to maternal PCBs and dioxins on the neonate's thyroid hormone status. *Epidemiology* 11 (3): 239.

W.H.O. (2011). Guidelines for Drinking-water Quality, fourth edition.

W.H.O. (2016). Dioxins and their effect on human health.

Xing, D., Zhao, H., and Li, Q.Q. (2008). Arabidopsis CLP1-SIMILAR PROTEIN3, an ortholog of human polyadenylation factor CLP1, functions in gametophyte, embryo, and postembryonic development. *Plant Physiology* 148: 2059–2069.

Yang, Y.Y., Zhao, J.L., Liu, Y.S. et al. (2018). Pharmaceuticals and personal care products (PPCPs) and artificial sweeteners (ASs) in surface and ground waters and their application as indication of wastewater contamination. *Science of the Total Environment* 616–617: 816–823.

Zennegg, M., Schluep, M., Streicher-Porte, M. et al. (2014). Formation of PBDD/F from PBDE in electronic waste in recycling processes and under simulated extruding conditions. *Chemosphere* 116: 34–39.

Zheng, J., He, C.T., Chen, S.J. et al. (2017). Disruption of thyroid hormone (TH) levels and TH-regulated gene expression by polybrominated diphenyl ethers (PBDEs), polychlorinated biphenyls (PCBs), and hydroxylated PCBs in e-waste recycling workers. *Environment International* 102: 138–144.

14

Zero-Valent Iron (ZVI) for Groundwater Remediation

Naresh K. Sethy[1], *Zeenat Arif*[1], *K.S. Sista*[2], *Pradeep Kumar*[1], *P.K. Mishra*[1], *and Rajesh Saha*[1]

[1] *Department of Chemical Engineering & Technology, Indian Institute of Technology (BHU), Varanasi, Uttar Pradesh, India*
[2] *Research and Development, Tata Steel, Jamshedpur, Jharkhand, India*

14.1 Introduction

Water, being one of the basic necessities of life, occupies prime importance, touching several billions of lives globally. Even though 71% of the Earth's surface is covered by water, only 3% of this is freshwater and rest is salty. Among freshwater, nearly 70% is stuck in a frozen state in the form of glaciers and ice caps, and only 30% is available on the subsurface and surface for use. The water below the surface is generally termed 'groundwater' and is one of the major sources of freshwater globally. India is the largest user of groundwater in the world, with consumption of about 250 cubic kilometres per year. Being a major source for agriculture and industries, about 75% of irrigation, 85% of rural, and 50% of urban water demands are met by groundwater resources (Ministry of Jal Shakti 2019). Groundwater is generally present in the form of water tables and aquifers. Unlike surface waters (lakes, ponds), the groundwater is not directly visible. Groundwater is considered to be a renewable source which restores its balance through the water cycle comprising various phenomena like evaporation, transpiration, condensation, precipitation, infiltration, percolation, and runoff but due to overpopulation, deforestation, rapid urbanization, and industrialization, groundwater is overexploited, leading to imbalance in the water cycle and continuous depletion of global groundwater resources. According to the data of the World Health Organization (WHO), the scarcity of water resources has created challenges for over 40% of the world's population, i.e. more than 2 billion people have no access or enough clean water. This demand for water is doubling every 15 years, and it is estimated that by 2050 the global population will cross 9.8 billion, leading to huge demand. On the other hand, rapid increase in industrial and municipal activities also contributes to augmenting the level of pollutants in groundwater, impacting not only its usage but also affecting the health of numerous people. So, it is of prime importance to remediate groundwater before its use for various useful purposes (Fu et al. 2014).

Groundwater Geochemistry: Pollution and Remediation Methods, First Edition.
Edited by Sughosh Madhav and Pardeep Singh.

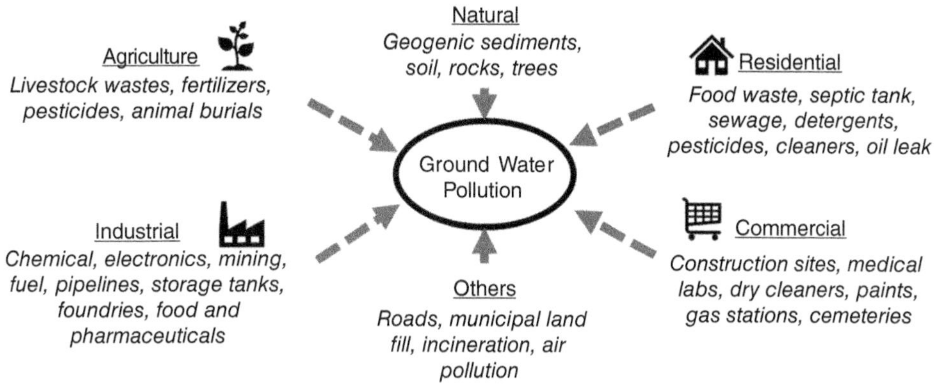

Figure 14.1 Sources of groundwater pollution.

Natural events and anthropogenic activities are major sources of groundwater pollution. A wide range of contaminants like physical, chemical (organic and inorganic), bacterial, and radioactive are prevalent in groundwaters. When man-made products such as gasoline, oil, road salts, and chemicals after spillage of improper usage come into contact with groundwater, it becomes unsafe and unfit for human use (Tratnyek and Johnson 2006). Even unaffected by human activities, groundwater contains some impurities due to leaching or seeping of geogenic matter from soil, rocks, etc. into groundwater. An image showcasing various sources of groundwater pollution is given in Figure 14.1.

14.1.1 Groundwater Remediation

Removal of polluting contaminants from groundwater to nullify its potential damage is otherwise termed as groundwater remediation. Pollutants like nitrates, chlorides, fluorides, arsenic, heavy metals, organic compounds, other pharmaceutical and industrial complex wastes, and microorganisms majorly affect the groundwater quality. Groundwater remediation techniques are mainly divided into two approaches: (i) *Ex-Situ* approach, in which the polluted groundwater is first pumped out and then subjected to treatment by use of physical, chemical, or biological techniques before reinjecting the treated water back to the source; and the (ii) *In-Situ* approach, which involves treatment of groundwater within the subsurface by use of physical, chemical, and biological treatment technologies (Tratnyek and Johnson 2006; Forbort 2009).

The in-situ approach is basically implemented in three modes: (i) containment; (ii) removal; and (iii) flushing. In the containment mode, contaminated groundwater movement is contained by use of active or passive barriers like soil barriers, frozen barriers, cutoff walls, bio barriers, etc. This mode of approach is not contaminant specific. In the second mode of approach removal of various contaminants is performed using physical, chemical, and biological techniques. In the third approach, flushing of contaminated groundwater from the region of concern to another area by use of natural of forced draft of water is done. In a similar way, the ex-situ approach also operates in (i) immobilization and

(ii) treatment modes. In the immobilization mode of operation, the targeted contaminant is allowed to trap or solidify onto a stable matrix which can be later disposed. In the treatment mode, various physical, chemical, and biological techniques are used. In both approaches, the treatment of groundwater carries good weightage as this mode really attacks the contaminant for removal. In the rest of the modes, the contaminant is either contained, immobilized, or diverted from the prime location but still remains a threat to the environment (IAEA 1999).

Physical treatment techniques like mechanical screening and filtration majorly target the removal of coarse to fine solid contaminants like gravel, grit, sludge particles, and other suspended matter. Air sparging or air stripping is another physical technique used for removal of a few volatile organic compounds (Reddy 2008; Zoltek and Melear 1978). Pump and treat is a conventional and widely used physical technique for removal of organics and hydrocarbons, which is generally assisted with chemical and biological techniques for multiple containment removal (Cheremisinoff 1997b; Reddy 2008). Chemical treatment techniques make use of active and reactive chemical elements or compounds to remove contaminants from groundwater through techniques like adsorption, ion exchange, coagulation, precipitation, oxidation, reduction, etc. (Bradl and Xenidis 2005; Ghosh et al. 2019; Shankar et al. 2014; Suarez 2009; Taylor 2015; Yin 1999). On the other hand, biological techniques such as bioaugmentation, bioventing, biosparging, phytoremediation, etc., make use of microorganisms, plants, and their extracts to effectively remove harmful containments from groundwater (Cheremisinoff 1997a; Tripathi and Ram 2018; Sims et al. 1992). Among the three, physical techniques are easy to use but are restricted to removal of very few contaminants and are costly. Biological techniques are eco-friendly, however take a longer time and have lower yields of removal. So, chemical techniques, even though they are a bit costlier, are largely preferred and widely used due to their high efficiency and higher yields along with faster response times (Lim et al. 2014; Singh et al. 2018). However, the suitability and integration of individual techniques depends on the onsite conditions as well.

Major contaminants in the groundwater comprise organic (VOCs [volatile organic compounds], pesticides, chlorinated solvents, benzene and pyrene derivatives), inorganic (aluminium, antimony, arsenic, barium, beryllium, cadmium, chlorides, chromium, copper, cyanide, dissolved solids, fluorides, iron, lead, manganese, mercury, nickel, nitrates, nitrites, selenium, silver, sodium, thallium, zinc, etc.), microbial (bacteria, viruses, etc.)

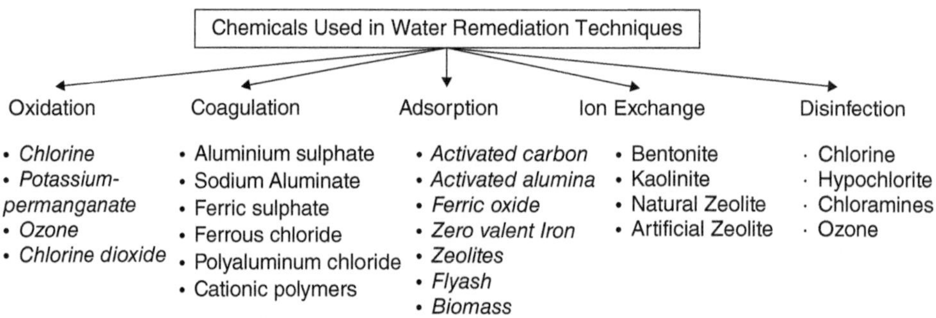

Figure 14.2 Chemicals used in water remediation.

and others (Diaz-Cruz and Barcelo 2008; Duan et al. 2020; Hasan et al. 2020; Lapworth et al. 2012; Sarkar and Paul 2016; Sizmur et al. 2017; Waller 1982). Treatment of most of the above contaminants by chemically dominant techniques like oxidation, adsorption, coagulation, precipitation, ion exchange, disinfection, etc. involves usage of various chemical species (Asere et al. 2019; Kumar and Jain 2013; Prabu and Parthiban 2013; Singh et al. 2018). Technique wise examples of various chemicals frequently used are showcased in Figure 14.2. One such prominent material which is of prime importance in present-day research due to its low cost, high availability, and multiple contaminant removal (chlorinated compounds, nitro compounds, heavy metals, dyes, aromatics) ability is zero-valent iron (ZVI) (Duan et al. 2020; Chen et al. 2019; Liu and Wang 2019; Shen et al. 2020; Summer et al. 2020; Qin et al. 2020). Focus on ZVI for removal of groundwater contaminants is dealt with in the subsequent sections.

14.1.2 Zero-Valent Iron (ZVI)

Iron is one of the rich elements in the Earth's crust typically present in the form of oxides. In its various forms of existence (Fe^{+3}, Fe^{+2}, Fe) iron is widely used in many applications such as construction, chemical, metallurgical, catalysts, environment remediation, food supplement, fuel, medicine, electronics, etc. Iron in the zero-valency state (Fe^0), which is prominently used in environmental remediation, especially water treatment applications, is ZVI. ZVI can be defined as the solid-state iron appearing in the form of lumps, chunks, wires, or powders typically ranging from millimetre to nanoscale dimension and comprising dominating composition of metallic iron besides other minor impurities (Neumann et al. 2013; Sista and Dwarapudi 2018). Synthesis of ZVI can be performed using various physical, chemical, biological, and mechanical techniques. Utilization of these techniques involves basically two approaches: (i) top-down approach, in which bulk iron or iron compounds are brought down to ZVI; and (ii) bottom-up approach, in which iron in the form of ions is agglomerated to form nano or micro ZVI. An overview on the synthesis of ZVI is elucidated in Figure 14.3 (Crane and Scott 2012; Li et al. 2006; Mukherjee et al. 2016; Sista and Dwarapudi 2018).

Most of the laboratory research on usage of ZVI for water remediation applications is focused on the synthesis of ZVI in nano scale by the bottom-up approach. Reduction of iron salts by use of chemicals like sodium borohydride and biological extracts from leaves,

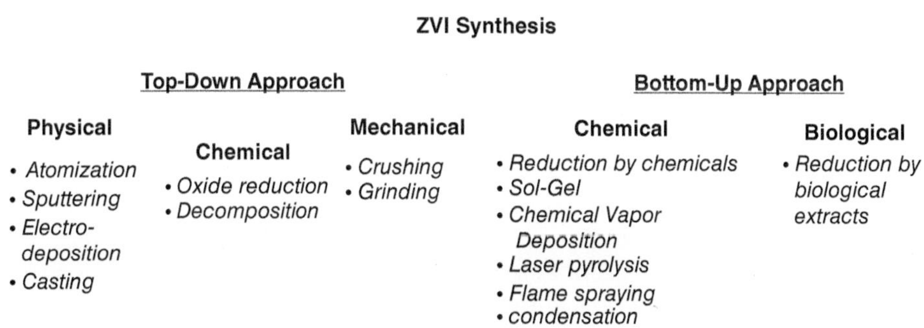

Figure 14.3 ZVI synthesis overview. *Sources:* Crane and Scott (2012) and Mukherjee et al. (2016).

roots, and stems of plants is widely used (Prabu and Parthiban 2013). On the other hand, most of the practical field applications involve usage of macro to micro scale ZVI produced from the top-down approach. ZVI produced from oxide reduction, Fe $(CO)_5$ decomposition, atomization, and electrodeposition is widely used (Hepure 2020; Hoganas 2018; Sista and Dwarapudi 2018). Due to difficulties in handling, nano scale ZVI is not preferred for field applications. However, due to their superior performance, usage of nano ZVI for groundwater remediation is also well reported (Tosco et al. 2014; Zhao et al. 2016).

Chemical composition, size, and surface area of ZVI stand very important for their performance in water purification. Many characterization techniques like x-ray diffraction (XRD), scanning electron microscopy (SEM), transmission electron microscopy (TEM), inductive coupled plasma (ICP), x-ray photoelectron spectroscopy (XPS), Mossbauer spectroscopy, Laser particle size analyser, UV–Vis spectroscopy, and BET surface area analyser are widely used to study the physical, chemical, morphological, and structural characteristics of ZVI (Li et al. 2006; Mukherjee et al. 2016). ZVI is used for removal of chlorinated methanes, chlorinated ethenes, chlorinated benzenes, trihalomethanes, polychlorinated hydrocarbons, pesticides, organic dyes, inorganic anions, heavy metals, radio actives, and other organic contaminants (Cook 2009; Li et al. 2006). Thus, ZVI is one of the potential agents for removal of multiple contaminants from groundwater. The mechanism by which the ZVI reacts to or attacks the contaminants is slightly complex and varies with the operation parameters and ambient conditions. Details on the mechanism are discussed in Section 14.2.

14.2 Removal Mechanism Using ZVI

Iron products adopt different principles of contaminant removal based on their oxidation state of appearance. For example, oxides like hematite having Fe in +3 form remove contaminants by only adsorption, whereas magnetite, having Fe in +2 form, removes by both adsorption as well as reduction. On the other hand, the zero-valent form of iron, Fe^0, implements one or the combination of adsorption, reduction, oxidation, precipitation, and co-precipitation principles to remove various contaminants (Naseri et al. 2017; Nicomel et al. 2016). This behaviour of ZVI is attributed to its corrosion in aqueous media and the behaviour of corrosion products formed on the surface of ZVI. Several reactions ZVI undergoes in an aqueous system are expressed in Table 14.1 (Guan et al. 2015; Noubactep 2008, 2009; Triszcz et al. 2009).

Among the Eqs. (1)–(8), the dominance of the single reaction or combination of reactions largely depends on the dissolved oxygen content, pH, water quality, presence of impurities, and other hydrodynamic conditions. Corrosion of iron under anaerobic conditions proceeds by Eqs. (1)–(2), while under aerobic conditions by Eqs. (3)–(6). Thus, it must be understood that oxidation (corrosion) of iron is feasible in both aerobic and anaerobic conditions due to the presence of water. On the other hand, if the pH of the aqueous media is very acidic under aerobic conditions, iron corrosion proceeds by generation of oxidizing species as in Eqs. (7)–(8). Major species generated from corrosion of iron responsible for various contaminant removal are: Fe^0, Fe^{+2}(aq), Fe^{+2}(s); H_2 – acts as good electron donors and thus promotes reduction of contaminants; H_2O_2 and OH radical – promote oxidation and oxide

Table 14.1 Possible reactions of and with ZVI.

Corrosion:	
$Fe^0 + 2H^+ = Fe^{+2} + H_2(g)$	(1)
$Fe^{+2} + 2H_2O = Fe(OH)_2 + 2H^+$	(2)
$2Fe^0 + O_2 + 2H_2O = 2Fe^{+2} + 4OH^-$	(3)
$3Fe^{+2} + O_2 + 2H_2O = Fe_3O_4 + 4H^+$	(4)
$4Fe^{+2} + O_2 + 10H_2O = 4Fe(OH)_3 + 8H^+$	(5)
$4Fe^{+2} + O_2 + 6H_2O = 4FeOOH + 8H^+$	(6)
$Fe^0 + O_2 + 2H^+ = Fe^{+2} + H_2O_2$	(7)
$Fe^{+2} + H_2O_2 = Fe^{+3} + OH^- + OH(radical)$	(8)
Reduction of contaminant:	
$Fe^0 + C^+ = Fe^{+2} + C$	(9)
$Fe^{+2}(aq) + C^+ = Fe^{+3}(aq) + C$	(10)
$Fe^{+2}(s) + C^+ = Fe^{+3}(s) + C$	(11)
$H_2 + C^+ = H^+ + C$	(12)
Oxidation of contaminant:	
$C^- + H_2O_2 = C$	(13)
$C^- + OH(radical) = C$	(14)
Precipitation of contaminant:	
$C + nOH^- = C(OH)_n$	(15)
Adsorption of contaminant:	
$Sorption\ site + C = C - Sorption\ site$	(16)
Co-precipitation of contaminant:	
$C + n\ Fe_x(OH)_y^{(3x-y)} = C - [Fe_x(OH)_y^{(3x-y)}]_n$	(17)

$Fe^{+2}(aq)$- released in bulk water, $Fe^{+2}(s)$- on the ZVI surface, C^+- containment in oxidized from, C^- – contaminant in reduced form, C-contaminant.
Source: Guan et al. (1994-2014), Noubactep (2009) and Triszcz et al. (2009).

species $Fe(OH)_2$, $Fe(OH)_3$, FeOOH, Fe_2O_3, Fe_3O_4 and many other forms; ($Fe(OH)_n^{n-2}$, $Fe(OH)_n^{n-3}$, $Fe(H_2O)_x^{2+}$, $Fe(H_2O)_x^{3+}$: $n \leq 3$, $x \leq 6$) – have active sorption sites and promote adsorption, precipitation, and co-precipitation. Possible reaction for contaminant removal for each of the principles is shown in Eqs. (9)–(17). Among them, removal by reduction, oxidation, and precipitation undergoes chemical transformation, whereas removal through adsorption and co-precipitation occurs by surface or structural entrapment (Noubactep 2009). Iron oxy-hydroxide present from longer duration, i.e. aged oxy-hydroxides, lead to entrapment onto the surface leading to adsorption, while freshly formed iron-oxy hydroxides after preliminary adsorption also enable structural entrapment of the contaminants while ageing leading to co-precipitation (Guan et al. 2015; Noubactep 2009). In practice, it not easy to differentiate between adsorption, precipitation, and co-precipitation, and a complex interplay

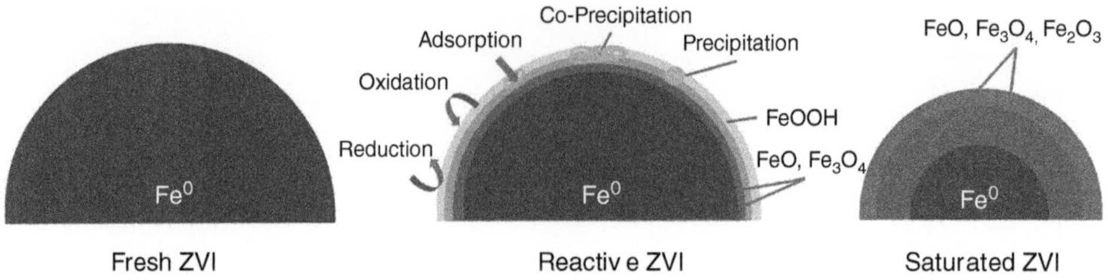

Figure 14.4 ZVI reaction modes.

and synergistic effect of these processes is expected to drive the contaminant removal. One another factor that governs the removal of contaminants by ZVI is the mass transport of contaminant towards ZVI and mass transport of active species from ZVI. So, the thickness of oxide scale and its properties like porosity, surface area, and density are also very important for water remediation using ZVI (Noubactep 2008). Fresh ZVI doesn't have any oxide scale and once it reacts with water it forms porous layers of various oxide and oxy-hydroxide scales; once the ZVI surface is saturated a hard shell of less porous oxide is formed, obstructing the ZVI performance. A pictographic illustration of ZVI reaction with water representing all the three modes of fresh, reactive, and saturated ZVI is shown in Figure 14.4.

Thus, ZVI promotes multiple contaminant removal by reduction (Eg: Cr[VI], nitrates, nitro compounds, chlorinated organic compounds), oxidation (Eg: phenols, dyes), adsorption-precipitation-co precipitation (Eg: As) and their combination (heavy metals, organics) (Fu et al. 2014; Mukherjee et al. 2016). In most of the cases the reduction and oxidation reactions of ZVI are supported by adsorption of the contaminant as well. Various research works performed on the usage of ZVI are discussed in Section 14.3.

14.3 Groundwater Contaminants and their Removal by ZVI

14.3.1 Arsenic

Arsenic, one of the most carcinogenic and toxic pollutants, is prevalent in groundwater in two different forms: arsenite (As[III]) and arsenate (As[V]). The WHO recommends the acceptable level of arsenic in drinking water to be less than 10 ppb (Xi et al. 2019). The removal mechanism of arsenic by ZVI features to be simple yet delivers to be complex as it involves multiple processes like adsorption, precipitation, co-precipitation, reduction, and oxidation (Biterna et al. 2010; Triszcz et al. 2009) as shown in the Eqs. (9)–(17) of Table 14.1. Various corrosion products of iron like ferrous and ferric oxides and hydroxides facilitate the arsenic removal by these processes (Mak et al. 2009).

14.3.2 Chlorinated Organic Compounds (COC)

Chlorinated organic compounds (COC) such as trichloroethylene, tertrachloroethylene, vinyl chloride, carbon tetrachloride, hexachlorocyclohexane, polychlorinated biphenyls, dichlorodiphenyltrichloroethane, etc. are formulated in daily use chemicals and are very toxic to environment (He et al. 2010). ZVI degrades these harmful contaminants by reduction mechanism as shown in the Eqs. (9)–(12) in Table 14.1, and is proven successful in degradation of trichloroethylene, trichloroethane, carbon tetrachloride, hexachlorocyclohexane, etc. (Román et al. 2013; Zhang et al. 2011). ZVI also gained interesting results for onsite remediation of COCs.

14.3.3 Dyes

Dyes, because of their widespread use, presence of toxic aromatic intermediates, and conventional removal difficulties are one of the primary concerning contaminants for removal from groundwaters (Oh et al. 1997). ZVI facilitates the removal or decolorization of dyes by both reduction and oxidation mechanisms as shown by Eqs. (9)–(14) in Table 14.1. Under anaerobic conditions, dyes like acid orange, acid blue, and reactive red are reduced by ZVI (Wang et al. 2010). Under aerobic conditions, strong oxidants generated by ZVI support the removal of dyes like acid black and brilliant blue (Chang et al. 2009).

14.3.4 Heavy Metals

Heavy metals are among the major contaminants of groundwater. Having no biodegradability, these heavy metals undergo accumulation in living tissue and concentrate through food chain. Heavy metals in their ionic forms such as Ni (II), Pb (II), Cu (II), Cd (I), Se (VI), Cr (VI), Zn (II), etc. are prevalent in the groundwaters. They undergo reduction in presence of ZVI as shown by Eqs. (9)–(12) in Table 14.1, which allows them to convert into neutral or non-toxic oxidized forms (Calabrò et al. 2012; Phanthasri et al. 2020; Shi et al. 2013; Xiao et al. 2011). Removal of Cr(VI) by ZVI has gained a lot of research interest, wherein the ZVI reduces the toxic Cr (VI) to Cr (III) followed by precipitation of Cr (III) onto iron oxy-hydroxides (Cissoko et al. 2009).

14.3.5 Nitrate

Nitrate is one of the most alarming groundwater pollutants known. Nitrogen containing sources like plant fertilizers, animal manure, septic systems, and other spent offs from industries contribute to nitrate pollution (Rocca et al. 2007). Because of its toxic effects, WHO have restricted the maximum concentration of nitrate–nitrogen (NO^{3-}–N) to 10 mg/L in drinking water beyond which it stands harmful to usage (Ahn et al. 2008). Unlike several disadvantages of conventional nitrate removal processes, removal of nitrate by reduction through ZVI stands unique and prominent. In the process of nitrate removal ZVI undergoes oxidation to its +2 and +3 oxidized forms, wherein the nitrate ion (NO_{3-}) gets reduced to N_2, NH_3, and NH_4^+ forms (Suzuki et al. 2012; Xu et al. 2012) as showcased in the Eqs. (9)–(12) of Table 14.1.

14.3.6 Nitro Amino Compounds (NACs)

Nitro amino compounds (NACs) are one of the potential and toxic carcinogens. Even at very low contamination levels, NACs like nitrobenzene, trinitrotoluene, dinitrotoluene, etc. stands highly toxic (Barreto-Rodrigues et al. 2009). Unlike conventional methods like biological processes which stands ineffective, removal of NACs through reaction with ZVI stands effective (Yin et al. 2012; Zhang et al. 2010b) and proceeds by mechanism shown in Eqs. (9)–(12) in Table 14.1. For example, nitro group compounds are effectively reduced by ZVI into aromatic amino compounds (Dong et al. 2010).

14.3.7 Phenolic Compounds

Discharge of toxic phenolic compounds like chlorophenols, nitrophenols, methyl phenols, etc. from domestic, agricultural, and industrial activities into groundwater showcases drastic and long-lasting effects on humans and animals (Britto et al. 2008). Removal of phenol compounds by ZVI is facilitated by Fentons reagent reaction as shown by Eq. (7) in Table 14.1, which further reacts with corrosion products of iron to form OH radical as shown by Eq. (8) in Table 14.1, which decomposes the phenolic compounds as shown in Eqs. (13)–(14) in Table 14.1. Further, ZVI also promotes the removal of phenolic compounds by adsorption and precipitation (Shimizu et al. 2012).

Works done by various researchers globally on usage of ZVI for groundwater treatment to remove various contaminants are compiled in the Table 14.2.

14.4 Advantages and Limitations of ZVI

Zero-valent iron is one the most studied materials for groundwater treatment due to its high reactivity, wide availability, multi-contaminant removal capability, and environmental compatibility. Besides, the technologies based on ZVI also add benefit in terms of low cost, easy handling and on-site flexibility (Crane and Scott 2012; Sen et al. 2017). Also, the steady and prolonged formation of iron oxides due to corrosion of iron drives most of the contaminant removal. Another added advantage with ZVI is the option of reusability by means of a simple thermochemical treatment. Besides several advantages, practical usage of ZVI also attracts a few demerits in terms of narrow pH range of usage, particle agglomeration, loss in surface area and reactivity, decrease in removal efficiency due to thick oxide scales, foreign particle interference, and sometimes elevated cost of operation (Lu et al. 2016). Further leaching of iron from ZVI during treatment also leads to iron contamination above the permissible limits. Overall, weighing up the pros and cons of the process, ZVI is one of the best agents for environmental remediation, especially for groundwater. A pictographic representation of various advantages and limitations of ZVI is shown in Figure 14.5

14.5 Recent Advances and Prospects in ZVI

Preliminary reports on the usage of zero-valent products for environmental applications goes back to 1972 (Sweeny and Fischer 1972). Two decades later, more detailed study on their usage, especially on usage of ZVI for remediation of groundwater, is reported (Gillham and O'Hannesin 1994, 1998; Tratnyek and Scherer 2003). Later on, widespread usage of

Table 14.2 Research works on usage of ZVI for multiple contaminant removal.

S. No	Contaminant	Reagent	ZVI properties	Conditions	Remarks	References
1. Arsenic						
1.1	As(III)	ZVI with $CuSO_4$	• Surface area = 0.588 m^2 /g • Shape = spherical	• Dosage of reagent = 0.5 mg/L • Aerobic condition	• Fe-Cu galvanic cell system accelerates arsenic removal efficiency • Arsenic removal rate was 99.8%	Xi et al. (2019)
1.2	As (V)	ZVI with Fe_3O_4	• 5 micro m ZVI • Surface area = 0.468 m^2 /g • Shape = spherical	• Dosage of reagent = 0.10 mg/L • Aerobic and anaerobic conditions	• Fe (II) was essential to activate ZVI and to remove As (V). • ZVI system showed high As (V) removal percentage	Xue et al. (2019)
1.3	As (III)	Nano-ZVI	• Size = 32 nm • Shape = spherical	• Dosage of reagent = 0.7 mg/L • pH (3, 7 and 9)	• ZVI nanoparticles were synthesized from leaf extracts. • As(III) oxidation is reported.	Rana et al. (2018)
1.4	As(V) and As(III)	Acid-purged ZVI	• Shape = rod shaped	• Aerobic condition • Dosage of reagent = 33 L/min • pH 6 ± 1.0	• As(V) reduced to As (III) and sparsely to As (0) • arsenic rejection efficiency >90% (~20 mg/L removed).	Abass et al. (2016)
1.5	As(III) and As(V)	Nano-ZVI with reduced graphene oxide	• Size = 23 nm • Shape = spherical	• Shape = spherical • pH < 9.1 and pH > 9.1	• The adsorption capacity of As (III) and As(V) was 35.83 mg/L and 29.04 mg/L, respectively	Wang et al. (2014)
1.6	As(III)	Nano-ZVI	• Size = 60 nm • Shape = spherical	• Dosage of reagent = 2–5 g/L • pH 3–9	• The role of intra-particle diffusion and reduction is affecting the distribution of arsenic in nano ZVI materials	Yan et al. (2012)
1.7	As(V)	ZVI (iron chips)	• Shape = Rectangular	• Dosage of reagent = 10micro g/L • pH 4–11	• Effect of anions (Cl^-, CO_3^{2-}, NO_{3-},PO_{4-}, SO_4^{2-}) is explored. • Cl^-, CO_3^{2-} and Mn slightly decreased As (III) removal. • NO_{3-}, PO_{4-} largely impacted As (V) removal.	Biterna et al. (2010)

(*Continued*)

Table 14.2 (Continued)

S. No	Contaminant	Reagent	ZVI properties	Conditions	Remarks	References
1.8	As(III) and As(V)	ZVI (wire)	• Shape = Rod shaped	• Dosage of reagent = 7micro g/L • pH 3, 7, and 9	• Effect of operating conditions on ZVI systems for arsenic removal observed.	Triszcz et al. (2009)
1.9	As(III) and As(V)	Nano-ZVI with activated carbon	• Surface area = 0.486 m^2 /g • Size = 48 nm • Shape = spherical	• Aerobic condition • pH = 6.5	• The adsorption capacity for arsenite and arsenate calculated was 18.2 and 12.0 mg/g, respectively. • PO_4^- decreased the removal of both arsenite and arsenate. • Ca^{2+}, Mg^{2+} enhanced adsorption	Zhua et al. (2009)
2. Chlorinated Organic Compounds (COC)						
2.1	2- Per Chloroethene (PCE)	ZVI (micro and nano powders)	• Shape = spherical	• Aerobic condition • Dosage of reagent = 24 l/min • pH 9	• ZVI can simulate and inhibit bacterial PCE dichlorination. • H_2 produced by ZVI is used by bacterial cultures.	Summer et al. (2020)
2.2	Polychlorinated aromatic pollutants (PCP)	Nano-ZVI	• Size = 26 nm • Shape = spherical	• Aerobic and anaerobic conditions • pH = 3–4	• Short-chain organic acids enhance iron nanoparticles to degrade pentachlorophenol.	Ou et al. (2016)
2.3	1,4-dichlorobenzene	ZVI with Granular Activated Carbon	• Surface area = 0.650 m2/g • Shape = spherical	• Aerobic conditions • Basic pH	• 1,4-DCB could be efficiently dechlorinated in both anode and cathode chamber.	Yang et al. (2010)
2.4	Carbon tetrachloride (CT)	ZVI with organic ligand	• Surface area = 0.325 m2/g	• Aerobic and anaerobic conditions • Temp is 700° c	• Organic ligands significantly enhanced the rates and the extents of CT removal. • The dichlorination of CT was 89.1% (with EDTA).	Zhang et al. (2011)
2.5	Trichloroethylene (TCE)	ZVI with Granular activated carbon	• Shape = spherical • Size micro level	• Aerobic conditions • Temp is 750° c • Dosage of reagent = 24 mg/L	• Reagent composites liberated a greater amount of Cl • The desorption limited TCE degradation rate constant decreased as the composite dosage was increased.	Tsenga et al. (2011)

2.6	Trichloroethylene (TCE)	Nano- ZVI with Cu(II)	• Size = 19.6 nm • Shape = spherical	• pH = 3–6	• TCE was significantly degraded (95%) in 10 minutes in nZVI Fenton system with 20 mMCu(II)	Choi and Lee (2012)
2.7	Vinyl chloride (VC) and 1,2 dichloroethane (1,2-DCA)	Surfactant stabilized Nano-ZVI coated with palladium	• Shape = triangular • Surface area = 0.200 m2/g	• pH = 3–9 • Aerobic conditions • Biotic and abiotic processes.	• Stabilized NZVI was shown capable of remediating VC and 1,2-DCA	Wei et al. (2012))
3. Dyes						
3.1	Congo red	Poly (Gg -AAm) /ZVI	• Surface area = 2.775 m2/g	• pH =7.0 • temperature = 45° c	• Adsorption capacity for CR at pH 7.0 was found to be 153.8 mg/L • Adsorption of CR onto [poly (Gg -AAm)/ZVI] was feasible, spontaneous, and endothermic.	Reddy et al. (2020)
3.2	Reactive blue 13	Nano-ZVI	• Size = 36 nm • Shape = Spherical	• PH = basic • Dosage of reagent = 1 g/L	• ZVI enhanced anaerobic bio reduction of reactive blue 13. • Sludge morphology and microbial community were changed by ZVI addition.	Li et al. (2013)
3.3	Azo dye direct sky blue 5B	ZVI	• Shape = Spherical	• PH = 3–11 • temperature = 320° c • Aerobic conditions	• Fenton's reaction promoted OH radical concentration • Degradation ratio increased with the increase of ZVI dosage and decrease of pH value	Chen et al. (2011)
3.4	Acid Orange 7 (AO7, monoazo), Reactive Red 120 (RR120, diazo), and Acid Blue 9(AB9,triphenyl-methane),	Nano- ZVI	• Size = 60 nm • Shape = semi spherical	• PH = 6.5 • Dosage of reagent = 30 g/L	• At dissolved oxygen of 6 mg L^{-1}, more than 99% of each dye was decolorized within 12 minutes and high COD removals were obtained (97% for AO7, 87% for RR120, and 93% for AB9).	Wang et al. (2010)

(*Continued*)

Table 14.2 (Continued)

S. No	Contaminant	Reagent	ZVI properties	Conditions	Remarks	References
3.5	C.I. Acid Orange 7	Nano-ZVI	• Shape = spherical • Surface area = $0.875\,m^2/g$	• PH = 9 • Aerobic conditions	• The decolorization rate increased with the increase of hydrogen peroxide concentration and power density but decreased with the increase of initial pH value.	Zhang et al. (2009)
4. Heavy metals						
4.1	Chromium (Cr^{6+})	Fe-C composite	• Size = micro level • Shape = rod shaped	• Aerobic and anaerobic conditions • pH = 3–9.5	• Cr(VI) was effectively removed under acid and anaerobic conditions • composite is synthesized by ball milling. • Cr(VI) removal is 94.01%.	Wang et al. (2020b)
4.2	Selenate (Se^{6+})	Zeolite-Supported nano-Iron chips	• Size =35 nm • Shape = spherical	• Aerobic conditions • Temp = 220° c • pH =8.5	• Zeolite performed as a supporter of the insoluble products • Removal efficiency was 85.2%.	Phanthasri et al. (2020)
4.3	Chromium (Cr^{6+})	Fe-Al bimetallic nano ZVI	• Shape = spherical	• PH = 6.5 • Dosage of reagent = 15 g/L • Aerobic and anaerobic conditions	• Removal efficiency for Cr^{6+} by Fe/Al was 1.47 g/L • Cr^{6+} removal mechanisms by include reduction, precipitation, and adsorption.	Ou et al. (2020)
4.4	Cadmium (Cd)	Nano ZVI-zeolite composite	• Size = 23.5 nm • Surface area = $1.258\,m^2/g$ • Shape = spherical	• pH = 2.5–8.5 • Aerobic and anaerobic conditions	• The nZVI-zeolite composite was suggested as an excellent adsorbent for PRB technology. • The best nZVI-zeolite composite showed removal efficiency of 20.6 mg/g for Cd.	Tasharrof et al. (2020)
4.5	Bromate (BrO_3^-)	Iron filling with activated carbon fibre (nZVI/ACF)	• Size = 23.5 nm • Surface area = $0.752\,m^2/g$	• pH = 6.5–9 • Aerobic conditions • Dosage of reagent = 12.5 g/L	• The nZVI/ACF plays a noble metal free electrode for electro-catalytic BrO_3^- reduction. • The BrO_3^- reduction could be performed in a wide dissolved oxygen and pH range	Yao et al. (2020)

4.6	Selenate (Se^{6+})	ZVI with NaClO and H_2O_2	• Size = micro level • Surface area = 0.356 m^2 /g	• pH = 9 • Aerobic conditions • Temp =320° c	• Minimal dosage of oxidants required was affected by the oxidant species. • More than 85% of Se^{6+} was reduced to Se^0 and Se^{2-}.	Li et al. (2018)
4.7	Lead (Pb^{2+}) and NO^-_3	Nano ZVI supported on chelating resin	• Size = 10–30 nm • Surface area = 0.864 m^2 /g • Shape = spherical	• pH = 4–8.5 • Aerobic and anaerobic conditions	• Removal efficiencies of Pb^{2+} and NO^-_3 can reach 87% and 97%, respectively.	Shi et al. (2013)
4.8	Nickel (Ni)	ZVI/Pumice granular mixtures	• Shape = rod shaped	• pH = 4–11 • Aerobic conditions • Dosage of reagent = 50 g/L	• The performance of the granular mixture is enhanced if the ZVI/ Pumice weight ratio is high (74% removal was take place in 50 mg/L nickel solution).	Calabrò et al. (2012)
4.9	Copper(II)	Nano ZVI immobilized in polymer nanofibres	• Size = 28 nm • Surface area = 0.635 m^2/g • Shape = Cylindrical	• Aerobic and anaerobic conditions • pH = 6.5–9	• The interaction between Cu(II) ions and these mult iwalled carbon nanotubes (MWCNT) reinforced ZVI. • Cu^{2+} chemisorption occurs via chemical reduction and deposition on the ZVI NP surfaces to form Fe/Cu alloy.	Xiao et al. (2011)
4.10	Zinc (Zn)	ZVI	• Surface area = 1.256 m^2/g • Shape = spherical	• pH = Acidic • Aerobic conditions	• Zinc removal enhanced by dissolved oxygen and ferric ion addition. • Zinc adsorbed on ZVI was eluted by a reducing agent such as citric acid	Kishimoto et al. (2011)
4.11	Chromium (Cr^{6+})	Bentonite-supported-nano ZVI	• Size = 10–30 nm • Shape = spherical	• pH = 3–9 • Aerobic conditions • Dosage of reagent = 15 g/L	• Reagent is used to remediate electroplating wastewater with removal efficiencies for Cr > 90%.	Shi et al. (2011))

(Continued)

Table 14.2 (Continued)

S. No	Contaminant	Reagent	ZVI properties	Conditions	Remarks	References
4.12	Hg(II), Zn(II) and hydrogen sulphide,	Nano-ZVI	• Size = 35–60 nm • Shape = spherical	• Aerobic and anaerobic conditions • pH = basic	• Zn(II) removal was achieved via sorption to the Fe_3O_4 shell followed by $Zn(OH)_2$ precipitation. • Hydrogen sulphide was immobilized on the nZVI surface as S_2^{2-} and S^{2-} species.	Yan et al. (2010)
4.13	Cobalt (Co^{2+})	Nano- ZVI	• Size = 20–80 nm • Shape = spherical	• pH = 3–11 • Aerobic conditions • Dosage of reagent = 1 g/L	• Effects of concentration, contact time, repetitive loading, pH and ageing on the extent of retardation of Co^{2+} ions investigated • The extent of Co^{2+} uptake increased with the increasing pH.	Uzum et al. (2008)
5. Nitrate						
5.1	Nitrate (NO^-_3)	Fe-Cu bimetallic rod	• Shape = rod shaped	• pH = 7 • Aerobic conditions	• The results showed that 98% nitrate removal efficiency • Optimization of operation conditions and the combination mode of Fe^0 @ Cu^0 with formic acid, Fe^{3+} and UV were investigated.	Wang et al. (2020a)
5.2	Nitrate (NO^-_3)	ZVI (rod)	• Shape = rod shaped • Surface area = 2.451 m^2/g	• Aerobic and anaerobic conditions • pH = 4–9	• Fe_3O_4 scale accelerated the electron transfer and the removal efficiency of NO^-_3 • The removal efficiency of NO_3– was up to 98.1%.	Xu et al. (2012)
5.3	Nitrate (NO^-_3)	Nano-TiO_2/Fe^0 composite	• Size = 20 nm • Shape = spherical	• pH = 11 • Aerobic conditions • Dosage of reagent = 10 g/L	• The composition (TiO2/nZVI ratio) greatly influences the conversion of nitrate to N_2 due to its reducing condition.	Pan et al. (2012)
5.4	Nitrate (NO^-_3)	ZVI (iron chips)	• Shape = rectangular	• Aerobic and anaerobic conditions • pH = 3–8.5	• Reduction efficiency was enhanced by either decreasing the aqueous pH or increasing the solid–liquid ratio. • Fe^0 coated with an iron corrosion product repeatedly reduced NO^-_3 to ammonium ion	Suzuki et al. (2012))

5.5	Nitrate ($NO^-{}_3$)	Nano- ZVI	• Size = 30–50 nm • Shape = spherical	• Aerobic and anaerobic conditions • pH = 3–9 • Dosage of reagent = 1 g/L	• The reduction efficiency of nitrate decreased quickly with increasing initial pH value. • The removal rate of nitrate reached 96.4% in 30 minutes	Zhang et al. (2010a)
6. Nitro Amino Compounds (NACs)						
6.1	N-Nitro sodimethylamine (NDMA)	ZVI	• Size = micro level • Shape = spherical • Surface area = 0.8656 m^2 /g	• pH = 11 • Aerobic conditions • Dosage of reagent = 12.5 g/L	• H_2 and Fe^{2+} generation by sulfidated ZVI was almost independent of nitrate concentration • sulfidation improved the efficiency of NDMA reduction by ZVI	Qin et al. (2020)
6.2	Nitro benzene (NB)	Biochar supported Nano ZVI (nZVI/BC)	• Size = 34 nm • Shape = spherical	• Aerobic and anaerobic conditions • pH = 6	• NB removal by nZVI/BC involves simultaneous adsorption and reduction mechanism. • nZVI/BC exhibited better catalytic activity, stability and durability than nZVI.	Wei et al. (2019)
6.3	Nitro benzene (NB)	ZVI (wire)	• Shape = rod shaped	• pH = 7 • Aerobic conditions • Dosage of reagent = 5 g/L	• The NB removal increased with the decrease of pH with removal efficiency of 88.7% @ pH -7. • Improvement of NB removal was observed in the present of Cl^-, $SO_4{}^{2-}$, and $NO_3{}^-$.	Yin et al. (2012)
6.4	Nitro benzene	ZVI	• Shape = spherical • Size = micro level	• pH = 3–11 • Aerobic conditions	• The final removal efficiency was 80.98%. • NB reduction rate constant linearly decreased with increased pH	Dong et al. (2010)

(*Continued*)

Table 14.2 (Continued)

S. No	Contaminant	Reagent	ZVI properties	Conditions	Remarks	References
6.5	2,4,6-tri nitro toluene (TNT)	Nano- ZVI	• Size = 43 nm • Shape = spherical	• pH = 4 • Aerobic and anaerobic conditions	• The results show that >99% TNT is degraded	Zhang et al. (2010b)
6.6	Nitro benzene (NB)	Nano- ZVI	• Size = 25 nm • Shape = spherical • Surface area = 1.325 m^2 /g	• pH = 7 • Aerobic conditions	• Reaction efficiency of 83% was achieved • Results confirm that non-structural Fe from Fe(III) was reduced predominantly to ZVI, which was responsible for the reduction of nitrobenzene to aniline.	Gu et al. (2010)
7. Phenol						
7.1	Ibuprofen and Phenol	Nano- ZVI	• Size = 36 nm • Shape = spherical	• pH = 4–6 • Aerobic conditions	• Very low generation of toxic 4-isobutylacetophenone as by-product • Technique reported can be to pollutant mixtures.	Minella et al. (2019)
7.2	phenol	ZVI with hydrogen peroxide	• Size = 26 micro m • Shape = cylindrical	• pH = 6 • Aerobic and anaerobic conditions	• 90% TOC conversions take place. • The presence of ferryl ion complexes of high oxidative capacity, also contributing the mineralization of the remaining organic intermediates.	Segura et al. (2012)
7.3	phenol	Nano- ZVI	• Size = 42 nm • Shape = spherical	• pH = 4 • Aerobic conditions • Dosage of reagent = 0.5 g/L to 500 mg/L	• The removal efficiency was increased with an increase in nZVI dosage and decreased with an increase in initial phenol concentration and initial pH.	Babuponnusami and Muthukumar (2012)

7.4	Phenol	ZVI (powder)	• Size = micro level • Shape = triangular	• pH = 2–8.1 • Aerobic and anaerobic conditions	• At solution pH 3, 91% of phenol removal was achieved • The contribution of the Fenton reaction was found to be 77% of overall Total organic carbon (TOC) reduction.	Shimizu et al. (2012)
8. Other compounds						
8.1	Dichloro diphenyl trichloro ethane (DDT)	ZVI with activated carbon	• Shape = spherical	• pH = 4–8.5 • Aerobic and anaerobic conditions	• DDT pre-adsorbed on activated carbon is rapidly dechlorinated by ZVI. • Hydrogen species (H*) act as active reductants.	Kopinke et al. (2020)
8.2	Thiamphenicol (TAP)	Amorphous ZVI microspheres (A-mZVI)	• Size = micro level • Shape = spherical	• pH = 6 • Aerobic conditions • Dosage of reagent = 10 g/L	• Amorphorization process promoted electron donating ability of A-mZVI. • A-mZVI shows higher apparent TAP reduction rate constant than its crystalline counterpart C-mZVI	Shen et al. (2020)
8.3	Sulfa diazine (SDZ)	Nano- ZVI with graphene (3D-GN@nZVI)	• Size = 22 nm • Shape = spherical • Surface area = 2.365 m^2 /g	• pH = 4.6 • Aerobic conditions • Temp =160° c	• Sulfadiazine was removed mainly by the attack of ˙OH radical. • Excellent catalytic properties of 3D-GN@nZVI observed	Yang et al. (2019)

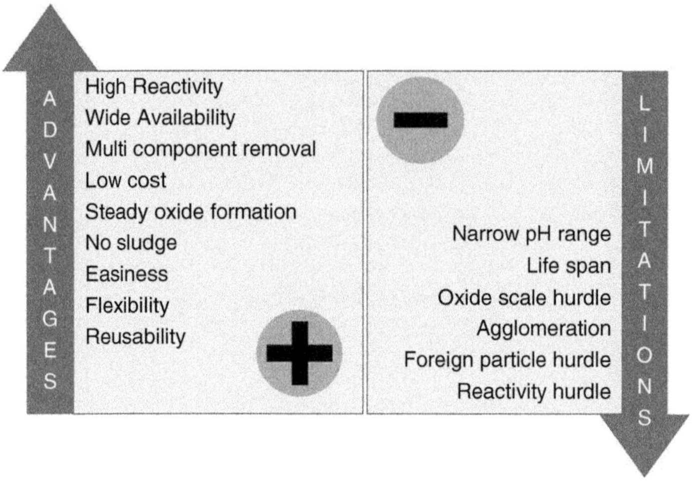

Figure 14.5 Advantages and Limitations of ZVI.

granular ZVI in permeable reactive barriers (PRB) for in situ removal of contaminants from groundwater is observed (Bastiaens et al. 2010). From the second half of twentieth century till now ZVI, due to its versatile properties, uniqueness, and applicability, has been widely used in one form or the other for removal of contaminants from groundwater. In recent years, ZVI has gained much attention and rather than using pristine ZVI, which accompanies certain demerits in terms of life span, oxide scale formation, reactivity, etc., various modifications in terms of size, surface, support, chemistry, and use of additional physical and chemical processes are being done by researchers globally to improve ZVI performance. Few researchers have subjected ZVI to pre-treatment techniques like acid washing, hydrogen reduction, and ultrasonication to refresh the ZVI surface and enhance its reactivity (Guan et al. 2015). Size modification of ZVI is achieved either by comminution or milling techniques, or alternately by use of chemical/biochemical techniques to synthesize nano-ZVI, which exhibits holistic superior properties compared to the bulk counterparts (Deshpande et al. 2020; El-sayed 2020; Maksoud et al. 2020; Rana et al. 2018). On the other hand, chemical modification of ZVI by doping other metal like Au, Co, Cu, Ni, Pd, Pt, etc. to form bimetallic ZVI for enhanced corrosion and contaminant removal is also reported (Ou et al. 2020; Scaria et al. 2020; Tso et al. 2020; Wang et al. 2020a). Similarly, coupling of ZVI with other active adsorbent materials like carbon, activated carbon, and iron oxides to achieve hybrid ZVI is also reported (Kopinke et al. 2020; Wang et al. 2020b; Zhao et al. 2020). Even though nano-ZVI has superior properties its practical usage attracts difficulties like, less stability, aggregation, separation difficulty, etc. In this regard, various works on immobilization of nano-ZVI on several kinds of supports such as bentonite, zeolite, kaolinite, titanium dioxide, chitosan, graphene, polyacrylic acid, and carboxymethyl cellulose to improve nano-ZVI control and stability are reported (Liu et al. 2020; Phanthasri et al. 2020; Tasharrof et al. 2020; Yao et al. 2020). Apart from the advancements that took place on the ZVI particle, little advancement on ZVI from a technology point of view is also observed. ZVI assisting with other physical processes like ultrasonication, UV illumination, gas bubbling, microwave energy, magnetic energy, etc. to enhance the active site formation, reactivity, and mass transport of active species is reported (Moradi and Eslami 2020;

Wang et al. 2020c; Xu and Huang 2019). Besides physical ones, chemically assisted processes like addition of oxidants, ligands (EDTA, Oxalate, etc.), anions (Fe^{+2}, Co^{+2}, Mn^{+2}, etc.) to improve oxidation, electron transfer, and complex formation of ZVI with contaminants is reported (Guan et al. 2015; Fu et al. 2014; Li et al. 2018; Xi et al. 2019; Xue et al. 2019). Thus, the future of ZVI technology is in the usage of modified ZVI particles along with proper support of physical, chemical, or a combination of technologies as well as integration with other water treatment technologies to achieve superior contaminant removal from groundwater.

14.6 Conclusions

- Groundwater remediation is one of the prime concerns of today's world.
- Remediation technologies like oxidation, disinfection precipitation, coagulation, adsorption, ion exchange, etc. make use of various chemicals like chlorine, ozone, aluminium sulphate, ferrous chloride, sodium aluminate, activated carbon, activated alumina, zeolites, iron, natural clay, etc. One such prominent material which is of prime importance due to its low cost, high availability, and multiple contaminant removal ability (chlorinated compounds, nitro compounds, heavy metals, dyes, aromatics) is zero-valent iron (ZVI).
- ZVI in the form of lumps or chunks or wires or powders with the dimension ranging from millimetre scale to nano scale is synthesized by various physical, chemical, biological, and mechanical techniques using a top-down or bottom-up approach.
- ZVI implements one or a combination of adsorption, reduction, oxidation, precipitation, and co-precipitation principles to remove various contaminants from groundwater like metalloids (As), heavy metals (Cr, Ni, Pb, Zn), nitrates, phosphates, chloro-organic compounds, nitro amino compounds, dyes, phenol compounds, etc.
- ZVI possesses several advantages of high reactivity, wide availability, multi component removal, no sludge formation, flexibility, low cost, reusability, steady oxide formation and technology easiness, and a few limitations like thick oxide formation, foreign particle hurdle, agglomeration, reactivity hurdle, life span, and narrow pH operation.
- Usage of ZVI for groundwater treatment has been widely reported for the past two decades and there has been much recent advancement in terms of size, surface, support, chemistry, and usage to form pretreated ZVI, micro ZVI, nano ZVI, immobilized nano ZVI, bimetallic ZVI, hybrid ZVI along with additional physical and chemical processes that are being attempted by researchers globally.
- ZVI, due to its versatile properties, uniqueness, and applicability, is being widely used in one form or the other for removal of contaminants from groundwater and the future of ZVI technology is in the usage of novel and modified ZVI particles, along with other integrated physical or chemical or combination of processes.

References

Abass, K.M., Maa, T., and Kong, S. (2016). A novel MD-ZVI integrated approach for high arsenic groundwater decontamination and effluent immobilization. *Process Safety and Environmental Protection* 102: 190–203.

Ahn, S.C., Oh, S.Y., and Cha, D.K. (2008). Enhanced reduction of nitrate by zero valent iron at elevated temperatures. *Journal of Hazardous Materials* 156: 17–22.

Asere, T.G., Stevens, C.V., and Laing, G.D. (2019). Use of (modified) natural adsorbents for arsenic remediation: a review. *Science of the Total Environment* 676: 706–720. https://doi.org/10.1016/j.scitotenv.2019.04.237.

Babuponnusami, A. and Muthukumar, K. (2012). Removal of phenol by heterogenous photo electro Fenton-like process using nano-zero valent iron. *Separation and Purification Technology* 98: 130–135.

Barreto-Rodrigues, M.T., Silva, F.T., and Paiva, T.C.B. (2009). Optimization of Brazilian TNT industry wastewater treatment using combined zero-valent iron and Fenton processes. *Journal of Hazardous Materials* 168: 1065–1069.

Bastiaens, L., Phillips, D.H., and Kalin, R.M. (2010). Ten year performance evaluation of a field-scale zero-valent iron permeable reactive barrier installed to remediate trichloroethene contaminated groundwater. *Environmental Science and Technology* 44: 3861–3869.

Biterna, M., Antonoglou, L., Lazou, E., and Voutsa, D. (2010). Arsenite removal from waters by zero valent iron: batch and column tests. *Chemosphere* 78: 7–12.

Bradl, H. and Xenidis, A. (2005). Remediation Techniques. In: *Heavy Metals in the Environment: Origin, Interaction and Remediation. Interface Science and Technology*, vol. 6 (ed. H.B. Bradl), 165–261. Elsevier https://doi.org/10.1016/S1573-4285(05)80022-5.

Britto, J.M., Oliveira, S.B., Rabelo, D., and Rangel, M.C. (2008). Catalytic wet peroxide oxidation of phenol from industrial wastewater on activated carbon. *Catalyst Today* 133: 582–587.

Calabrò, P., Moraci, N., and Suraci, P. (2012). Estimate of the optimum weight ratio in zero-valent iron/pumice granular mixtures used in permeable reactive barriers for the remediation of nickel contaminated groundwater. *Journal of Hazardous Materials* 207–208: 111–116.

Chang, S.H., Chuang, S.H., Li, H.C. et al. (2009). Comparative study onthe degradation of I.C. Remazol brilliant blue R and I.C. acid black 1 by fentonoxidation and Fe0/air process and toxicity evaluation. *Journal of Hazardous Materials* 166: 1279–1288. https://doi.org/10.1016/j.jhazmat.2008.12.042.

Chen, B., Wang, X., Wang, C., and Li, S. (2011). Degradation of azo dye direct sky blue 5B by sonication combined with zero-valent iron. *Ultrasonics Sonochemistry* 18: 1091–1096.

Chen, Q., Fan, G., Na, W. et al. (2019). Past, present, and future of groundwater remediation research: a scientometric analysis. *Environmental Research and Public Health* 16: 3975. https://doi.org/10.3390/ijerph16203975.

Cheremisinoff, N.P. (1997a). Groundwater contamination. In: *Groundwater Remediation and Treatment Technologies* (ed. N.P. Cheremisinoff), 127–168. William Andrew Publishing https://doi.org/10.1016/B978-081551411-4.50006-5.

Cheremisinoff, N.P. (1997b). Pump-and-treat remediation technology. In: *Groundwater Remediation and Treatment Technologies* (ed. N.P. Cherminisoff), 203–258. William Andrew Publishing https://doi.org/10.1016/B978-081551411-4.50008-9.

Choi, K. and Lee, W. (2012). Enhanced degradation of trichloroethylene in nano-scale zero-valent iron Fenton system with Cu(II). *Journal of Hazardous Materials* 211–212: 146–153.

Cissoko, N., Zhang, Z., Zhang, J.H., and Xu, X.H. (2009). Removal of Cr(VI) from simulative contaminated groundwater by iron metal. *Process Safety and Environment Protection* 87: 395–400. https://doi.org/10.1016/j.psep.2009.07.001.

Cook, S.M. (2009). Assessing the Use and Application of Zero-Valent Iron Nanoparticle Technology for Remediation at Contaminated Sites. https://clu-in.org/download/studentpapers/Zero-Valent-Iron-Cook.pdf.

Crane, R.A. and Scott, T.B. (2012). Nanoscale zero-valent iron: future prospects for an emerging water treatment technology. *Journal of Hazardous Materials* 211: 112–125.

Deshpande, B.D., Agrawal, P.S., Yenkie, M.K.N., and Dhoble, S.J. (2020). Prospective of nanotechnology in degradation of waste water: a new challenges. *Nano-Structures & Nano-Objects* 22: 100442. https://doi.org/10.1016/j.nanoso.2020.100442.

Diaz-Cruz, M.S. and Barcelo, D. (2008). Trace organic chemicals contamination in groundwater recharge. *Chemosphere* 72: 333–342. https://doi.org/10.1016/j.chemosphere.2008.02.031.

Dong, J., Zhao, Y., Zhao, R., and Zhou, R. (2010). Effects of pH and particle size on kinetics of nitrobenzene reduction by zero-valent iron. *Journal of Environmental Sciences* 22: 1741–1747.

Duan, C., Ma, T., Wang, J., and Zhou, Y. (2020). Removal of heavy metals from aqueous solution using carbon-based adsorbents: a review. *Journal of Water Process Engineering* 37: 101339. https://doi.org/10.1016/j.jwpe.2020.101339.

El-Sayed, M.A. (2020). Nanoadsorbents for water and wastewater remediation. *Science of the Total Environment* https://doi.org/10.1016/j.scitotenv.2020.139903.

Forbort, J. (2009). Treatment for organic contaminants: biological treatment. In: *Groundwater Treatment Technology*, 3e (ed. E.K. Nyer), 127–195. Wiley https://doi.org/10.1002/9780470432839.ch3.

Fu, F., Dionysioub, D.D., and Liu, H. (2014). The use of zero-valent iron for groundwater remediation and wastewater treatment: a review. *Journal of Hazardous Materials* 267: 194–205.

Ghosh, S., Debsarkar, A., and Dutta, A. (2019). Technology alternatives for decontamination of arsenic-rich groundwater – a critical review. *Environmental Technology and Innovation* 13: 277–303. https://doi.org/10.1016/j.eti.2018.12.003.

Gillham, R.W. and O'Hannesin, F.H. (1994). Enhanced degradation of halogenated aliphatics by zero-valent iron. *American Journalism* 32: 958–967.

Gillham, R.W. and O'Hannesin, F.H. (1998). Long-term performance of an in situ "Iron Wall" for remediation of VOCs. *American Journalism* 36: 164–170.

Gu, D., Jia, H., Li, H., and Boyd, S.A. (2010). Synthesis of highly reactive Subnano-sized zero-valent iron using smectite clay templates. *Environmental Science & Technology* 40: 4258–4263.

Guan, X., Sun, Y., Qin, H., and Dong, H. (2015). The limitations of applying zero-valent iron technology in contaminants sequestration and the corresponding countermeasures: the development in zero-valent iron technology in the last two decades (1994-2014). *Water Research* 75: 224–248.

Hasan, H.A., Muhammad, M.H., and Ismail, N.I. (2020). A review of biological drinking water treatment technologies for contaminants removal from polluted water resources. *Journal of Water Process Engineering* 33: 101035. https://doi.org/10.1016/j.jwpe.2019.101035.

He, F., Zhao, D., and Paul, C. (2010). Field assessment of carboxymethyl cellulose stabilized iron nanoparticles for in situ destruction of chlorinated solvents in source zones. *Water Resources* 44: 2360–2370.

Hepure. (2020). Contaminants treated with ZVI products. https://hepure.com/products/what-can-be-treated-with-zvi

Hoganas. (2018). Green contaminants removal for water and soil treatment. https://www.hoganas.com/en/powder-technologies/water-and-soil-treatment. https://doi.org/10.1016/j.scitotenv.2020.137802

IAEA. (1999). Technical options for the remediation of contaminated groundwater. International Atomic Energy Agency 1088.

Kishimoto, N., Iwano, S., and Narazaki, Y. (2011). Mechanistic consideration of zinc ion removal by zero-valent iron. *Water, Air, and Soil Pollution* 221: 183–189.

Kopinke, F., Sühnholz, S., Georgi, A., and Mackenzie, K. (2020). Interaction of zero-valent iron and carbonaceous materials for reduction of DDT. *Chemosphere* 253: 126712. https://doi.org/10.1016/j.chemosphere.2020.126712.

Kumar, S. and Jain, S. (2013). History, introduction, and kinetics of ion exchange materials. *Journal of Chemistry* https://doi.org/10.1155/2013/957647.

Lapworth, D.J., Baran, N., Stuart, M.E., and Ward, R.S. (2012). Emerging organic contaminants in groundwater: a review of sources, fate and occurrence. *Environmental Pollution* 163: 287–303. https://doi.org/10.1016/j.envpol.2011.12.034.

Li, X.Q., Elliott, D.W., and Zhang, W.X. (2006). Zero-valent iron nanoparticles for abatement of environmental pollutants: materials and engineering aspects. *Critical Reviews in Solid State and Materials Sciences* 31: 111–122.

Li, W.W., Zhang, Y., Yang, Y.L. et al. (2013). Synergetic decolorization of reactive blue 13 by zero-valent iron and anaerobic sludge. *Bioresource Technology* 149: 38–43.

Li, Y., Guo, X., Dong, H. et al. (2018). Selenite removal from groundwater by zero-valent iron (ZVI) in combination with oxidants. *Chemical Engineering Journal* 345: 432–440.

Lim, K.T., Shukor, M.Y., and Wasoh, H. (2014). Physical, chemical, and biological methods for the removal of arsenic compounds. *BioMed Research International* https://doi.org/10.1155/2014/503784.

Liu, Y. and Wang, J. (2019). Reduction of nitrate by zero valent iron (ZVI)-based materials: a review. *Science of the Total Environment* 671: 388–403. https://doi.org/10.1016/j.scitotenv.2019.03.317.

Liu, K., Li, F., Cui, J. et al. (2020). Simultaneously removal of Cd (II) and As (III) by graphene-like biochar supported zero-valent iron from irrigation waters under aerobic conditions: synergistic effects and mechanisms. *Journal of Hazardous Materials* 395: 122623. https://doi.org/10.1016/j.jhazmat.2020.122623.

Lu, H.J., Wang, J.K., Ferguson, S. et al. (2016). Mechanism, synthesis and modification of nano zerovalent iron in water treatment. *Nanoscale* 00: 1–13.

Minella, M., Bertinetti, S., Hanna, K., and Vione, D. (2019). Degradation of ibuprofen and phenol with a Fenton-like process triggered by zero-valent iron (ZVI-Fenton). *Environmental Research* 179: 108750. https://doi.org/10.1016/j.envres.2019.108750.

Mak, M.S.H., Rao, P.H., and Lo, I.M.C. (2009). Effects of hardness and alkalinity on the removal of arsenic(V)from humic acid-deficient and humic acid-rich groundwater by zero-valent iron. *Water Resources* 43: 4296–4304.

Maksoud, M.I.A.A., Elgarahy, A.M., Farrell, C. et al. (2020). Insight on water remediation application using magnetic nanomaterials and biosorbents. *Coordination Chemistry Reviews* 403: 213096. https://doi.org/10.1016/j.ccr.2019.213096.

Ministry of Jal Shakti, Government of India. (2019). Empowering Village Communities for A Sustainable Water Future – A Resource Book for Jaldoots. http://cgwb.gov.in/06112019%20Jaldoot%20Resource%20Book.pdf.

Moradi, M. and Eslami, A. (2020). Reductive/oxidative degradation of tetrachloroethene and its transformation products using combination of permeable ZVI column and UV/Fe/peroxydisulfate system: RSM design and synergy study. *Journal of Water Process Engineering* 36: 101288. https://doi.org/10.1016/j.jwpe.2020.101288.

Mukherjee, R., Kumar, R., Sinha, A. et al. (2016). A review on synthesis, characterization, and applications of nano zero valent iron (nZVI) for environmental remediation. *Critical Reviews in Environmental Science and Technology* 46: 443–466.

Naseri, E., Mwakabona, H.T., Noubactep, C. et al. (2017). Making Fe0-based filters a universal solution for safe drinking water provision. *Sustainability* 9: 1224. https://doi.org/10.3390/su9071224.

Neumann, A., Kaegi, R., Voegelin, A. et al. (2013). Arsenic removal with composite iron matrix filters in Bangladesh: a field and laboratory study. *Environmental Science and Technology* 47 (9): 4544–4554. https://doi.org/10.1021/es305176x.

Nicomel, N.R., Leus, K., Folens, K. et al. (2016). Technologies for Arsenic Removal from water: current status and future perspectives. *International Journal of Environmental Research and Public Health* 13 (62): 1–24. https://doi.org/10.3390/ijerph13010062.

Noubactep, C. (2008). A critical review on the process of contaminant removal in Fe–H_2O systems. *Environmental Technology* 29: 909–920.

Noubactep, C. (2009). An analysis of the evolution of reactive species in Fe^0/H_2O systems. *Journal of Hazardous Materials* 168: 1626–1631.

Oh, S.W., Kang, M.N., Cho, C.W., and Lee, M.W. (1997). Detection of carcinogenic amines from dyestuffs or dyed substrates. *Dyes and Pigments* 33: 119–135.

Ou, Y., Wei, C.Y., and Shih, Y. (2016). Short-chain organic acids increase the reactivity of zerovalent iron nanoparticles toward polychlorinated aromatic pollutants. *Chemical Engineering Journal* 284: 372–379.

Ou, J.H., Sheu, Y.T., Sun, Y.J., and Kao, C.M. (2020). Application of iron/aluminum bimetallic nanoparticle system for chromium-contaminated groundwater remediation. *Chemosphere* 256: 127158. https://doi.org/10.1016/j.chemosphere.2020.127158.

Pan, J.R., Huang, C., Hsieh, W., and Wu, B. (2012). Reductive catalysis of novel TiO2/Fe0 composite under UV irradiation for nitrate removal from aqueous solution. *Separation and Purification Technology* 84: 52–55.

Phanthasri, J., Grisdanurak, N., Wantala, K. et al. (2020). Role of zeolite-supported nanoscale zero-valent iron in selenate removal. *Water, Air, and Soil Pollution* 231: 199. https://doi.org/10.1007/s11270-020-04587.

Prabu, D. and Parthiban, R. (2013). Synthesis and characterization of Nanoscale Zero Valent Iron (NZVI) nanoparticles for environmental remediation. *Asian Journal of Pharmanutical Technology* 3: 181–184.

Qin, H., Guan, X., and Tratnyek, P.G. (2020). Effects of sulfidation and nitrate on the reduction of N-Nitrosodimethylamine by zerovalent iron. *Environmental Science and Technology* 53: 9744–9754.

Rana, A., Kumari, S., and Tyagi, M. (2018). Leaf-extract mediated zero-valent iron for oxidation of Arsenic (III): preparation, characterization and kinetics. *Chemical Engineering Journal* 347: 91–100.

Reddy, K.R. (2008). Physical and chemical groundwater remediation technologies. In: *Overexploitation and Contamination of Shared Groundwater Resources*, NATO Science for Peace and Security Series C: Environmental Security (ed. C.J.G. Darnault), 257–274. Springer https://doi.org/10.1007/978-1-4020-6985-7_12.

Reddy, G.S.M., Bhaumik, M., Maity, A., and Ray, S.S. (2020). Removal of Congo red from aqueous solution by adsorption using gum ghatti and acrylamide graft copolymer coated with zero valent iron. *International Journal of Biological Macromolecules* https://doi.org/10.1016/j.ijbiomac.2020.01.099.

Rocca, C.D., Belgiorno, V., and Meric, S. (2007). Heterotrophic/autotrophic denitrification (HAD) of drinking water: prospective use for permeable reactive barrier. *Desalination* 210: 194–204.

Román, S., Alonso, M.L., Bartolomé, L. et al. (2013). Relevance study of bare and coated zero valent iron nanoparticles for lindane degradation from its byproduct monitorization. *Chemosphere* 93: 1324–1332.

Sarkar, A. and Paul, B. (2016). The global menace of arsenic and its conventional remediation – a critical review. *Chemosphere* 158: 37–49. https://doi.org/10.1016/j.chemosphere.2016.05.043.

Scaria, J., Nidheesh, P.V., and Kumar, M.S. (2020). Synthesis and applications of various bimetallic nanomaterials in water and wastewater treatment. *Journal of Environmental Management* 259: 110011. https://doi.org/10.1016/j.jenvman.2019.110011.

Segura, Y., Martínez, F., Melero, J.A. et al. (2012). Enhancement of the advanced Fenton process (Fe0/H2O2) by ultrasound for the mineralization of phenol. *Applied Catalysis B: Environmental* 113–114: 100–106.

Singh, N.B., Nagpal, G., Agrawal, S., and Rachna (2018). Water purification by using adsorbents: a review. *Environmental Technology and Innovation* 11: 187–240. https://doi.org/10.1016/j.eti.2018.05.006.

Shankar, S., Shanker, U., and Shikha (2014). Arsenic contamination of groundwater: a review of sources, prevalence, health risks, and strategies for mitigation. *The Scientific World Journal* https://doi.org/10.1155/2014/304524.

Shen, W., Wang, X., Jia, X. et al. (2020). Amorphization enables highly efficient anaerobic thiamphenicol reduction by zero-valent iron. *Applied Catalysis B: Environmental* 264: 118550. https://doi.org/10.1016/j.apcatb.2019.118550.

Shi, L.N., Zhang, X., and Chen, Z. (2011). Removal of Chromium (VI) from wastewater using bentonite-supported nanoscale zero-valent iron. *Water Research* 45: 886–892.

Shi, J., Yi, Z., He, H., and Li, A. (2013). Preparation of nanoscale zero-valent iron supported on chelating resin with nitrogen donor atoms for simultaneous reduction of Pb^{2+} and NO_3. *Chemical Engineering Journal* 230: 166–171.

Shimizu, A., Tokumura, M., Nakajima, K., and Kawase, Y. (2012). Phenol removal using zero-valent iron powder in the presence of dissolved oxygen: roles of decomposition by the Fenton reaction and adsorption/precipitation. *Journal of Hazardous Materials* 201: 60–67.

Sims, J.L., Suflita, J.M., and Russell, H.H. (1992). *In-Situ Bioremediation of Contaminated Groundwater. Groundwater Issue*. United States Environmental Protection Agency https://www.epa.gov/sites/production/files/2015-06/documents/insitubio_contam_gw.pdf.

Sista, K.S. and Dwarapudi (2018). Iron powders from steel industry by-products. *ISIJ International* 58: 999–1006.

Sizmur, T., Fresno, T., Akgul, G. et al. (2017). Biochar modification to enhance sorption of inorganics from water. *Bioresource Technology* 246: 34–47. https://doi.org/10.1016/j.biortech.2017.07.082.

Suarez, G. (2009). Treatment for organic contaminants: physical/chemical methods. In: *Groundwater Treatment Technology*, 3e (ed. E.K. Nyer), 39–126. Wiley https://doi.org/10.1002/9780470432839.ch2.

Summer, D., Schöftner, P., Watzinger, A., and Reichenauer, T.G. (2020). Inhibition and stimulation of two perchloroethene degrading bacterial cultures by nano- and micro-scaled zero-valent iron particles. *Science of the Total Environment* 722: 137802.

Suzuki, T., Moribe, M., Oyama, Y., and Niinae, M. (2012). Mechanism of nitrate reduction by zero-valent iron: equilibrium and kinetics studies. *Chemical Engineering Journal* 183: 271–277.

Sweeny, K.H. and Fischer, J.R. (1972). Reductive degradation of halogenated pesticides. *Water Pollution by Organic Pesticides* 56: 267–279.

Tasharrof, S., Rouzitalab, Z., Maklavany, D.M. et al. (2020). Adsorption of cadmium using modified zeolite-supported nanoscale zero-valent iron composites as a reactive material for PRBs. *Science of the Total Environment* 736: 139570. https://doi.org/10.1016/j.scitotenv.2020.139570.

Taylor, P.A. (2015). Physical, chemical, and biological treatment of groundwater at contaminated nuclear and NORM sites. *Environmental Remediation and Restoration of Contaminated Nuclear and Norm Sites*: 237–256. https://doi.org/10.1016/B978-1-78242-231-0.00010-7.

Tosco, T., Papini, M.P., Viggi, C.C., and Sethi, R. (2014). Nanoscale zerovalent iron particles for groundwater remediation: a review. *Journal of Cleaner Production* 77: 10–21.

Tratnyek, P.G. and Johnson, R.L. (2006). Nanotechnologies for environmental cleanup. *Applications Feature* 1: 44–48.

Tratnyek, P. and Scherer, M. (2003). Permeable reactive barriers of zero valent metals in groundwater pollution remediation. *Environmental and industrial application* 25: 371–417.

Tripathi, S.M. and Ram, S. (2018). Bioremediation of groundwater: An overview. *International Journal of Applied Engineering Research* 13 (24): 16825–16832. https://www.ripublication.com/ijaer18/ijaerv13n24_19.pdf.

Triszcz, J.M., Porta, A., and Einschlagc, F.S.G. (2009). Effect of operating conditions on iron corrosion rates in zero-valent iron systems for arsenic removal. *Chemical Engineering Journal* 150: 431–439.

Tsenga, H.H., Su, J.G., and Liang, C. (2011). Synthesis of granular activated carbon/zero valent iron composites for simultaneous adsorption/dechlorination of trichloroethylene. *Journal of Hazardous Materials* 192: 500–506.

Tso, C.P., Kuo, D.T.F., and Shih, Y.H. (2020). Removal of Hexabromocyclododecane by carboxymethyl cellulose stabilized Fe and Ni/Fe bimetallic nanoparticles: the particle stability and reactivity in water. *Chemosphere* 250: 126155. https://doi.org/10.1016/j.chemosphere.2020.126155.

Uzum, C., Shahwan, T., Lieberwirth, I., and Hallam, K.R. (2008). Application of zero-valent iron nanoparticles for the removal of aqueous Co^{2+} ions under various experimental conditions. *Chemical Engineering Journal* 144: 213–220.

Wang, K.S., Lin, C.L., Wei, M.C. et al. (2010). Effects of dissolved oxygen on dye removal by zero-valent iron. *Journal of Hazardous Materials* 182: 886–895.

Wang, C., Zhanga, Z., Zhanga, J., and Chena, S. (2014). Removal of As(III) and As(V) from aqueous solutions using nanoscale zero valent iron-reduced graphite oxide modified composites. *Journal of Hazardous Materials* 268: 234–241.

Wang, W., Zhao, P., Hu, Y. et al. (2020c). Application of weak magnetic field coupling with zero-valent iron for remediaiton of groundwater and waste water: A review. *Journal of Cleaner Production* 262: 121341. https://doi.org/10.1016/j.jclepro.2020.121341.

Wang, B., An, B., Liu, Y. et al. (2020a). Selective reduction of nitrate into nitrogen at neutral pH range by iron/ copper bimetal coupled with formate/ferric ion and ultraviolet radiation. *Separation and Purification Technology* 248: 117061. https://doi.org/10.1016/j.seppur.2020.117061.

Wang, W., Hu, B., Liang, Z. et al. (2020b). Cr(VI) removal by micron-scale iron-carbon composite induced by ball milling: the role of activated carbon. *Chemical Engineering Journal* 389: 122633. https://doi.org/10.1016/j.cej.2019.122633.

Wei, Y., Yang, S.W., Che, C.H., and Huang, D.H. (2012). Biodegradable surfactant stabilized nanoscale zero-valent iron for in situ treatment of vinyl chloride and 1,2-dichloroethane. *Journal of Hazardous Materials* 211–212: 373–380.

Wei, G., Zhang, J., Luo, J. et al. (2019). Nanoscale zero-valent iron supported on biochar for the highly efficient removal of nitrobenzene. *Frontiers of Environmental Science & Engineering* 4: 61. https://doi.org/10.1007/s11783-019-1142-3.

Waller and Roger, M. (1982). *Groundwater and the Rural Homeowner*. Pamphlet: U.S. Geological Survey https://www.usgs.gov/special-topic/water-science-school/science/contamination-groundwater?qt-science_center_objects=0#qt-science_center_objects.

Xi, Y., Zou, J., Luo, Y., and Lin, G. (2019). Performance and mechanism of arsenic removal in waste acid by combination of CuSO4 and zero-valent iron. *Chemical Engineering Journal* 375: 121–128.

Xiao, S., Ma, H., Shen, M. et al. (2011). Excellent copper(II) removal using zero-valent iron nanoparticle-immobilized hybrid electrospun polymer nanofibrous mats. *Colloids and Surfaces A: Physicochemical and Engineering Aspects* 381: 48–54.

Xu, L. and Huang, Y. (2019). A simple and novel method to enhance As (V) removal by zero valent iron and activated iron media through air injection at intervals. *Chemosphere* 222: 421–425.

Xu, J., Hao, Z., Xie, C. et al. (2012). Promotion effect of Fe2+ and Fe3O4 on nitrate reduction using zero-valent iron. *Desalination* 284: 9–13.

Xue, G., Wang, Q., and Qian, Y. (2019). Simultaneous removal of aniline, antimony and chromium by ZVI coupled with H_2O_2: implication for textile wastewater treatment. *Journal of Hazardous Materials* 368: 840–848. https://doi.org/10.1016/j.jhazmat.2019.02.009.

Yin, Y. (1999). Insitu Chemical Treatment. Technology evaluation report. Groundwater Remediation Technologies Analysis Centre https://clu-in.org/download/toolkit/inchem.pdf.

Yin, W., Wu, J., Wang, X. et al. (2012). Experimental study of zero-valent iron induced nitrobenzene reduction in groundwater: the effects of pH, iron dosage, oxygen and common dissolved anions. *Chemical Engineering Journal* 184: 198–204.

Yan, W., Herzing, A.A., Kiely, C.J., and Zhang, W. (2010). Nanoscale zero-valent iron (nZVI): aspects of the core-shell structure and reactions with inorganic species in water. *Journal of Contaminant Hydrology* 118: 96–104.

Yan, W., Vasic, R., and Koel, B.E. (2012). Intraparticle reduction of Arsenite (As(III)) by nanoscale zerovalent iron (nZVI) investigated with in situ X-ray absorption spectroscopy. *Environmental Science & Technology* 46: 7018–7026.

Yang, J., Cao, L., Guo, R., and Jia, J. (2010). Permeable reactive barrier of surface hydrophobic granular activated carbon coupled with elemental iron for the removal of 2,4-dichlorophenol in water. *Journal of Hazardous Materials* 184: 782–787.

Yang, Y., Xu, L., Li, W., and Yang, J. (2019). Adsorption and degradation of sulfadiazine over nanoscale zero-valent iron encapsulated in three-dimensional graphene network through oxygendriven heterogeneous Fenton-like reactions. *Applied Catalysis B: Environmental* 259: 118057. https://doi.org/10.1016/j.apcatb.2019.118057.

Yao, F., Yang, Q., Sun, J. et al. (2020). Electrochemical reduction of bromate using noble metal-free nanoscale zero-valent iron immobilized activated carbon fiber electrode. *Chemical Engineering Journal* 389: 123588. https://doi.org/10.1016/j.cej.2019.123588.

Zhang, H., Zhang, J., Zhang, C., and Zhang, D. (2009). Degradation of C.I. acid Orange 7 by the advanced Fenton process in combination with ultrasonic irradiation. *Ultrasonics Sonochemistry* 16: 325–330.

Zhang, J., Hao, J.U., Zhang, Z., and Xu, X. (2010a). Kinetics of nitrate reductive denitrification by nanoscale zero-valent iron. *Process Safety and Environmental Protection* 88: 439–445.

Zhang, X., Lin, Y., and Chen, Z. (2010b). Degradation of 2,4,6-trinitrotoluene (TNT) from explosive wastewater using nanoscale zero-valent iron. *Chemical Engineering Journal* 158: 566–570.

Zhang, J., Deng, B., Guo, J., and Lan, Y. (2011). Ligand-assisted degradation of carbon tetrachloride by microscale zero-valent iron. *Journal of Environmental Management* 92: 1328–1333.

Zhao, X., Liu, W., Cai, Z. et al. (2016). An overview of preparation and applications of stabilized zero-valent iron nanoparticles for soil and groundwater remediation. *Water Research* 100: 245–266.

Zhao, M., Zhang, C., Yang, X. et al. (2020). Preparation of highly conductive pyrogenic carbon-supported zero-valent iron for enhanced Cr(VI) reduction. *Journal of Hazardous Materials* 396: 122712. https://doi.org/10.1016/j.jhazmat.2020.122712.

Zhua, H., Wua, X., and Wang, H. (2009). Removal of arsenic from water by supported nano zero-valent iron on activated carbon. *Journal of Hazardous Materials* 172: 1591–1596.

Zoltek, J. and Melear, E.L. (1978). Physical and chemical methods. *Journal – Water Pollution Control Federation* 50 (6): 1066–1075. http://www.jstor.org/stable/25040271.

15

Various Purification Techniques of Groundwater

Dan Bahadur Pal[1], Amit Kumar Tiwari[1], and Deen Dayal Giri[2]

[1] *Department of Chemical Engineering, Birla Institute of Technology, Mesra, Ranchi, Jharkhand, India*
[2] *Department of Botany, Maharaj Singh College, Saharanpur, Uttar Pradesh, India*

15.1 Introduction

Groundwater is a crucial substance for all life existing on the earth. It is freely available for human utilization. In last few centuries after industrialization groundwater has been frequently recorded to be polluted, and in the twenty-first century consistent access to affordable clean water has become of prime importance and a big challenge for all the countries across the world (Xiaolei et al. 2013). There are different pollutants responsible for the degrading quality of groundwater. Among these, heavy metals are the most important, which seriously affect groundwater quality and pose threats to human health even when present in trace levels Singh et al. 2010). Heavy metals commonly present in chemical industries' discharge are cadmium (Cd), iron (Fe), copper (Cu), chromium (Cr), arsenic (As), and lead (Pb). These heavy metals cause very harmful effects in humans when present in groundwater (Kumar et al. 2014). For example, the presence of Cd in ground water is highly hazardous. The hazardous heavy pollutants may enter into the groundwater bodies through agricultural practices and related activities (Bhanjana et al. 2017). Minimization of such entry of undesirable metals to groundwater is a very important and ever-challenging task compared to extremely toxic heavy metals containing industrial waste. The toxic metals adversely damage terrestrial and aquatic life (Wang et al. 2010). Abundantly present toxic arsenic in the upper Earth surface is known for different health issues in several states of India like Uttar Pradesh, Assam, Jharkhand, West Bengal, and the adjoining country of Bangladesh (Siddiqui and Chaudhry 2018). The arsenic enters in organisms via drinking water, either from natural sources such as arsenic-rich rocks and soils or from anthropogenic sources. The anthropogenic sources of groundwater pollution include industry effluent of electroplating, paint, cement, and waste of metallurgy, electronics, petroleum refining, mining, ceramic and glass processing, as well as manufacturing and production of medicines, dye and pesticides, wood preservatives, and leaching of landfill (Kord et al. 2013). Application of heavy metal-containing pesticides, herbicides,

Groundwater Geochemistry: Pollution and Remediation Methods, First Edition.
Edited by Sughosh Madhav and Pardeep Singh.

and fertilizers in agriculture and forestry ultimately enters into the potable water sources (Hashemi et al. 2018). There are about 200 different water-soluble minerals of heavy metals that are toxic in their different oxidation states. The heavy metals present in groundwater is used as drinking water for a majority of the population in developing countries without any filtration, so people easily interact with the cell surface and it is more deadly than non-potable surface water. Long-term exposure of heavy metals in the drinking water at extremely low doses can significantly harm skin and vital organs like the central nervous system and liver, and may even cause cancer of different organs (Kumar et al. 2014; Gaurav et al. 2017). Heavy metal accumulations are able to demineralise bones and cause renal dysfunction. It is necessary to prevent the industrial discharge of heavy metals into the rivers, lakes, and ponds and such discharges should be monitored for metal concentration to make sure they are well below permissible limits (Hashemi et al. 2018).

The groundwater presence of the heavy metals cadmium in trace quantities is toxic, causing severe diseases of the kidney, especially to the proximal tubular cells, and bone demineralisation either directly or indirectly due to renal dysfunction (Hashemi et al. 2018). Cadmium-containing effluents therefore are required to be treated so their final concentration is well below permissible limit of less than or equal to 0.003 mg/L, before release from industries. Cadmium is present in the effluents of different chemical industries, including petrochemical complexes, metal and cement production, pigment and battery production, etc. (Hashemi et al. 2018).

There are a range of different methodologies for heavy metal removal from groundwater including membrane separation, filtration, reverse osmosis, flocculation, coagulation, electro-coagulation, precipitation, and adsorption (Siddiqui and Chaudhry 2018). Among these, desorption is one of the most common and commercially adopted techniques mainly due to efficiency, simple procedure, and low operational cost (Gaurav et al. 2017; Pal et al. 2018). The available industrial adsorbents and activated carbons are attractive options because of their unique pores along with great chemical and mechanical resistance (Cecen and Aktas 2011).

The present chapter briefly describes various sources of heavy metals polluting groundwater and the availability of techniques to purify the heavy metal-polluted groundwater.

15.2 Sources of Heavy Metals in Groundwater Pollution

Heavy metals of pedogenic or lithogenic origin are relatively less mobile compared to heavy metals in soil due to anthropogenic sources, and hence anthropogenic heavy metals of the soils are more bioavailable (Kaasalainen and Yli-Halla 2003). The solid matter containing heavy metals at contaminated sites is originated from a wide variety of anthropogenic sources. The main sources are metal mine tailings, coal combustion residues, fertilizers application in agriculture fields, compost, animal manures, biosolids, pesticides, improperly disposed metal wastes in landfills, petrochemicals, leaded gasoline and lead-based paints, and atmospheric deposition (Zhang et al. 2010). Some of the major heavy metal pollution sources are discussed in the following sections.

15.2.1 Agriculturally Important Chemicals

Agriculture activities can be considered the first main human influence on groundwater (Scragg 2006), because plants need both macro- and micronutrients for their growth, development, and yield. But many agriculture fields are deficient in important micronutrient heavy metals such as Co, Cu, Fe, Mn, Mo, Ni, and Zn, which are important for healthy plant growth (Letham et al. 2005). During agricultural practice, these elements are supplied to crops in soil or as a foliar spray. There are many pesticides which are traditionally used in agriculture and horticulture that contain substantial amount of heavy metals. About 10% of the approved chemicals for their use as insecticides and fungicides in different countries contain compounds of Cu, Hg, Mn, Pb, and Zn. For example, bordeaux mixture and other copper-based pesticides and fungicidal sprays contain compounds like copper sulphate and copper oxychloride (Jones 1983).

15.2.2 Biosolids

The application of different biosolids like livestock manures, composts, and municipal sewage sludge to land unintentionally accumulate various heavy metals like As, Zn, Cd, Hg, Cr, Cu, Pb, Ni, Se, Ti, Mo, Sb, etc. in the soil (Basta et al. 2005). Manures derived from poultry, cattle, and pig are generally applied in agriculture crops and pastures in the form of solids or slurries and considered as precious fertilizers (Sumner 2000). However, in the food of pig and poultry Cu and Zn is added as dietary supplements for development. The arsenic content of poultry manure has the potential to cause metal contamination of the soil when applied continuously and not consciously (Sumner 2000).

15.2.3 Community and Manufacturing Waste

Wastewater generated from the community and the manufacturing sector have generally been used in land for several decades across the world (Reed et al. 1995). According to an estimate the total area irrigated by such wastewater is about 20 million hectares. In the majority of Asian and African cities, about half of the total vegetables supplied to urban areas are irrigated by wastewater (Bjuhr 2007).

15.2.4 Mining and Industrial Release

In many countries, mining, grinding, and extraction of metal from ores release a wide variety of metal pollutants that contaminate groundwater. Generally mine tailings containing high concentrations of metal are directly released into adjoining natural low lands or nearby wetlands (DeVolder et al. 2003). Mining and smelting of Pb and Zn ores have degraded the quality of groundwater at level posing risk to soil, aquatic ecosystems, and human health at several localities. The reclamation methods applied to remediate these sites are quite lengthy, expensive, and sometimes it may even not even able to restore soil productivity.

15.2.5 Airborne Sources of Metal

The main airborne sources of metals are stack or duct emissions of air, gas, vapour streams, and fugitive discharge from storage areas or piles of waste. The particulate matter in these gaseous sources contains metals because at high temperature processing, As, Cd, and Pb are volatilized and converted to their oxides, which mix and get attached to the fine particulate matter (Smith et al. 1995).

15.3 Heavy Metal Effects on Environment and Human Health

Human exposed to high level of heavy metals are common with rapid urbanization and large-scale industrialization across the world (Asaduzzaman et al. 2017). Our dependence on electronic equipment and discarding of obsolete electronics in improper ways has become responsible for health effects in many instances. Harmful chemicals are released into the environment during manual dismantling, burning to recover precious heavy metals, and dumping of residues. The important heavy metals in electronics are arsenic (As), aluminium (Al), copper (Cu), chromium (Cr), cadmium (Cd), lead (Pb), zinc (Zn), manganese (Mn), nickel (Ni), mercury (Hg), and cobalt (Co). All the heavy metals have their own chemical composition, reaction properties, distribution, metabolism, excretion, and transmission (Sankhla et al. 2016). Heavy metals influence a number of diverse systems and organs, resulting in both acute and chronic effects on children's health, ranging from minor upper respiratory irritation to chronic respiratory, cardiovascular, nervous, urinary, and reproductive disease.

15.3.1 Lead

This metal enters into the ecosystem both naturally as well as from emission sources. Atmospheric emission of vehicular exhaust, lead paint chips, pesticides, fertilizers, acid-lead batteries, and burning of printed circuit boards are some common emission sources from which lead is transported via air. The main effects of lead on the human body are brain disorders, kidney damage, hypertension, anaemia, and reproductive system effects (Kakker and Jaffery 2004). Lead is a human carcinogen, as per the EPA (Environmental Protection Agency). Acute high levels of exposure are one of the causes of miscarriage in pregnant women, whereas in men it can damage testicular spermatogenesis. Lead affects the process of erythropoietin and harem biosynthesis. Due to permeable blood brain barrier and higher gastrointestinal uptake, kids are more susceptible to lead compared to adults. In young children it alters metabolites of vitamin D, causing behavioural disturbances, slow growth, and learning and concentration difficulties.

15.3.2 Mercury

The sources of mercury in the environment include industrial smoke, mining, and burning of fossil fuel, as well as e-waste burning. Mercury can have adverse effects on aquatic and terrestrial life, including humans, through the contaminated air, soil, water, and food chain. Build-up of mercury in different environmental components at especially higher trophic level is called biomagnification. The acute exposure of mercury vapours may cause 'pink' disease or acrodynia. It is also responsible for minimata disease. High levels of exposure to mercury can permanently damage the kidneys, brain, and developing foetus. The nervous system is very sensitive to all forms of mercury. As per the EPA, mercuric chloride and methyl mercury are possible human carcinogens. Short-term exposure to mercury leads to lung damage, increase in blood pressure, eye irritation, and skin rashes. In adults it causes Parkinson neuropathy, allergic dermatitis, respiratory failure, memory impairment, tunnel vision, proteinuria renal syndrome, acute renal failure, deafness, tremor and ataxia, headache, impaired sexual function, fatigue, and depression (Kakker and Jaffery 2004). Mercury could be responsible for late walking and talking, reduced neurological development, autism, and intellectual disorders in children. It has been proven through many studies that it binds to the DNA (deoxyribonucleic acid), leading to strand break *in vitro*, without causing point mutations and producing negative effects on sperm cells.

15.3.3 Cadmium

Cadmium and its compounds are highly toxic in nature. Industrial sources and electronic wastes are important sources of cadmium pollution in the environment. This heavy metal occurs in almost all soils, surface waters, and in plants. It is considered a potential threat to wildlife species when not properly recycled as it may leach out into the soil, harming soil micro flora (Sharma et al. 2012). Cadmium and its compounds are known human carcinogens, as per the EPA (Environmental Protection Agency). In long-term exposure it causes kidney disorders, lung damage, and fragile bones. In higher plants and animals, cadmium causes decreased fidelity of DNA synthesis, gene mutations, and chromosomal abnormalities. Higher blood cadmium is associated with more coughing and wheezing (Zeng et al. 2016).

The mechanisms of cadmium carcinogenesis are complex. It may modulate gene expression by altering signal transduction, interfere with enzymes of the cellular antioxidation and generation of reactive oxygen species, and may promote DNA damage by inhibiting DNA repair enzymes. According to Zhang et al. (2017), cadmium exposure induces the expression of proteins involved in transcription- and translation-related processes. Lethal levels of cadmium trigger distinct protein expression patterns in earthworms (Prasad et al. 2020). It may affect apoptosis induction in cells or play a role in the disruption of cadherin-mediated cell adhesion (Malaguarnera 2012). High-level exposure of cadmium in humans can cause them to develop itai-itai disease (Aoshima 2012).

15.3.4 Barium

Aerial concentration of barium is increasing, mainly due to industrial emissions. There is not any established role of barium in human nutrition. The short-term exposure of the element causes abdominal cramps, diarrhoea, vomiting, difficulty breathing, and fluctuation in blood pressure. Large doses of barium result in altered heart rhythm, high blood pressure, paralysis, and even death. Vasoconstriction by direct stimulation of arterial muscle is the basic reason for high blood pressure.

15.3.5 Arsenic

Arsenic is naturally present in trace quantities in soil, water, and air. The anthropogenic sources of arsenic are industrial smoke emission and burning of waste electronic equipment. The other sources are volcanoes, microorganisms, and pesticides. Arsenic is able to alter the genetic material of fish residing in arsenic-polluted water. Arsenic is toxic to plants and animals and is even carcinogenic to humans (Pal et al. 2017). The toxic effects on humans include skin lesions to cancer of the brain, kidney, liver, and stomach (Murcott 2012). According to the EPA, short-term exposure may lead to nausea and vomiting, damage to blood vessels, diminished RBC (red blood cell) and WBC (white blood cell) production, abnormal heartbeat, and a sensation of pinching pins and needles in hands and feet, whereas skin pigmentation and appearance of small corns and warts on palms develop in long-term exposure. Arsenic is able to induces chromosomal aberrations, micronuclei formation, aneuploidy, and sister-chromatid exchanges. The effects of inorganic arsenic are more pronounced compared to the organic form in organisms including humans (Ng 2005).

A brief description of the effects of heavy metals and their compound pollutants on water quality and human health is presented in Table 15.1.

Table 15.1 Chemicals and heavy metal pollutants in groundwater.

Metal	Resources	Effects on human health and others
Al	Naturally present in rocks and mine drainage.	Decreases transparency of water due to its precipitation during leading to turbid or discoloured water.
As	Natural processes as well as industrial waste of pesticide industries, smelting of Cu, Pb, and Zn ore, e-waste burning.	Acute and chronic toxicity skin, liver, and kidney, cancer, decreases blood haemoglobin.
Cd	Trace amount is present in rocks, coal, and petroleum. In acidic water, dissolves and percolates to the groundwater. Its atmospheric sources are manufacturing and mining, waste, landfill leachate, metal plating, water pipes, batteries, paints and pigments, plastic stabilizers.	It can replace zinc in the body to alter physiology. It may cause high BP (blood pressure), liver, bone softening, kidney damage. It can damage testicular tissue leading to impotency and red blood cells to anaemia. It is toxic to aquatic life. A carcinogen. Long-term exposure causes itai-itai disease, having severe pain in the joints and spine.
Cr	It enters into the atmosphere from fossil-fuel burning and waste incineration, old mining runoff and mineral leaching to groundwater, cement-plant discharge	Cr^{+6} is toxic whereas Cr^{+3} is a nutritionally important element. Cr^{+6} causes liver and kidney damage, internal haemorrhaging, respiratory damage, dermatitis, and ulcers on the skin at high concentration exposure. Inhaling Cr^{+6} can cause asthmatic bronchitis and lung cancer.
Cu	Metal plating, commercial and domestic waste, mining, and mineral leaching.	It causes stomach and intestinal distress, liver and kidney damage, anaemia in high doses.
F	Naturally present in water or as an additive to community water supplies	It is used to prevent tooth decay in children but high levels can stain or mottle teeth, and cause calcification of the bones and joints.
Fe	Natural mineral of sediment and rocks, mining, commercial waste, corrosion.	Impart an acidic astringent taste to water, brownish colour to laundered clothing

(*Continued*)

Table 15.1 (Continued)

Metal	Resources	Effects on human health and others
Pb	Manufacturing, mining, plumbing, gasoline, coal, water preservative.	Affects red blood cells, impediment of normal physical and mental development of children. Neurotoxic affecting kidneys and the reproductive system.
Hg	Manufacturing waste, mining, smelting, fossil-fuel combustion, pesticides, coal, batteries, lamps, and switches.	Acute and chronic toxicity lead to kidneys and nervous system disorders. Hg is methylated in water bodies by microbial activity. This toxic metal enters the human body through the food chain to affects the CNS (central nervous system), kidneys, and immune system in long-term exposure. It impairs foetus growth and harms infants of exposed mother.
Ni	Naturally occurs in soils and groundwater. Used in electroplating, stainless steel and alloy products, mining, and refining.	In laboratory animals it has been shown to damage the heart and liver when exposed to large amounts for long period.
Ag	Ore mining and processing, product manufacturing, and discarding. Used in photography, electric and electronic equipment, electroplating, alloy, solder.	Chronic exposure may cause blue-grey coloration of the skin, mucous membranes, eyes, and organs.
Zn	Occurs naturally in water of the area of its mining. It is part of industrial waste, metal plating, plumbing, and sludge.	Helps in wound healing and no ill health effects until exposure to very high doses. Toxic to plants at high levels. Imparts undesirable taste to water.

15.4 Methods for Purification of Groundwater

There are many techniques which can be used to purify the heavy metal polluted groundwater. These techniques include chemical precipitation, ion exchange, membrane filtration, coagulation and flocculation, electrochemical treatment, and adsorption. The different techniques are described here briefly and only the process of adsorption is detailed here.

15.4.1 Chemical Precipitation

This is the most extensively used method for effective groundwater treatment in industries due to its ease and cheap operation (Jalil and Alsewailem 2009). The conventional chemical precipitation techniques are done by hydroxide and sulphide precipitation. Constraint of hydroxide precipitation generates huge volumes of small density sludge and the presence of composite agents inhibit metal hydroxide precipitation (Gu and Evans 2008). Colloidal precipitates formed during sulphide precipitation pose difficulty in separation due to settling or filtration issues. Further, under acidic conditions, sulphide precipitants generate toxic hydrogen sulphide fumes.

15.4.2 Ion Exchange

Groundwater polluted with heavy metal can be treated using ion-exchange procedures because of their fast kinetics with high treatment capability and high removal efficiency (Liu et al. 2008). Both synthetic and natural resins are used in the ion-exchange process. The resins are selected based on their ability to exchange cations of the metals present in the groundwater. Once capability of the resin is achieved, the resin is rejuvenated. The performance and economics of ion exchange resin are highly related to the capability of resin. The heavy metals are generally found in the form of HM^{2+} ions in the solution regardless of its complexes with chlorine, which makes its separation complex with metal ion exchange resin. Resin exchange resembles the sorption process, as in both cases a dissolved metal species is taken up by a solid; however, resin ion exchange is a stoichiometric technique, while sorption is a non-stoichiometric technique. Each ion that has been separated from the solution is replaced by an equivalent quantity of other ionic species of the same sign from the ion exchange material. In sorption, the metal ion is taken up without being replaced by an equivalent species. Though this difference is clear, it is difficult to describe between exchange and sorption, since each ion exchange method is accompanied by electrolytic sorption or desorption on the common sorbents such as alumina and activated carbon acts as ion exchanger.

15.4.3 Membrane Filtration

The membrane filtration procedure is an effective and efficient environmentally friendly technology for water treatment without phase-transition procedure in heavy metal removal that could be operated in small spaces and can be easily scaled up. The most common membrane methods to remove metals from the groundwater are ultrafiltration (Amana et al. 2008), reverse osmosis (Kaczala et al. 2009), nanofiltration (Saeed et al. 2005), and electro-dialysis (Jai et al. 2007). Some processes of membrane filtrations are described here briefly.

15.4.3.1 Ultrafiltration

In ultrafiltration membrane separation technology, an ultrafiltration membrane is used, which is highly efficient in removing macromolecular organic compounds at relatively low operation pressure and energy consumption compared to nanofiltration and reverse osmosis membranes (Valipour et al. 2016). The most common problem associated with ultrafiltration is membrane fouling. Water flux and separation efficiency of membrane decreased due to the adsorption and deposition of pollutants in the membrane pores or on the membrane surface (Geng et al. 2017).

15.4.3.2 Reverse Osmosis

The process of reverse osmosis separation is suitable for a wide range of applications involving removal of salt and/or dissolved solids from a solution. Thus, the reverse osmosis process may be used for saline and brackish water desalination for producing good quality water for drinking as well as commercial applications. The process could be applied for the production of ultrapure water used for boiler feed and other industrial purposes.

The RO (reverse osmosis) membrane systems are useful for groundwater purification for drinking and for the purpose of water reuse. This technique is utilized for removing huge amount of pollutants from groundwater by pushing the water under pressure through a semipermeable membrane. A semipermeable membrane permits only a few atoms or molecules to pass through it. Air molecules are passed through it but microbes and pests bigger than pores are screened at the door. The reverse osmosis technique of osmosis takes place in reverse. The osmosis takes place naturally across the semipermeable membrane spontaneously without energy input. The water moves from its higher potential to lower potential, i.e. from pure water to polluted water having high concentrations of different ions or molecules of pollutant. The reverse osmosis pure water is moved apart from the membrane from polluted water solution by applying energy to the saltier solution. The semipermeable membrane permits the passage of water molecules but not others of dissolved salts, organics, bacteria, and pyroxene. The reverse osmosis process demineralises groundwater and produces clear water by holding back major pollutants.

15.4.3.3 Nanofiltration

The main advantage of membrane separation techniques is its operation without heating, so its energy consumption is usually lower compared to other conventional thermal separation techniques. The membrane filtration process is associated with drawbacks such as high power consumption for pumping pressures, the restoration of the membranes, the need for pre-treatment, and limited retention of salts and univalent ions. The nanofiltration membranes are expensive and sensitive to free chlorine and most importantly membrane fouling.

15.4.4 Coagulation and Flocculation

Occurrence of coagulation and flocculation takes place in succession by overcoming the forces stabilizing the suspended particles and allowing their collision for growth of floc. This process has been used to purify the water. Amudaa et al. (2009) were able to remove tungsten from groundwater employing precipitation, coagulation, and flocculation using ferric chloride. The process requires only tanks for performing coagulation and flocculation, but a major disadvantage is the high operational cost compared to other techniques. The requirement of considerable quantities of coagulant and flocculants to achieve desired level of flocculation is its drawback. The process produces certain quantities of physicochemical sludge requiring external processing. Such processing costs can escalate in the case of treatment of large volumes of groundwater. The coagulation technique is under the inhibitory control of some inhibitors that limit clot formation to a permissible level, thus avoiding the thrombus propagation. This delicate balance is interrupted whenever the procoagulant activity of the coagulation factors is increased, or the activity of naturally occurring inhibitors is decreased (Previtali et al. 2011).

15.4.5 Electrochemical Treatment

The electrochemical technologies have gained importance globally in the past three decades due to the stringent ecological regulations in reference to the wastewater discharge. In the electrochemical treatment of groundwater electricity is passed through electrodes, and groundwater acts as electrolyte in an electrochemical reactor. The most emphasized techniques are electrocoagulation, electroflotation, and electrodeposition. Oliveira et al. (2008) reported clarification of groundwater by electroflotation using aluminium electrodes. Electrochemical heavy metal removal from groundwater is rapid and well-controlled, requiring only a few chemicals, with good reduction yields and less sludge production. The technique needs high initial capital investment and continuous, expensive electricity supply to operate, which is the major restriction of the technique.

15.4.6 Adsorption

The adsorption process is well recognized for economic effective removal of heavy metal from polluted water. The adsorption process can be operated with the ease of flexible design and operation and generates high-quality metal pollution-free effluent. The adsorbent could be reversibly regenerated by suitable desorption procedure, in some cases. The different adsorption parameters like efficiency, selectivity, equilibrium time, regeneration, and stability usually depend on the characteristics of the material used to prepare adsorbents. Heavy metals present in groundwater can be adsorbed on solid matrices of activated carbon and metal oxides, as well as on the variety of waste agricultural products. Better adsorbent material can be selected depending upon the nature and type of metal pollution present in the groundwater. Study of adsorption isotherms and kinetic parameters is required before selecting the adsorbent. The different adsorbent adsorbs the metals following different kinetic parameters are discussed below.

15.4.6.1 Adsorption Isotherm

There are three classical models, namely Langmuir model, Freundlich model, and Temkin model, which are generally fit to explain the adsorption isotherm of adsorbent material.

15.4.6.1.1 Langmuir Model The Langmuir isotherm model is expressed in the form of this equation (Langmuir 1918):

$$q_e = \frac{q_m b_0 C_e}{1 + b_0 C_e} \tag{15.1}$$

where q_m denotes the maximum adsorption capacity of the adsorbent, i.e. how many miligrams of metals are adsorbed on one gramme of adsorbent, b_0 is called Langmuir isotherm constant (L/mg) and Ce represents equilibrium concentration of metal in aqueous solution (mg/L). The linear form of the Langmuir isotherm is as follows:

$$\frac{C_e}{q_e} = \frac{1}{q_m b_0} + \frac{tC_e}{q_m} \tag{15.2}$$

This equation is used to calculate the value of Langmuir parameters q_m and b_0 from the slope and intercept of the linear plot drawn between Ce/qe and Ce.

15.4.6.1.2 Freundlich Model As per the Freundlich model, sorption energy exponentially decreases upon the completion of the sorption centres of the adsorbent (Freundlich 1906). This isotherm is represented as an empirical equation:

$$q_e = K_f C_e^{1/n} \tag{15.3}$$

where Freundlich constant (K_f) represents the bonding energy (mg/g) (L/g) and n denotes heterogeneity factor. Freundlich equilibrium constants could be determined from the linear plot between $\ln q_e$ and $\ln C_e$ using the following equation:

$$\ln q_e = \ln K_f + \frac{1}{n} \ln C_e \tag{15.4}$$

Freundlich isotherm is applied widely in heterogeneous systems, especially organic compounds and highly interactive species on activated carbon and molecular sieves. The slope value 1/n varies between 0 and 1 and represents measure of adsorption intensity or surface heterogeneity. A slope above the value of one indicates a co-operative adsorption (Foo and Hameed 2010).

15.4.6.1.3 Temkin Model The Temkin model generalized and linearized forms of the model are given by the following equations (Temkin and Pyzhev 1940):

$$q_e = \frac{RT}{b_T} \ln K_T C_e \tag{15.5}$$

$$q_e = \frac{RT}{b_T} \ln K_T + \frac{RT}{b_T} \ln C_e \tag{15.6}$$

where K_T is Temkin isotherm constant (L/g), R represents the ideal gas constant (8.314 J/mol K), T denotes temperature (K), and b_T is a constant related to the heat of sorption (J/mol). A plot between q_e and ln C_e gives a straight line from which the values of K_T and b_T are evaluated from the slope and intercept.

15.4.7 Adsorption Kinetics Analysis

The study of reaction kinetics is necessary for the design of sorption systems and reaction rate controlling step, as chemical reaction occurs. The sorption depends on physiochemical characteristic of the adsorbent and system environmental conditions like temperature (Nadeem et al. 2006). The amount of metals adsorbed q_t at time t is calculated from the equation:

$$q_A = \frac{(C_{0-} C_e)V}{W} \tag{15.7}$$

where qt: sorption capacity at time t (mg/g), C_0 is the initial metal ion concentration (mg/L), C_e is the metal ion concentration at equilibrium (mg/L), V is the volume of solution (L), W is the weight of the adsorbent (g), and q_{As} ion is the adsorption capacity at equilibrium (mg/g) (Liu 2010).

Pseudo-first-order, pseudo-second-order, intra-particle, and Elovich models were used to fit the adsorption data, and four kinetic fitting curves were obtained.

The pseudo-first order model of Lagrange is generally expressed as follows (Farhan et al. 2012):

$$\frac{dq_t}{dt} = K\left(q_e - q_t\right) \tag{15.8}$$

After integration and applying the boundary conditions qt = 0 to qt = qt at T = 0 to t = t; the integrating form of the equation becomes:

$$\log\left(q_e - q_t\right) = \log q_e + \frac{k_1}{2.303}t \tag{15.9}$$

A plot log (qe-qt) versus t gives a straight line from which k_1 and q_e can be evaluated from the slope and intercept. Its analytical equation can be expressed as:

$$q_t = q_e\left[1 - \left(\exp k_1 t\right)\right] \tag{15.10}$$

The pseudo-second order model is based on sorption capacity of the solid phases, which has been applied for analysing chemisorption kinetics rate. It is given by:

$$\frac{dq_t}{dt} = k_2\left(q_e - q_t\right)^2 \tag{15.11}$$

The boundary conditions $q_t = 0$ and $q_t = q_t$ at T = 0 to t = t, respectively. The integrated form of equation becomes:

$$\frac{t}{q_t} = \frac{1}{q_e^2 k_2} + \frac{t}{q_e} \tag{15.12}$$

A plot t/qt versus t gives a straight line from which k_2 and qe can be evaluated:

$$q_t = \frac{q_e^2 k_2 t}{1 + q_e k_2 t} \tag{15.13}$$

The intra-particle diffusion equation (Furusawa and Smith 1974) can be written as:

$$q_t = t^{0.5} k_{ipd} + C \tag{15.14}$$

where q_t is the amount of dye adsorbed onto the adsorbent at time *t* (mg/g), *C* is the intercept, and K_{ipd} is the intra-particle diffusion rate constant (mg/g/min).

Elovich model equation (Zhang et al. 2019):

$$q_t = \frac{1}{2}\log\left(\alpha^2\right) + \frac{1}{2}\log t \tag{15.15}$$

where α (mg/g/min) and β (mg/g) are the rate constants related to the adsorption and desorption processes, respectively, and the values can be obtained from the plot of qt versus lnt.

15.5 Summary

Heavy metals released from various sources enter the groundwater. The frequently encountered metals in groundwater are arsenic, cadmium, mercury, lead, chromium, etc. These metals significantly affect different components of aquatic and terrestrial ecosystems, including humans, when present in the groundwater. Polluted precious groundwater could be restored to their use in industries, irrigation, and consumable drinking water standards by the various technologies available, however it is very costly in terms of work force and energy consumption. The restoration of groundwater quality is of prime importance for life on the planet. So, the major emphasis should be to prevent the pollution sources to reduce the groundwater contamination by anthropogenic means. Continuous monitoring of water quality parameters of groundwater should be monitored in heavy metal contamination-prone areas to prevent toxic hazards to human.

Acknowledgement

The authors acknowledge Birla Institute of technology, Mesra, Ranchi, Jharkhand, and Indian Institute of Technology (BHU), Varanasi, for characterization and raw materials, respectively, and NPIU (TEQIP-III), Govt. of India for the financial support.

Financial and Ethical Disclosures

There are no conflicts of interest between authors. There is no involvement of human or animal cell in this work. All the co-authors have seen the final manuscript and agreed with the submission. The manuscript has not been published elsewhere. The article is not consideration under any other journal in full or in part. No data or figures have been fabricated or manipulated. The funding agencies (NPIU/AICTE New Delhi, India) have been duly acknowledged. All the authors agree to transfer copyright to book publishers. All the further responsibilities will be undertaken by the corresponding author.

References

Amana, T., Kazi, A.A., Usman Sabri, M., and Banoa, Q. (2008). Potato peels as solid waste for the removal of heavy metal copper (II) from waste water/industrial effluent. *Colloids and Surfaces* 63: 116–121.

Amudaa, O.S., Adelowoa, F.E., and Ologunde, M.O. (2009). Kinetics and equilibrium studies of adsorption of chromium (VI) ion from industrial wastewater using Chrysophyllum albidum (Sapotaceae) seed shells. *Colloids and Surfaces* 68: 184–192.

Aoshima, K. (2012). Itai-Itai disease: cadmium induced renal tubular osteomalacial. *Nihon Eiseigaku Zasshi* 67 (4): 455–463.

Asaduzzaman, K., Khandaker, M.U., Binti Baharudin, N.A. et al. (2017). Heavy metals in human teeth dentine: a bio-indicator of metals exposure and environmental pollution. *Chemosphere* 176: 221–230.

Basta, N.T., Ryan, J.A., and Chaney, R.L. (2005). Trace element chemistry in residual-treated soil: key concepts and metal bioavailability. *Journal of Environmental Quality* 34 (1): 49–63.

Bhanjana, G., Dilbaghi, N., Singhal, N.K. et al. (2017). Copper oxide nanoblades as novel adsorbent material for cadmium removal. *Ceramics International* 43 (8): 6075–6081.

Bjuhr, J. (2007). Trace Metals in Soils Irrigated with Waste Water in a Periurban Area Downstream Hanoi City, Vietnam, Seminar Paper, Institutionen f¨or markvetenskap, Sveriges lantbruksuniversitet (SLU), Uppsala, Sweden.

Cecen, F. and Aktas, O. (2011). *Activated Carbon for Water and Wastewater Treatment: Integration of Adsorption and Biological Treatment*. Weinheim: Wiley.

DeVolder, P.S., Brown, S.L., Hesterberg, D., and Pandya, K. (2003). Metal bioavailability and speciation in a wetland tailings repository amended with biosolids compost, wood ash, and sulfate. *Journal of Environmental Quality* 32 (3): 851–864.

Farhan, A.M., Salem, N.M., Dujaili, A.H., and Awwad, A.M. (2012). Biosorption of Cr(VI) ions from electroplating wastewater by walnut Shell powder. *American Journal of Environmental Engineering* 2 (6): 188–195.

Foo, K.Y. and Hameed, B.H. (2010). Insights into the modeling of adsorption isotherm systems. *Chemical Engineering Journal* 156 (1): 2–10.

Freundlich, H.M.F. (1906). Über die adsorption in losungen. *Zeitschrift für Physikalische Chemie* 57: 385–470.

Furusawa, T. and Smith, J.M. (1974). Intraparticle mass transport in slurries by dynamic adsorption studies. *AIChE Journal* 20 (1): 88–93.

Gaurav, N., Sivasankari, S., Kiran, G.S. et al. (2017). Utilization of bioresources for sustainable biofuels: a review. *Renewable and Sustainable Energy Reviews* 73: 205–214.

Geng, Z., Yang, X., Boo, C. et al. (2017). Self-cleaning anti-fouling hybrid ultrafiltration membranes via side chain grafting of poly(aryl ether sulfone) and titanium dioxide. *J. Membr. Sci.* 529: 1–10.

Gu, X.Y. and Evans, L.J. (2008). Surface complexation modelling of Cd(II), Cu(II), Ni(II), Pb(II) and Zn(II) adsorption onto kaolinite. *Geochimica Cosmochimica Acta* 72: 267–276.

Hashemi, Z., Sharfiifard, H., and Lashanizadegan, A. (2018). Grape stalks biomass as raw material for activated carbon production: synthesis, characterization and adsorption ability. *Materials Research Express* 5: 660.

Jai, P.H., Wook, J.S., Kyu, Y.J. et al. (2007). Removal of heavy metals using waste eggshell. *Journal of Environmental Sciences* 19: 1436–1441.

Jalil, S.A. and Alsewailem, F.D. (2009). Saudi Arabian clays for lead removal in wastewater. *Applied Clay Science* 42: 671–674.

Jones, H.G. (1983). Estimation of an effective soil water potential at the root surface of transpiring plants. *Plant, Cell & Environment* 6 (8): 671–674.

Kaasalainen, M. and Yli-Halla, M. (2003). Use of sequential extraction to assess metal partitioning in soils. *Environmental Pollution* 126 (2): 225–233.

Kaczala, F., Marques, M., and Hogland, W. (2009). Lead and vanadium removal from a real industrial wastewater by gravitational settling/sedimentation and sorption onto Pinus sylvestris sawdust. *Bioresource Technology* 100: 235–243.

Kakker, P. and Jaffery, F.N. (2004). Biological markers for metal toxicity. *Environmental Toxicology and Pharmacology* 19: 335–349.

Kord, M.F., Bazrafshan, E., Farzadkia, M., and Amini, S. (2013). Arsenic removal from aqueous solutions by Salvadora Persica stem ash. *Journal of Chemistry* 2013: 740847.

Kumar, S., Bhanjana, G., Jangra, K. et al. (2014). Utilization of carbon nanotubes for the removal of rhodamine B dye from aqueous solutions. *Journal of Nanoscience and Nanotechnology* 14 (6): 4331–4336.

Langmuir, I. (1918). The adsorption of gases on plane surface of glass, mica and platinum. *Journal of the American Chemical Society* 40 (9): 1361–1403.

Letham, D.L.D., Pence, N.S., Lasat, M.M., and Kochian, L.V. (2005). Molecular and physiological investigations of *Thlaspi caerulescens L. Roots and Soil Management: Interactions between Roots and the Soil* 48: 95–106.

Liu, B. (2010). Uncertain risk analysis and uncertain reliability analysis. *Journal of Uncertain Systems* 4 (3): 163–170.

Liu, Z.R., Zhou, L.M., Wei, P. et al. (2008). Competitive adsorption of heavy metal ions on peat. *Journal of China University Mining and Technology* 18: 255–260.

Malaguarnera, M. (2012). Cadmium and cancer risk. ICS 2012, Cadmium Symposium Sassari, Italy, Abstract book: pp, 39.

Murcott, S. (2012). *Arsenic contamination in the world.* IWA publishing.

Nadeem, M., Mahmood, A., Shahid, S.A. et al. (2006). Sorption of lead from aqueous solution by chemically modified carbon adsorbents. *Journal of Hazardous Materials* 138 (3): 604–613.

Ng, J.C. (2005). Environment contamination of arsenic and its toxicological impact on humans. *Environmental Chemistry* 2 (3): 146–160.

Oliveira, L., Franca, A.S., Alves, T.M., and Rocha, S.D.F. (2008). Evaluation of untreated coffee husks as potential biosorbents for treatment of dye contaminated waters. *Journal of Hazardous Materials* 155: 507–512.

Pal, D.B., Giri, D.D., Singh, P. et al. (2017). Arsenic removal from synthetic waste water by CuO nano-flakes synthesized by aqueous precipitation method. *Desalination & Wastewater Treatment* 62: 355–359.

Pal, D.B., Lavania, R., Srivastava, P. et al. (2018). Photo-catalytic degradation of methyl tertiary butyl ether from wastewater using CuO/CeO_2 composite nanofiber catalyst. *Journal of Environmental Chemical Engineering* 6: 2577–2587.

Prasad, N., Thakur, P., and Pal, D.B. (2020). Cadmium removal from aqueous solution by jackfruit seed bio-adsorbent. *Springer Nature Applied Sciences* 2: 1018.

Previtali, E., Bucciarelli, P., Passamonti, S.M., and Martinelli, I. (2011). Risk factors for venous and arterial thrombosis. *Blood Transfusion* 9: 120–138.

Reed, S.C., Crites, R.W., and Middlebrooks, E.J. (1995). *Natural Systems for Waste Management and Treatment*, 2e. New York, NY: McGraw-Hill.

Saeed, A., Iqbal, M., and Akhtar, M.W. (2005). Removal and recovery of lead (II) from single and multimetal (Cd, Cu, Ni, Zn) solutions by crop milling waste (black gram husk). *Journal of Hazardous Materials* 117: 65–73.

Sankhla, M.S., Kumari, M., Nandan, M. et al. (2016). Effect of electronic waste on environmental and human health – a review. *IOSR Journal of Environmental Science Toxicology and Food Technology* 10 (9): 98–104.

Scragg, A. (2006). *Environmental Biotechnology*, 2e. Oxford, UK: Oxford University Press.

Singh, A., Sharma, R.K., Agrawal, M., and Marshall, F.M. (2010). Health risk assessment of heavy metals via dietary intake of foodstuffs from the wastewater irrigated site of a dry tropical area of India. *Food and Chemical Toxicology* 48: 611–619.

Sharma, P., Fulekar, M.H., and Bhagwan, P. (2012). E-waste a challenge for tomorrow. *Research Journal of Recent Sciences* 1 (3): 86–93.

Siddiqui, S.I. and Chaudhry, S.A. (2018). A review on graphene oxide and its composites preparation and their use for the removal of As(III) and As (V) from water under the effect of various parameters: application of isotherm, kinetic and thermodynamics. *Process Safety and Environmental Protection* 119: 138–163.

Smith, L.A., Means, J.L., Chen, A. et al. (1995). *Remedial Options for Metals-Contaminated Sites.* Boca Raton, FL: Lewis Publishers.

Sumner, M.E. (2000). Beneficial use of effluents, wastes, and biosolids. *Communications in Soil Science and Plant Analysis* 31 (11–14): 1701–1715.

Temkin, M.I. and Pyzhev, V. (1940). Kinetics of ammonia synthesis on promoted iron catalyst. *Acta Physicochimica USSR* 12: 327–356.

Valipour, A., Ayyaru, S., and Ahn, Y. (2016). Application of graphene-based nanomaterials as novel cathode catalysts for improving power generation in single chamber microbial fuel cells. *Journal of Power Sources* 327: 548–556.

Wang, Y.Q., Wang, G.Z., Wang, H.Q. et al. (2010). Chemical template synthesis of micro-nanoscale magnesium silicate hollow spheres for wastewater treatment. *Chem. A Eur. J.* 16: 3497–3503.

Xiaolei, Q., Alvarez, P.J.J., and Li, Q. (2013). Applications of nanotechnology in water and wastewater treatment. *Water Research* 47: 3931–3946.

Zeng, X., Xijin, X., Xiangbin, Z. et al. (2016). Heavy metals in PM2.5 and in blood, and children's respiratory symptoms and asthma from an e-waste recycling area. *Environmental Pollution* 210: 346–353.

Zhang, M.K., Liu, Z.Y., and Wang, H. (2010). Use of single extraction methods to predict bioavailability of heavy metals in polluted soils to rice. *Communications in Soil Science and Plant Analysis* 41 (7): 820–831.

Zhang, L., Duan, X., He, N. et al. (2017). Exposure to lethal levels of benzo[a]pyrene or cadmium trigger distinct protein expression patterns in earthworms (Eisenia fetida). *Science of the Total Environment* 595: 733–742.

Zhang, H., Omer, A.M., Hu, Z. et al. (2019). Fabrication of magnetic bentonite/carboxymethyl chitosan/sodium alginate hydrogel beads for Cu (II) adsorption. *International Journal of Biological Macromolecules* 135: 490–500.

16

Various Remediation Measures for Groundwater Contamination

Ankita Ojha[1,2] *and Dhanesh Tiwary*[1]

[1] *Department of Chemistry, IIT (BHU), Varanasi, Uttar Pradesh, India*
[2] *Department of Chemistry, Maharaja College, VKSU, Arrah, Bihar, India*

16.1 Introduction

Earth has been labelled the 'blue planet' because 71% of its surface is covered by water. Water is not only on the surface but is also present under the surface as 'groundwater' in aquifers and in the atmosphere as water vapour. However, from Table 16.1 it can be clearly seen that groundwater holds a very small percentage of total distribution of water in ecological resources (UNEP Report 2003). More than two billion of the world's population and 40% of agriculture depend on groundwater for their functioning. The ever-increasing population is creating thrust on the natural resources and water is the most basic natural resource that cannot be compromised. With such an increasing need for human beings, groundwater contamination is a rising issue in parallel that may possess serious threats for living beings consuming groundwater directly through various sources or entering into the food chain either through plants or other sources. It not only affects the area that is owned by a person but also tends to spread to other regions. The contamination of the groundwater is multi-fold process originating from multiple sources. There are both minor reasons, such as from installation of wells improperly or placing water sources close to sewers, as well as major reasons, such as fertilizers in the agriculture run-off and industrial dumps, for groundwater contamination.

This distribution of water on Earth can be categorized in Table 16.1.

Chemical substances such as polychlorinated hydrocarbons, biphenyls, polycyclic aromatic hydrocarbons, radionuclides, petrochemicals, heavy metals or transition metals (Ni, Co, Cr, Cu, Hg, Pb), and radioactive wastes are continuously released from industries. Leaky tanks, oil spills, and landfill leachates as well as the water immiscible organic compounds (non-aqueous phase liquids, NAPLs) are few of the other varieties of contamination being added to groundwater.

Groundwater Geochemistry: Pollution and Remediation Methods, First Edition.
Edited by Sughosh Madhav and Pardeep Singh.

Table 16.1 Sources of water and their distribution on Earth.

Type of source	Percentage
Ocean	97.2
Glaciers and other ice	2.15
Groundwater	0.61
Freshwater lakes	0.009
Inland sea	0.008
Soil moisture	0.005
Atmosphere	0.001
River	0.0001

Source: Nace (1967) and The Hydrologic Cycle (Pamphlet), USGS (1984).

The groundwater remediation is based on the needs of humans and its purpose is to make the water fit for consumption. The contamination can either be treated through a clean-up process or through containment at the site of contamination. The treatment method is decided on the basis of some considerations. The treatment of contaminated groundwater is a task in itself because even though enough contaminants are present in toxicity range, still their concentration is quite low. Besides, the groundwater is at depth of 100 m below the surface. The first thing that needs to be decided is whether the source is to be removed or the cleanup is needed for the groundwater. Excavation and elimination of wastes or the removal of contaminated soil can be one of the processes for source removal, including the vacuum extraction method. The containment of the source is either done through the chemical fixation method or by physically encapsulating the contaminants. There are chances for the elimination of potential contaminants from the waste source to the groundwater through the containment process. The cleanup of groundwater may be either through conventional methods (physical methods, chemical treatments, and biological processes) or nanotechnological advancements. General approaches for groundwater remediation include the assessment and remediation of the sites that are contaminated. This should be done in order to avoid unnecessary delays (Sharma and Reddy 2004).

There has been an increasing number of remedial measures that are used for assessment of pollution or contamination of groundwater in the past few decades. But, it should also be kept in mind that even though each technique is an advance in itself, that doesn't make it qualify for overall remediation of contaminants. While the contaminant removal can be extremely difficult and not very economical, it still is the need of the time. On the other hand, it is widely accepted that each contaminant needs its own type of remedial measure which may also vary according to the geological conditions of the treatment site. The chapter, therefore, tracks such protocols and sets of points that are needed for consideration before the application of a technique at a particular site on the basis of geological, geochemical, and biological aspects of the site of treatment as well as understanding the nature of the contaminant and planning the treatment method accordingly without any sort of risk to the environment (Figure 16.1).

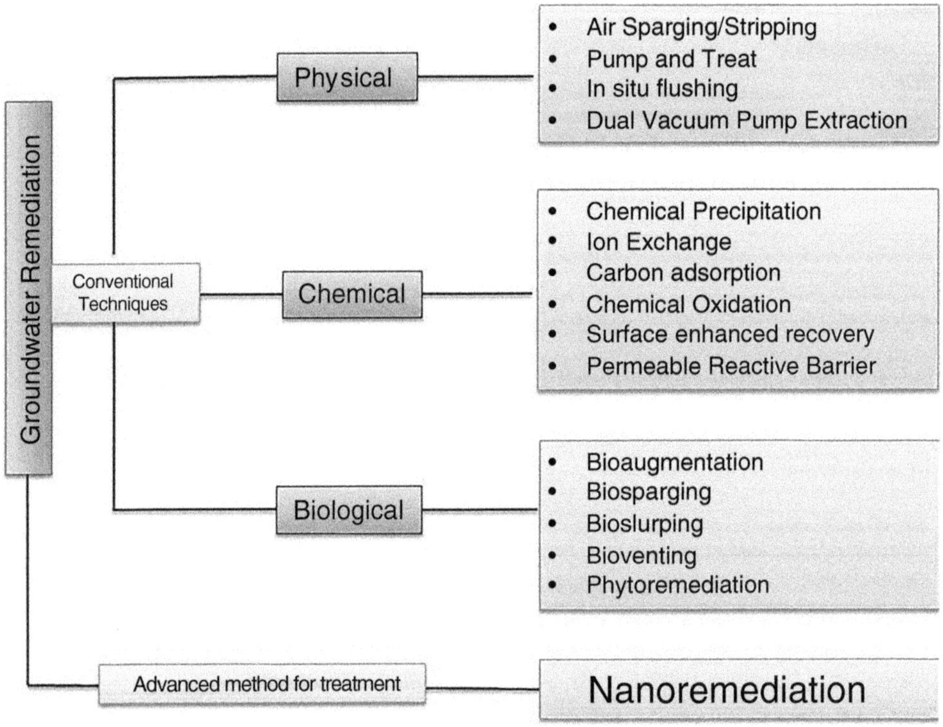

Figure 16.1 Summary of treatment methods.

16.1.1 Determining the Remedial Measure

The first step in the strategy for planning the remediation is *site characterization*. Assessment and collection is done for the data that represent the type of contaminant and its distribution for the particular site which is under investigation. This lays the foundation for determining the kind of remedial measures that can be applied. In addition to this, the data also helps to design, implement, and monitor the strategy for remediation. The site characterization is a well-tailored process for meeting site-specific requirements. This process is very expensive and takes a long time; hence, the characterization should be optimized. Geology (stratigraphy and formations), hydrogeology, hydraulic features, and site contamination properties (type, concentration, and contaminant distribution) must be taken into account. Cone penetrometer and sensor-based devices have helped a lot in site characterization. Chemical tracers are often used for identification and quantification of contaminated zones on the basis of their contaminant affinity. Upon the identification of the contaminant and characterization of the site, an assessment of risk is done, which is a systematic method for evaluation for the determination of the potential risk of the contaminant to human health as well as the ecosystem. It is then used for the plan-of-action for remedial measures while taking into account the goals that are necessary for reduction of potential risks. It can be categorized into four steps: identification of hazards, assessment of exposure to human beings, measuring the toxicity, and then finally characterizing the risk. The final step is the application of remedial measures, taking technical, commercial, and regulatory considerations into account. The remedial measure is decided on the basis of flexibility (site-specific soil and groundwater properties), its compatibility for the particular geological condition, speed, and cost of application. The method is either in-situ (because they are economical, little site disturbances and enhanced safety of on-site workers) or ex-situ (where subsurface conditions need not be considered, easy to control and monitor) (Reddy 2008).

16.2 Conventional Techniques for Groundwater Contamination Remediation

16.2.1 Physical Methods of Treatment

Treatment of contaminated groundwater begins at the physical level, primarily involving the removal of large particles or obstructions that prevent the easy flow of the water. There are several ways of achieving the purified groundwater using some of the physical methods. These are either air stripping, which is an in-situ method of treatment, or pump-treat method, which is an ex-situ method of remediation. With the advent of time these techniques have evolved as per the site and contaminant requirement. The modification may be as simple as advancement in the method of device designing or combination of other technology with the existing one. As mentioned previously, the geological and hydrogeological structures largely determine the mode of treatment.

16.2.1.1 Air Sparging or Air Stripping

One of the most common and basic methods of groundwater remediation is the use of the *air strip* water cleaning process. This technique is known as *air sparging technique* or in situ *volatilization* techniques. Mostly applied for petrochemical compounds, chlorinated solvents, or hydrocarbon contamination, this technique is mainly beneficial for volatile organic compounds (VOCs) (Marley et al. 1992). The method involves the application of compressed air to a manifold system, which further compresses the air into an array of air injection wells. These wells send this air to the surface below the lowest contamination zone. The air rises towards the surface due to buoyancy and will remove or degrade the contaminants through a series of mechanisms categorized as: mass transfer, mass transformations, and mass transport mechanisms. The mass transfer mechanism comprises processes such as adsorption, desorption, volatilization, and dissolution. Mass transformations occur through biodegradation, and mass transport consists of diffusion, advection (horizontally flowing), and dispersion (Reddy and Adams). This technique is highly effective for BTEX compounds, i.e. benzene, toluene, xylene, and ethyl benzene. Some of the advantages of air stripping include the in situ treatment and therefore minimal perturbation at the operation sites. Its simple and easy installation process combined with cost effectiveness promotes its significance more than other methods. It reduces the time of operation by simply skipping the removal, treatment, storage, or discharge processes, where it is one to years under optimum conditions. On the contrary, it holds certain limitations in the applications as there is a chance of contaminant spray to the clean areas due to the application of pressure or formations which are low-permeable, reducing its advantage. The vapours that are generated are very harmful and may migrate to the undesirable places, such as basements of buildings, subways, and tunnels, leading to toxicity due to inhalation. Other disadvantages of air sparging techniques are the formation of complex intermediate physicochemical conditions within the subsurface environment. To avoid such adversities and hurdles for effective implementation of this process, it is highly recommended to have an idea about the area under study (Reddy et al. 1995).

16.2.1.2 Pump and Treat Method

Another most widely used groundwater remediation physical method is the pump and treat method, which physically removes the water from the subsurface. Pump and treat is a simple equipment-based technique where the removal of source zone is done for free-phase contamination. Conventionally this technique involves pumping the contaminated water to the surface for treatment purposes (Brusseau 2019). Extraction wells are used to remove the contaminated water and clean water is brought to the flow associated with pumping removes. The water that is being pumped from the subsurface is subjected to various processes such as air stripping, carbon-based adsorption, or biological treatments. In general, it can be stated that treatment is done above the earth surface and then the clean water is moved to the supply system or re-injected. The general process lying behind the operation is largely affected by dissolution of free products, their desorption, and kinetics of dissolution. However, there are certain limits for the application of this technique for the groundwater remediation. In the zones having low permeability the release of water is through diffusion, which is rather a very slow process and hence increases the time required for contamination treatment. The solute adsorption and desorption can also limit the volume of water and contaminants. Presence of NAPLS (such as hydrocarbons, phthalates, esters, and biphenyls) can also affect the treatment technique to a great extent. The residual contamination sometimes lingers due to rebound, and biofouling of extractions may also occur (USEPA 1996). Immiscible liquids are tough to displace or push out and may serve as long-term contaminant sources. The working of pump and treat method decreases drastically with time and thus, even though there is continuous treatment, the residual contamination may still be there. Besides all these limitations, the pump and treat method still stands as one of the best physical remediation methods and its effectiveness can be enhanced by certain advanced processes such as ISCO (in-situ chemical oxidation), which is discussed in detail later. One such method is the containment of the source zone, i.e. the area in which disposal and spilling of contaminants takes place. Untreated contaminant source zones act as long-term continuous sources of pollution and extend the site clean-up time (Cheremisinoff 1997). The discussed limitations have restricted the use of this conventional method for the free product treatment or controlling the migration of containment plumes (Reddy 2008).

16.2.1.3 In-situ Flushing

The process involves solution flushing through pumping in the groundwater via injection wells. The solution flows downwards and desorbs, solubilizes contaminants, and flushes them out of the groundwater. After their solubilization, the solution is flushed out through the extraction wells. The contaminated solution is then treated using conventional wastewater treatment methods and further pumped back to the injection well for recycling purposes (USEPA 1991; Roote 1997). This method helps in low exposure to the contaminants for the on-site workers as well as the environment. It is simple, easy to operate, and can be applied for organic as well as inorganic contaminants. The disadvantage related to this technique includes large functioning time in case of soil heterogeneity or permeability or the presence of organic materials within the soil. The solubilized contaminants may travel beyond the extraction wells and contamination may spread unintentionally around the wells. The solution, soil, type of contaminants, and interaction with groundwater may alter the working time of the technique. The process may sometimes be expensive owing to the solutions used, which are higher in cost or long time of remediation (Reddy 2008).

16.2.1.4 Dual-phase Vacuum Extraction (DPVE) Technique

This remediation technology is also termed multi-phase extraction and it is a high-vacuum system technology that removes both contaminated groundwater and soil vapour. A high-vacuum extraction well that is equipped with a screened section is the constituent of DPVE systems is installed in zone of contaminated soils and groundwater. The water table is depressed using fluid/vapour extraction systems and water flows easily to the extraction well. DPVE treats the contaminants both from above and below the water table. On lowering the water table in the well, the unsaturated soil is exposed (area termed capillary fringe in the smear zone). This area, which is a cone of depression and an associated cone of ascension, contains many undissolved chemicals and these contaminants are removed by the vapour extraction technique. Liquid phase organic contaminants and groundwater are further separated and then treated. As the capillary fringe is the most contaminated area, it is good that DPVE can be combined with other techniques for the treatment and also reduces the cleanup time. This technique is mostly effective for the soil where there is more silt or clay, as they require high vacuum, and higher vacuum increases the mass transfer of the volatile impurities. The technology however holds certain limitations. The geology of the site and the chemical or physical characteristics of contaminants are highly effective in influencing this technology. This technique most of the time needs certain complementary technologies (pump-and-treat or air sparging) for the recovery of groundwater from high-yielding aquifers. The process is sometimes believed to be limited due to depth as the vacuum limitation is around 1 bar, i.e. the vacuum is unable to lift up the water table in the well for a depth more than 10 m. In such cases air entrainment is applied for extraction purposes. For long chain hydrocarbons, the combination of DPVE with bioremediation (bioventing or bioaugmentation), air sparging, or bioventing is a very important need. Besides these limitations, DPVE has been reported to be of great use for the removal of VOCs, semi-volatile organic compounds (SVOCs) and fuels from groundwater and soil (Athens and Goodrich 2000).

16.2.2 Chemical Methods of Treatment

Chemical methods of treatment are more expensive and take more time for their execution, but some contaminants can only be treated through this method. They aim at the addition of such solvents or chemicals that are capable of removing or degrading the contaminants in the groundwater. Various remediation techniques for the treatments include carbon adsorption, chemical precipitation method, ion-exchange techniques, chemical oxidation processes, etc. All these techniques are most common chemical treatment processes for a range of contaminants such as heavy metals, radionuclides, hardness, petroleum product leakage, VOCs, and other organic contaminants. This technique converts contaminants into inactive states or mineralized forms. The reactions may be oxidation, reduction, redox, or neutralization reactions based on the contaminants and the chemical used for the purpose (Hashim et al. 2011). The methods of treatment are chosen on the basis of type of chemical/ions needed to be treated. Heavy metal and radionuclides treatments are some of the chemicals that are tough for the treatment and chemical remediation stands as one of the efficient technology for treatment of such contaminants. While chemical precipitation focuses on dealing with heavy metals and their removal as phosphates, sulphides, hydroxides, and carbonates, ion exchange works on a variety of metal ions ranging from alkali and alkaline earth metals and heavy metals to radioactive metal ions. Addition of chemicals may also have side effects on the treatment process, especially when we are working with the removal of contamination, sidewise increasing other pollutants. Chemical oxidation methods are used for NAPLs and adsorption techniques are adopted by carbon-based adsorbents for VOCs and NVOCs (non-volatile organic compounds) (Cheremisinoff and Goessmann 1992). The materials used are mostly metallic oxides or polymeric entities. The remediation depends on a variety of factors such as texture of soil, geochemistry, and biogeochemistry of the subsurface, type of contaminant being worked with, and the biological profile of the site. The efficiency of the method largely depends on the operative conditions and is mainly done with the application of solutions of acids, bases and salts, chelating agents, or surfactants. The generation of waste sludge after the treatment is another issue that needs to be addressed while working with chemical treatment techniques (Koul and Taak 2018). The biggest limitation that is attached with the application of chemical remediation method is its limited field data availability; most of the methods are laboratory treatments and still need pilot-scale action plans. The contaminated groundwater mixing with the chemicals is highly uneven (Cheremisinoff and Goessmann 1992).

16.2.2.1 Chemical Precipitation

The chemical precipitation is commonly employed in groundwater treatment for the removal of hardness (presence of Ca^{2+}, Mg^{2+}) and heavy metals (Hg^{2+}, Cd^{2+} and some other transition metal ions). The process in general involves the addition of chemical reagents to the stream of water in a well-stirred reaction vessel in either a batch-wise method or steady flow method. The processes have been performed, both as lab as well as pilot-level experiments and proved their effectiveness for industrial wastes and acid mine drainages, which are the prime sources of heavy metal contamination. Most of the metal ions can be easily converted to their insoluble salts or complexes by certain chemical reactions that occur between these reagents and dissolved metal ions in the stream. Some of the common coagulants that have been used are calcium carbonate and granulated lime for the removal of arsenic and nickel (Lee et al. 2007) and benzenedaimidoethathiol dianion or Metx (commercial name) for removal of most of the soft heavy metals from the samples collected from acid mine drainage (Matlock et al. 2002). The insoluble salts or complexes (precipitates) that are generated due to the chemical treatment are then removed either by settling by themselves or mechanical treatment and/or filtering before use.

Some of the common contaminants that are treated are: precipitation of Pb by limestone; sulphide for the treatment of Hg; removing Cd(II) by changing the pH; and so on. Use of sodium sulphide salts for sulphide precipitation is the most used method for chemical precipitation but they may prove toxic at acidic pH. They may be fatal if accidently released or leak out, so it may be replaced by dithiocarbonate precipitation (Evanko and Dzombak 1997). The process limitations is many fold as there is parallel generation of toxic sludge, which needs proper disposal and the treated water needs the adjustments in pH after the process is complete. Since most of the metals are precipitated as hydroxides there is a chance that any change in pH may lead to the return of the metal ions in the system again. Similar to the process of the pump and treat method, non-termination of the contamination source can result in superfluous treatment requirements. The metals that are held by complexing agents such as EDTA (ethylenediaminetetraacetic acid) or cyanides are not easy to precipitate out. The system of treatment needs well-qualified operators with proper system control and large amounts of reagents that overall lead to high cost of remediation and hence reduce the efficacy of this method (Salami et al. 2013).

16.2.2.2 Ion Exchange

This technique is mainly applied for remediation of ionic contaminants such as cations like hardness (Ca^{2+}, Mg^{2+}), heavy and transition metal ions (Pb^{2+}, Hg^{2+}, Ni^{2+}, Cu^{+}, Cd^{2+}), alkali and alkaline earth metals, and radioactive wastes (strontium, caesium), as well as anions like perchlorates, pertechnates, sulphates, and arsenide. Most of the remediation techniques are not very beneficial in the treatment of ionic impurities. The conversion of metal ion salts to their hydroxide was not very effective because even if not completely soluble, it's sparingly soluble (Papelis et al. 1988; Tripp and Clifford 2006). The ion-exchange process for remediation of groundwater contamination remediation is usually done by passing the water downward. Pressure is applied on water either through a fixed bed of granular medium (which is composed of cation and anion exchange media) or spherical beads. The cations present in water are replaced with non-harmful cations in the bed and anions are displaced by other anions from the exchange bed. Most of the time this ion-exchange media used for remediation is made up of zeolites (both natural and synthetic) and synthetic resins. Solvent-impregnated resins are an important class of ion-exchange resins for groundwater remediation. They give the dual advantage of liquid–liquid extraction combined with ion exchange (Vilensky et al. 2002).

Perchlorate is one the important constituent of solid rocket fuels and hence they exist as the single largest source of groundwater contamination by perchlorate ions. Various techniques such as reverse osmosis, carbon adsorption, and chemical reduction have been applied for its remediation, but they have been showing very strong affinity for strong-base anion exchange resins (SBA) (Freeman 1961). Besides perchlorates, nitrates and sulphates are also some of the basic anions that are treated in groundwater by ion-exchange resins. Materials of ion-exchange resins come from natural sources such as seaweeds, algae, humic acid, and peats as well as laboratory synthesized products such as zeolites and resins. Some of the common examples of resins used are Duolite GT-73, Amberlite IRC-748, and Amberlite IRA-96, which are embedded in polystyrene-divnylbenzene matrix (Vilensky et al. 2002). Sodium phlogopite, modified duolite and amberlite, and newly synthesized silicotitanate and sodium titanate are some of the exchange resins that have been studied for removal of radioactive wastes such as strontium and caesium from the groundwater (Bortun et al. 1997; Marinin and Brown 2000). Prussian blue entrapped in alginate beads and modified hydroxyapatite for the removal for Cs^+ (Metwally et al. 2017) and separation of uranium from iron using Dowex-1X8 anion resins have been done in treatment through ion-exchange methods (Pulhani et al. 2012).

16.2.2.3 Carbon Adsorption

Activated carbon and its resins can adsorb a range of organic compounds in water. There are many sources of activated carbon and they may range from natural sources, such as wood, fruit pits, or cellulose-based material, as well as made from the available sources of carbon (peat, lignite, etc.). Activation of carbon is done by subjecting it to the heat treatment (800–1000 °C) in a furnace in the atmosphere of steam cutting off the air supply or insufficient oxygen is given to open up the pores (micropores and macropores depending on the heat situations) and enhance the area for adsorption. The carbon after adsorption may be reactivated again by heating the same way, but the activity will not be the same as previous time. Bituminous coal is one of the most common adsorbents that is being used for the activated carbon treatment method of groundwater remediation (Hyman and Dupont 2001). Carbo-Iron, PlumeStop, CAT-100, and BOS-200 are some of the commercially available activated carbon used for the treatment which can adsorb chlorinated solvents or petroleum-based compounds (Fan et al. 2017). The adsorption of organic compounds on the surface of activated carbon is mostly physiosorption but may also be chemical sometimes. While the micropores are suitable for the adsorption of VOCs, macropores are more beneficial for the use in adsorption of large molecules. Activated carbon can adsorb a range of VOCs (petrochemicals as well as aromatics) from groundwater. The organic compounds tend to attach with the graphite-like surface of activated carbon. Granular activated carbons (GACs) have been utilized for the reduction of organic compounds in groundwater (Stenzel and Merz 1989). Recently, colloidal activated carbon (CAC) has been used for in-situ remediation by injecting suspensions into the subsurface. This helps in preventing the spread of hydrophobic pollutants by providing a sorption barrier (Georgi et al. 2015). There are certain points that must be kept in mind while working on the carbon adsorption for the groundwater contamination. Filtering out of oil and grease must be done prior to the treatment as they interfere in the process of adsorption. Suspended particles must be pre-treated before the application for they may clog the pores and reduce the surface area. The life of the carbon beds may be increased by pre-stripping the VOCs and the bed must be soaked, washed, and then started just before the process starts (Hyman and Dupont 2001).

16.2.2.4 Chemical Oxidation

In situ chemical oxidation, or ISCO, is a process in which the organic contaminants are rapidly destroyed or converted to nontoxic molecules such as water and carbon dioxide upon addition of chemical oxidants (can be liquid or gases) by injection or mixing. These oxidants may range from air or oxygen/ozone to the liquids such as permanganate solution, peroxides, and persulphates. The introduction of oxygen or ozone can be done directly by injection into soil or groundwater after on-site generation through air and electricity. This technique is very effective for a range of contaminants as well as geological conditions. The process potentially enhances aerobic degradation that has been going on naturally in soil and groundwater. It has been found very effective in the treatment of dense non-aqueous phase liquids (DNAPLs). However this technique needs highly specialized, well managed personnel and compatible devices (Siegrist 2011). The oxidants generate reactive species that mineralize or transform contaminants such as chlorinated hydrocarbons (trichloroethylene), petroleum fuels (benzene, ether and toluene), phenols and polycyclic aromatic hydrocarbons (anthracene, phenanthrene, and naphthalene), and some pesticides and explosives (trinitrotoluene, RDX). One of the examples for this is ISCO of RDX-contaminated groundwater in Nebraska Ordnance Plant by using permanganate (Albano 2009). Lab experiments have revealed that these reactions are electron transfer reaction or free radical processes and may be simple or complex in nature with varying intermediates following second-order kinetics. The activation of free radical generations is primarily controlled by factors such as pH, salinity, temperature, and contaminants (Siegrist et al. 2011). This process holds great potential for treatment, but at sites with small target zones and lower concentration of contaminants. Synergizing of ISCO with other remedial measures such as biostimulation or surface enhanced recovery can be of great benefit. 'Rebounding' in the contaminants occurring after the completion of oxidation may also occur; it is rare but must be taken into account while planning for the process. This process can also perturb the subsurface conditions, although sometimes temporarily, due to change in redox environment or pH or decrement in microbial activity. Some of the key areas that need to be looked into are deep insightful study of degradation steps of a particular contaminant, understanding the chemistry of oxidant on the subsurface, the effects of oxidants on biogeochemistry of the subsurface, deliverance of oxidants in the zone of low permeability, and chances of rebound once the treatment is over (Siegrist et al. 2010; Tsitonaki et al. 2010).

16.2.2.5 Surfactant Enhanced Recovery (SER)

The pump and treat method is quite ineffective for organic compounds that are hydrophobic in nature because of their low solubility, which makes it difficult to treat. A new technique involving the use of surfactants has been introduced for such contaminants in the past few years. In a way this technique is a somewhat modified version of the pump-treat method with slight variations. This method is applicable for the treatment of DNAPLs such as trichloroethylene mostly found in the areas with oil spills and leakage trapped in the system due to capillary forces in the groundwater by increasing the solubility and mobility of the contaminants. It is one of the alternatives for the pump-treat method in treating aquifer contamination (Lee 2010). The surfactant promotes the extraction of these contaminants from the subsurface or aquifers by the reduction of interfacial tension, which in turn enhances the vertical mobility. The value of the interfacial tension should be greater than 1 dyne/cm. Some of the common surfactants that are used in SER are Rexophos 25/97 and Alkasurf NP10 for the treatment of DNAPLs such as nonylphenol and ethoxylate. It had been concluded from a series of experiments that flooding of surfactant has been a very efficient method for the DNAPL-contaminated groundwater, that too very rapidly (Fountain et al. 1996). While solubilization promotes the dissolution of the contaminant, mobilization leads to untrapping of it. The process of SER involves the injection of surface-active agents (surfactants) into the groundwater, which is contaminated. An extraction pump is then used to remove the groundwater from the injection point in a downstream way. The groundwater is further subjected to the treatment above the ground, contaminants are separated from it, and then the water is reused. The surfactants used are chemical, but they are food-grade, biodegradable, as well as non-toxic (NFESC Technical Report 2003). During the solubilization the surfactant molecules form micelles (colloidal agglomerates in water) above the critical micellar concentration (CMC). The DNAPLs are solubilized by getting incorporated within the micelle forming a micro-emulsion. The DNAPLs' aqueous solubility increases manifolds and promotes the site groundwater remediation (Londergan et al. 2001). The factor that controls the SER activity is the geological heterogeneity, which reduces the effectiveness of process. Certain challenges in designing of the SER include DNAPL location, optimized surfactant solution for a given composition of DNAPL, and the variety of soil, along with hydraulic properties' determination for the aquifer.

16.2.2.6 Permeable Reactive Barriers

Permeable reactive barriers (PRBs) are emplacing reactive media for the interception of contaminated subsurface, providing a pathway for flowing through reactive media and conversion of contaminants into environmentally suitable forms to achieve remediation down the gradient (USEPA 1989). PRBs consist of permanent or replaceable reactive media that are placed in the subsurface in the path of flow of contaminated groundwater. A passive treatment system is created through which contaminated water flows under the natural gradient. Further on moving through the PRBs, the contaminants change into non-toxic, less harmful, and immobile species making the groundwater contamination free. The PRB techniques are a barrier only for contaminants, not the groundwater. They are more permeable compared to the neighbouring aquifer materials for the contaminants to be treated and the groundwater flows steadily without tempering hydrogeology of groundwater (USEPA 1989). PRBs hold several potential advantages compared with the other conventional treatment methods like the pump-treat method. The degradation or immobilization of the contaminants is in-situ, which prevents their exposure on the surface, prevents cross media contamination, and cuts off the need of storage, transportation, treatment, or disposal above the surface. Since a natural gradient of groundwater is used as contaminant carrier, there is no need for extra energy input to carry out the process. But the reaction media needs periodic replacement or repairing for effective functioning of the system as the precipitants or microbes may cause clogging. The treatment processes mainly relies on degradation of the contaminant rather than only changing the phase and also control the migration of the pollutants (Thiruvenkatachari et al. 2008).

PRBs can either be continuous type or funnel-gate type. The continuous PRBs are made up of single reactive zones which are placed across the contaminant plume, and the funnel-gate system is made up of a reactive permeable gate that is placed between two impermeable walls in such a way that it directs contaminant plume at the reactive zones. These configurations are decided on the basis of hydrological properties of the site and cost of reactive materials. Continuous PRB operating is cheaper than funnel-gate configurations. Therefore, it becomes necessary to control the ratio of cost of reactive material and cost of barrier construction. Funnel-gate configuration is used in the case of reactive zones that require less material as it uses high-cost reactive materials along with consideration of target contaminant and treatment level to be achieved. An alternate for these expensive barriers is use of multiple caisson that is a cylindrical, load-bearing, and hollow enclosure used for excavation (Gavaskar et al. 1998). This can be installed from ground surface and the excavation is done with the need for personnel to enter. The installation of conventional PRBs needs certain degrees of excavation that restrict PRBs to very shallow depth, reaching up to 20 m only. Slurry injection techniques (producing a treatment zone within the boundary of containment through boring a series of holes and then reactive material which is either a chemical or a particulate mixture in the treatment zones) or hydrofracturing technology (fracturing of a surface by application of water or air under very high pressure, that develops laterally from the point of initiation and then injection of reactive media is done) may be used for overcoming the limitations of depth (Vidic 2001). The operating cost and maintenance can be relatively minimized by placing the reactor cells into the subsurface containing reactive media and contaminant plume capture into the reactor for treatment purposes. This method was labelled as passive groundwater capture (ITRC 2005). The technique for construction of reactive barriers mainly depends upon several factors such as depth of the barriers, geological constructions and geochemistry, soil excavation, space that is available for handling, and disposal of the contaminants and soil, along with considerations for the health and safety of the human resources working at the site.

The contaminant treatment can be done by three possible ways: degradation (either chemical or biological), precipitation (immobilization of contaminant by formation of insoluble compound) and sorption (adsorption or complex formation). The degradation can be attained either through reduction or oxidation. The most common reactive material used is zero-valent iron (ZVI) and other materials that are used for PRBs are organics (activated carbon, sawdust, sludge, or leaf), complexing agents for alkalines such as ferrous sulphate, hydroxyapatite, polymers, zeolites, clays, and some of the metal oxides (Thiruvenkatachari et al. 2008). ZVI is applied for the reductive treatment of contaminants such as halogenated hydrocarbons (through precipitation and degradation) and inorganic contaminants such as U(VI) and Cr(VI). Precipitation may lead to potential risk of dissociation and remobilization of the contaminants (Scherer et al. 2000). Certain limitations that are associated with ZVI are no effects on DNAPLs, reduced lifetime due to formation of surface coatings at the site, effect of pH may affect the functioning of the barriers, and less information on the performance of material in the long run. The technology may be combined with bioremediation or ISCO for enhanced decontamination of groundwater. ISCO combined with PRB gives such advantages as: less treatment time, the complete degradation of the contaminants, injection of oxidants to a great depth, and treatment of recalcitrant organic compounds. On the other side, there are also certain limitations for this combination. The operational history is little known, oxidant selection needs to done properly by considering site characteristics and contaminants to be dealt with, toxic by-products or explosive gases may also be generated during the process, and other factors such as pH, temperature, etc. are some of the points that need to be taken care of. Sequential reactive media are used in cases where more than one class of contaminants are dealt with. Chlorinated solvents or hydrocarbons are denser than water, while petroleum is lighter than water. These two kinds of contaminants when present together at a single site create trouble when a single reactive medium is used, as chlorinated solvents are more in oxidized forms and petroleum compounds are in relatively reduced forms. Therefore, introduction of two or more than two reactive barriers comes as a solution for such cases (Fiorenza et al. 2000).

16.2.3 Biological Methods of Treatment

The use of biological methods for treatment of contaminants for groundwater is termed bioremediation. It is viable and more efficient for the treatment of petrochemical wastes. Bioremediation not only involves the application of biological process or biostimulation for coordinating with existing microbes in soil for degradation of contamination, but also monitors and verifies the natural attenuation of these contaminants. The biological method of treatments involves a variety of processes. This method mainly implies use of organic matter, microorganisms (mainly bacteria and fungi), and plants to clean contaminated groundwater. Bioaugmentation, bioventing, bioslurping, phytoremediation, and biosparging are some of the ways to use biological material for breaking down certain chemicals and compounds that are produced in industrial waste in groundwater. Biological methods are more convenient to use because contaminated water is not to be removed for treatment purposes. The treatment may be in-situ, which is cost-effective but difficult to operate as stimulant and microbial mixing is a tedious job, or ex-situ, which provides good control over the process but is expensive and increases the contaminant exposure. Environmental conditions such as pH, salinity, temperature, humidity, redox conditions, and nutrient concentration controls these biodegradation processes to a great extent. Overall this technique is well-tailored, site-specific, and cost-effective method for remediation of groundwater (Brown et al. 2017).

16.2.3.1 Bioaugmentation

Bioaugmentation is a process of adding cultured microbes into the subsurface for the degradation of contaminants of soil and water. In general it can be stated that it is a method of adding biocatalysts that facilitate degradation of targeted pollutants. The addition of inocula, which consists of strains of *Dehalococcoides* for the degradation of chlorinated ethenes, is one example of bioaugmentation (Lee et al. 2008). This is a highly popular remediation technique mainly due to its low cost of application. Pre-grown culture enhances the microbial concentration at the site and thereby promotes more contaminant cleanup and reduces the time and cost of cleanup. Although there may be the possibility sometimes that some indigenous species of microbes that are present in very small quantities and therefore may not be able to prevent the contaminant spread or they do not have tendency to degrade a contaminant, bioaugmentation creates a pathway such that a sufficient number of microbes with desirable requirements can be cultured and used at the site for treatment (Atlas 1991). The microbial inocula are prepared in laboratory after obtaining microbes from soil or groundwater either from the site where the process of bioaugmentation is to be done or from any other site where the biodegradation of the chemicals of interest is known to be occurring and may produce the desired result. These microbes present in the soil or groundwater are isolated and then added to media containing the contaminants that are to be degraded. The microbes that have the potential to metabolize (degrade) the contaminants tend to grow and the rest perish. The microbial population of interest is then isolated. Bioaugmentation hold the high advantage of being used at the sites with chlorinate solvents (perchloroethylene and trichloroethylene) for their reductive dechlorination. In a way, it can be stated that this method introduces highly concentrated and specialized populations of chemical specific microbes to the contaminated area. However, every solution comes with a problem. In the case of bioaugmentation the underlying problem is that introduction of a microorganism at the point of contamination may fail because of two reasons. On the first note, the microbe that been introduced may not be able to create its own niche at the site due to varying environment. Experimentally it has been observed that these microbes do not survive beyond a few weeks when introduced into a new environment. On the second point of observation, deliverance of these microbes to the site of contamination is a tough task, as they might be strongly sorbed by the solid surfaces of the containers carrying them (Gentry et al. 2004).

16.2.3.2 Bioventing

Bioremediation is mainly controlled by the oxygen availability for aerobic microbial metabolic activities. The natural, in-situ process of aerobic biodegradation of contaminants in water by supplying air or oxygen to the microorganisms that are present in soil subsurface is known as bioventing. This is one of the first, an on-site remediation method that includes the biodegradation of organic compounds in groundwater applied at a very large scale. The metabolic activity of indigenous bacteria or archaeabacteria is enhanced by pumping of air or oxygen (direct injection) to the unsaturated zone or sometimes, by the addition of nutrients through extraction wells. Bioventing is low-rate treatment method, hence oxygen is supplied adequately to the microbial environment to flourish (Das and Dash 2014). The bioventing methods mostly helps in degradation of VOCs that are residuals of adsorbed fuels and assists their degradation as these vapours move up through the microbial active soils. It is more common and easy to perform as the hardware used is readily available. Mid-weight or lighter petroleum products such as jet fuels, diesel, gasoline, etc., are treated by this technique. The pumping of warm and humid air is beneficial in the case of areas with colder or drier conditions (Brown et al. 2017). Bioventing can be combined with the soil vapour extraction (SVE) technique, in a way that extracted contaminants in vapour phase are treated above the surface and in situ remediation is also done (Maier 2009). The cost of application of bioventing technology depends on several factors, such as type of contaminant and its concentration, permeability of the soil, spacing in the well and number of the wells, rate of pumping of oxygen or air, and off-gas processes. Most of the materials used are cheaper compared to other techniques and fewer personnel are required for the operation (Hinchee et al. 1991). Maximum output of the process can be achieved by lowering the air flow rates, lengthening of flow paths, and optimization of conditions for microbial growth by adding moisture and nutrition. Bioventing techniques not only effectively degrade petrochemical products but are also great when it comes to dealing with volatile chlorinated contaminants. The air withdrawn, also known as off-gas, is treated further by adsorption through activated carbon or degraded biologically through biological filters or treated thermally. This method reduces the remediation cost by lowering or eliminating the requirement of off-gas treatment, if passed through uncontaminated soil that works as a biofilter (Lee and Swindoll 1993).

16.2.3.3 Biosparging

This is an in-situ remediation method that exploits the indigenous microbial flora for the bioremediation of organic contaminants in the subsurface. This is done by injecting air as well as nutrients in the aquifer below the contamination zone at a very low rate. The air that is injected promotes the biodegradation of organic contaminants aerobically. In a way, this techniques can be considered as the air sparging type method, but the only point that distinguishes these two processes is that there is emphasis on aerobic biodegradation-based removal, that too at a very lower rate of injection (<5 ft^3/minutes) compared to typical air-sparging techniques where the rate is around 10–20 ft^3/minute and the process is mainly volatilization. The targeted contaminants of this technique are BTEX compounds, which are from petroleum and oil spills or leakages (Kao et al. 2008). The method may be applied for a range of products, such as mid-weight petroleum products such as diesel or jet fuel, lighter products such as gasoline, as well as heavier products such as lubrication oils and greases. The prime reason of using biosparging is the need to avoid SVE for the reduction of its cost for installation and operation. The methodology is affected by the geological formations and conditions of the process such as rate of flow, pressure of injection, and turning system on/off (Johnson et al. 2001). Biosparging of propane was done for the biodegradation of 1,4 dioxane with *Rhodococcus ruber* ENV425 and was shown to be highly effective with the application of propane (Lippincott et al. 2015). The biosparging technique has two advantages over the conventional air stripping technique: (i)the amount of air required in biosparging is much less than that of the amount required for air sparging technique; and (ii) the VOC concentrations are highly reduced in the off-gas, which on the other hand need SVE for the typical air sparging technique and hence no further treatment is required. There are some limitations for this technique, which can be stated as follows. Flushing is very difficult in the zones with less permeability, and the process cannot be operated below 30 ft or 9 m and the process becomes extremely difficult in the case of multiphasic components. Biosparging is restricted by some of the site conditions such as presence of free products. In the presence of free products there may be chance of mounding, thus free product may migrate and contamination could spread (Salami et al. 2013; Wilson et al. 1996).

16.2.3.4 Bioslurping

A combination of bioventing method and vacuum-enhanced free fuel pumping technique with SVE for recovering free products that are lighter than water, termed as light non-aqueous phase liquids or LNAPLs, to elimination of these organic compounds from the groundwater from the vadose zone and also bioremediating the soil is termed bioslurping. This technique uses a 'slurp' tube, which goes into the layer containing the LNAPLs. The pumps draw liquid (groundwater and LNAPLs) along with the soil gas in the upward direction. The extraction wells are made up of a PVC (polyvinyl chloride) riser (unslotted), which is connected to a PVC (slotted) well screen adjusted in interception with the water table. The piping of the well is around 2 in. in diameter (AFCEE/ERT report 1997). The pump takes up the LNAPLs (jet fuels, crude oil, petroleum diesel, or gasoline) from the subsurface and capillary fringe, brings them up to the surface, and then separates them from water and air. The process is termed as bioslurping, because on the introduction of air or oxygen in the unsaturated zone of contaminated soil the biological degradation of these LNAPLs occurs.

Once the LNAPLs are removed this system is reverted back to the bioventing technique to wrap up the remediation. Cadotte et al. (2007) studied the life cycle assessment of bioslurping and other bioremediation techniques for diesel. They found that bioslurping was the method with high input of materials and more output of emissions. Non-halogenated VOCs and semi-VOCs are target contaminants of this technique. This technique however is less effective in the case of soil having low permeability. Excess soil moisture negatively impacts on the permeability of the soil and hence reduces the oxygen transfer. Introduction of modified Fenton reaction in addition to bioslurping was found to be an effective remedial measure for overcoming this obstacle (Dae-Eun et al. 2005). The bioslurping technique comes with certain restrictions that must be taken care of while performing the remedial operation. The most important regulatory factor is the soil moisture, as discussed above. While too much moisture in the soil may reduce the permeability, low moisture may result in the inhibition of microbial activity. Besides this, temperature regulates this too. Low temperature can result in slow remediation process. Fluctuations in the water table and water level may lead to difficulty in aerating and lower the degradation efficiency. Mixing of phase occurs as a result of fuel, water, and air in one stream at the subsurface which need separators for treatment (Place et al. 2003).

16.2.3.5 Phytoremediation

In general, phytoremediation can be termed as restoration of the environment using plants and their related microorganisms for groundwater and soil treatment by removal, containment, inactivation, or degradation of harmful or toxic contaminants followed by revitalizing the groundwater or soil for human needs. The plants that are grown at the site are harvested from time to time and the contaminants are destroyed. From the wetlands to the modern-day urban green roofs, the application of phytoremediation can be seen in range. In this method of treatment of contaminated groundwater, such trees and plants whose roots absorb contaminants from groundwater are planted. This technique is used in areas where the roots of plants can reach up to the groundwater. *Populus* and *Salix* trees are the traditional remediation measure and planted as live cuttings, whips (unrooted or bare-rooted), or bare-rooted trees. The common mechanism for the phytoremediation is the reduction of the contaminants present in soil or water. The reduction of contaminants can be achieved by certain ways:

- Through the root zone due to co-metabolism with the microbes or enzymatic reactions, termed rhizodegradation.
- Contaminant uptake by the plant through tissues and then phytodegradation.
- Translocation of contaminants to leaves and transpiration known as phytovolatalization.
- Contaminant immobilization near the root zone through photosequestration.
- Uptake and sequestration of unmodified contaminants through phytoaccumalation.

The trees are planted in trenches or open holes in such a way that the roots are in contact with the zone of partial saturation (capillary fringe) of the groundwater (Gatliff et al. 2016).

The common contaminants that have been treated by phytoremediation include agricultural chemicals (fertilizers and pesticides), heavy metals (cadmium, lead), explosives (trinitroluene), petrochemical compounds, and VOCs (Banuelos 1994; Rock 1996; Chaney et al. 1997). The phytoremediation process depends on various factors as in the level and extent of soil contamination, the bioavailability of contaminants by the roots, ability of interception by plants, and the accessibility of the contaminant as well as its availability (Vangronsveld et al. 2009). The use of plant and soil amendment combination to reduce the soil metal bioavailability is termed phytostabilization and the real-time decontamination of trace elements by plants is named phytoextraction (Vassilev et al. 2004). Methyl tert-butyl ether (MTBE) and tert-butyl alcohol (TBA) from gasoline-contaminated groundwater sources was treated by conifers (*Pinus*) (Arnold et al. 2007); Break fern or Chinese Ladder fern for the accumulation of arsenic and transgenic cotton wood trees for taking up of selenium are some examples of plant-based remediation measures (Salami 2011; Salami et al. 2013). *Salsola kali, Prosopis, and Brassica juncea* are the plant species that are known for uptake of a considerable amount of heavy metals (Akpor and Muchie 2010). Phytoremediation, in general, can be concluded to be holding a very important role in the treatment of groundwater contamination as it enhances the degradation as well as the removal of organic as well as inorganic (metal ions) contaminants from the groundwater.

The phytoremediation is a cost-effective process, even cheaper than the conventional in-situ and ex-situ processes with an easiness to monitor the decontamination. The process is self-sustaining in itself as the microbes that degrade the contaminants multiply in the presence of contaminants and reduce when the concentrations reduce. Valuable metals can be recovered and reused by this method known as 'phytomining'. It is one the least harmful methods as it includes natural processes and helps in environmental protection as well, but it has certain limitations of its own. Leaching of contaminants may happen due to the low biomass and slow growth process, and the plant survival may be affected due to the soil conditions and toxic contaminants. There are chances of bioaccumulation of toxic wastes, especially metal ions or metals which slightly enter the food chain, and its biomagnification can cause damaging effects on the human system. The lack of basic information about some of the biochemical processes may affect the process. When the contaminated groundwater is at the level deeper than the zone of rooting, then there must be criteria for the selection of plants to be grown in that region. However, the most beneficial part of this treatment method is the genetic modification of plants as per the requirement on the basis of site and type of contaminant being considered for remediation (Vassilev A, Schwitzguébel). A summary of conventional groundwater remediation modes can be found in Table 16.2.

16.3 Nanotechnological Advancements in the Groundwater Contamination Remediation: Nanoremediation

The conventional groundwater remediation techniques are quite expensive, less efficient, and more tedious. Therefore, there is a challenging task in the field of modern research and there is a need for advanced materials. Modern research in the field of groundwater remediation has shifted attention towards the application of nanostructured materials. These materials can be used either as adsorbent or catalysts for the removal of toxic and harmful contaminants from groundwater. The production and application of such devices and systems for the groundwater remediation with desirable design and structures constitute an important part of nanotechnological application for contamination treatment and the technique is labelled as nanoremediation. The application of nanotechnology in environmental remediation can be stated in three possible roles: conversion of pollutants into eco-friendly or sustainable products; removal of the contaminant; and sensing the

Table 16.2 Summary table for the conventional groundwater remediation modes.

Method	Contaminant	Advantage	Disadvantage	References
Physical Methods				
Air sparging	VOCs	Simple, cost-effective, and in-situ	Harmful vapour generation, formation of complex intermediates	Reddy et al. (1995), Marley et al. (1992)
Pump and treat	Free products	Easy monitoring and less chances of leakage of contaminants	Efficiency decrease with the time, costly, interfered by NAPLs	Reddy (2008), Cheremisinoff (1997)
DPVE	VOCs, SVOCs, and fuels	Simultaneously treats soil and groundwater, enhanced biodegradation	Limited by depth	Athens and Goodrich (2000)
In-situ flushing	Organic and inorganic contaminants	Simple, easy to use	Dislocation of solubilized contaminants	Reddy (2008)
Chemical Methods				
Chemical precipitation	Hardness, heavy metals, transition elements	Best method for the treatment of metal ion impurities	Sludge generation, pH dependent	Salami et al. (2013)
Ion exchange	Cation and anions, radioactive wastes	Effective for a large number of cations and even anions, in-situ process	Biofouling and clogging of the resins can reverse the effect	Papelis et al. (1988), Vilensky et al. (2002)
Carbon adsorption	VOCs, metal ions, big molecules	Hydrophobic pollutants spreading is controlled	Oil and grease, suspension may reduce the activity	Georgi et al. (2015), Fan et al. (2017)
Chemical oxidation	DNAPLs, phenols, aromatics, chlorinated hydrocarbons	Can easily be combined with other techniques for better results	Highly specialized personnel, rebounding of contaminants may occur	Siegrist et al. (2011b), Albano (2009)
SER	DNAPLs	Food grade surfactants used, chemicals used are biodegradable and non-toxic	Challenges due to soil geochemistry and hydraulic properties	Fountain et al. (1996), Londergan et al. (2009a)
PRB	Organic, inorganic	Economical, easily modified as per need, no separate disposal needed	Long treatment time, shallow depth	Fiorenza et al. (2000), Scherer et al. (2000)

(*Continued*)

Table 16.2 (Continued)

Method	Contaminant	Advantage	Disadvantage	References
Biological Methods				
Bioaugmentation	Chlorinated solvents	In-situ treatment of contaminants	Deliverance of microbes at the site of contamination is risky	Gentry et al. (2004), Atlas (1991)
Bioventing	VOCs, diesel, gasoline	Less oxygen required as compared, can be combined with SVE	Low permeability of soil may affect the functioning	Brown et al. (2017), Maier (2009), Das and Dash (2014)
Biosparging	BTEX	Turns off need of SVE, cost reduction, less air needed	Difficult in multiphasic components	Salami et al. (2013)
Bioslurping	LNAPLs (diesel, gasoline)	aerobic biological degradation	Reduced by humidity and less permeability	Place et al. (2003), Salami et al. (2013)
Phytoremediation	Organic contaminants, MTBE, heavy metals, and VOCs	Environment friendly, natural, in situ	Bioaccumulation of toxic metals	Vangronsveld et al. (2009), Banuelos (1994)

pollutants (Masciangioli and Zhang 2003). Nanotechnological advancements are less expensive and time-consuming compared to conventional methods. Past years have shown the boom in the area of application of nanotechnology for the environmental pollution remediation, especially treatment of water. Nanomaterials are the materials having dimensions between 1 and 100 nm categorized as natural and man-made. The small size and large surface area (which is sometimes also modified as per the requirement) makes them achieve wider distribution and a range of contaminants for treatment purposes.

Natural nanoparticles have biogenic origin (produced in the living systems) and man-made are designed in laboratories according to the desirable properties required. There is also a category of incidental nanomaterials that are due to pyrogenic activities (combustion) or geological activities. Mostly they are classified as inorganic or carbon-based nanomaterials on the basis of their composition (Tratnyek and Johnson 2006; Nowack and Bucheli 2007).

Amongst the range of nanomaterials available, some of the nanomaterials that are helpful in site remediation of contaminated groundwater include carbon-based nanomaterials (CNM) such as fullerenes and carbon nanotubes (CNTs), nanoparticles of metals such as nano zero-valent iron (nZVI), bimetallic nanoparticles, and some of the nanomaterials as semiconductor photocatalysts (that oxidize toxic contaminants into nontoxic ones such as titania) (Matlochová et al. 2013). Of all the mentioned nanomaterials, nZVI is the most widely studied and applied for treatment purposes. Nanoremediation not only treats the solubilized contaminants but is also effective against NAPLs with the application of some nano-sized metal oxides. They are used in cleaning of oil spills and also reduce these contaminants. Similar to the conventional methods, the nanotechnological methods is also based on the concept of adsorptive or reactive mechanisms and in-situ or ex-situ set up. The adsorptive methods imply sequestration of the contaminants (hardness, heavy metals, or radionuclides) and the reactive method deals with the decomposition of contaminants (organic especially petrochemicals or aromatics). In-situ and ex-situ are the methods that have been discussed in detail earlier (Tratnyek and Johnson 2006).

nZVI are prepared by the chemical reduction of Fe^{2+} and Fe^{3+} by the borohydrides and ranges between 10 and 100 nm diameter. They have a core shell structure with metallic iron at the centre and bivalent and trivalent iron form the shell. This has a large surface area compared to other nanoparticles in use, and thereby provides a more reactive site for contamination treatment (Rajan 2011). The properties of formation of microemulsions and magnetic behaviour of iron nanoparticles are exploited for this purpose. Nano-zero valent iron can be used for a wide range of organic contaminants, especially TCE (trichloroethylene), pesticides, and PCBs (polychlorinated biphenyls) (Zhang 2003), as well as arsenic removal from the groundwater (Yean et al. 2005). Elemental iron in combination with other metals such as Au, Ag, Ni, Pt, and Pd form bimetallic nanoparticles with enhanced redox reaction, and hence act as very active catalysts for contaminant breakdown (Nutt et al. 2006). CNTs (known for their high chemical stability, strong adsorption behaviour, easy modification of functionality, and thermal tolerance) bind strongly to the atoms or molecules by their graphite sheet surfaces and were used for removal of heavy metals such as Cr(III), Pb(II), and Zn (II), arsenides, some of the organic compounds such as dioxins, and biological impurities (Rajan 2011). There are certain regulatory instructions for the applications of CNTs. The CNTs must be removed after treatment and before the discharge of water to prevent their bioaccumulation. The surfaces must be functionalized in such a way that they become more soluble. This will promote the reuse of these nanomaterials and also make them biocompatible with certain species.

The limited knowledge about the potential accumulation or nanotoxicity of these particles somewhat limits the use of nanotechnology in remediation. They might create negative health impacts if they are taken through food or inhaled directly or absorbed through the skin. According to some researchers, nanotubes of carbon produce the same effects as asbestos sheets in mice, i.e. they are airborne and when deposited in lungs cause lesions and inhibit the respiratory processes (Lam et al. 2003). In some cases, there was change in the structure of the microbial community after the use of nZVI (Macé et al. 2006). The nanotechnology, even though it has been proven to be highly effective and great results were achieved, needs a lot of deep insight and observation for its impact on the surroundings.

16.4 Conclusion

With the rising pressure of population growth, the thrust on the groundwater utilization is a matter of serious concern. While most of us are dependent on it for survival there is been growing limitations on its use because of the limited availability. Besides, contaminated groundwater will create pressure on the health infrastructures of the countries. The above discussed methods ranging from physical to chemical and biological and then advanced nanotechnologies have been helpful in combatting these contaminants to a great extent. However, each technology comes with its own limitations and challenges. Therefore, it becomes very necessary to prevent the contamination in order to attain clean groundwater and living with the fact that 'prevention is better than cure'. Moreover, site analysis, risk assessment, and application of suitable remediation methods are very important tasks and have the power to decide the fate of many lives that are dependent on these aquifers. As it can be seen from the above findings, selection of remediation technique is a very complex process and needs to be site-specific, taking all the geological, hydrological, and environmental features into consideration, and thus must be analysed properly before application of a particular technique. Improper implementation of any technique may create a serious threat by reversing the positive effect and threaten many lives. Nanotechnological advancements are quite a growing field and the coming era is an era of nanotechnology, so it becomes important to explore more and more in this field. Combination of nanotechnology with conventional techniques may also shoot up the remediation processes. Most of the techniques in the conventional modes have been found to be time-consuming or costly. Combining some techniques together (sequentially or concurrently) may also reduce the cost of operation of remedial measures. Chemical methods of treatments needs to be advanced, as using lots of chemicals for treatment can reverse the purpose of treatment. Phytoremediation serves the dual purpose of treatment and removal of pollutant using the most eco-friendly pathway, i.e. plantation. Bioremediation techniques, when combined with pump and treat or air sparging, showed efficient results and were quite promiscuous. Even when we are having such a variety of treatment techniques that can deal with a range of contaminants, still we need to focus on preventing contamination by keeping a check on the sources of these contaminations. Hazardous wastes and leachates must be monitored and sensing applications must be used to control their activity.

References

AFCEE/ERT (1997). Report for Engineering Evaluation and Cost analysis of Bioslurper initiative, A005. 25 March, 1997.

Akpor, O.B. and Muchie, M. (2010). Remediation of heavy metals in drinking water and wastewater treatment systems: processes and applications. *International Journal of the Physical Sciences* 5 (12): 1807–1817.

Albano, J.A. (2009). In Situ Chemical Oxidation of RDX-Contaminated Groundwater with Permanganate at the Nebraska Ordnance Plant. Dissertations & Theses in Natural Resources. 30.

Arnold, C.W., Parfitt, D.G., and Kaltreider, M. (2007). Phytovolatilization of oxygenated gasoline-impacted groundwater at an underground storage tank site via conifers. *International Journal of Phytoremediation* 9 (1): 53–69.

Athens, N. and Goodrich, D.M. (2000). Soil and Groundwater Decontamination System with Vacuum Extraction. US Patents, Patent Number: 6,158, pp. 924.

Atlas, R.M. (1991). Bioremediation: using nature's helpers-microbes and enzymes to remedy mankind's pollutants. In: Proceedings of the Biotechnology in the Feed Industry. Alltech's Thirteenth Annual Symposium (eds. T.P. Lyons and K.A. Jacques), 255–264. Nicholasville, Kentucky: Alltech Technical Publications.

Banuelos, G.S. (1994). Extended abstract: managing high levels of B and Se with trace element accumulator crops. Symposium. In: Emerging Technologies in Hazardous Waste Management VI, 1994 Book of Abstracts, September 19–21, 1994, Atlanta, GA, vol. 2 (eds. D.W. Tedder and American Chemical Society. Division of Industrial and Engineering Chemistry), 1344–1347. Washington, DC: American Chemical Society.

Bortun, A.I., Bortun, L.N., and Clearfield, A. (1997). Evaluation of synthetic inorganic ion exchangers for cesium and strontium removal from contaminated groundwater and wastewater. *Solvent Extraction and Ion Exchange* 15 (5): 909–929.

Brown, L.D., Cologgi, D.L., and Ulrich, A.C. (2017). Bioremediation of oil spills on land. In: Oil Spill Science and Technology, 2e), Chapter 12 (ed. P.K. Arora), 699–729. Singapore: Springer.

Brusseau, M.L. (2019). Soil and groundwater remediation. *Environmental and Pollution Science*: 329–354.

Cadotte, M., Deschênes, L., and Samson, R. (2007). Selection of a remediation scenario for a diesel-contaminated site using LCA. *The International Journal of Life Cycle Assessment* 12 (4): 239–251.

Chaney, R.L., Malik, M., Li, Y.M. et al. (1997). Phytoremediation of soil metals. *Current Opinion in Biotechnology* 8 (3): 279.

Cheremisinoff, N.P. (1997). Pump-and-treat remediation technology. In: Groundwater Remediation and Treatment Technologies (ed. N.P. Cherminisoff), 203–258. William Andrew Publishing.

Cheremisinoff, P.N. and Goessmann, G. (1992). Comparing groundwater remediation options. *Remediation Journal* 2 (2): 153–169.

Dae-Eun, K., Seung-Won, S., Min-Kyoung, K., and Kong, S.-H. (2005). Enhanced Bioslurping System for Remediation of Petroleum Contaminated Soils.

Das, S. and Dash, H.R. (2014). Microbial bioremediation: a potential tool for restoration of contaminated areas. In: Microbial Biodegradation and Bioremediation (ed. S. Das) Chapter 1, 1–21. Oxford: Elsevier.

Evanko, C.R. and Dzombak, D.A. (1997). E Series: TE-97-01 Remediation of Metals-Contaminated Soils and Groundwater.

Fan, D., Gilbert, E.J., and Fox, T. (2017). Current state of in situ subsurface remediation by activated carbon-based amendments. *Journal of Environmental Management* 204: 793–803.

Fiorenza, S., Oubre, C.L., and Ward, C.H. (2000). Sequenced Reactive Barriers for Groundwater Remediation. New York, NY: Lewis Publishers.

Fountain, J.C., Starr, R.C., Middleton, T. et al. (1996). A controlled field test of surfactant-enhanced aquifer remediation. *Ground Water* 34 (5): 910–916.

Freeman, D.H. (1961). Thermodynamics of binary ion exchange systems. *The Journal of Chemical Physics* 35 (1): 189.

Gatliff, E., Linton, P.J., Riddle, D.J., and Thomas, P.R. (2016). Phytoremediation of soil and groundwater. In: Bioremediation and Bioeconomy (ed. M.N.V. Prasad), 589–608. Elsevier.

Gavaskar, A.R., Gupta, N., Sass, B.M. et al. (1998). Permeable Barriers for Groundwater Remediation: Design, Construction, and Monitoring. Columbus, Ohio: Battelle Press.

Gentry, T.J., Rensing, C., and Pepper, I.L. (2004). New approaches for bioaugmentation as a remediation technology. *Critical Reviews in Environmental Science and Technology* 34: 447–494.

Georgi, A., Schierz, A., Mackenzie, K., and Kopinke, F.-D. (2015). Colloidal activated carbon for in-situ groundwater remediation – transport characteristics and adsorption of organic compounds in water-saturated sediment columns. *Journal of Contaminant Hydrology* 179: 76–88.

Hashim, M.A., Mukhopadhyay, S., Sahu, J.N., and Sengupta, B. (2011). Remediation technologies for heavy metal contaminated groundwater. *Journal of Environmental Management* 92: 2355–2388.

Hinchee, R.E., Ong, S.K., and Hoeppel, R. (1991). A Treatability Test for Bioventing, in Proceedings of the 84th Annual Meeting and Exhibition, Air and Waste Management Association, Vancouver, BC, 91–19.4.

Hyman, M. and Dupont, R.R. (2001). Groundwater remediation using carbon adsorption. In: Groundwater and Soil Remediation: Process Design and Cost Estimating of Proven Technologies, 109–135. Reston, VA: ASCE Press https://doi.org/10.1061/9780784404270.ch04.

ITRC (2005). Permeable Reactive Barriers: Lessons Learned/New Directions. Technical/Regulatory Guidelines. Washington, DC: ITRC.

Johnson, P.C., Johnson, R.L., Bruce, C.L., and Leeson, A. (2001). Advances in in situ air sparging/biosparging. *Bioremediation Journal* 5 (4): 251–266.

Kao, C.M., Chen, C.Y., Chen, S.C. et al. (2008). Application of in situ biosparging to remediate a petroleum-hydrocarbon spill site: field and microbial evaluation. *Chemosphere* 70 (8): 1492–1499.

Koul, B. and Taak, P. (2018). Chemical methods of soil remediation. In: Biotechnological Strategies for Effective Remediation of Polluted Soils, 77–84. Singapore: Springer.

Lam, C.-W., James, J.T., McCluskey, R., and Hunter, R.L. (2003). Pulmonary toxicity of single-wall carbon nanotubes in mice 7 and 90 days after intratracheal instillation. *Toxicological Sciences* 77 (1): 126–134.

Lee, K.S. (2010). Numerical simulation of surfactant-enhanced remediation of heterogeneous aquifer contaminated with nonaqueous phase liquids. *Journal of Material Cycles and Waste Management* 12 (3): 193–203.

Lee, M.D. and Swindoll, C.M. (1993). Bioventing for in situ remediation. *Hydrological Sciences -Journal- des Sciences Hydrologiques* 38: 4–8.

Lee, M., Paik, I.S., Kim, I. et al. (2007). Remediation of heavy metal contaminated groundwater originated from abandoned mine using lime and calcium carbonate. *Journal of Hazardous Materials* 144 (1–2): 208–214.

Lee, P.K., Macbeth, T.W., Sorenson, K.S. Jr. et al. (2008). Quanepsying genes and transcripts to assess the in situ physiology of Dehalococcoides spp. in a trichloroethene-contaminated groundwater site. *Applied and Environmental Microbiology* 74: 2728–2739.

Londergan, J.T., Meinardus, H.W., Mariner, P.E. et al. (2009). DNAPL removal from a heterogeneous alluvial aquifer by surfactant-enhanced aquifer remediation. *Groundwater Monitoring & Remediation* 21 (4): 57–67.

Lippincott, D., Streger, S.H., Schaefer, C.E. et al. (2015). Bioaugmentation and propane biosparging for in situ biodegradation of 1,4-dioxane. *Groundwater Monitoring & Remediation* 35 (2): 81–92.

Macé, C., Desrocher, S., Gheorghiu, F. et al. (2006). Nanotechnology and groundwater remediation: a step forward in technology understanding. *Remediation Journal* 16 (2): 23–33.

Maier, R.M. (2009). Microorganisms and organic pollutants. In: Environmental Microbiology, 2e), Chapter 20 (eds. R.M. Maier, I.L. Pepper and C.P. Gerba), 387–420. Academic Press.

Marinin, D.V. and Brown, G.N. (2000). Studies of sorbent/ion-exchange materials for the removal of radioactive strontium from liquid radioactive waste and high hardness groundwaters. *Waste Management* 20: 545–553.

Marley, M.C., Hazebrouck, D.J., and Walsh, M.T. (1992). The application of in situ air sparging as an innovative soils and ground water remediation technology. *Ground Water Monitoring & Remediation* 12 (2): 137–145.

Masciangioli, T. and Zhang, W.-X. (2003). Peer-reviewed: environmental technologies at the nanoscale. *Environmental Science and Technology* 37: 102A.

Matlochová, A., Plachá, D., and Rapantová, N. (2013). The application of nanoscale materials in groundwater remediation. *Polish Journal of Environmental Studies* 22 (5): 1401–1410.

Matlock, M.M., Howerton, B.S., and Atwood, D.A. (2002). Chemical precipitation of heavy metals from acid mine drainage. *Water Research* 36 (19): 4757–4764.

Metwally, S.S., Ahmed, I.M., and Rizk, H.E. (2017). Modification of hydroxyapatite for removal of cesium and strontium ions from aqueous solution. *Journal of Alloys and Compounds* 709: 438–444.

Nace, R.L. (1967). Are we running out of water? Report (USGS). *Geological survey circular* 536: 1–12.

NFESC Technical Report TR-2219-ENV, Surfactant-Enhanced Aquifer Remediation (SEAR) Implementation Manual, NAVAL FACILITIES ENGINEERING COMMAND, Washington, DC 20374–5065. Intera Inc. (formerly Duke Engineering & Services) and Naval Facilities Engineering Service Center, April 2003.

Nowack, B. and Bucheli, T.D. (2007). Occurrence, behavior and effects of nanoparticles in the environment. *Environmental Pollution* 150: 5.

Nutt, M.O., Heck, K.N., Alvarez, P., and Wong, M.S. (2006). Improved Pd-on-Au bimetallic nanoparticle catalysts for aqueous-phase trichloroethene hydrodechlorination. *Applied Catalysis B: Environmental* 69 (1–2): 115–125.

Papelis, C., Hayes, K.F., and Leckie, J.O. (1988). HYDRAQL: A Program for the Computation of Chemical Equilibrium Composition of Aqueous Batch Systems Including Surface -Complexation Modeling of Ion Adsorption at the Oxide/Solution Interface. Stanford, CA: Stanford University.

Place, M., Hoeppel, R., Chaudhry, T., et al. (2003). Application Guide For Bioslurping Principles And Practices Of Bioslurping Addendum: Use Of Pre-Pump Separation For Improved Bioslurper System Operation. Nov. 2013.

Pulhani, V.A., Dafauti, S., and Hegde, A.G. (2012). Separation of uranium from iron in ground water samples using ion exchange resins. *Journal of Radioanalytical and Nuclear Chemistry* 294: 299–302.

Rajan, C.S. (2011). Nanotechnology in groundwater remediation. *International Journal of Environmental Science and Development* 2 (3): 182–187.

Reddy, K.R. (2008). Physical and Chemical Groundwater Remediation Technologies, NATO Science for Peace and Security Series (ed. C.J.G. Darneault), 257–274. Netherlands: Springer.

Reddy, K.R., Kosgi, S., and Zhou, J.A. (1995). Review of in-situ air sparging for the remediation of VOC-contaminated saturated soils and groundwater. *Hazardous Waste & Hazardous Materials* 12 (2): 97–118.

Rock, S. (1996). Phytoremediation of organic compounds: mechanisms of action and target contaminants. In: Workshop on Phytoremediation of Organic Wastes, December 17–19 (eds. W.W. Kovalick and R. Olexsey). Ft. Worth, TX. An RTDF meeting summary.

Roote, D.S., (1997). In-situ Flushing, Technology Overview Report, Groundwater Remediation Technologies Analysis Center.

Salami, L. (2011). Remediation of Contaminated groundwater: An Overview. A seminar paper presented at Lagos State University, August, 2011.

Salami, L., Susu, A.A., Olafadehan, O.A., and Odunlami, M.O. (2013). Remediation of contaminated groundwater: an overview. *International Journal of Chemical Engineering Research* 5 (1): 19–33.

Scherer, M.M., Richter, S., Valentine, R.L., and Alvarez, P.J.J. (2000). Chemistry and microbiology of permeable reactive barriers for in situ groundwater clean-up. *Critical Reviews in Microbiology* 26 (4): 221–264.

Sharma, H.D. and Reddy, K.R. (2004). Geoenvironmental Engineering: Site Remediation, Waste Containment, and Emerging Waste Management Technologies. In: Environmental and Pollution Science, 329–354. Hoboken, New Jersey: John Wiley & Sons.

Siegrist, R.L., Crimi, M.L., Petri, B. et al. (2010). In Situ Chemical Oxidation for Groundwater Remediation: Site Specific Engineering and Technology Application. ER-0623 Final Report (CD-ROM, Version PRv1.01 October 29, 2010). Prepared for the ESTCP, Arlington, VA, USA.

Siegrist, R.L., Crimi, M., Simpkin, T.J. et al. (2011). ISCO status and future directions. In: In Situ Chemical Oxidation for Groundwater Remediation, vol. XLIII (eds. R. Siegrist, M. Crimi and T. Simpkin). New York: Springer-Verlag.

Stenzel, M.H. and Merz, W.J. (1989). Use of carbon adsorption processes in groundwater-treatment. *Environmental Progress* 8 (4): 257–264.

Thiruvenkatachari, R., Vigneswaran, S., and Naidu, R. (2008). Permeable reactive barrier for groundwater remediation. *Journal of Industrial and Engineering Chemistry* 14: 145–156.

Tratnyek, P.G. and Johnson, R.L. (2006). Nanotechnologies for environmental cleanup. *Nano Today* 1: 44.

Tripp, A.R. and Clifford, D.A. (2006). Ion exchange for the remediation of perchlorate-contaminated drinking water. *Journal AWWA* 98 (4): 105–114.

Tsitonaki, A., Petri, B., Crimi, M. et al. (2010). In situ chemical oxidation of contaminated soil and groundwater using Persulfate: a review. *Critical Reviews in Environmental Science and Technology* 40 (1): 55–91.

National water summary 1984: Hydrologic events, selected water-quality trends, and ground-water resources. Water Supply Paper, U.S. Geological Survey, 1985.

UNEP, (2003). Groundwater and its susceptibility to degradation: a global assessment of the problem and options for management, early warning and assessment report series, UNEP/DEWA/RS. 03-3, joint publication from United Nations Environment Programme, Department for International Development and the Natural Environment Research Council (NERC).

USEPA (1989). Evaluation of Groundwater Extraction Remedies, vol. 1 and 2. Washington, DC: EPA Office of Emergency and Remedial Responses.

USEPA, (1991). Engineering Bulletin: In Situ Soil Flushing, EPA/540/2-91/021, and Washington, D.C.

USEPA (1996). Pump-and-Treat Ground-Water Remediation: A Guide for Decision Makers and Practitioners, EPA/625/R-95/005. Washington, D.C.: Office of Research and Development.

Vangronsveld, J., Herzig, R., Weyens, N. et al. (2009). Phytoremediation of contaminated soils and groundwater: lessons from the field. *Environmental Science and Pollution Research* 16: 765–794.

Vassilev, A., Schwitzguébel, J.P., Thewys, T. et al. (2004). The use of plants for remediation of metal contaminated soils. *Scientific World Journal* 4: 9–34.

Vidic, R.D. (2001). Permeable Reactive Barriers: Case Study Review, Technology Evaluation Report TE01-01. Groundwater Remediation Technologies Analysis Center.

Vilensky, M.Y., Berkowitz, B., and Warshawky, A. (2002). In situ remediation of groundwater contaminated by heavy- and transition-metal ions by selective ion-exchange methods. *Environmental Science & Technology* 36: 1851–1855.

Wilson, D.J., Norris, R.D., and Clarke, A.N. (1996). Groundwater cleanup by in-situ sparging. X. Air channelling model for biosparging of non-aqueous phase liquid. *Separation Science and Technology* 31 (10): 1357–1376.

Yean, S., Cong, L., Yavuz, C.T. et al. (2005). Effect of magnetite particle size on adsorption and desorption of arsenite and arsenate. *Journal of Materials Research* 20 (12): 3255–3264.

Zhang, W.X. (2003). Nano-scale iron particles for environmental remediation: an overview. *Journal of Nanoparticle Research* 5: 323–332.

17

Various Remediation Measures for Groundwater Contamination

Paramdeep Kaur, Abhishek Dawar*, and Baljinder Singh*

Department of Biotechnology, Panjab University, Chandigarh, India
**Abhishek Dawar and Paramdeep Kaur have equal contributions.*

17.1 Introduction

Groundwater is the main source of potable water for humankind, animals, and plants. The majority of the world's population depends upon groundwater for water supply; for example, nearly 19% of the 610 billion m^3 annual total water supply of China accounts for groundwater (Jiang 2015), whereas Denmark's population is entirely dependent upon groundwater (Schmoll et al. 2006). Groundwater used for irrigation accounts for 40% of the world's food (Thiruvenkatachari et al. 2008). Contamination of groundwater is an emerging environmental concern as it makes the groundwater unsuitable for consumption and poses a great threat to human and animal health. Chemical compounds released from various anthropogenic activities such as industrial or mining activities, accidental spills, and agricultural activities account for the major contaminants in groundwater. Due to inappropriate methods of disposal of these contaminants, they are either directly released into aquifers or infiltrate through the ground to reach the groundwater (Babiker et al. 2004; Sethi and Molfetta 2019). Various inorganic contaminants (such as arsenic, fluorides, and nitrates), organic compounds, pesticides, and pharmaceuticals associated with high exposure risks are known to contaminate groundwater (Menchen et al. 2017; Schweitzer and Noblet 2018; Sharma et al. 2019). Nitrates, which are one of the most common groundwater pollutants, have shown to cause serious health conditions such as blue baby syndrome, or methemoglobinaemia, in newborn babies (Bouchard et al. 1992; Knobeloch et al. 2000). Petroleum hydrocarbons, which often contaminate groundwater due to accidental leaks, are mutagenic and carcinogenic in nature (Logeshwaran et al. 2018). Moreover, polluted groundwater also affects the economy of industries related to fisheries because fish are heavily impacted by the chemical contaminants.

There are various precautionary principles that have been taken to protect the groundwater from various contaminants or chemicals. Therefore, in many countries

Groundwater Geochemistry: Pollution and Remediation Methods, First Edition.
Edited by Sughosh Madhav and Pardeep Singh.

groundwater quality inspection programmes are introduced to determine the hydrological system and development of aquifer vulnerability maps. Regular inspection of groundwater quality has to be done to determine the trends. The levels of contaminants have to be compared with World Health Organisation (WHO) guidelines for potable quality of water (Edition 2011).

It is very difficult to rarefy the pollutants present in groundwater than surface water (Zeidan n.d.).Various physical, chemical, and biological methods (e.g. pump and treat, adsorption, air sparging, bioremediation, chemical oxidation, and reduction) have been employed to remediate contaminated groundwater. The very first attempts of groundwater treatment at large scale began in the 1980s with the application of the pump and treat method; since then there have been many advancements to make the remediation process more efficient (Voice et al. 1992; Gehringer et al. 1995). The focus has recently shifted towards in-situ remediation of groundwater due to the associated advantages. Application of innovative technologies such as permeable reactive barriers, slow release materials, and optimisation of the conventional methods have made groundwater remediation more dependable.

17.2 Ex-Situ and In-Situ Remediation

17.2.1 Pump and Treatment (Ex-Situ Treatment)

Pump and treat (P and T) is an ex-situ groundwater remediation system (O'connor et al. 2018) initiated in 1987 (Guo et al. 2019); this system is considered one of the most effective and cost-efficient techniques for containment of widespread groundwater contaminated by various chemical and other non-biodegradable materials (Hoffman 1993). The functional implementation of the pump and treat system for groundwater remediation can be characterized into three categories (Truex et al. 2015). Groundwater extraction from aquifers has to be done for hydraulic gradient manipulation and contaminant mass removal. The above function is carried out by the pump portion of P and T. This is later followed by the 'treat' portion of PandT which is an aboveground part; in this, specified treatment goals have to reached through routine engineering designs (Truex et al. 2017).

17.2.2 In-Situ Remediation

In-situ remediation has become the method of choice for groundwater remediation due to its various advantages over the ex-situ methods. In-situ remediation is cost effective and these techniques have a very high decontamination rate. These is no need to displace groundwater and it is also noticed that during in-situ remediation the risk of cross contamination is far lower unlike other techniques, like ex-situ, where pollutant materials are exposed to other sites. It involves the application of various biological methods and chemical methods such as biosparging (Angmo et al. 2020), bioventing (Werkneh and Rene 2019), ion exchange (Zhang et al. 2019), activated carbon adsorption (Perrich 2018), chemical oxidation and reduction (Devi et al. 2016), and flocculation (Teh et al. 2016), for the containment of harmful hazardous compounds and chemicals. Therefore, development of in-situ remediation technologies is the current focus for efficient clean-up of groundwater and one of the most promising technologies used in in-situ remediation is permeable reactive barriers.

17.2.3 Permeable Reactive Barriers (PRB)

Permeable reactive barriers (PRB) or permeable reactive treatment zones are widely known for its cost effectiveness. This technique helps in capturing and removing the contaminants passively and releases uncontaminated water (Gillham et al. 2010). There are three different primary removal methods in permeable barrier methods (Tratnyek et al. 2003):

1) Chemical reaction
2) Reactions involving biological mechanisms
3) Precipitation and sorption

There are various reactive materials used in this process, for example zero-valent iron (Kornilovych et al. 2018), peat moss (sphagnum) (Faisal et al. 2018), biological barriers (Gibert et al. 2019), surfactant modified zeolites (Shang et al. 2017), etc.

17.3 Methods of Groundwater Remediation

17.3.1 Chemical Oxidation and Reduction

17.3.1.1 In-situ Chemical Oxidation

In-situ chemical oxidation (ISCO) is one of the most popular methods for treatment of contaminated groundwater. There are a large number of organic chemicals or contaminants that are really sensitive to chemical reactions or oxidative degradation, when they react with different oxidants. The most commonly used oxidants for groundwater remediation include ozone (Andaluri and Suri 2017), sodium permanganate (Zhang and Dong 2019), persulphate (Chen et al. 2016; Zhong et al. 2017), and hydrogen peroxide (Yang et al. 2020). The purpose of this method is to inject an oxidizing compound into the contaminated area so that the oxidizing compound reacts with organic contaminants and transforms them into oxidant-specific by-products and carbon dioxide. There are different mechanisms that are followed during degradation like free radical or electron transfer process, which may involve different pathways and various intermediates. Oxidizing agents that are selected for the treatment of organic contaminants will react with almost every organic contaminant present. Therefore, a high oxidant demand is required for the of groundwater containing an exorbitant amount of organic contaminants (Filler et al. 2009; Siegrist et al. 2011).

This process of oxidation of contaminants takes place in the aqueous phase, therefore the hydrophobic contaminants that remain sorbed to soil or contaminants present in the non-aqueous phase easily escape the action of these oxidizing chemicals. Consequently, contaminant rebound that is leaching back of the contaminants into the aqueous phase is a common problem associated with in-situ chemical oxidation. Therefore, subsequent ISCO treatments are often required. Combinations of surfactants and oxidants have been used to overcome this problem as the surfactant causes desorption and emulsification of contaminants, making them available to the oxidants and thus leading to optimisation of the process (Besha et al. 2018). Persulphate, when used in combination with surfactant Lauryl Betaine (LB), has proved to be highly effective for the oxidation of polycyclic aromatic hydrocarbon (PAH) compounds, especially phenanthrene (Bouzid et al. 2017). The use of C_{12}-MADS (sodium dodecyl diphenyl ether disulfonate) surfactant has also shown to enhance the oxidation of PAHs and the oxidant utilization efficiency (Wang et al. 2017). Surfactant Enhanced In-situ Chemical Oxidation (S-ISCO®) is a proprietary technology that is being widely accepted for groundwater remediation. It utilizes a plant-based, biodegradable surfactant/co-solvent mixture VeruSOL for improving the action of oxidants. Groundwater treatment performed for coal tar clean up in Queens, New York, using sodium persulphate and VeruSOL resulted in 90% reduction of the total SVOCs (semi-volatile organic compounds) and VOCs (volatile organic compounds) with a 95% decrease in naphthalene contamination (Dahal et al. 2016).

Another novel technique, polymer-amended in situ chemical oxidation (PA-ISCO) is also being employed to increase the interaction between the contaminants and the oxidants by using water-soluble polymers and polyphosphates that enhance the viscosity of ISCO solution formulations and help to mitigate the problems associated with subsurface permeability heterogeneity and poor ISCO fluid delivery. Polyphosphates act as anti-coagulants and particle stabilizers and reduce the accumulation of MnO_2 particles, which are by-products of the reactions between permanganate and various contaminants. PA-ISCO based treatment using sodium permanganate as oxidant, xanthan gum biopolymer as viscosifier, and sodium hexametaphosphate (SHMP) as an anti-coagulant has shown 103% improvement in efficiency compared to permanganate-only treatment (Silva et al. 2017).

The use of micro-nano-bubbles (MNBs) is another innovative approach to increase the effectiveness of in situ oxidation while using ozone as the oxidant. Ozone is a highly efficient oxidant but its activity is often limited due to its moderately low stability in the aqueous solution. MNBs with diameters less than 100 μm present rapid mass transfer rates and increase the availability of the gas at the site for longer periods (Hu and Xia 2018). Groundwater remediation at a trichloroethylene (TCE) contaminated site in Japan using ozone MNBs along with H_2O_2 (mass ratio of ozone: H2O2 of 1 : 1) lead to 99% removal of TCE in groundwater in just six days, bringing the TCE levels below the local permissible limit (Xia et al. 2018).

17.3.1.2 In-Situ Chemical Reduction

In-situ chemical reduction (ISCR) is also very similar to ISCO: in ISCR a reductant or reducing agent in the liquid form is introduced into the contaminated zone that causes the transformation of the contaminants into a less pertinacious form, and it has been noted that ISCR is more effective than ISCO. Examples of various reductants that have been used for groundwater treatment include polysulphides (Storch et al. 2002), zero-valent metals (ZVMs) (Nidheesh et al. 2018), iron minerals (Berens et al. 2020), and bimetallic materials (Han and Yan 2014). Either zero-valent or ferrous (+2) iron are the most widely used reducing agents for groundwater treatment and have shown the tendency to degrade both inorganic and organic contaminants (Brown et al. 2006). ISCR using commercial zero-valent iron microparticles (ZVIm) of groundwater of Bailin's landfill (Sabiñanigo, Spain), where the waste generated by a lindane factory was dumped for years, resulted in efficient removal of persistent chlorinated organic compounds (COCs) dissolved in groundwater (Dominguez et al. 2018). Inorganic contaminants such as chromium, dichromates, and nitrates (Ludwig et al. 2007; Vilardi et al. 2017) have also been successfully removed from groundwater using iron-based reduction methods.

17.3.2 Ion Exchange Methods

Ion exchange is the process that is used mainly for purification of aqueous solution, using resin, because it contains an ionic functional group to which exchangeable ions are attached (Srikanth et al. 2010). During this process when water is forcefully passing downward under pressure through a particular granular medium, anions are displaced by certain anions from the solution and cations are displaced by certain cations. Zeolites (Fedoročková et al. 2016) and synthetic resins (Park et al. 2020) are widely used for ion exchange methods (Hyman and Dupont 2001). The application of ion exchange resins is also very effective and economical in removal of heavy metals like mercury, zinc, nickel, lead, chromium (Cavaco et al. 2007), and radionuclides (Yang et al. 2016) from groundwater.

17.3.3 Flocculation

Flocculation is the process in which aggregates are formed when the solution is colloidally unstable (Slomkowski et al. 2011). Flocculation is induced by flocculating agents, which are necessary to remove suspended particles from liquid (Krupińska 2011). Examples of flocculating agents are Chitson (Hughes et al. 1990), Isinglass (Walker et al. 2007), Moringa Olifera seeds (Horse radish tree) (Teixeira et al. 2012), Gelatine, Guar Gum (Nakamura et al. 2020), Alginates (Brown seaweed extract) (Tian et al. 2019), etc. In this process particles of a size below 10 μm are sedimented with the help of certain coagulants and flocculants, which are used by different procedures (Jekel and Gimbel 2000).

17.3.4 Adsorption

The rapid growth of industrialization in our country and all over the world has caused some serious threats to environment and their natural resources like groundwater, etc. As these industries directly pour their wastewater into rivers, lakes, ponds, etc., the chances of percolation of polluted water to groundwater are really very high. Therefore, there are various steps taken to control the pollutants. One of the most popular methods is using adsorbents such as active carbon, diatomite, polyaluminium chloride, and polyferricsulfate to treat groundwater. This method brings a significant change by removing odour and taste causing compounds and other contaminants. Activated carbon has a really good capacity of adsorption because of its highly porous nature and extremely large surface area (Bin Jusoh et al. 2005). Diatomite (hydrated silica) is effectively used in adsorption of contaminants from contaminated zones in groundwater treatment because they are nanostructures, highly porous, and have extremely large surface area (Sheng et al. 2008; Sharipova et al. 2017; Garcia-Alonso et al. 2018). Polyaluminium chloride is effectively used as a coagulant in water treatment plants (Gadekar and Ahammed 2020) and also used in deflourination of water (Cho et al. 2020). Poly ferric sulphate (PFS) is an inorganic coagulant. PFS plays an extraordinary role in decreasing the chemical oxygen demand (COD) and biochemical oxygen demand, and it also helps in removing all kinds of oil presence with more than 60% of removal rate (Liu et al. 2013) and it is also noticed that water that is treated with PFS is safe for human consumption (Hendrich et al. 2001; Liu et al. 2013).

17.3.5 Bioremediation

Bioremediation is a process in which microorganisms play a crucial role in degradation of the toxic contaminants present in subsurface of groundwater, and this is one of the most effective and viable technologies. In-situ bioremediation is primarily used for organic contaminants such as chlorinated solvents typically following the principle of oxidation–reduction manipulation, in which microbial catalysed biochemical reaction leads to the degradation of chemicals (Faris and Vlassopoulos 2003). Inorganic contaminants can also be treated with the help of bioremediation approaches; for example, nitrates and sulphates, which are used in in-situ bioremediation, are really very effective for treating contaminants present in groundwater (Brusseau 2019).

With the activity of microorganisms, we can reduce a vast number of environmental pollutants such as petroleum hydrocarbons, following the oxidation of these pollutants resulting in the innocuous final compounds such as CO_2 and H_2O, whereas pollutants like chlorinated solvents and nitrates are prone to undergo reduction because they are highly oxidized. Such redox reactions are mediated by microorganisms which maybe anaerobic or aerobic and they flourish on the contaminant degradation process for growth by obtaining its food and energy (Scherer et al. 2000). In-situ bioremediation of various organic contaminants in groundwater requires aerobic conditions in the aquifer substrate but the concentration of dissolved oxygen in groundwater (below 3 mg/L) is not enough to support efficient bioremediation processes (Lee et al. 2014). Therefore, different techniques such as biosparging and bioventing are often used to obtain the suitable conditions (Vidali 2001). Biosparging is a technology in which organic matter present in a saturated zone is degraded by indigenous microorganisms. In this process the biological activity of indigenous microorganisms increased by injecting oxygen and nutrients into the saturated zone. Whereas in the case of bioventing, the unsaturated zone is overflowed by oxygen either through injection or extraction to increase concentration of oxygen. The most commonly used method is direct air injection as in this method the flow rates of oxygen can be easily controlled and provide only enough supply to sustain microbial activity. A recent development for maintaining aerobic conditions for groundwater treatment is the application of ORCs (oxygen-releasing compounds) such as magnesium peroxide (MgO_2) and calcium peroxide (CaO_2), which slowly decompose to oxygen when they encounter water. The use of ORCs has proved to be more economical, environmental, and effective (Mosmeri et al. 2018; Mosmeri et al. 2019).

Furthermore, to improve the aerobic biodegradation oxygen-releasing alginate beads (ORABs) have been brought into application. They help in overcoming limitations associated with the fast release of oxygen, provide suitable conditions for ORCs, and enhance their activity. In a study, the encapsulation of calcium peroxide within alginate beads for biodegradation of 1,4-dioxane by *Mycrobacterium* sp. PH-06 in batch and column conditions elucidated maximum degradations of 99 and 68.1%, whereas without ORABs 34.3 and 18% of 1,4-dioxane were degraded, respectively (Lee et al. 2014).

Bioaugmentation is another method of bioremediation in which mixed cultures and selected strains are added to the wastewater reactors to increase catabolism of a specific compound. Bioaugmentation is also very helpful in solving pragmatic problems like WWTP (wastewater treatment plants) and enhancing removal efficiency (Herrero and Stuckey 2015; Huang and Ye 2020). Large-scale bioaugmentation of a site contaminated with total petroleum hydrocarbons (TPH) at a petroleum facility has shown to decrease the concentrations of TPHs from 1564 to 89 $mg\,L^{-1}$ in 32 days (Poi et al. 2018). The bioaugmented columns (inoculated with Mn-oxidizing strain Pseudomonas sp. QJX-1) compared to non-bioaugmented controls have shown improvement in As(III) and Sb(III) removal while studying the treatment of groundwater contaminated with various inorganic contaminants such as Fe(II), Mn(II), As(III), and Sb(III).The other advantages of using bioaugmentation included reduced start-up time and anti-shock load capacity (Bai et al. 2016).

17.4 Conclusion

Groundwater contamination is a serious threat to the environment. Therefore, it is very important to develop efficient methods for groundwater remediation. Since the initial initiatives taken for groundwater remediation in 1950s, there has been great advancement in the techniques used for the treatment of groundwater. The focus has shifted from ex-situ remediation to in-situ remediation and various physical, chemical, and biological methods are being employed for groundwater remediation. Developmental and innovative changes have increased the effectiveness of the existing methods of remediation, whereas various new methods of remediation have also been introduced. Different methods of groundwater remediation have proven to be good for different kinds of contaminants. Although there has been great development, combinations of groundwater treatment methods are required for efficient clean-up.

References

Andaluri, G. and Suri, R. (2017). Removal of 1, 4-dioxane and volatile organic compounds from groundwater using ozone-based advanced oxidation process. *Ozone: Science and Engineering* 39: 423–434.

Angmo, D., Dutta, R., Kumar, S.I., and Sharma, A. (2020). Natural biological treatment of effluent and sludges to combat the burden of waste. In: *Earthworm Assisted Remediation of Effluents and Wastes* (eds. S.A. Bhat, A.P. Vig, F. Li and B. Ravindran), 107–122. Singapore: Springer.

Babiker, I.S., Mohamed, M.A., Terao, H. et al. (2004). Assessment of groundwater contamination by nitrate leaching from intensive vegetable cultivation using geographical information system. *Environment International* 29: 1009–1017.

Bai, Y., Chang, Y., Liang, J. et al. (2016). Treatment of groundwater containing Mn (II), Fe (II), As (III) and Sb (III) by bioaugmented quartz-sand filters. *Water Research* 106: 126–134.

Berens, M.J., Hofstetter, T.B., Bolotin, J., and Arnold, W.A. (2020). Assessment of 2, 4-dinitroanisole transformation using compound-specific isotope analysis after in situ chemical reduction of iron oxides. *Environmental Science and Technology* 54: 5520–5531.

Besha, A.T., Bekele, D.N., Naidu, R., and Chadalavada, S. (2018). Recent advances in surfactant-enhanced in-situ chemical oxidation for the remediation of non-aqueous phase liquid contaminated soils and aquifers. *Environmental Technology and Innovation* 9: 303–322.

Bin Jusoh, A., Cheng, W., Low, W. et al. (2005). Study on the removal of iron and manganese in groundwater by granular activated carbon. *Desalination* 182: 347–353.

Bouchard, D.C., Williams, M.K., and Surampalli, R.Y. (1992). Nitrate contamination of groundwater: sources and potential health effects. *Journal-American Water Works Association* 84: 85–90.

Bouzid, I., Maire, J., Brunol, E. et al. (2017). Compatibility of surfactants with activated-persulfate for the selective oxidation of PAH in groundwater remediation. *Journal of Environmental Chemical Engineering* 5: 6098–6106.

Brown, R.A., Lewis, R., Fiacco JR, R., and Leahy, M. The technical basis for in situ chemical reduction (ISCR). Proceedings, Fifth International Conference on Chlorinated and Recalcitrant Compounds, Monterey, CA, USA, May, 2006. 22–25.

Brusseau, M.L. (2019). Soil and groundwater remediation. In: *Environmental and Pollution Science* (eds. M.L. Brusseau, I.L. Pepper and C.P. Gerba), 329–354. Amsterdam: Elsevier.

Cavaco, S.A., Fernandes, S., Quina, M.M., and Ferreira, L.M. (2007). Removal of chromium from electroplating industry effluents by ion exchange resins. *Journal of Hazardous Materials* 144: 634–638.

Chen, J., Qian, Y., Liu, H., and Huang, T. (2016). Oxidative degradation of diclofenac by thermally activated persulfate: implication for ISCO. *Environmental Science and Pollution Research* 23: 3824–3833.

Cho, D.-W., Han, Y.-S., Lee, J. et al. (2020). Water defluorination using granular composite synthesized via hydrothermal treatment of polyaluminum chloride (PAC) sludge. *Chemosphere* 247: 125899.

Dahal, G., Holcomb, J., and Socci, D. (2016). Surfactant-oxidant co-application for soil and groundwater remediation. *Remediation Journal* 26: 101–108.

Devi, P., Das, U., and Dalai, A.K. (2016). In-situ chemical oxidation: principle and applications of peroxide and persulfate treatments in wastewater systems. *Science of the Total Environment* 571: 643–657.

Dominguez, C.M., Romero, A., Fernandez, J., and Santos, A. (2018). In situ chemical reduction of chlorinated organic compounds from lindane production wastes by zero valent iron microparticles. *Journal of Water Process Engineering* 26: 146–155.

Edition, F. (2011). Guidelines for drinking-water quality. *WHO Chronicle* 38: 104–108.

Faisal, A., Sulaymon, A., and Khaliefa, Q. (2018). A review of permeable reactive barrier as passive sustainable technology for groundwater remediation. *International journal of Environmental Science and Technology* 15: 1123–1138.

Faris, B. and Vlassopoulos, D. (2003). A systematic approach to in situ bioremediation in groundwater. *Remediation Journal: The Journal of Environmental Cleanup Costs, Technologies and Techniques* 13: 27–52.

Fedoročková, A., Sučik, G., and Raschman, P. (2016). Activated zeolite and magnesite as potential reactive materials for passive acidic groundwater treatment technology. *Solid State Phenomena* 244. Trans Tech Publ,: 221–227.

Filler, D.M., Van Stempvoort, D.R., and Leigh, M.B. (2009). Remediation of frozen ground contaminated with petroleum hydrocarbons: feasibility and limits. In: *Permafrost Soils*, 279–301. Berlin, Heidelberg: Springer.

Gadekar, M.R. and Ahammed, M.M. (2020). Use of water treatment residuals for colour removal from real textile dye wastewater. *Applied Water Science* 10: 1–8.

Garcia-Alonso, J.A., Zurita-Martinez, F., Guzmán-González, C.A. et al. (2018). Nanostructured diatomite and its potential for the removal of an antibiotic from water. *Bioinspired, Biomimetic and Nanobiomaterials* 7: 167–173.

Gehringer, P., Eschweiler, H., and Fiedler, H. (1995). Ozone-electron beam treatment for groundwater remediation. *Radiation Physics and Chemistry* 46: 1075–1078.

Gibert, O., Assal, A., Devlin, H. et al. (2019). Performance of a field-scale biological permeable reactive barrier for in-situ remediation of nitrate-contaminated groundwater. *Science of the Total Environment* 659: 211–220.

Gillham, R.W., Vogan, J., Gui, L. et al. (2010). Iron barrier walls for chlorinated solvent remediation. In: *In Situ Remediation of Chlorinated Solvent Plumes* (eds. H.F. Stroo and C.H. Ward), 537–571. New York: Springer.

Guo, Z., Brusseau, M.L., and Fogg, G.E. (2019). Determining the long-term operational performance of pump and treat and the possibility of closure for a large TCE plume. *Journal of Hazardous Materials* 365: 796–803.

Han, Y. and Yan, W. (2014). Bimetallic nickel–iron nanoparticles for groundwater decontamination: effect of groundwater constituents on surface deactivation. *Water Research* 66: 149–159.

Hendrich, S., Fan, M., Sung, S.-W. et al. (2001). Toxicity evaluation of polymeric ferric sulphate. *International Journal of Environmental Technology and Management* 1: 464–471.

Herrero, M. and Stuckey, D. (2015). Bioaugmentation and its application in wastewater treatment: a review. *Chemosphere* 140: 119–128.

Hoffman, F. (1993). Ground-water remediation using "smart pump and treat". *Groundwater* 31: 98–106.

Hu, L. and Xia, Z. (2018). Application of ozone micro-nano-bubbles to groundwater remediation. *Journal of Hazardous Materials* 342: 446–453.

Huang, H. and Ye, L. (2020). Biological technologies for cHRPs and risk control. In: *High-Risk Pollutants in Wastewater* (eds. H. Ren and X. Zhang), 209–236. Elsevier.

Hughes, J., Ramsden, D., and Symes, K. (1990). The flocculation of bacteria using cationic synthetic flocculants and chitosan. *Biotechnology Techniques* 4: 55–60.

Hyman, M. and Dupont, R.R. Groundwater and soil remediation: process design and cost estimating of proven technologies. 2001. American Society of Civil Engineers.

Jekel, M. and Gimbel, R. 2000. Water, 4. Treatment by flocculation and filtration. Ullmann's Encyclopedia of Industrial Chemistry.

Jiang, Y. (2015). China's water security: current status, emerging challenges and future prospects. *Environmental Science and Policy* 54: 106–125.

Knobeloch, L., Salna, B., Hogan, A. et al. (2000). Blue babies and nitrate-contaminated well water. *Environmental Health Perspectives* 108: 675–678.

Kornilovych, B., Wireman, M., Ubaldini, S. et al. (2018). Uranium removal from groundwater by permeable reactive barrier with zero-valent iron and organic carbon mixtures: laboratory and field studies. *Meta* 8: 408.

Krupińska, I. (2011). Potassium permanganate as an oxidizing and flocculating agent in groundwater treatment. *Civil and Environmental Engineering Reports* 6: 75–88.

Lee, C.-S., Le Thanh, T., Kim, E.-J. et al. (2014). Fabrication of novel oxygen-releasing alginate beads as an efficient oxygen carrier for the enhancement of aerobic bioremediation of 1, 4-dioxane contaminated groundwater. *Bioresource Technology* 171: 59–65.

Liu, C., He, Y., Li, F., and Wang, H. (2013). Preparation of poly ferric sulfate and the application in micro-polluted raw water treatment. *Journal of the Chinese Advanced Materials Society* 1: 210–218.

Logeshwaran, P., Megharaj, M., Chadalavada, S. et al. (2018). Petroleum hydrocarbons (PH) in groundwater aquifers: an overview of environmental fate, toxicity, microbial degradation and risk-based remediation approaches. *Environmental Technology and Innovation* 10: 175–193.

Ludwig, R.D., Su, C., Lee, T.R. et al. (2007). In situ chemical reduction of Cr (VI) in groundwater using a combination of ferrous sulfate and sodium dithionite: a field investigation. *Environmental Science and Technology* 41: 5299–5305.

Menchen, A., De Las Heras, J., and Alday, J.J.G. (2017). Pesticide contamination in groundwater bodies in the Júcar River European Union pilot basin (SE Spain). *Environmental Monitoring and Assessment* 189: 146.

Mosmeri, H., Gholami, F., Shavandi, M. et al. (2018). Application of magnesium peroxide (MgO 2) nanoparticles for toluene remediation from groundwater: batch and column studies. *Environmental Science and Pollution Research* 25: 31051–31061.

Mosmeri, H., Gholami, F., Shavandi, M. et al. (2019). Bioremediation of benzene-contaminated groundwater by calcium peroxide (CaO2) nanoparticles: continuous-flow and biodiversity studies. *Journal of Hazardous Materials* 371: 183–190.

Nakamura, A., Ozaki, M., and Murakami, K. (2020). Elucidation of the aggregation mechanism of bentonite with cationic guar gum as flocculant and application to filtration. *Colloids and Surfaces A: Physicochemical and Engineering Aspects* 596: 124660.

Nidheesh, P., Khatri, J., Singh, T.A. et al. (2018). Review of zero-valent aluminium based water and wastewater treatment methods. *Chemosphere* 200: 621–631.

O'connor, D., Hou, D., Ok, Y.S. et al. (2018). Sustainable in situ remediation of recalcitrant organic pollutants in groundwater with controlled release materials: a review. *Journal of Controlled Release* 283: 200–213.

Park, M., Daniels, K.D., Wu, S. et al. (2020). Magnetic ion-exchange (MIEX) resin for perfluorinated alkylsubstance (PFAS) removal in groundwater: roles of atomic charges for adsorption. *Water Research* 181: 115897.

Perrich, J.R. (2018). *Activated Carbon Adsorption for Wastewater Treatment*. CRC press.

Poi, G., Shahsavari, E., Aburto-Medina, A. et al. (2018). Large scale treatment of total petroleum-hydrocarbon contaminated groundwater using bioaugmentation. *Journal of Environmental Management* 214: 157–163.

Scherer, M.M., Richter, S., Valentine, R.L., and Alvarez, P.J. (2000). Chemistry and microbiology of permeable reactive barriers for in situ groundwater clean up. *Critical Reviews in Microbiology* 26: 221–264.

Schmoll, O., Howard, G., Chilton, J. et al. (2006). *Protecting groundwater for health: managing the quality of drinking-water sources*. World Health Organization.

Schweitzer, L. and Noblet, J. (2018). Water contamination and pollution. In: *Green Chemistry*. Elsevier.

Sethi, R. and DI Molfetta, A. (2019). *Groundwater Contaminants. Groundwater Engineering*. Springer.

Shang, H., Javadi, S., and Zhao, Q. (2017). Organic surfactant modified zeolite as a permeable reactive barrier component – a laboratory study. *Geotechnical Frontiers* 2017: 443–449.

Sharipova, A.A., Aidarova, S.B., Bekturganova, N.Y. et al. (2017). Triclosan adsorption from model system by mineral sorbent diatomite. *Colloids and Surfaces A: Physicochemical and engineering aspects* 532: 97–101.

Sharma, B.M., Bečanová, J., Scheringer, M. et al. (2019). Health and ecological risk assessment of emerging contaminants (pharmaceuticals, personal care products, and artificial sweeteners) in surface and groundwater (drinking water) in the Ganges River Basin, India. *Science of the Total Environment* 646: 1459–1467.

Sheng, G., Hu, J., and Wang, X. (2008). Sorption properties of Th (IV) on the raw diatomite—effects of contact time, pH, ionic strength and temperature. *Applied Radiation and Isotopes* 66: 1313–1320.

Siegrist, R.L., Crimi, M., and Simpkin, T.J. (2011). *In Situ Chemical Oxidation for Groundwater Remediation*. Springer Science and Business Media.

Silva, J.A., Crimi, M., Palaia, T. et al. (2017). Field demonstration of polymer-amended in situ chemical oxidation (PA-ISCO). *Journal of Contaminant Hydrology* 199: 36–49.

Slomkowski, S., Alemán, J.V., Gilbert, R.G. et al. (2011). Terminology of polymers and polymerization processes in dispersed systems (IUPAC recommendations 2011). *Pure and Applied Chemistry* 83: 2229–2259.

Srikanth, M., Sunil, S., Rao, N. et al. (2010). Ion-exchange resins as controlled drug delivery carriers. *Journal of Scientific Research* 2: 597–597.

Storch, P., Messer, A., Palmer, D., and Pyrih, R. Pilot test for in situ geochemical fixation of chromium (VI) using calcium polysulfide. Proceedings of the Third International Conference on Remediation of Chlorinated and Recalcitrant Compounds: pts 1a–2b, 2002.

Teh, C.Y., Budiman, P.M., Shak, K.P.Y., and Wu, T.Y. (2016). Recent advancement of coagulation–flocculation and its application in wastewater treatment. *Industrial and Engineering Chemistry Research* 55: 4363–4389.

Teixeira, C.M.L.L., Kirsten, F.V., and Teixeira, P.C.N. (2012). Evaluation of Moringa oleifera seed flour as a flocculating agent for potential biodiesel producer microalgae. *Journal of Applied Phycology* 24: 557–563.

Thiruvenkatachari, R., Vigneswaran, S., and Naidu, R. (2008). Permeable reactive barrier for groundwater remediation. *Journal of Industrial and Engineering Chemistry* 14: 145–156.

Tian, Z., Zhang, L., and Ni, C. (2019). Preparation and flocculation properties of modified alginate amphiphilic polymeric nano-flocculants. *Environmental Science and Pollution Research* 26: 32397–32406.

Tratnyek, P.G., Scherer, M.M., Johnson, T.L., and Matheson, L.J. (2003). Permeable reactive barriers of iron and other zero-valent metals. *Environmental Science and Pollution Control Series* 1: 371–422.

Truex, M.J., Johnson, C.D., Becker, D.J. et al. (2015). *Performance Assessment for Pump-and-Treat Closure or Transition*. Richland, WA (United States): Pacific Northwest National Lab (PNNL).

Truex, M., Johnson, C., Macbeth, T. et al. (2017). Performance assessment of pump-and-treat systems. *Groundwater Monitoring and Remediation* 37: 28–44.

Vidali, M. (2001). Bioremediation. An overview. *Pure and Applied Chemistry* 73: 1163–1172.

Vilardi, G., Di Palma, L., and Verdone, N. (2017). Competitive reaction modelling in aqueous systems: the case of contemporary reduction of dichromates and nitrates by nZVI. *Chemical Engineering Transactions* 60: 175–180.

Voice, T.C., Pak, D., Zhao, X. et al. (1992). Biological activated carbon in fluidized bed reactors for the treatment of groundwater contaminated with volatile aromatic hydrocarbons. *Water Research* 26: 1389–1401.

Walker, S.L., Camarena, M.C.D., and Freeman, G. (2007). Alternatives to isinglass for beer clarification. *Journal of the Institute of Brewing* 113: 347–354.

Wang, L., Peng, L., Xie, L. et al. (2017). Compatibility of surfactants and thermally activated persulfate for enhanced subsurface remediation. *Environmental Science and Technology* 51: 7055–7064.

Werkneh, A.A. and Rene, E.R. (2019). *Applications of Nanotechnology and Biotechnology for Sustainable Water and Wastewater Treatment. Water and Wastewater Treatment Technologies* (eds. X. Bui, C. Chiemchaisri, T. Fujioka and S. Varjani), 405–430. Singapore: Springer.

Xia, Z., Hu, L., Kusaba, S., and Song, D. (2018). Remediation of TCE contaminated site by ozone micro-nano-bubbles. In: *The International Congress on Environmental Geotechnics*, 796–803. Springer.

Yang, H., Luo, M., Luo, L. et al. (2016). Highly selective and rapid uptake of radionuclide cesium based on robust zeolitic chalcogenide via stepwise ion-exchange strategy. *Chemistry of Materials* 28: 8774–8780.

Yang, Z.-H., Verpoort, F., Dong, C.-D. et al. (2020). Remediation of petroleum-hydrocarbon contaminated groundwater using optimized in situ chemical oxidation system: batch and column studies. *Process Safety and Environment Protection* 138: 16–26.

Zeidan, B. A (n.d.). Groundwater Quality Management.

Zhang, M. and Dong, J. (2019). Phase-transfer catalysis enhanced remediation of trichloroethylene polluted groundwater by potassium permanganate. *Environmental Technology* 41: 1–12.

Zhang, X., Ye, C., Pi, K. et al. (2019). Sustainable treatment of desulfurization wastewater by ion exchange and bipolar membrane electrodialysis hybrid technology. *Separation and Purification Technology* 211: 330–339.

Zhong, H., Tian, Y., Yang, Q. et al. (2017). Degradation of landfill leachate compounds by persulfate for groundwater remediation. *Chemical Engineering Journal* 307: 399–407.

18

Exploration of Water Resources Using Remote Sensing and Geographic Information System

Chandrashekhar Azad Vishwakarma[1], Vikas Rena[2], Deepali Singh[2], and Saumitra Mukherjee[2]

[1] *TERI School of Advanced Studies, New Delhi, India*
[2] *School of Environmental Sciences, Jawaharlal Nehru University, New Delhi, India*

18.1 Introduction

18.1.1 Remote Sensing and Geographic Information System

Lillesand et al. (2004) defined the term remote sensing (RS) as 'the science and art of obtaining information about an object, area and phenomenon through analysis of data acquired by any device that is not in contact with the object, area, or phenomenon under investigation'. The concept of sensing an object by a satellite sensor is similar to that of our eyes. A range of wavelengths enters our eyes and sends a signal to the brain to interpret the object. For example, one reading the black area of pages in the form of letters, results into formation of words and, further collection of information by the collation and identification of words and sentences. The sensor in satellite RS also works in the same way, except the range of wavelength, which is not fixed like that for our eyes. The sensors of satellites gather information by emitted and reflected electromagnetic energy from various features on the Earth surface (Figure 18.1). They can work over a wide range of electromagnetic spectrum, on spaceborne or airborne platforms, and are used to analyse, map, and monitor Earth's resources (Lillesand and Kiefer 2007). RS technology plays a significant role by collecting information of inaccessible areas and generating information with spatial, spectral, and temporal variability. It is a useful tool to assess, monitor, and manage the groundwater resources (Chowdhury et al. 2009).

In combination with RS, the geographic information system (GIS) is a powerful tool to analyse the spatial data and is used as a decision support system in many fields, including the environment and engineering (Chowdhury et al. 2009). GIS is defined by Chang (2016) as 'a computer system for capturing, storing, querying, analyzing and displaying geospatial data'. The geospatial data is also known as geographically referenced data. It entails both location and the characteristics of the surface features such as forests, agriculture lands, and built-up areas on the Earth's surface (Figure 18.2) (Chang 2006).

Groundwater Geochemistry: Pollution and Remediation Methods, First Edition.
Edited by Sughosh Madhav and Pardeep Singh.

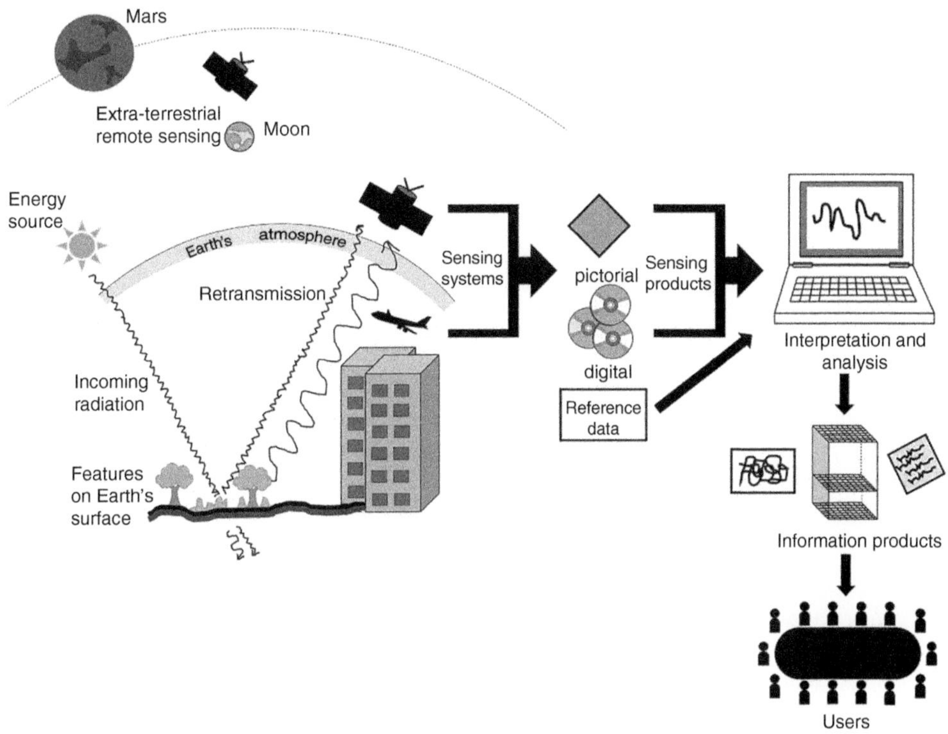

Figure 18.1 Pictorial representation of remote sensing data acquisition and analysis.

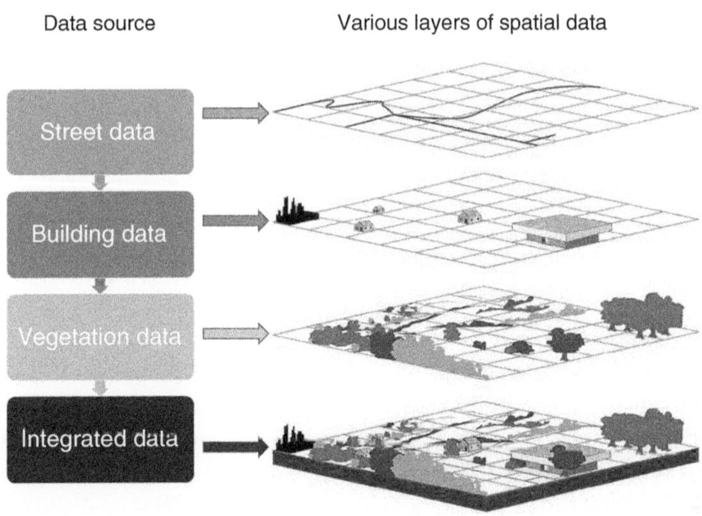

Figure 18.2 Schematic diagram of geographic information system. *Source:* Chang, K.-T. (2006). Introduction to geographic information systems: McGraw-Hill Higher Education Boston.© 2006, McGraw Hill.

The application of this technique is not only important in identification of groundwater potential zones but also in conservation and monitoring of the sources of water. The analysis and identification of suitable aquifers can also be determined using traditional standard methods like test drilling and stratigraphic analysis (Chowdhury et al. 2009). This kind of analysis is, however, very expensive, time consuming, and requires human resources. RS technology is an economic alternative for hydrogeological mapping (Sander et al. 1996). The availability and movement of groundwater in a particular area is influenced by many factors like lithology, topography, geological structures, depth of weathering, secondary porosity, extent of fractures, climatic conditions, and the correlation between these factors (Jaiswal et al. 2003). Therefore, while it is clear that direct measurement of hydrological properties is not possible by RS data, it may derive direct and indirect measurements of the different surface features, collected from satellite imagery in the form of slope, elevation, landforms, soils, land use/land cover, vegetation, and water bodies that further act as indicators of groundwater existence (Sander et al. 1996; Chowdhury et al. 2009). The delineation of groundwater potential zones and groundwater modelling need large volumes of data. Therefore, GIS has emerged as a powerful decision-making tool. Jaiswal et al. (2003) explained the role of GIS in combining with various factors for delineation of groundwater potential zones. Thus, for groundwater potential zone and groundwater modelling, the integrated approach of RS and GIS has become a precious tool (Chowdhury et al. 2009).

The integrated approach of two techniques for groundwater modelling and its availability was elucidated by many researchers. Several studies have used RS and GIS techniques to delineate groundwater potential zones and found successful results (Karanth and Seshu Babu 1978; Chi and Lee 1994; Krishnamurthy and Srinivas 1995; Kamaraju et al. 1996; Krishnamurthy et al. 1996; Sander et al. 1996; Khan and Mohd 1997; Ravindran 1997; Edet et al. 1998; Saraf and Choudhury 1998; Krishnamurthy et al. 2000; Shahid et al. 2000; Jaiswal et al. 2003; Sikdar et al. 2004; Sener et al. 2005; Solomon and Quiel 2006). Chowdhury et al. (2009) has explained the integrated approach to use these two spatial data management techniques for development of groundwater strategies in a hard rock terrain. RS and GIS have importance in various applications like assessment and exploration of the groundwater resources, selection of artificial recharge sites, GIS-based subsurface flow and pollution modelling, assessment of groundwater pollution and its planning for protection, analysis of hydrogeological data and process monitoring, and assessment of natural recharge distribution sites (Jha et al. 2007).

18.1.2 Hydro-geomorphological Unit and Groundwater

The presence and movement of groundwater below the Earth's surface is governed by many factors like physiography, drainage, geology, structure, and hydrogeology (Rao et al. 2001). False Color Composite (FCC) of a satellite imagery is helpful in identification of geomorphological units by visual interpretation. Various studies have been carried out by Krishnamurthy and Srinivas (1995), Ghayoumian et al. (2007), Murthy (2000), Sener et al. (2005), Sreedevi et al. (2005), Gogu et al. (2001), Bonham-Carter (1996), and Chopra and Sharma (1993) for identification of groundwater potential zones by analysing different geomorphic and landform features (Gaur et al. 2011). Use of satellite data in identification of different landforms with favourable groundwater zones has been reported from different parts of India (Dhakate et al. 2008).

Mukherjee et al. (2007) have revealed that the occurrence and amount of groundwater in hard rock terrain is related to lithological, structural, and geomorphological factors. The network of joints, fractures, and faults in the weathered zones is a good signature of groundwater movement and storage. Data from multiple types of sensor can be used for groundwater restoration (Mukherjee 2008). Roy (2009) used integrated approach of RS and GIS in Sikkim, which has Himalayan rocky terrain. Through lineament studies, many fractured aquifers were identified and it was found that the principal mode of recharge of groundwater in Sikkim is direct infiltration of rainfall through joints, fractures, and weathered zones of the rocks and through soil covers (Roy 2009).

18.1.3 Effect of Vegetation and Deforestation on Hydrology

Forestry and plantation programmes are carried out throughout the world. They help in creating a clean and hygienic environment by removal of carbon dioxide and addition of oxygen in nature. They help in maintenance of climate at local and regional scale, as well as play a significant role in precipitation and groundwater recharge. But the rate of deforestation is several million hectares more than afforestation, which has declined by 8.4 million hectares or 0.2% per year between 1990 and 2005 (FAO 2005; van Dijk and Keenan 2007). Therefore, afforestation needs to be promoted for water-related problems such as to stop or mitigate river flooding, landslides, salinity, soil erosion, river pollutant loads, and global warming (van Dijk and Keenan 2007).

The calculation of Normalized Difference Vegetation Index (NDVI) is a prevalent method for identification of vegetation using remotely sensed data. Vegetation index is the algorithm used to simplify the data in multiple reflectance bands into a single value, which may be used to derive vegetation parameters such as biomass, leaf area index (LAI), productivity, and percent vegetation ground cover (Tucker 1979). The NDVI is calculated by subtraction of red reflectance value from near-infrared reflectance value, which is divided by the sum of red and near-infrared reflectance (Jackson et al. 2004). The map generated by NDVI using multispectral sensor may be used to infer the impact of urbanization in groundwater recharge in the area (Mukherjee 2008).

18.1.4 Land Use Land Cover Change Analysis and Groundwater

Although the terms 'land use' and 'land cover' are frequently used interchangeably, they actually convey different meanings. The term 'land cover' refers to the area covered by natural features on the surface of the earth, i.e. the biotic and abiotic components of the Earth's surface, and the term 'land use' describes how the land is being used by humans (NRSC 2012). RS uses multiple bands, spanning different ranges in the electromagnetic (EM) spectra, in delineation of various land use/land cover classes. Land cover/land use mapping can be performed by both visual interpretation and digital analysis techniques in the GIS platform (Singh and Dubey 2012). Supervised and unsupervised classification are the two most commonly known digital techniques used for classification of land use/land cover in various GIS interfaces such as ArcGIS, ENVI, ERDAS Imagine, etc. In supervised classification the user must have prior knowledge about various classes in the land cover area whereas, no prior information about the land cover area is required in case of unsupervised classification (NRSC 2012).

The effect of land use change can be seen on biodiversity, water and radiation budgets, trace gas emissions, and many other factors, which directly or indirectly alter the global biosphere (Riebsame et al. 1994). The change in land use patterns can transform ecosystems and their functions, especially the hydrological performances. A study conducted in watersheds of a major human habitation zone in the Himalayas shows that these watersheds were more efficient in water conservation and with reduced percentage of runoff (Heilongtan 2007). Although the changes in land use due to construction are necessary, they adversely affect the quality and quantity of groundwater (Mahamuni and Upasani 2011).

18.1.5 Identification of Water-Ice Using Extraterrestrial Remote Sensing

Just as RS is used for groundwater and natural resource exploration on Earth, it has also been used in space. Several space agencies of leading nations have been working towards this goal for the past 50 years. This field has seen exponential growth in the last decade due to the discovery of water-ice on the Moon and Mars. Various regions of the electromagnetic (EM) spectra including visible (McEwen et al. 2007), IR (infrared) (Murchie et al. 2007), microwave (Seu et al. 2007; Plaut et al. 2009), and gamma rays (Boynton et al. 2002) have been exploited to infer natural resources on planetary bodies.

Water acts as a solvent in various cellular reactions for life on Earth and, is therefore, a requirement for habitability on any planet. Thus, central to the primary objective of recent planetary exploration missions oriented towards the search for traces of life forms, the detection of liquid water continues to remain the prime focus. Additionally, its potential is being explored with regards to an alternate fuel source for longer journeys as well as sample return missions from other planets (Elitzur et al. 2016; Doyle and Peck 2018; Doyle and Peck 2019)

18.2 Methods for Groundwater Mapping

RS and GIS have become useful tools in natural resource management, environment impact assessment (EIA), and land planning. Groundwater potential zone identification is a challenging task in hard rock terrain and drought-prone and lateritic areas (Mukherjee et al. 2007; Roy et al. 2020). For evaluation of groundwater potential zones, various thematic layers like geology, slope, rainfall, elevation, NDVI, drainage density, land use land cover (LULC), lineament density, Topographic Wetness Index (TWI), proximity to surface water bodies, hydraulic conductivity (K), and soil texture are generated by using field data coupled with remotely sensed data in GIS platform (Arulbalaji et al. 2019). The layers can further integrated using weightage overlay, analytical hierarchy process (AHP), fuzzy logic, artificial neural network (ANN), and other statistical methods.

18.2.1 Weighted Overlay

Weighted overlay is the basic method of multi-criteria decision analysis (MCDA) for groundwater potential and recharge zone mapping. Based on literature review, expert opinion, and local hydro-geologic conditions, weightage is assigned to different thematic layers considering the relative importance of each layer separately. All the datasets are further resampled using the nearest-neighbour algorithm for the same resolution (Zadeh 1965). On the basis of their histograms, the groundwater potential map generated using this approach may be classified from not suitable to most suitable sites. These maps are further validated with the groundwater yield data. Significantly reliable results of groundwater potential zones may be used for effective planning for the study area (Jaiswal et al. 2003).

18.2.2 Frequency Ratio (FR) Model

The frequency ratio (FR) is a simple geospatial assessment tool which has been used in various studies for delineation of groundwater potential zones (Ozdemir 2011; Pourtaghi and Pourghasemi 2014; Moghaddam et al. 2015; Balamurugan et al. 2016; Mousavi et al. 2017). Ozdemir (2011) used the FR method for groundwater spring potential mapping (Moghaddam et al. 2015). This bivariate statistical technique was used to calculate the relationship between dependent variables, i.e. spring location, and independent variables, i.e. spring formation factors (Ozdemir 2011). Using these two variables, a relationship was developed between occurrence and non-occurrence of springs in an area with their corresponding factors (Ozdemir 2011). FR model was used to derive a quantitative relationship between spring occurrence and responsible factors. The FR was computed for a class of spring affecting factors as below:

$$FR = \frac{S/TS}{C/TP} \tag{18.1}$$

Where, S is the total number of springs in each class, TS represents the total number of springs, C is the number of pixels in each class, and TP is the total number of pixels. In this model, the FR value of each class of thematic layer was considered as weight of that particular class of thematic layer, which determined the groundwater potentiality (Balamurugan et al. 2016). The FR value was derived from all thematic layers with the relationship between training springs using Eq. (18.1). All the thematic layers were integrated on the basis of FR value and a groundwater spring potential map was generated using Eq. (18.2) in raster calculator tool ArcGIS software.

$$\begin{aligned}\textit{Spring potential map} = {} & FR_{LULC} + FR_{NDVI} + FR_{Elevation} + FR_{Slope\ degree} + FR_{Aspect} + \\ & FR_{Drainage\ density} + FR_{Drainage\ buffer} + FR_{TWI} + FR_{SPI} + FR_{Slope\ length} + \\ & FR_{Plan\ curvature} + FR_{Fault\ buffer} + FR_{Lithology} + FR_{Road\ buffer}\end{aligned} \tag{18.2}$$

18.2.3 Analytical Hierarchy Process (AHP)

Analytical hierarchy process (AHP) is a frequently used technique for natural resource management, suitability analysis, and site selection, under multi-criteria decision analysis. It is a less expensive, multi-parametric evaluation technique most suitable for places where good quality data is lacking. Various thematic layers or feature classes are given weightage or ranking (1–9) based on literature review and expert opinion, similar to the weighted overlay method. Relative importance of various thematic layers is assigned by using the AHP method of MCDA (Gupta and Srivastava 2010). The weights can also be assigned using the PriEsT AHP tool with proxy data to assess the dominant and sensitive factors governing groundwater potential zones (Berhanu and Hatiye 2020). Normalization of assigned weights during weightage overlay is carried out using hierarchical ranking in pairwise comparison matrix of AHP. Consistency Index and Consistency Ratio are calculated to validate the consistency of assigned weights with the desired results using the methodology followed by Saaty (1980). Further, reclassification of the thematic layers is done based on the normalized assigned weights using weightage overlay model. Reclassified thematic layers are then used in GIS environment and overlay analysis is carried out to delineate groundwater potential zones.

18.2.4 Fuzzy Logic

The AHP method of MCDA, although widely used in groundwater potential zone mapping, is also sometimes criticized due to inherent uncertainties and imprecision, associated with the perception of the decision maker during mapping (Chen et al. 2011). During the analysis of uncertainty, fuzzy logic in MCDA provides membership or non-membership function to each criteria, for more flexibility in decision making. AHP combined with fuzzy membership function sets a good framework for assessment and reliability of criteria functions. Therefore, in this technique, integration of AHP with fuzzy set theory, FAHP(Fuzzy Analytical Hierarchy Process), has improved the groundwater potential zone mapping by drawing specific conclusions despite ambiguity, vagueness, and imprecise information (Mallick et al. 2019).

This technique works by supporting a feature object that can be used as membership value between 0 (Non-member) and 1 (full member), explaining the degree of membership function. When fuzzy membership functions are assigned to spatial variance, their trend results in fuzzy boundaries for each potential zone being mapped. Standardization of data includes four scales, i.e. ordinal, nominal, interval and ratio. Following the objective and hypothesis, linear and categorical type of fuzzy membership functions may be used.

18.2.4.1 Pairwise Comparison Matrix Formation

- Fuzzy geometric mean and fuzzy weighting of criterion can be determined by following Buckley's method (Buckley and Uppuluri 1987).
- Weighted linear combination aggregation method is used to calculate GWPZ map.

$$\mathrm{GWPI} = \sum_{w=1}^{m}\sum_{i=1}^{n}\left(\mathrm{wj} \times \mathrm{xi}\right) \tag{18.3}$$

where, GWPI = groundwater potential index,
x_i = FMF thematic maps,
w_j = normalized weight of the jth theme,
m = total number of themes, and n = total number of classes in a theme.

Further sensitivity analysis is carried out to assess the effective weight of each criterion with the assigned weight following the formula mentioned below:

$$\textit{Effective weight} = \frac{\left(\textit{Themew} \times \textit{Themes}\right)}{\textit{GWPI}} \times 100 \tag{18.4}$$

where, w and s are the weight and scaled values of the theme (theme) assigned to each grid, respectively, and GWPI is the groundwater potential index (as calculated from Eq. 18.3).

Relative operating curve (ROC) analysis is a method used to assess the accuracy or qualitative validation of groundwater potential zone. The X-axis of the curve encompasses the false-positive values, whereas the true-positive values are plotted on Y-axis. The area under the ROC curve indicates accuracy in groundwater potential zone mapping. The relationship between area under the curve and accuracy of the prediction process is classified as excellent (0.9–1), very good (0.8–0.9), good (0.7–0.8), average (0.6–0.7), and poor (0.5–0.6) (Naghibi et al. 2015).

18.2.5 ANN Method

Artificial Neural Network (ANN) is a computational mechanism which may be used to derive information from one multivariate space to another, given a set of data that represents that mapping (Garrett 1994). A support vector machine (SVM) is a tool based on nonlinear transformations of covariates into a higher dimensional feature space (Vapnik 2013). ANN and SVM are often used in combination for groundwater-productivity potential mapping. In this method, the thematic layers are selected for data mining and spatial database preparation. Back-propagation algorithm is adopted for ANN, whereas polynomial kernel is used for SVM during groundwater productivity potential mapping (Lee et al. 2018). The back-propagation algorithm is a trained set of associated inputs and outputs. The training sets of the spatial database are based on GPP-prone and non-GPP-prone locations. Further, quantitative calculation of training sets along with selected thematic layers are used in SVM and ANN data mining models for the preparation of a groundwater potential map (Lee et al. 2012). Later, factor analysis of the weights assigned to each thematic layer and well yield data are used for validation of groundwater potential map.

18.2.6 Techniques for Water-Ice Detection on Extraterrestrial Planets

The following are a few techniques which have been successfully employed for water-ice detection in extraterrestrial RS.

18.2.6.1 Morphological Analysis

This type of analysis, on the planetary bodies, aims at identifying features that are indicative of water activity using the visible region of the EM spectra. For instance, several studies on Mars have reported the presence of valley networks with dendritic patterns (Hynek et al. 2010), fluvio-lacustrine basins (Fassett and Head III 2008; Goudge et al. 2015), desiccation polygons (El-Maarry et al. 2013), and layered ejecta craters (Mukherjee et al. 2020). These features, similar to those on Earth, suggest that Mars was hydrologically active in the past.

18.2.6.2 Reflectance Spectroscopy

Reflectance spectroscopy uses several bands in the visible, near-infrared (NIR), and thermal infrared (TIR) region, termed as multispectral or hyperspectral and thermal RS, respectively. It works on the principle that surface features differ from each other in the way they reflect or absorb light, which in turn reflects their composition and their physical as well as chemical state. The difference in reflectance properties helps in identification of minerals by recording their spectral signatures.

Aqueous alteration history may be determined by detection of hydrated minerals such as phyllosilicates, carbonates, sulphates, chlorides, etc. as they are products of water-rock interactions. Studies have reported many such minerals on Mars using CRISM and OMEGA hyperspectral sensors (Bishop et al. 2009; Ehlmann et al. 2009; Osterloo et al. 2010; Wray et al. 2016). Planetary studies make use of band ratioing techniques as well as continuum removal to enhance features in a certain wavelength range, to detect specific minerals. The following is the CRISM adjusted H_2O ice index formula:

$$\mathrm{BD1500\_2} = 1 - \left(\frac{R1525}{a \times R1367 + b \times R1808} \right) \tag{18.5}$$

where, R(λ) is the apparent reflectance at wavelength λ in mm, BD1500_2 is band ratio for water ice on surface or in atmosphere. a = 1-b; b = (λc- λs)/(λl- λs) λc –central wavelength, λs and λl are shorter and longer wavelength points along the continuum (Viviano-Beck et al. 2014)

Different band ratios are assigned colours in the RGB channels to develop FCC. The reflectance curve obtained is then matched with spectral library of minerals archived from Earth and Apollo missions, to identify secondary hydrous minerals.

18.2.6.3 Radar

Radar is one of the most popularly employed techniques for below ground detection of water-ice, successfully used on both the Moon and Mars. Some of the payloads using radar RS include SHAllow RADar (SHARAD) aboard the Mars Reconnaissance Orbiter (MRO) and Mini-SAR on Chandrayaan-1. Both these instruments work in the microwave region of the EM spectra, although the principles behind their operation differ significantly.

SHARAD works like a sonar and reflects some of the energy back when it encounters an interface with different dielectric constant. The dielectric constant is dependent upon the composition of the object as well as its porosity. As the transmitted waves reach back to the sounder, they experience a time delay (Seu et al. 2007). This helps in developing a 'radargram' which may be compared with a simulated radargram generated using topographic information to detect the time delay. The time delay radargram is then converted to 'depthgram' by using most probable values of dielectric constant (such as 3.2, in the case of ice) to represent radar signals in terms of distance. If the conversion of time to depth brings subsurface reflection in alignment with the surroundings, it leads to identification of the composition of the object (Plaut et al. 2009).

Chandrayaan-1 Mini-SAR and Mini-RF of Lunar Reconnaissance Orbiter (LRO) have contributed significantly in studying the scattering properties of the materials in the lunar polar regions thereby settling the long-standing debate on whether water-ice exists on Moon (Spudis et al. 2013; Mishra et al. 2014; Putrevu et al. 2016; Verma et al. 2018). Hybrid polarimetry data is used to calculate Stokes vector (S) from which child parameters such as Circular Polarization Ratio (CPR) can be derived. CPR, which is the ratio of same sense circular (SC) polarization to opposite sense circular (OC) polarization, is used to determine evidences of possible water-ice deposits in cold, dark, permanently shadowed regions where CPR ratio of more than one suggests presence of water-ice deposits (Spudis et al. 2009).

18.2.6.4 Gamma-Ray and Neutron Spectroscopy

This technique relies on protons and alpha particles, present in the cosmic rays, for excitation. When these particles strike the surface of a planetary body, they generate neutrons from other nuclei, forming a chain reaction. While some neutrons on de-excitation emit gamma rays, some are able to escape the surface and can be detected at the sensor. The flux of these leakage neutrons is a function of composition of the surface. For instance, hydrogen, if present below the surface, is very effective at capturing neutrons, since it is almost the same size as a neutron (Boynton et al. 2002). This approach indirectly hints at the presence of water-ice and has been used on both Mars and the Moon. The ambiguity in results may be due to lack in clarity of the chemical state in which hydrogen detections have been made. This uncertainty has therefore resulted into limitation of this technique to mid-latitudes and poles (Wray et al. 2016).

18.3 Conclusion

With the advancement of technology, management and restoration of water resources has become more convenient for the science and administrative community. The responsible factors for groundwater availability are found in the form of various data types. The integration and compilation of all the data need some specific method which can generate water potential sites. Weightage overlay, FR, AHP, fuzzy logic-AHP, and ANN are some of the crucial methods for groundwater potential mapping. The methods involve preparation of various thematic layers and assigning weights based on their relative importance. Statistical methods like factor analysis and analysis of pairwise comparison matrix are used for geostatistic spatial database preparation and further, generate a potential map. The potential map can also be validated with relative operating curve (ROC), consistency ratio, consistency index, and field-based well yield data.

The scope of RS is not limited to Earth. It has crossed the dimensions of Earth and now it is used to infer natural resources on other planetary bodies. Most significant amongst them is water, since it is pivotal in the search for past life on other planets. Scientists have utilized the entire electromagnetic spectrum, from gamma ray to microwave, to devise direct and indirect methods for detection of water-ice, at the surface as well as subsurface level. These techniques are used to make qualitative and quantitative assessments of water-ice in the area with the help of elemental, morphological, and mineralogical detections. Mostly they are used in combination with one another since ground truthing of all the areas is negligible to absent. The field of RS is constantly being probed to develop new techniques for exploration of natural resources in both terrestrial and extraterrestrial environments for the benefit of the human community.

References

Arulbalaji, P., Padmalal, D., and Sreelash, K. (2019). GIS and AHP techniques based delineation of groundwater potential zones: a case study from southern Western Ghats, India. *Scientific Reports* 9 (1): 1–17.

Balamurugan, G., Seshan, K., and Bera, S. (2016). Frequency ratio model for groundwater potential mapping and its sustainable management in cold desert, India. *Journal of King Saud University-Science* https://doi.org/10.1016/j.jksus.2016.08.003.

Berhanu, K.G. and Hatiye, S.D. (2020). Identification of groundwater potential zones using proxy data: case study of Megech Watershed, Ethiopia. *Journal of Hydrology: Regional Studies* 28: 100676.

Bishop, J.L., Parente, M., Weitz, C.M. et al. (2009). Mineralogy of Juventae Chasma: sulfates in the light-toned mounds, mafic minerals in the bedrock, and hydrated silica and hydroxylated ferric sulfate on the plateau. *Journal of Geophysical Research, Planets* 114 (E2): E00D09.

Bonham-Carter, G.F. (1996). *Geographic Information Systems for Geosciences*. Ontario, Canada: Modeling With GIS. Pergamon/Love Printing Service Ltd., 398 pp.

Boynton, W.V., Feldman, W., Squyres, S. et al. (2002). Distribution of hydrogen in the near surface of Mars: evidence for subsurface ice deposits. *Science* 297 (5578): 81–85.

Buckley, J.J. and Uppuluri, V. (1987). *Fuzzy Hierarchical Analysis Uncertainty in Risk Assessment, Risk Management, and Decision Making*, 389–401. Springer.

Chang, K.-T. (2006). *Introduction to Geographic Information Systems*. McGraw-Hill Higher Education Boston.

Chang, K.T. (2016). *Geographic information system*, 1–10. International Encyclopedia of Geography: People, the Earth, Environment and Technology.

Chen, V.Y., Lien, H.-P., Liu, C.-H. et al. (2011). Fuzzy MCDM approach for selecting the best environment-watershed plan. *Applied Soft Computing* 11 (1): 265–275.

Chi, K. and Lee, B. (1994). *Extracting potential groundwater area using remotely sensed data and GIS techniques*. Paper presented at the Proceeding of the regional seminar on integrated applications of remote sensing and GIS for land and water resources management, Bangkok (Bangkok: Economic and Social Commission for Asia and the Pacific).

Chowdhury, A., Jha, M., Chowdary, V., and Mal, B. (2009). Integrated remote sensing and GIS-based approach for assessing groundwater potential in West Medinipur district, West Bengal, India. *International Journal of Remote Sensing* 30 (1): 231–250.

Chopra, R. and Sharma, P.K. (1993). Landform analysis and ground water potential in the Bist Doab area Punjab, India. *International Journal of Remote Sensing* 14 (17).

Dhakate, R., Singh, V., Negi, B. et al. (2008). Geomorphological and geophysical approach for locating favorable groundwater zones in granitic terrain, Andhra Pradesh, India. *Journal of Environmental Management* 88 (4): 1373–1383.

van Dijk, A.I. and Keenan, R.J. (2007). *Planted Forests and Water in Perspective*. Elsevier.

Doyle, K. and Peck, M.A. (2018). *Water electrolysis for propulsion of a crewed mars mission*. Paper presented at the 2018 AIAA Aerospace Sciences Meeting.

Doyle, K.P. and Peck, M.A. (2019). Water electrolysis propulsion as a case study in resource-based spacecraft architecture (February 2020). *IEEE Aerospace and Electronic Systems Magazine* 34 (9): 4–19.

Ehlmann, B.L., Mustard, J.F., Swayze, G.A. et al. (2009). Identification of hydrated silicate minerals on Mars using MRO-CRISM: geologic context near Nili Fossae and implications for aqueous alteration. *Journal of Geophysical Research, Planets* 114 (E2): E00D08.

Elitzur, S., Rosenband, V., and Gany, A. (2016). Combined energy production and waste management in manned spacecraft utilizing on-demand hydrogen production and fuel cells. *Acta Astronautica* 128: 580–583.

El-Maarry, M.R., Pommerol, A., and Thomas, N. (2013). Analysis of polygonal cracking patterns in chloride-bearing terrains on Mars: indicators of ancient playa settings. *Journal of Geophysical Research, Planets* 118 (11): 2263–2278.

Edet, A., Okereke, C., Teme, S., and Esu, E. (1998). Application of remote-sensing data to groundwater exploration: a case study of the Cross River State, Southeastern Nigeria. *Hydrogeology Journal* 6 (3): 394–404.

FAO. (2005). Food security in the context of economic and trade policy reforms: insights from country experiences.

Fassett, C.I. and Head, J.W. III (2008). Valley network-fed, open-basin lakes on Mars: distribution and implications for Noachian surface and subsurface hydrology. *Icarus* 198 (1): 37–56.

Garrett, J. (1994). Where and why artificial neural networks are applicable in civil engineering. *Journal of Computing in Civil Engineering* 8: 129–130.

Gaur, S., Chahar, B.R., and Graillot, D. (2011). Combined use of groundwater modeling and potential zone analysis for management of groundwater. *International Journal of Applied Earth Observation and Geoinformation* 13 (1): 127–139.

Ghayoumian, J., Mohseni Saravib, M., Feiznia, S. et al. (2007). Application of GIS techniques to determine areas most suitable for artificial groundwater recharge in a coastal aquifer in southern Iran. *Journal of Asian Earth Science* 30: 364–374.

Gogu, R.C., Carabin, G., Hallet, V. et al. (2001). GIS based hydrogeological databases and groundwater modeling. *Hydrology Journal* 9: 555–569.

Goudge, T.A., Aureli, K.L., Head, J.W. et al. (2015). Classification and analysis of candidate impact crater-hosted closed-basin lakes on Mars. *Icarus* 260: 346–367.

Gupta, M. and Srivastava, P.K. (2010). Integrating GIS and remote sensing for identification of groundwater potential zones in the hilly terrain of Pavagarh, Gujarat, India. *Water International* 35 (2): 233–245.

Heilongtan, K. (2007). Land use change and its impact on hydro-ecological linkages in Himalayan watersheds. *Tropical Ecology* 48 (2): 151–161.

Hynek, B.M., Beach, M., and Hoke, M.R. (2010). Updated global map of Martian valley networks and implications for climate and hydrologic processes. *Journal of Geophysical Research, Planets* 115 (E9): E09008.

Jackson, T.J., Chen, D., Cosh, M. et al. (2004). Vegetation water content mapping using Landsat data derived normalized difference water index for corn and soybeans. *Remote Sensing of Environment* 92 (4): 475–482.

Jaiswal, R., Mukherjee, S., Krishnamurthy, J., and Saxena, R. (2003). Role of remote sensing and GIS techniques for generation of groundwater prospect zones towards rural development--an approach. *International Journal of Remote Sensing* 24 (5): 993–1008.

Jha, M.K., Chowdhury, A., Chowdary, V., and Peiffer, S. (2007). Groundwater management and development by integrated remote sensing and geographic information systems: prospects and constraints. *Water Resources Management* 21 (2): 427–467.

Kamaraju, M., Bhattacharya, A., Reddy, G.S. et al. (1996). Ground-water potential evaluation of West Godavari District, Andhra Pradesh state, India – a GIS approach. *Groundwater* 34 (2): 318–325.

Karanth, K. and Seshu Babu, K. (1978). Identification of major lineaments on satellite imagery and on aerial photographs for delineation for possible potential groundwater zones in

Penukonda and Dharmavaram taluks of Anantapur district. Paper presented at the Proc. of Joint Indo US Workshop on Remote Sensing of Water Resources, NRSA, Hyderabad.

Khan, S. and Mohd, A. (1997). *GIS approach for water resources assessment in parts of Rajasthan and Haryana using remote sensing techniques*. Paper presented at the Proceedings of natural symposium on remote sensing for natural resources (Pune, India, 4–6 December 1996). Indian society of remote sensing (ISRS) National natural resources management system (NNRMS), Ahmedabad, India.

Krishnamurthy, J. and Srinivas, G. (1995). Role of geological and geomorphological factors in ground water exploration: a study using IRS LISS data. *International Journal of Remote Sensing* 16 (14): 2595–2618.

Krishnamurthy, J., Venkatesa Kumar, N., Jayaraman, V., and Manivel, M. (1996). An approach to demarcate ground water potential zones through remote sensing and a geographical information system. *International Journal of Remote Sensing* 17 (10): 1867–1884.

Krishnamurthy, J., Mani, A., Jayaraman, V., and Manivel, M. (2000). Groundwater resources development in hard rock terrain-an approach using remote sensing and GIS techniques. *International Journal of Applied Earth Observation and Geoinformation* 2 (3–4): 204–215.

Lee, S., Song, K.-Y., Kim, Y., and Park, I. (2012). Regional groundwater productivity potential mapping using a geographic information system (GIS) based artificial neural network model. *Hydrogeology Journal* 20 (8): 1511–1527.

Lee, S., Hong, S.-M., and Jung, H.-S. (2018). GIS-based groundwater potential mapping using artificial neural network and support vector machine models: the case of Boryeong city in Korea. *Geocarto International* 33 (8): 847–861.

Lillesand, T.M. and Kiefer, R.W. (2007). *Remote Sensing and Image Interpretation*, 5e. Wiley India Pvt. Limited.

Lillesand, T.M., Kiefer, R.W., and Chipman, J.W. (2004). *Remote Sensing and Image Interpretation*. New York: John Wiley and Sons.

Mahamuni, K. and Upasani, D. (2011). Springs: A common source of a common resource. Paper presented at the 13th Biennial Conference of the International Association for the Study of the Commons (IASC), Sustaining Commons: Sustaining our Future.

McEwen, A.S., Eliason, E.M., Bergstrom, J.W. et al. (2007). Mars reconnaissance orbiter's high resolution imaging science experiment (HiRISE). *Journal of Geophysical Research, Planets* 112 (E5): E05S02.

Mishra, P., Kumar, S., and Singh, D. (2014). An approach for finding possible presence of water ice deposits on lunar craters using MiniSAR data. *IEEE Journal of Selected Topics in Applied Earth Observations and Remote Sensing* 8 (1): 30–38.

Mallick, J., Khan, R.A., Ahmed, M. et al. (2019). Modeling groundwater potential zone in a semi-arid region of Aseer using fuzzy-AHP and geoinformation techniques. *Water* 11 (12): 2656.

Moghaddam, D.D., Rezaei, M., Pourghasemi, H. et al. (2015). Groundwater spring potential mapping using bivariate statistical model and GIS in the Taleghan watershed, Iran. *Arabian Journal of Geosciences* 8 (2): 913–929.

Mousavi, S.M., Golkarian, A., Naghibi, S.A. et al. (2017). GIS-based groundwater spring potential mapping using data mining boosted regression tree and probabilistic frequency ratio models in Iran. *AIMS Geosciences* 3: 91–115.

Mukherjee, S. (2008). Role of satellite sensors in groundwater exploration. *Sensors* 8 (3): 2006–2016.

Mukherjee, S., Sashtri, S., Gupta, M. et al. (2007). Integrated water resource management using remote sensing and geophysical techniques: Aravali quartzite, Delhi, India. *Journal of Environmental Hydrology* 15 (10): 1–10.

Mukherjee, S., Singh, D., Singh, P., and Roy, N. (2020). Morphological and morphometric analysis of a topographic depression near Huygens basin, Mars: identification of a putative endorheic playa. *Geomorphology* 351: 106912.

Murchie, S., Arvidson, R., Bedini, P. et al. (2007). Compact reconnaissance imaging spectrometer for Mars (CRISM) on Mars reconnaissance orbiter (MRO). *Journal of Geophysical Research, Planets* 112 (E5): E05S03.

Murthy, K.S.R. (2000). Groundwater potential in a semi-arid region of Andhra Pradesh—a GIS approach. *International Journal of Remote Sensing* 21 (9): 1867–1884.

Naghibi, S.A., Pourghasemi, H.R., Pourtaghi, Z.S., and Rezaei, A. (2015). Groundwater qanat potential mapping using frequency ratio and Shannon's entropy models in the Moghan watershed, Iran. *Earth Science Informatics* 8 (1): 171–186.

NRSC. (2012). Manualof National Land Use/Land Cover Mapping (Second Cycle) Using Multi-temporal Satellite Data. Department of Space, Hyderabad.

Osterloo, M.M., Anderson, F.S., Hamilton, V.E., and Hynek, B.M. (2010). Geologic context of proposed chloride-bearing materials on Mars. *Journal of Geophysical Research, Planets* 115 (E10): E10012.

Ozdemir, A. (2011). GIS-based groundwater spring potential mapping in the Sultan Mountains (Konya, Turkey) using frequency ratio, weights of evidence and logistic regression methods and their comparison. *Journal of Hydrology* 411 (3–4): 290–308.

Pourtaghi, Z.S. and Pourghasemi, H.R. (2014). GIS-based groundwater spring potential assessment and mapping in the Birjand township, Southern Khorasan Province, Iran. *Hydrogeology Journal* 22 (3): 643–662.

Plaut, J.J., Safaeinili, A., Holt, J.W. et al. (2009). Radar evidence for ice in lobate debris aprons in the mid-northern latitudes of Mars. *Geophysical Research Letters* 36 (2): L02203.

Putrevu, D., Das, A., Vachhani, J. et al. (2016). Chandrayaan-2 dual-frequency SAR: further investigation into lunar water and regolith. *Advances in Space Research* 57 (2): 627–646.

Rao, N.S., Chakradhar, G., and Srinivas, V. (2001). Identification of groundwater potential zones using remote sensing techniques in and around Guntur town, Andhra Pradesh, India. *Journal of the Indian Society of Remote Sensing* 29 (1–2): 69.

Ravindran, K. (1997). *Drainage morphometry analysis and its correlation with geology, geomorphology and ground water prospects in Zuvari basin, South Goa, using remote sensing and GIS*. Paper presented at the Proceedings of natural symposium remote sensing for natural resources with special emphasis on water management (Pune, India, 4–6 December 1996). Indian society of remote sensing (ISRS) National natural resources management system (NNRMS), Ahmedabad, India.

Riebsame, W., Parton, W., Galvin, K. et al. (1994). Integrated modeling of land use and cover change. *Bioscience* 44 (5): 350–356.

Roy, I. (2009, 27 November 2009). Proceedings of the Workshop on Integrated water resource management.

Roy, S., Hazra, S., Chanda, A., and Das, S. (2020). Assessment of groundwater potential zones using multi-criteria decision-making technique: a micro-level case study from red and lateritic zone (RLZ) of West Bengal, India. *Sustainable Water Resources Management* 6 (1): 4.

Saaty, T. (1980). *The Analytic Hierarchy Process: Planning, Priority Setting, Resource Allocation.* New York, NY: McGraw-Hill. Inc.

Sander, P., Chesley, M.M., and Minor, T.B. (1996). Groundwater assessment using remote sensing and GIS in a rural groundwater project in Ghana: lessons learned. *Hydrogeology Journal* 4 (3): 40–49.

Saraf, A. and Choudhury, P. (1998). Integrated remote sensing and GIS for groundwater exploration and identification of artificial recharge sites. *International Journal of Remote Sensing* 19 (10): 1825–1841.

Sener, E., Davraz, A., and Ozcelik, M. (2005). An integration of GIS and remote sensing in groundwater investigations: a case study in Burdur, Turkey. *Hydrogeology Journal* 13 (5–6): 826–834.

Seu, R., Phillips, R.J., Biccari, D. et al. (2007). SHARAD sounding radar on the Mars reconnaissance Orbiter. *Journal of Geophysical Research, Planets* 112 (E5): E05S05.

Singh, V. and Dubey, A. (2012). Land use mapping using remote sensing and GIS techniques in Naina-Gorma Basin, part of Rewa District, MP, India. *International Journal of Emerging Technology and Advanced Technology* 2: 151–156.

Shahid, S., Npath, S., and Rpay, J. (2000). Groundwater potential modeling in softrock using a GIS. *International Journal of Remote Sensing* 21: 1919–1924.

Sikdar, P., Chakraborty, S., Adhya, E., and Paul, P. (2004). Land use/land cover changes and groundwater potential zoning in and around Raniganj coal mining area, Bardhaman District, West Bengal-a GIS and remote sensing approach. *Journal of Spatial Hydrology* 4 (2): 1–24.

Solomon, S. and Quiel, F. (2006). Groundwater study using remote sensing and geographic information systems (GIS) in the central highlands of Eritrea. *Hydrogeology Journal* 14 (6): 1029–1041.

Spudis, P., Nozette, S., Bussey, B. et al. (2009). Mini-SAR: an imaging radar experiment for the Chandrayaan-1 mission to the Moon. *Current Science* 96 (4): 533–539.

Spudis, P., Bussey, D., Baloga, S. et al. (2013). Evidence for water ice on the moon: results for anomalous polar craters from the LRO mini-RF imaging radar. *Journal of Geophysical Research, Planets* 118 (10): 2016–2029.

Sreedevi, P.D., Subrahmanyam, K., and Ahmed, S. (2005). Integrating approach for delineating potential zones to explore for groundwater in the Pageru river basin,Cuddapah district, Andhra Pradesh, India. *Hydrogeology Journal* 13: 534–543.

Tucker, C.J. (1979). Red and photographic infrared linear combinations for monitoring vegetation. *Remote Sensing of Environment* 8 (2): 127–150.

Vapnik, V. (2013). *The Nature of Statistical Learning Theory*. Springer science and business media.

Verma, N., Mishra, P., Purohit, N., and Singh, D. (2018). Detection of Possible Water-Ice Deposits on Lunar Surface Using Conformity Coefficient: An Application of MiniSAR Data. Paper presented at the IGARSS 2018–2018 IEEE International Geoscience and Remote Sensing Symposium.

Viviano-Beck, C.E., Seelos, F.P., Murchie, S.L. et al. (2014). Revised CRISM spectral parameters and summary products based on the currently detected mineral diversity on Mars. *Journal of Geophysical Research, Planets* 119 (6): 1403–1431.

Wray, J.J., Murchie, S.L., Bishop, J.L. et al. (2016). Orbital evidence for more widespread carbonate-bearing rocks on Mars. *Journal of Geophysical Research, Planets* 121 (4): 652–677.

Zadeh, L.A. (1965). Fuzzy sets. *Information and Control* 8 (3): 338–353.

19

Recent Trends in Groundwater Conservation and Management

Amit Kumar Tiwari and Dan Bahadur Pal

Department of Chemical Engineering, Birla Institute of Technology, Mesra, Ranchi, Jharkhand, India

19.1 Introduction

Water is one of the major resources for life; for a very long time, surface water and groundwater have been used as common sources of water for agricultural work, supply for drinking, and industrial consumption. Dense and rising population, growth of urbanization, and fast-growing industrialization are important factors for maximizing the rate of farm production, which promotes high demands for water. In the view of development, groundwater has become a spinal cord of Indian agricultural and drinking water security. Groundwater also helps in the control of ecological balances and environmental conditions by maintaining the minimum required water level in the water bodies. As per the reports of various governments, per day consumption of groundwater almost doubled from 34 billion gallons to about 80 billion gallons in 40 years (1950–1990) (Sarkar 2014). During the same period, the capacity of surface-water reservoirs was enhanced almost three times, but at present have been been depleted in large scale. With this depletion, we are unable to meet the water demands from surface water resources and now groundwater is considered and utilized as a main resource. The surface area of India is around 2.45% of total surface area of the world and 4% water resources; around 16% population of total world's population is residing in India. The total precipitated water in the country is approximately 4000 km^3/year. Available water is estimated around 1869 km^3 from different sources, such as surface water and rechargeable groundwater; from this available water we can utilize only 60% for beneficial uses. Therefore, the current total usable water resource in the country is only 1122 km^3 (Singh and Singh 2002). Freshwater is an eternal resource; however, the supply of clean freshwater is steadily decreasing due to mismanagement or undue uses worldwide. The demand for water is already more than the supply in many parts of the country, as population continues increasing, which is creating a huge demand for clean water. These problems may be due to lack of proper awareness regarding the global importance of water conservation for entire ecosystem, but recently various awareness programs

Groundwater Geochemistry: Pollution and Remediation Methods, First Edition.
Edited by Sughosh Madhav and Pardeep Singh.

have been implemented. During the twentieth century, 50% or more of the wetlands of the world have been lost along with their environmental and economical values. Freshwater ecosystems which were full of biodiversity are also declining at a very fast rate compared to other ecosystems such as the marine ecosystem or land ecosystem (Singh and Chaturvedi 2017).

Subsurface water and groundwater are considered freshwater which is situated in between the pore spaces of soils and rocks. This water always flows within the aquifers below the water table (Srivastava et al. 2015). As is previously stated, about 16% of world's population is resides in India; but in comparison to population, our water resources are much less. Only 4% freshwater resources of the world are available in India, which are unevenly scattered over vast areas of the country. Because of increasing demands for water by increasing population and due to uncontrolled pollution, the quality of existing water resources is also diminishing. India's continuous industrial and agricultural developments also create huge additional requirement of water for consumption, while the quantity and supply of freshwater is still unsettled. It was found that the maximum urban areas suffer from water deficiency; nearly 40% water requirement of urban India is currently fulfilled by groundwater (according to Tata Institute of Social Sciences). Finally, it is noted that the water tables of a number of cities are at a horrific stage; their tables are falling at the rate of 2–3 m/year (Central Ground Water Board [CGWB] 2014). The scarcity of water has many negative impressions on the environment, lakes, rivers, wetlands, and other freshwater resources are highly affected.

Groundwater is most important for arid zones and semi-arid zones, while surface water has more importance for humid areas. However, from the reports of uses of water resources it is revealed how groundwater is important worldwide. Most of the European cities depend on groundwater; the biggest cities of Africa and Asia are based on surface water but millions of rural people in these countries are groundwater dependent. Peoples of other countries such as in sub-Saharan Africa are bound to survive on untreated groundwater supplies from protected wells with hand pumps, which may likely to be their best solution for many years to come because they do not have any other means of improved water supply (Gandhi and Namboodiri 2009). India has very good relief features such as landforms, climatic zones, hydrological cycle, and vegetation types compared to many other countries.

Rivers, lakes, ponds, and tanks are known as four common sources of surface water. India has around 10 360 rivers and their tributary rivers are flowed; the length of each tributary is more than 1.6 km. In all these river basins estimated mean annual flow is about 1869 km^3 (Upadhyay 2019). Apart from various technical conditions (topographical and hydrological) and due to other constraints, we are able to utilize only 32% (690 km^3) of the available surface water. The quantity and flow of water in a river depends on the size of its catchment area and rainfall or hydrological cycle inside the catchment area. Highly diversified hydrogeological set-up is found in India, this versatility is because it is a vast country. Due to diversified geological formations (lithological and chronological variations), climatological dissimilarities, complex tectonic framework, and various hydrochemical conditions, Indian sub-continent is very much tangled. Three major geological formations of India are (i) consolidated rocks: these formations are basically igneous and metamorphic rocks composed of granite, charnockite, quartz and Phyllite, slate, basalt, etc., and other associated igneous materials; (ii) semi-consolidated rocks: these rock formations are made from limestones, sandstones, pebbles, and boulder conglomerates; and (iii) unconsolidated rocks: such rocks are formed by boulders, pebbles, graded sands, and silt-clay. These rocks are of major potential for aquifer zones. A large distinguishable part of India is occupied by major geological rocks such as metamorphic, igneous rocks (basaltic), gondwana, and vindhyan rocks.

19.2 Occurrence of Groundwater

As reported in a survey of Basic Ground Water Hydrology, US, around 14% of all freshwater is groundwater (Jha 2007). If we consider the water locked in the form of ice in glacial areas, then around 94% of the earth's freshwater can be considered as groundwater. As we know that the groundwater is almost hidden from view, there are many miconceptions on its occurrence. It is prevalent perception that only trained people can find groundwater and they are the only people who know everything about it; the truth is that groundwater can be obtained by any person simply by digging just about anywhere, but the yield and quality of the obtained water may depend on the location. Actually, the upper surface of the soils is generally saturated with water up to a fixed depth. Groundwater is generally found either in unconsolidated rocks or consolidated rocks. After the falling of rainwater on the ground, this water works as a source of moisture for the vegetative and upper surfaces and then the extra water begins to infiltrate the earth. During infiltration water penetrates the subsurface due to the force of gravity, and it passes through various layers or zones as given in Figure 19.1. The rate of infiltration depends on the soil type, density of vegetation, the amount of rainfall, and duration of rainfall. The excessive rain always affects the infiltration rate and an excess amount of water creates surface runoff (overland flow).

Various geological formations of the earth are sources of groundwater. On the uppermost part of the Earth, different types of rocks are established, and these rocks are different from each other by origin, type, or on the basis of their age. These rocks have coarse or fine orifices and are called pores or voids. The water-holding capacity of a rock depends on the size of voids or the number of voids in a given specific volume of rock; this is known as porosity of the rock and it can be presented in percentage. More pores results in higher porosity, which is responsible for excessive stored water. The porosity ranges of some common geological materials are shown in Table 19.1 (Dürr et al. 2005).

Those rocks that are very near to the ground surface have empty spaces that are partially filled with water and to some extent filled with air. This zone is known as 'unsaturated zone or vadose zone', and may vary in depth (from 0 to 10 m or more) (Panda and Chidambaram 2019). Residence time of water in this zone may be based on the thickness of the unsaturated or vadose zone. At high depths,

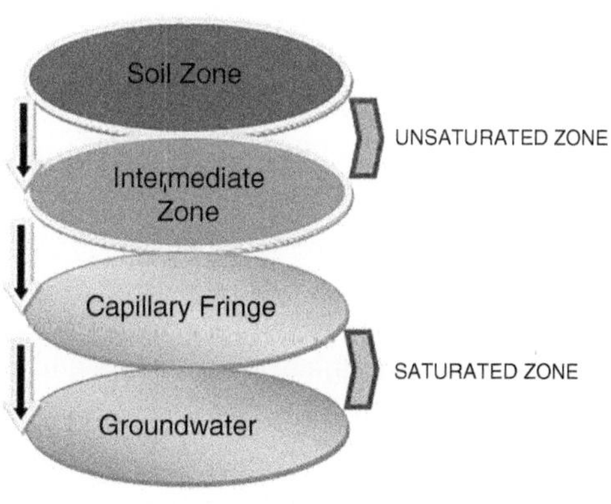

Figure 19.1 Water zones.

Table 19.1 Porosity % of consolidated rocks and unconsolidated rocks (Dürr et al. 2005).

S. No.	Unconsolidated rocks		S. No.	Consolidated rocks	
a	Name of Sediments	Porosity (%)	b	Name of Sediments	Porosity (%)
1.	Glacial till	20–30	1.	Fresh granite and gneiss	0.012
2.	Sand and gravel	20–30	2.	Shale	1–10
3.	Gravel	25–35	3.	Weathered granite and gneiss	1–15
4.	Sand	25–45	4.	Limestone and dolomite	1–20
5.	Silt	35–50	5.	Sandstone	5–30
6.	Clay	45–55	6.	Karst limestone	5–30
			7.	Fractured basalt	5–30
			8.	Vesicular basalt	10–40
			9.	Tuff	10–60

Source: (Dürr et al. 2005).

where all the blank spaces are totally filled with water, these zones are known as 'saturated zones'. If a deep well is dug in this zone, water will come out from the ground and settle down at the depth below and then all the blank spaces will be covered by water. This level of water saturation is known as the water table, and it builds on the upper surface of the saturated zone; groundwater refers to the saturated zone below water table. On the other hand, we can also say that all the natural water which is found below the Earth's surface is termed 'sub-surface water'.

19.3 Factors Affecting Availability of Groundwater

19.3.1 Impact of Climate Change

Many temporal processes including hydrological processes are collective effects of climatic activities. The climatic changes that naturally occur from time to time in space are directly associated with the changes and variables of the Earth's processes (Velasco et al. 2015). Global climatic changes always affect the natural processes, anthropogenic activities, and hydrologic cycle (Dragoni and Sukhija 2008). Changes in temperature and radiation balance in the atmosphere is linked with the hydrologic cycle. Various changes like increase in evapotranspiration rate, rate of atmospheric vaporization, changes in precipitation patterns, etc., are associated with hydrological cycle changes and responsible for global warming (Bates et al. 2008; IPCC 2013). Therefore, such types of changes always affect the hydrological dynamics of subsurface and are causes of changes in groundwater quantity, discharge, replenishment, storage, and quality in several water bodies (Green et al. 2007). It is very important to understand the influence of climatic changes on the complicated processes affecting the sustainability and availability of resources of surface water and groundwater (Dragoni and Sukhija 2008). It is well known that the climate changes have deadly effects on surface-water resources; during the various studies on climate change and future projections of climatic changes, it was found that the freshwater resources are unsafe and limited. The human activities and climatic stresses are threatening availability and sustainability of groundwater in many aquifers, and finally depletion of the subsurface water and groundwater resource are effects of these threats (Bovolo et al. 2009; Green et al. 2011). Efficient resource management is essential to understand the probable effects of overall environmental changes on accessibility and sustainability of groundwater resources.

In the twenty-first century, projections of climate change are rational and indicate that there is a higher chance of global warming because of the rise in temperature of air shift in frequencies of precipitation and its intensity, etc. (Mearns et al. 2007). These projections also include increased warming at high latitudes compared to the tropics; precipitation will increase in high latitudes and some parts of the tropics; and in the subtropical and lower mid-latitude regions decreased precipitation may occur. Increasing air temperature will be responsible for high evapotranspiration and uneven distribution of precipitation; sea level will increase due to rise in global average temperature. Overall climatic changes will play an effective role in the hydrological cycle and therefore resources of surface water and groundwater will be affected (Cisneros et al. 2014; Gurdak et al. 2009). Now, understanding on the potential effects of climate change is much improved and it is inculcated that the climate induced changes may directly or indirectly affect the groundwater systems.

19.3.2 Surface Water Precipitation and Evapotranspiration

Natural greenhouse emissions are presently a big problem, and these problems have been more or less induced by human activities (IPCC 2007). The endemic variation of forecast of rainfall suggests that numerous negative and positive changes may affect both surface water and groundwater processes. Increasing trends in air temperature will also affect the rate of evaporation and transpiration, which will create direct problems for the sustainability of surface water and groundwater resources for both global and regional areas (Barnett et al. 2008). Higher air temperatures will be the reason for increase in evaporation and transpiration, reduction in overflow and soil moisture, and even a little change in rainfall or evaporation/transpiration rates may lead to bigger changes in moisture content of soil and also reduction in groundwater recharge in many arid zones and semi-arid zones (Earman et al. 2006). In some regions due to increase in precipitation, groundwater recharge will also increase. It has been predicted that some portion of land at sub-tropical and mid-latitude areas will ultimately suffer from drought (Miller et al. 2009). Different places within the particular aquifers will experience diversified changes induced by climate. The extraction of groundwater may rise to satisfy the lack of surface water in the summer, because the demands of water may be at very high levels during this season.

19.3.3 Vadose Zone

It is anticipated that the climatic changes will have deep effects on the soil and overall hydrological conditions. Factors like moisture of soil and temperature of the Earth are crucial for interplay between earthborne climates, geobiochemical response, and soil and environment. According to Fan et al. (2007), variability in hydrology of vadose zone, low level of water tables, and resources of groundwater are seriously affected by the moisture content of soil and its temperature. Changes in the moisture content of soil will also affect the atmospheric processes, especially cumulus formation and amount of rainfall (Jungkunst et al. 2008). Climatic variables are spatiotemporal patterns in precipitation, evapotranspiration, and surface water conditions; these are the factors that may influence the level of moisture in soil. As suggested by Seneviratne et al. (2010), various other factors like soil texture, land slope, land use and other chemical, biological, and physical factors are too much important in this type of climatic changes.

19.3.4 Zone of Saturation

Numerous aquifers have well-water storage ability and they are potentially not so sensitive to climatic changes as surface water bodies. Available groundwater in these aquifers is capable of reducing droughts and can provide strong support to surface water to fulfill the water demand, especially in drought-prone areas (semi-arid and arid) (Miyakoshi et al. 2005; Taniguchi 2002). Continuous and severe dryness can change the hydrological behaviour of aquifers. Existing weaknesses of water resources are strongly associated with mutability of climate and are responsible for variable precipitation (Kundzewicz et al. 2007). Groundwater variability level is also more strongly correlated with precipitation than the degree or intensity of heat, but intensity of heat is very important for shallow or petty aquifers.

19.3.5 Recharging of Aquifers

Groundwater recharge is a very sensitive process for which all the factors such as climatic, local geological, type of soils, topography, vegetation, surface-water hydrology, and land make their own contributions. Groundwater recharge is generally affected by the occurrence of variation in precipitation, temperature, and evapotranspiration (Green et al. 2011). The understanding of the recharge process and techniques of recharge system has gradually improved but unfortunately the knowledge of recharge mechanisms and recharge rates is still very poor among the rural population. Even excessive rainfall and runoff water still cannot be utilized or stored in reservoirs by rural people, whereas it can be converted into a part of the soil-water systems or it can be stored in rivers, oceans, etc. As suggested by Holman et al. (2006), climatic changes are anticipated to have various effects on mechanisms of groundwater recharge and the rate of recharge, which will be positive for all the aquifer systems.

19.3.6 Groundwater Discharge

Depletion of groundwater occurs due to slower rates of recharge than the discharge rates of a given aquifer. This depletion may arise from straight effects of environmental changes or uncontrolled human actions (massive use of land, uncontrolled irrigation, infinite urbanization, etc.). Over the past century, it was found that rise in sea level is a contributor to changes in global groundwater discharge. Global warming is also a valid reason for rise in sea level, because large amounts of water were not properly stored in superficial reservoirs (Sahagian et al. 1994).

19.3.7 Groundwater Storage and Flow

Groundwater storage is playing a very important role in successfully dealing with climatic effects on groundwater resources. In many areas, groundwater is being stored from time to time, but the practice should be continued because it is the only solution for modulating drought effects on surface water resources. As predicted, due to climate change surface-water storage will become more limited (Mall et al. 2006). The stored groundwater will be reduced in many regions due to severe droughts; climate change will be the main cause of this scenario (Toure et al. 2016). However, some regions of the country may have credible and qualified groundwater storage. Global population growth will likely be the reason for groundwater depletion because higher population will create abstraction for groundwater recharge, which will affect the sustainability and availability of groundwater (Loaiciga 2003). Predicted growth in population and progressive demand of water is directly associated with reduction in water storage and water resources (Taniguchi et al. 2008).

19.4 Hydrogeological Setup of India

India contains diversified land formations of hard and fissured structures; these formations includes crystal, basalt, and other unified sedimentations such as carbonate rocks. Apart from this, the Himalayan terrain is covered by alluvial formations from west to east (Rajasthan to Brahmaputra valley). This combination of rugged topography and fissured rock formations is responsible for extirpation of aquifers.

The major rock formations found in the Indian subcontinent include metamorphics, igneous rocks, gondwana, and vindhyan rocks, out of which few rocks are covered by quaternary to recent sedimentary deposits. These rocks are unevenly spread in different parts of the country; quantification of distribution is not easy. With due regard for formation features and hydraulic phenomena to store and circulate groundwater in all the litho units, they can be categorized in to two broad groups: (i) porous formations, which can be further reclassified into unconsolidated and semi-consolidated formations – these types of formations have only primary level of porosity – and (ii) fissured formation or consolidated formations, which have mostly secondary or derived porosity. Two most important factors that control the occurrence and distribution of groundwater are physiographic and geomorphologic settings (Arora and Tiwari n.d.). On the basis of these two important factors, the entire country can be broadly divided into five distinct regions:

i) Northern mountainous terrain and hilly area;
ii) Indo-gangetic area;
iii) Peninsular shield area;
iv) Coastal area;
v) Cenozoic fault basin and low rainfall area.

The different rock formations act as distinct aquifer systems due to different hydrogeological properties. Indian aquifer systems can be divided into 14 main groups. A few very useful and important groups are alluvial aquifers; laterite aquifers; sandstone and shale aquifers; limestone aquifers; basalt aquifers; and crystalline aquifers.

19.5 Hydrological Conditions and Farming Systems

India has various diversified hydrological setups and environmental conditions, which range from alluvial plains to hard rock aquifers. These aquifers are distributed from Mawsynram, Meghalaya, recording rainfall of 11 872 mm/year, to Thar Desert of Rajasthan, with the recorded rainfall of less than 250 mm/year. These diversities are very challenging for management of water resources in India (Arora and Tiwari n.d.). India is very fortunate with a large amount of annual rainfall and total average annual river flow of 1953 km^3, but the amount of available annual utilizable surface water is estimated to be around 690 km^3 (Kumar et al. 2005).

19.6 Groundwater Resource Management

The groundwater resource management in India is a very complicated task, because it is directly connected to the relationship between the physical environment and human society. The irregular distribution of available groundwater resources and improper utilization indicate that a single strategy is not sufficient to solve the problem. On the other hand, each circumstance requires a different solution, which is a basic need on the ground of landscape setup, climatological conditions, hydrogeological settings, groundwater availability, socioeconomic setup, and water utilization pattern in a specific region (Varua et al. 2017). Scientific management of groundwater resources may the combination of (i) supply-side measures with the aim of enhancement of extraction of groundwater depending on availability and (b) demand-side measures with the objective of control, protection, and conservation of existing water bodies. Different options falling under these categories are mentioned in Sections 19.6.1 and 19.6.2.

19.6.1 Measures of Supply Side

When we think of supply measures, considering the environmental, social, and economical factors of the country, various aims come to mind, such as how we can increase the groundwater availability. These measures are considered 'structural measures', which is a mixture of technical developments and scientific advancement of groundwater resources. Additional groundwater resource development and advancement through suitable means like artificial recharge and rainwater harvesting may be good solutions for water scarcity problems. Complete knowledge of the hydrological conditions and control of hydrogeological factors is required for effective management of water supply; it is also necessary to have an idea about the behavior of different aquifers and groundwater levels under abstractions (Arora and Tiwari n.d.). Interaction between surface water and groundwater, changes in rate of drift, and changes in rate of recharge are also important points in this matter. The following points may be useful for various regions:

i) Scientific development of groundwater resources;
ii) Harvesting of rainwater and artificial recharge.

19.6.2 Measures of Demand Side

Groundwater resource management also needs to concentrate on the reasonable and controlled exploitation of the available resources; apart from various scientific developments it should be ensured that these water resources will be available for the long term. During demand-side management, few important considerations should be kept in mind such as need-based allocation, involvement of stakeholders in various aspects of planning, ownership, pricing, regulatory measures, implementation, management, and proper monitoring of projects.

19.7 Groundwater Development Prospects in India

It is estimated that the role of groundwater in the agricultural economy of India is continuously and rapidly increasing, and has been since the 1970s. During the last two decades, there has been an approximate 105% rise in groundwater-irrigated lands; it was the most important pillar for the Green Revolution. After testing of available groundwater resources in different geographical regions of the country, it was found that the share of alluvial areas is about 44% and eastern plain regions contribute around 43%. An unexpected increase in mechanized deep bore well and shallow wells with pump set is illustrating how fast groundwater irrigation is going to be more extensive (Varua et al. 2017). According to the last few census records the number of wells was <1 million in the last 40 years, but now it is more than 19 million; this tremendous increase is noticed from 2000.

It is true that groundwater irrigation is necessity for poverty alleviation, as the small and marginal farmers have limited land resources (20–38%) for cultivation, therefore they require other improved resources like mechanized well, tube well, etc. This situation leads to overexploitation of groundwater resources; therefore, an extensive accelerated plan for groundwater development is required on a priority basis for our country. It is true that due to availability of huge water resources in the country we have further scope for groundwater development and management plans. For the implementation of groundwater development and management plans, India can be categorized in different parts or aquifers such as coastal areas, waterlogged areas, deep aquifers, flood plain aquifers, and alluvial areas (Indo-Gangetic Plains).

19.8 Groundwater Conservation

Groundwater conservation is a big need for our secured future; it may support our life and livelihood systems in different ways, such as supply of required water for drinking, irrigation, electric generation, animal husbandry, etc. Limited and controlled use of water resources may be good, but some other ways of groundwater conservations could also be useful solutions of the problems. Some important goals of groundwater conservation are:

i) To ensure availability and sustainability of freshwater for the future by controlling withdrawal and replacement rate.

Table 19.2 Methods for conservation of water.

S.No.	Method	S.No.	Method
1.	Proper distribution of groundwater	**6.**	Increasing forest cover
2.	Rational use of groundwater	**7.**	Flood management
3.	Renovation of traditional water sources	**8.**	Use of rainwater effectively
4.	Recharge of groundwater	**9.**	Make effective use of soil water reserves
5.	Use of modern irrigation methods	**10.**	Take measures to avoid runoff

ii) To ensure conservation of energy by controlled consumption, because around 15% of total generated electricity is used for water management.
iii) To minimize human water use by changing habitats and reducing construction of new dams and other water diversion infrastructure.

Water conservation can be done by limited usage of water for different purposes like domestic usage, industries, agriculture, etc., and groundwater can also be preserved by maintaining groundwater level by recharging it. A few easy methods for conservation of water are suggested in Table 19.2, which can be adopted by common people.

Due to increasing demand for water for various purposes and declining strata of water, rainwater harvesting is now a worldwide matter of interest. Rainwater harvesting can be done with very little effort and minimum cost involvement. To minimize the water crisis, rainwater harvesting and recharge of groundwater by artificial approach are now much more popular and accepted as cost-effective methods in the entire world. No specific knowledge is required to store rainwater, but some expertise is required for the use of artificial recharging techniques because these are highly site-specific techniques. For this purpose, it is required to ensure the availability and suitability of land or area as a space for subsurface water storage. During selection of the area, the availability of surplus water from runoff in monsoons should also be taken into consideration, as it will ensure successful implementation of the schemes. Different organizations like MoWR (Ministry of Water Resources), CGWB, and many other organizations in India are continuously working on rainwater harvesting and preparing artificial recharge schemes; these schemes are suitable for different areas and also implemented in various parts of the country (Varua et al. 2017). After implementation and monitoring schemes, it was noticed that the rate of recharging groundwater was increased, groundwater level is rising, and water runoff is reduced.

Now, it is required to convert institutional efforts and processes into public campaign through society-based plans on collection of rainwater including artificial recharge, which would be helpful in making people and stakeholders aware of their responsibilities. It would also be helpful in the enhancement of the capacity of conservation of the schemes.

19.9 Conclusions

As per available literature on groundwater, it is established that the available water resources are very much affected by climatic changes. Climatic changes have a big effect on surface water resources, but there is little evidence about the effects of environmental changes on groundwater resources. Present groundwater resources and environmental systems are not in the equilibrium position, because these environmental changes are a problem for the amount of groundwater in several aquifers. The stagnation of aquifer recharge and water storing capacity are totally based on climate. Recharging of groundwater resources is really a tough job due to unpredicted and variable precipitation patterns. Continuous changes in quantity of surface water, and increased rate of evapotranspiration due to increase in air temperature, are also valid reasons for reduction of groundwater level. A deep knowledge is required to create long-term understanding on the effects of climatic changes on groundwater resources, and it would be better for proper planning and implementation of management schemes for retaining the groundwater resources. More easy and improved techniques should be designed to properly understand and establish the relationship between groundwater resources and environmental changes.

Acknowledgement

I would like to express my special thanks to faculty members of Department of Chemical Engineering, Birla Institute of Technology Mesra (Ranchi) for their critical reading and valuable suggestions for this chapter. I'm thankful to HOD, Department of Chemical Engineering, and Birla Institute of Technology Mesra (Ranchi) for providing valuable technical support and enough time for writing this chapter.

References

Arora, T. and Tiwari, V.M. Hydrological studies in India during last decade: a review. *Proceedings of the Indian National Science Academy* 85: 381–401.

Barnett, T.P., Pierce, D.W., Hidalgo, H.G. et al. (2008). Human-induced changes in the hydrology of western United States. *Science* 319 (5866): 1080–1083.

Bates, B.C., Kundzewicz, Z.W., Wu, S., and Palutikof, J.P. (2008). Climate change and water, Technical Paper IV of the Intergovernmental Panel on Climate Change, IPCC Secretariat, Geneva, 210 p.

Bovolo, C.I., Parkin, G., and Sophocleous, M. (2009). Groundwater resources, climate and vulnerability. *Environmental Research Letters* 4 (3): 035001.

Central Groundwater Board (CGWB) (2014). *Dynamic Groundwater Resources of India (as on 31st March, 2011)*. Faridabad: Ministry of Water Resources, River Development and Ganga Rejuvenation, Government of India.

Cisneros, B.E.J., Oki, T., Arnell, N.W. et al. (2014). Freshwater resources. In: *Climate Change 2014: Impacts, Adaptation, and Vulnerability, Part A: Global and Sectoral Aspects. Contribution of Working Group II to the Fifth Assessment Report of the Intergovernmental Panel on Climate Change* (eds. C.B. Field, V.R. Barros, D.J. Dokken, et al.), 229–269. Cambridge, United Kingdom and New York, NY, USA: Cambridge University Press.

Dragoni, W. and Sukhija, B.S. (2008). Climate change and groundwater – a short review. In: *Climate Change and Groundwater*, vol. 288 (eds. W. Dragoni and B.S. Sikhija), 1–12. London: Geological Society, Special Publications.

Dürr, H.H., Meybeck, M., and Dürr, S.H. (2005). Lithologic composition of the Earth's continental surfaces derived from a new digital map emphasizing riverine material transfer. *Global Biogeochemical Cycles* 19: GB4S10. https://doi.org/10.1029/2005GB002515.

Earman, S., Campbell, A.R., Newman, B.D., and Phillips, F.M. (2006). Isotopic exchange between snow and atmospheric water vapor: estimation of the snowmelt component of groundwater recharge in the southwestern United State. *Journal of Geophysical Research* 111: D09302.

Fan, Y., Miguez-Macho, G., Weaver, C.P. et al. (2007). Incorporating water table dynamics in climate modelling: water table observations and equilibrium water table simulations. *Journal of Geophysical Research* 112: D10125.

Gandhi, V.P. and Namboodiri, N.V. (2009). Groundwater Irrigation in India: Gains, Costs and Risks, Working Paper No.2009-03-08, Indian Institute of Management, Ahmedabad.

Green, T.R., Bates, B.C., Charles, S.P., and Fleming, P.M. (2007). Physically based simulation of potential effects of carbon dioxide: altered climates on groundwater recharge. *Vadose Zone Journal* 6 (3): 597–609.

Green, T.R., Taniguchi, M., Kooi, H. et al. (2011). Beneath the surface of global change: impacts of climate change on groundwater. *Journal of Hydrology* 405: 532–560.

Gurdak, J.J., Hanson, R.T., and Green, T.R. (2009). Effects of climate variability and change on groundwater resources of the United States, http://pubs.gov/fs/2009/3074.

Holman, I.P. (2006). Climate change impacts on ground-water recharge: uncertainty, shortcomings, and the way forward. *Hydrogeology Journal* 14 (5): 637–647.

IPCC, Climate Change 2007: Synthesis Report, Contribution of Working Groups I, II and III to the Fourth Assessment Report of the Intergovernmental Panel on Climate Change [Core Writing Team, R.K. Pachauri and A. Reisinger (eds.)], IPCC, Geneva, Switzerland, 104 p. (2007).

IPCC, Climate change 2013: The physical science basis, Contribution of Working Group I to the IPCC Fifth Assessment Report, Cambridge University Press, Cambridge, United Kingdom (2013).

Jha, B.M. (2007). *Management of Ground Water Resources for Ensuring Food Security in India*. New Delhi: National Ground Water Congress.

Jungkunst, H.F., Flessa, H., Scherber, C., and Fiedler, S. (2008). Groundwater level controls CO O and CH fluxes of three different hydromorphic soil types of a temperate forest ecosystem. *Soil Biology and Biochemistry* 40 (8): 2047–2054.

Kumar, R., Singh, R.D., and Sharma, K.D. (2005). Water resources of India. *Current Science* 89 (5): 794–811.

Kundzewicz, Z.W., Mata, L.J., Arnell, N.W. et al. (2007). Freshwater resources and their management. In: *Climate Change 2007: Impacts, Adaptation and Vulnerability* (eds. M.L. Parry, O.F. Canziani, J.P. Palutikof, et al.), 173–210. Cambridge University Press.

Loaiciga, H.A. (2003). Climate change and ground water. *Annals of the Association of American Geographers* 93 (1): 30–41.

Mearns, L.O., Hulme, M., Carter, T.R. et al. (2007). Climate scenario development. In: *Climate Change 2007: The Physical Science Basis, Contribution of Working Group 1 to the Fourth Assessment Report of Intergovernmental Panel on Climate Change* (eds. S. Solomon, D. Qin, M. Manning, et al.), 739–768. Cambridge, United Kingdom, and New York, USA: Cambridge University Press.

Miller, N.L., Dale, L.L., Brush, C.F. et al. (2009). Drought resilience of the California Centra Valley surface-ground-water-conveyance system. *Journal of the American Water Resources Association* 45 (4): 857–866.

Miyakoshi, A., Taniguchi, M., Okubo, Y., and Uemura, T. (2005). Evaluation of subsurface flow for reconstructions of climate change using borehole temperature and isotope data in Kamchatka. *Physical Earth Planet International* 152 (4): 335–342.

Mall, R.K., Gupta, A., Singh, R. et al. (2006). Water resources and climate change: an Indian perspective. *Current Science* 90 (12): 1610–1626.

Panda, B. and Chidambaram, S. (2019). Influence of the vadose zone on groundwater pollution – a review. *The International Journal of Civil, Environmental and Agricultural Engineering* 1 (1): 41–44.

Sahagian, D.L., Schwartz, F.W., and Jacobs, D.K. (1994). Direct anthropogenic contributions to sea level rise in the twentieth century. *Nature* 367: 54–57.

Sarkar, A. (2014). Economic implications of groundwater markets in India. *Nirma University Law Journal* 4 (1): 49–63.

Seneviratne, S.I., Corti, T., Davin, E.L. et al. (2010). Investigating soil moisture-climate interactions in a changing climate – a review. *Earth-Science Reviews* 99 (3–4): 125–161.

Singh, D.K. and Singh, A.K. (2002). Groundwater situation in India: problems and perspective. *International Journal of Water Resources Development* 18 (4): 563–580. https://doi.org/10.1080/07900620220000017400.

Singh, J.S. and Chaturvedi, R.K. (2017). Diversity of ecosystem types in India: a review. *Proceedings of the Indian National Science Academy* 83 (3): 569–594.

Srivastava, S.K., Chand, R., Raju, S.S. et al. (2015). Unsustainable groundwater use in Punjab agriculture: insights from cost of cultivation survey. *Indian Journal of Agricultural Economics* 70 (3): 365–377.

Taniguchi, M. (2002). Estimations of the past groundwater recharge rate from deep borehole temperature data. *Catena* 48 (1–2): 39–51.

Taniguchi, M., Burnett, W.C., and Ness, C.D. (2008). Integrated research on subsurface environments in Asian urban areas. *The Science of the Total Environment* 404 (2–3): 377–392.

Toure, A., Diekkruger, B., and Mariko, A. (2016). Impact of climate change on groundwater resources in Klela Basin, Southern Mali. *Hydrology* 3 (2): 17. https://doi.org/10.3390/hydrology3020017.

Upadhyay, P., India's Water Crisis: A Permanent Problem Which Needs Permanent Solutions. (September 11, 2019). Available at SSRN: https://ssrn.com/abstract=3451715 or doi:https://doi.org/10.2139/ssrn.3451715.

Varua, M.E., Basant, M., Ward, J., and Seema, D. (2017). Groundwater conservation attitudes, behaviour and water management: the case of farmers in rural India. In: *Transactions on Ecology and the Environment in Water Resources Management IX*, vol. 220 (eds. C.A. Brebbia and Z. Boukalova), 141–150. WIT Press https://doi.org/10.2495/WRM170141.

Velasco, E.M., Gurdak, J.J., Dickinson, J.E. et al. (2015). Interannual to multidecadal climate forcings on groundwater resources of the U.S. West Coast. *Journal of Hydrology: Regional Studies*, G Model EJRH-159: 16.

20

Groundwater Vulnerability Assessment Using Random Forest Approach in a Water-Stressed Paddy Cultivated Region of West Bengal, India

Rabin Chakrabortty, Paramita Roy, Indrajit Chowdhuri, and Subodh Chandra Pal

Department of Geography, The University of Burdwan, Bardhaman, West Bengal, India

20.1 Introduction

The most reliable source of drinking water is groundwater, which is easily available and also used for various purposes of human beings, such as irrigation and industrial activities. Approximately 1.5 billion people of the world depend upon groundwater directly or indirectly for their domestic and agricultural purposes (Mukherjee et al. 2012). Now, overexploitation of groundwater has led to growing demand, more polluted nature, and low cost of abstraction over the world (Menon 2007). In India, only 2% of groundwater is used for domestic and drinking purposes, but 89% of groundwater is used for agriculture and 9% for industrial activity (Pahuja et al. 2010; Suhag 2019). Approximately 85% of rural water demand, 50% of urban water demand, and about 60% of irrigated water are supplied by groundwater in India (Sekhri 2013). Freshwater is under threat due to the increasing demand of the population and industrial activities, and on the other hand, available groundwater is icontaminated by various pollutant and geogenic processes (Shrestha et al. 2017).

Rapid growth of irrigation has been observed since the Green Revolution and parallel groundwater extraction has increased in order to survive the growing population by increasing food production (Ahmed et al. 2014; Rosegrant et al. 2009). Groundwater is the only source for irrigation of high yielding and highly consumed Boro paddy cultivation in India (Castaño et al. 2010). Groundwater irrigation helped the socioeconomic development of people in different ways through high agricultural production, food security, development of the rural economy, and reduction of the poverty level in rural areas (Dutta 2018).

Geogenic and anthropogenic processes reduce the quality of groundwater and are a major challenge for the sustainable management of water resources in different parts of the world (Saha and Alam 2014). Groundwater quality generally depends upon topography,

Groundwater Geochemistry: Pollution and Remediation Methods, First Edition.
Edited by Sughosh Madhav and Pardeep Singh.

geology, aquifer characteristics, climate, and various hydrochemical processes, and these factors have led the geogenic process (Kumar et al. 2017; Santos et al. 2008). Precipitation, dissolution, absorption, cation exchange, oxidation, and biochemical reaction are the major processes to change the geohydrochemical characteristic of groundwater (Matthess 1982). In India the groundwater contamination has taken place due to the high concentration of arsenic (As), fluoride (F), nitrate (NO3), and salinity in the land (Alagumuthu and Rajan 2008).

Conversely, anthropogenic activity is one of the main causes of physical and chemical contamination of groundwater by urbanization, industrialization, and modern agricultural systems (Rahman 2008). In urban areas, there is leaching from landfills and polluted wetlands, seepage from open land or stream sewage, and in rural areas, agricultural practices of chemical fertilizer, pesticide, and poor sanitization of human wastage, etc. are contaminating groundwater (Chae et al. 2004; Saha et al. 2008). Pesticides like DDT (dichloro-diphenyl-trichloroethylene) and HCH (hexachlorocyclohexane) have been used extensively in agricultural sector of India for the last five years (Saha and Alam 2014). It is quantified that the annual use of chlorinated pesticides is approximately 25 000 metric tons in India (Sankararamakrishnan et al. 2005). As a result, long-term association with groundwater-contaminated harmful substances has an impact on human health, such as skin lesions, skin cancer, lung disease, obstetric problems, and neurological effects (Kumar et al. 2015).

In order to ensure a continuous supply of subsurface water and mitigate adverse effects, it is important to develop different strategies and guidance for the assessment and quality of groundwater. In this regard, the assessment of groundwater vulnerability (GV) is the most feasible step toward the achievement of the objective. GV assessment is knowing which area is more vulnerable to pollution than others, by means of hydrological factors and from a rigorous contamination point of view (El-Naqa et al. 2006; Shirazi et al. 2012). Generally, the GV map is used to denote the potential area of groundwater contamination by the various pollutants (Das and Pal 2019b).

Until now, a number of vulnerability assessment techniques or spatial mapping have been developed; there are mainly three types, viz. statistical, process-based, and index-based methods (Kumar et al. 2015). Statistical techniques are mainly based on the relationship between the spatial hydrological variables and contamination location in the groundwater (Arabameri et al. 2019; Khosravi et al. 2018; Machiwal et al. 2018). Different simulation models have been used in the process-based method where the models forecast the pollutant movement in groundwater (Fusco et al. 2020; Tiktak et al. 2006). The index-based method is comparatively the easiest and most precise method for GV, as there are no complexities in computational and data processing (Kumar et al. 2015; Sahoo et al. 2016).

In recent years, machine learning methods have been used to address various issues such as landslides, floods, and GV assessments, etc. Multivariate statistical methods, together with machine learning methods, are commonly used in hydrology studies (Rodriguez-Galiano et al. 2014). The machine learning methods like artificial neural network (ANN), random forest (RF), decision tree (DT), and maximum entropy (ME) have been used for susceptibility and vulnerability mapping (Rahmati et al. 2016). RF is an ensemble learning technique, so this is the same basic algorithm that generates multiple, repeated predictions that are averaged to produce a unique model (Breiman 2001). This model was used in this study.

The Bankura district, located in the lower Bengal Gangetic plains of India, which is prone to drought, has been selected as a study region in which the contaminations have become an acute problem, as well as the lowering of groundwater levels due to the extraction of water for Boro paddy cultivation. In this area, low-lying aquifers are at risk of pollution as a result of the rapid injection of chemical fertilizers and pesticides into agriculture and, at times, unplanned urban solid and liquid urban waste, which is reliable for contamination.

The objective of the study is to assess the vulnerability and mapping of groundwater in the Bankura district of India using RF and geo-hydrological attributes in the geospatial environment.

20.1.1 Study Area

Bankura district is the focus area for GV assessment, positioned in the eastern slope of Chhota Nagpur plateau and the western part of West Bengal in India. The study area covers 6882 km^2 and located between the latitude of 22°38′ to 23°38’ N and longitude of 86°36′ to 87°46′ E (Figure 20.1). Physiographically, the district is associated with undulating plateau upland, granite gneiss, skeleton and lateritic, and eastern flowing river system. The district

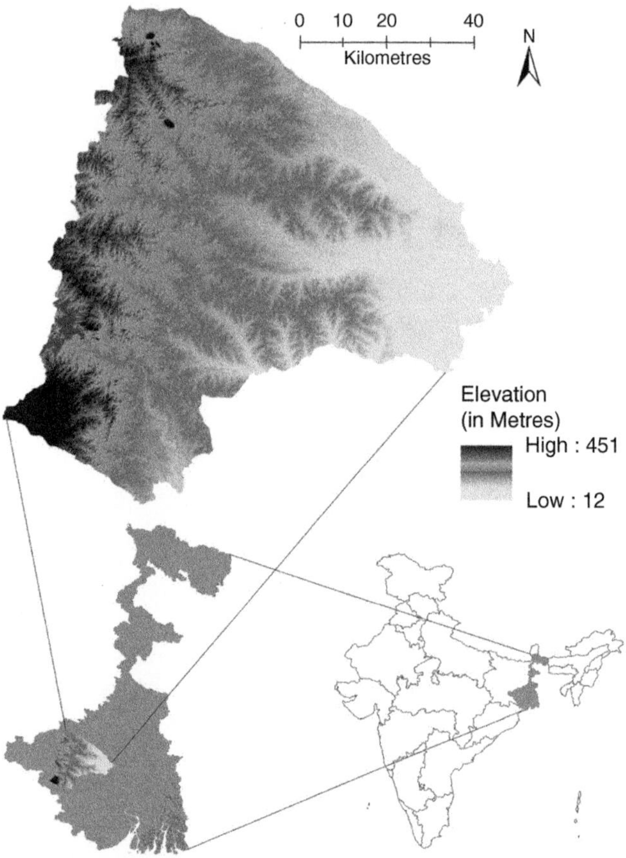

Figure 20.1 Location of the study area.

Figure 20.2 Field photographs.

receives high temperature (35–40 °C), medium rainfall (100–140 cm), and 80% of rainfall is received during the monsoon months (June–September). Tropical dry deciduous forest is located in different parts which cover 21.27% of the total land area. This region is known as a major drought-prone and water-scarce area where major wells are going dry in summer and winter season (Figure 20.2). The groundwater level varies from 1.5 to 22 m in the district but the artesian aquifer is generally obtained between the depth spans of 30–75 m. The groundwater of Bankura is contaminated by different materials but extensive occurrence of fluoride contamination takes place resulting from rock–water interaction (Chakrabarti and Bhattacharya 2013).

20.2 Materials and Methods

20.2.1 Database

Different types of data have been considered for this study to estimate the GV in Bankura district of West Bengal, India. The details database is shown in Table 20.1.

Table 20.1 Database and its sources.

Parameters	Data Type	Data Source	Data Details
Elevation	Raster Grid	ALOS PALSAR DEM, (Alaska Satellite Facility)	12.5 m spatial resolution
Slope gradient (degree)	Raster grid		12.5 m spatial resolution
Geology (detailed lithology and deposits)	Line, point, and polygon coverage	Geological Survey of India	Different unit of lithology
Geomorphology	Line, point, and polygon coverage	Geological Survey of India	Different spatial geomorphological unit
Drainage density	Polygon coverage buffer	ALOS PALSAR DEM, (Alaska Satellite Facility)	12.5 m spatial resolution
Topographical Wetness Index (TWI)	Raster grid		12.5 m spatial resolution
LULC	Spatial/Raster grid	Sentinel 2A (USGS [US Geological Survey])	10 m spatial resolution
pH	Point-specific information	C.G.W.B. Eastern Region	Groundwater geo-chemistry data
EC			
TDS			
SSP			
SAR			

20.2.2 Methodology

The detailed methods for estimating the GV are shown in Figure 20.3.

20.2.2.1 Random Forest

The RF model is an ensemble learning of multivariate statistical technique (Trigila et al. 2015; Were et al. 2015). The algorithm of RF is composed of an ensemble of classification and simple regression trees (Breiman 2001). RF uses the training dataset array class as DTs. Random sampling methodology is used in the design of the final sample. In addition, the client must be sensitive to the following two essential properties of the RF: t (tree number) and m (variable number). The calibration of RF model has been built based on bagging techniques to select training datasets randomly (Breiman 2001; Trigila et al. 2015). In the incremental phase of forest learning, each DT is loaded with a random sample chosen from the training dataset (Hosseinalizadeh et al. 2019). The algorithm for a RF of mean square error is given by the following equation:

$$\varepsilon = \left(v_{observed} - v_{response}\right)^2 \tag{20.1}$$

Where, ε is the mean–square error from the algorithm, $v_{observed}$ is the variables from the observed data, and $v_{response}$ is the variables from the result.

The predictions from RF trees is deliberated as follows:

$$S = \frac{1}{K}\Sigma K^{th} v_{response} \tag{20.2}$$

Where, S is the prediction of random-forest, and K applies to the individual trees in the forest.

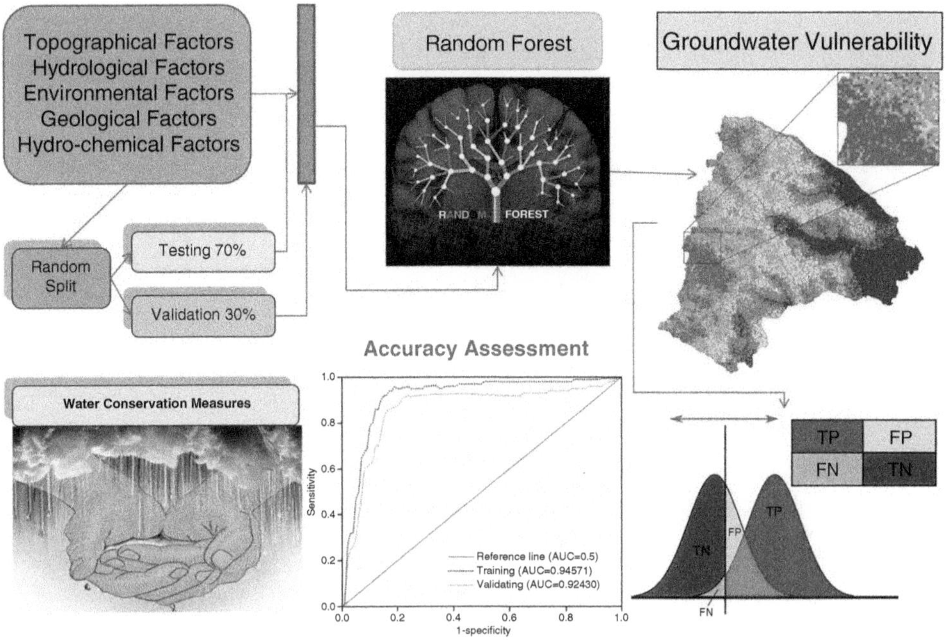

Figure 20.3 Methodology flowchart.

20.3 Results

20.3.1 Groundwater Vulnerability Causal Factors

20.3.1.1 Topographical Factors

Slope and elevation are considered topographical factors for estimating the GV. These factors were estimated considering the ALOSPALSER DEM (digital elevation model) from Alaska Satellite Facility as a raster grid format. The elevation of this region ranges from 12.0 to 451 m. The very high (183.0–451.0) and high (129.0–183.0) elevation zones are mainly located in the western part of this region (Figure 20.4). Most of the areas of this region are associated with moderate to very low elevation zone and located between the middle to the eastern portion. The amount of slope of this region varies from 0 to 51° and is unevenly distributed in the overall region. The very high (14.4–51.0) and high (6.4–14.4) slope gradients are mainly located in the western portion and there are some residual hills located there. For these reasons, there is a variation of slope gradient within the western portion and regional rugged topography has been initiated.

20.3.1.2 Hydrological Factors

Hydrological parameter is an important element regarding the occurrences of the water and its quantity. Here for this purpose, the drainage density and topographical wetness index (TWI) has been selected purposely for estimating the GV in this region. These components are estimated from the ALOSPALSER DEM with 12.5 m spatial resolution. The methods for estimate the DD and TWI are:

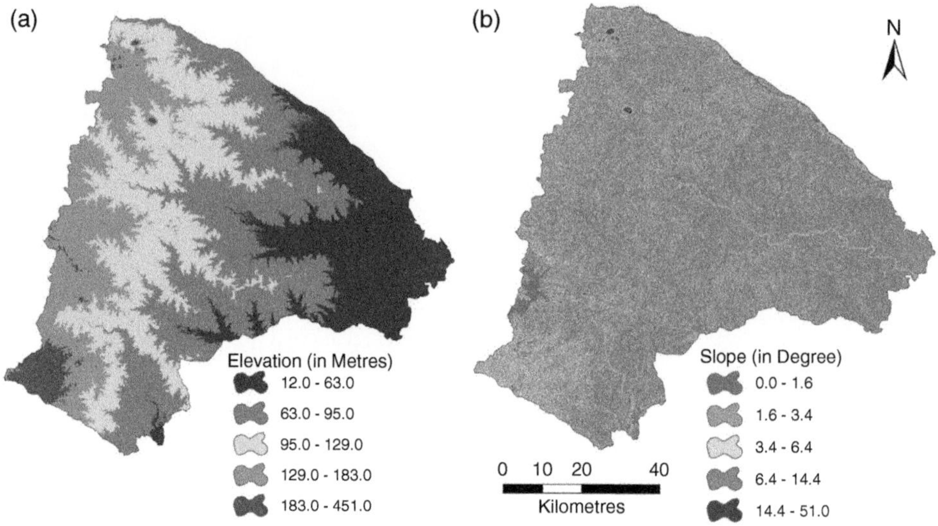

Figure 20.4 Topographical factors: elevation (a) and slope (b).

$$DD = \frac{L_n}{A_b} \tag{20.3}$$

Where, L_n is the length of the stream and A_b is the respective area.

$$TWI = In\left(\frac{as}{tanB}\right) \tag{20.4}$$

Where A means the identified catchment area, a = A/L, L = length of the column, *tanB* represents slope.

The values of drainage density of this region vary from 0.0 to 49.1. The very high (30.4–49.1) and high (21.5–30.4) drainage density is mainly associated with the major streamline and its surrounding regions (Figure 20.5). The TWI of this region is varying from 6.9 to 28.8 and this element is unevenly distributed in the entire region. The very high (18.0–28.0) and high (14.7–18.0) TWI of this region is mainly located in the major water bodies and their surrounding environment (Chowdhuri et al. 2020). Apart from this, most of the region is associated with moderate to very low TWI.

20.3.1.3 Environmental Factor

The land use and land cover are some of the important elements that largely influence the nature of water resources both in surface and subsurface. They are directly and indirectly controlling the nature of runoff of a particular region. The diversified LULC (land use land cover) are associated in this region, which have an important role in the nature of runoff and storage of subsurface water resources (Figure 20.6). The rapid LULC transformation is one of the key factors for various environmental hazards in various parts of the world.

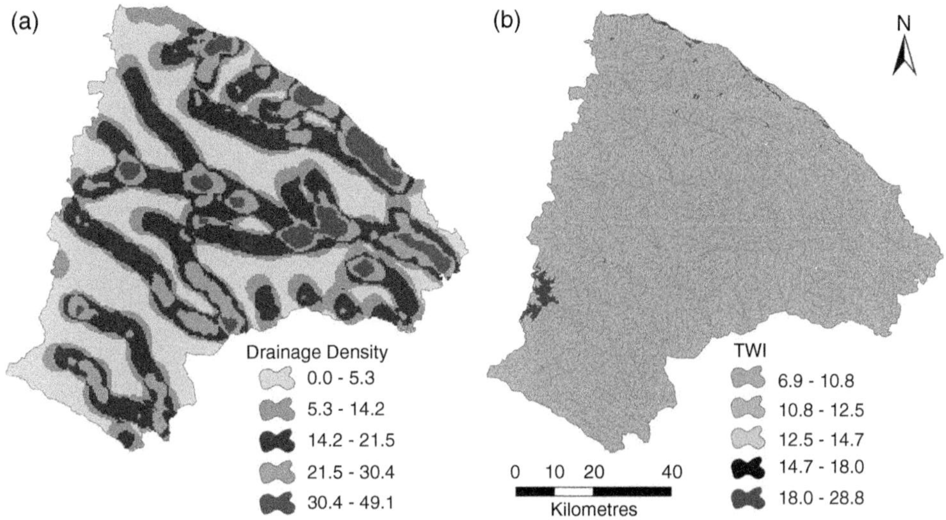

Figure 20.5 Hydrological factors: drainage density (a) and topographical wetness index (b).

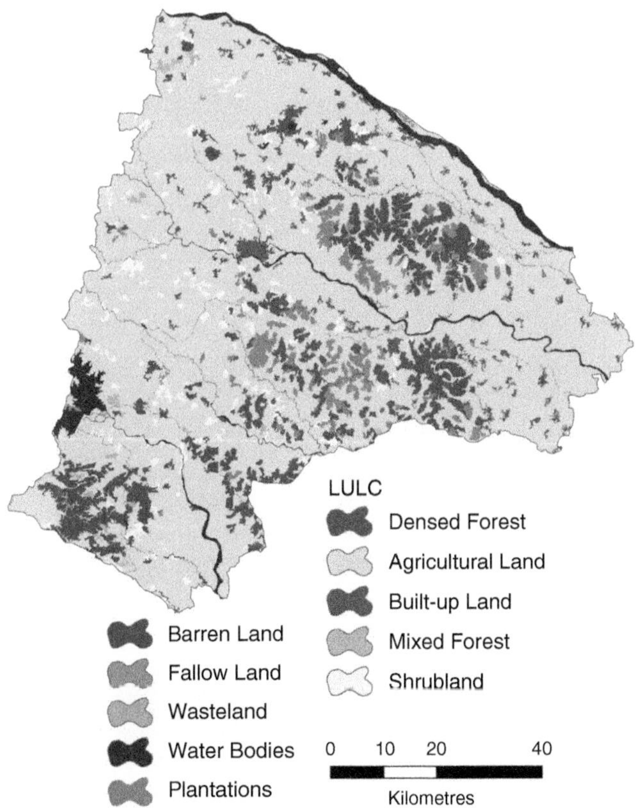

Figure 20.6 Environmental factor (LULC).

20.3.1.4 Geological Factors

Geomorphology, geology, and the orientation of the lineaments are the key elements for accumulation and storage of subsurface water resources. Different types of geomorphologic features are situated in this region; these are associated from dissected hills to alluvial plain. In terms of geomorphological characteristics, most of the region is associated with pediment pediplain complex (Figure 20.7). Here, the water bodies and alluvial plain are most suitable for optimal recharge of subsurface water. The undulating rugged topography is not suitable for recharging the subsurface water table. Different geological formations are also situated in this region which have a dissimilar impact on water resources; the

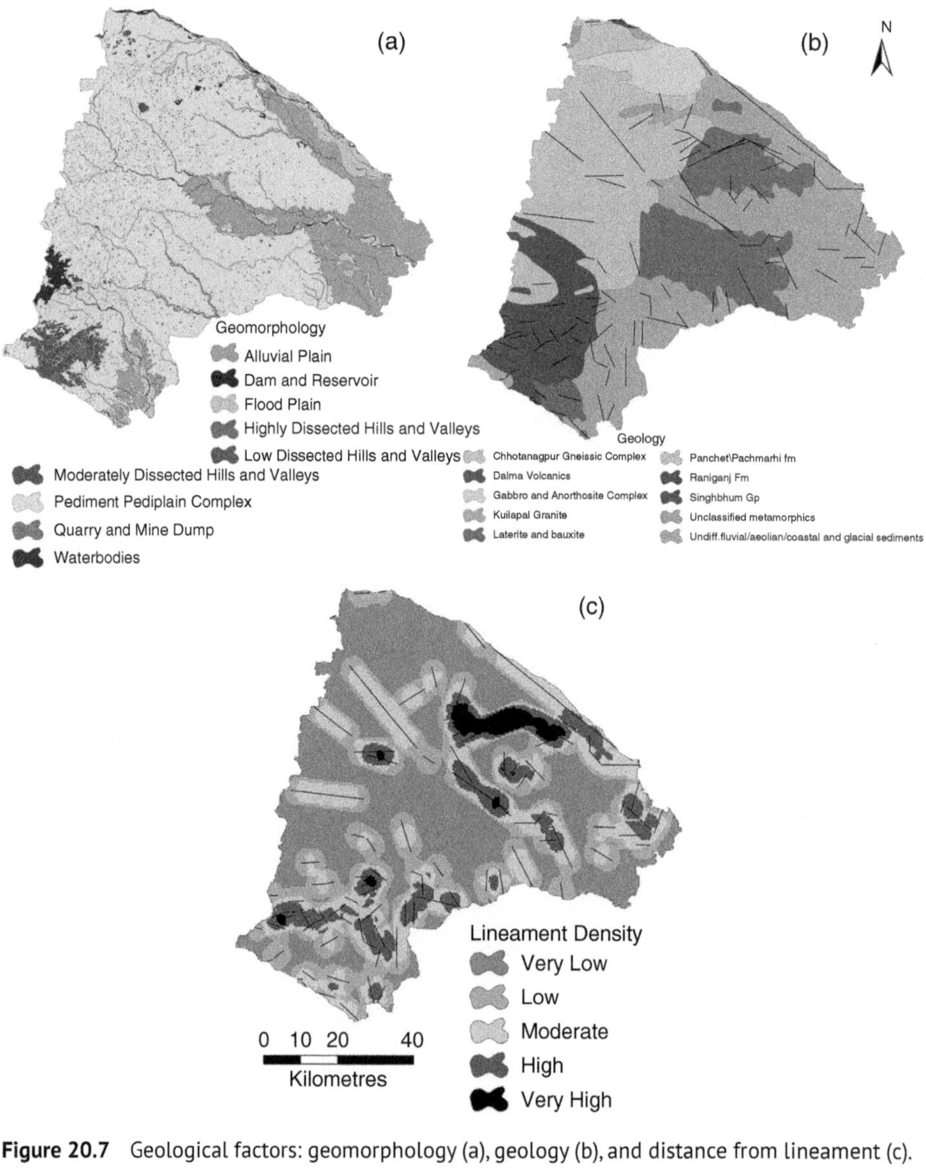

Figure 20.7 Geological factors: geomorphology (a), geology (b), and distance from lineament (c).

Table 20.2 Details of the geological formation.

Formation	Age
Unclassified metamorphics	Archaean–Proterozoic
Chhotanagpur Gneissic Complex	Proterozoic (undifferentiated)
Singhbhum Gp	Palaeoproterozoic
Raniganj Fm	Permian
Fluvial sediments	Quaternary
Laterite and bauxite	Cenozoic
Panchet\Pachmarhi fm	Triassic
Dalma Volcanics	Archaean–Proterozoic
Gabbro and Anorthosite Complex	Proterozoic
Dalma Volcanics	Archaean–Proterozoic
Kuilapal Granite	Mesoproterozoic

detailed descriptions of the different geological formations are shown in Table 20.2. Apart from this, there is a significant impact of major and minor lineaments associated in this region regarding the storage of water resources. The spatial variation in lineament density was considered for the acquisition of specific knowledge about its major position in hydrological behaviors in a specific region (Das et al. 2019). The very high and high lineament density are located in the southwestern and western part of this region. Most of the areas of this region are associated with moderate to very low lineament density.

20.3.1.5 Hydrochemical Factors

Domestic water that is used exclusively for drinking and other domestic uses, and must be free of toxic substances and contaminants. The existence of some cations and anions in freshwater under acceptable ranges is important for human health. So estimation of these elements is essential for estimating the GV in this region. The water pH of this region ranges from 6.90 to 8.19 and most of the areas are associated with very high (7.94–8.19) to moderate (7.49–7.72) pH zones (Figure 20.8). Electrical conductivity (EC) of this region ranges from 120.05 to 944.10 μS/cm and most of the areas are associated with moderate (1532.71–2239.03) to very low (120.05–944.10) EC zones. Total dissolved solids (TDS) of this region range from 51.01 to 683.85 mg/L; most of the areas are associated moderate (299.18–388.52) to very low (51.01–209.84) TDS zones. Soluble sodium percentage (SSP) ranges from 19.90 to 69.17 and most of the areas are associated with moderate (33.43–40.77) to very low (19.90–27.44) SSP zones. Sodium adsorption ratio (SAR) ranges from 1.04 to 11.89 and most of the areas are associated with moderate (3.63–5.12) to very low (1.04–2.44) SAR zones.

20.3.2 Groundwater Vulnerability (GV) Assessment

The GV raster has been estimated with the help of the RF model and its spatial distribution has been created in GIS (geographical information system) platform. This GV raster has been reclassified into various qualitative classes following Jenks natural breaks

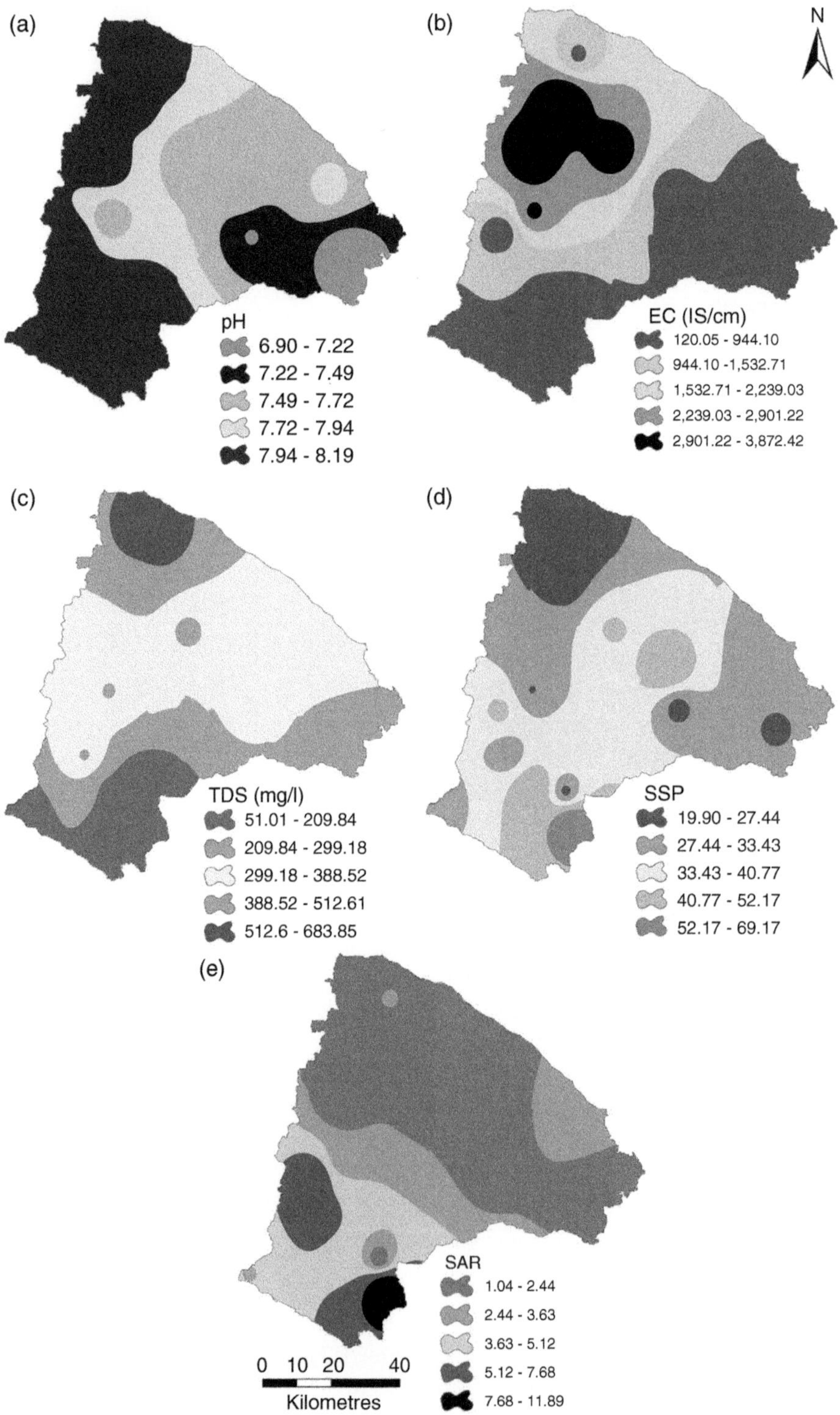

Figure 20.8 Hydrochemical factors: pH (a), EC (b), TDS (c), SSP (d) and SAR (e).

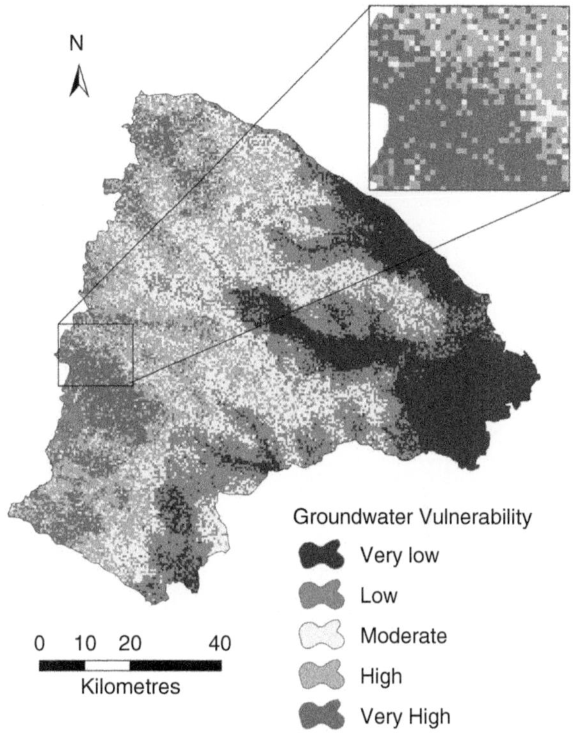

Figure 20.9 Groundwater vulnerability assessment.

optimization method. This method, also considered a differentiation of natural breaks in Jenks, is a way of information clustering intended to decide the optimal value structure for various groups. This is achieved by attempting to limit individual class's mean error from either the category mean, whereas increasing that class's deviation again from values of the another classes (Jenks 1967). In other terms, the approach helps to eliminate the variability within groups and to optimize the difference among groups (McMaster 1997). Very high GV is mainly located in the western part of this region where significant absence of sufficient recharge capacity and undulating rugged topography are associated. High GV is mainly located in the western and some extent of the middle part of this region where some residual hills are situated with significant undulating topography. Moderate GV areas are located only the middle part of this region. Low GV areas are located into the middle and eastern part of this region. Very low GV areas are located only the eastern part of this region where there are no such variations of elevation and slope gradient. Much of this region is associated with very low, low, and moderate GV zones (Figure 20.9). The entire region is dominated by subsistence agricultural practices where the paddy is most important crop. The eastern part of this region is the one of the most fertile regions of Bengal delta. The western part suffered acute water crisis but the eastern part facilitated the adequate availability of the water resources. This characteristic is also capable of making this eastern part an important agricultural region. To make this analysis relevant the actual information in different aquifers has been collected from field investigation. The characteristics of two selected litholog are shown in Figure 20.10, which was collected from Sihas village (23.012794 N, 87.593948 E) and Joyrambati village (22.9253 N, 87.6149 E).

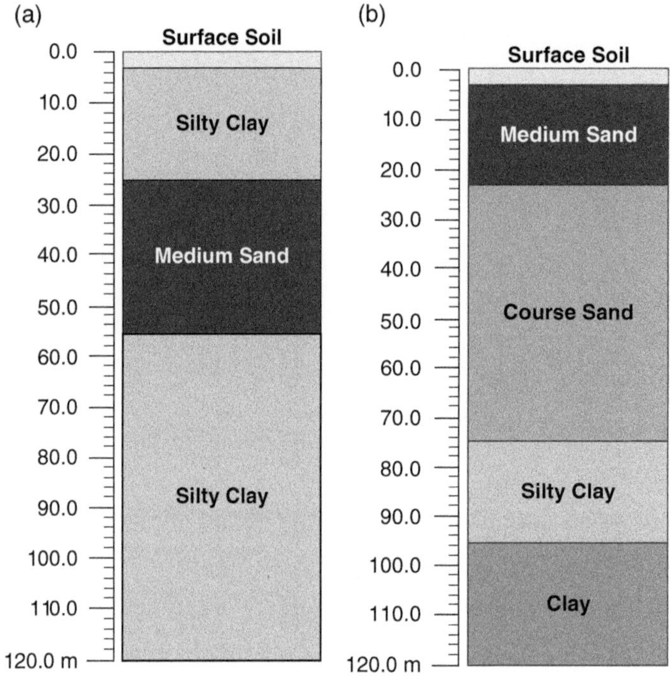

Figure 20.10 (a) Plot of the litholog data in Sihas village (23.012794 N, 87.593948 E) and (b) Joyrambati village (22.9253 N, 87.6149 E).

20.3.3 Accuracy Assessment

Verification is a crucial element of any scientific analysis. Groundwater recharge, which is specifically related to groundwater depletion and contamination of water, was measured through area of analysis. The receiver operating characteristics (ROC) area under curve (AUC) has been recommended for measuring which model is more fitting and correlated with better accuracy. ROC is one of the effective approaches which can estimate the projected model's accuracy. This method is now being considered in various scientific fields for that reason.

$$S_{AUC} = \sum_{k=1}^{n} \left(X_{k+1} - X_k \right) \left(S_k + 1 - S_{k+1} - \frac{S_k}{2} \right) \tag{20.5}$$

Where, S_{AUC} is the AUC, X_k is the 1-specificity, and S_k is the sensitivity of the ROC.

$$SPF = \frac{TN}{FP + TN} \tag{20.6}$$

SPF (specificity) indicates non-flood area. This method is applicable to predicting the GV for classifying non-vulnerable areas.

We used the ROC curve for validation purposes with consideration of the primary information about status of groundwater in recent years. Different cutting thresholds are deemed to define the optimum solution by the classifier's output (Chakrabortty et al. 2020;

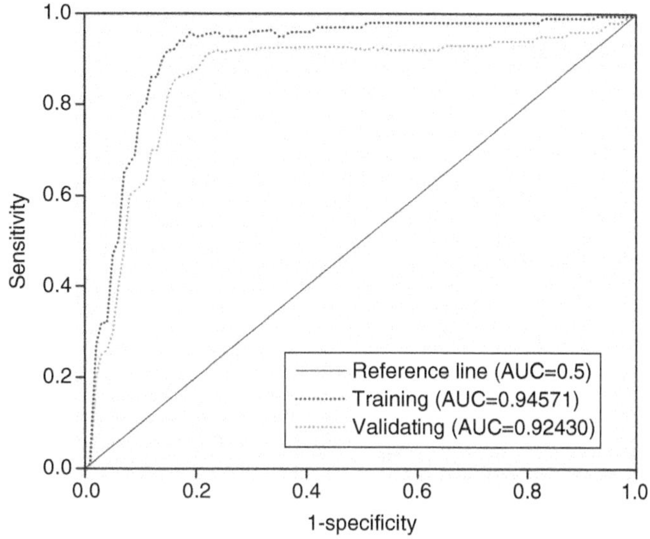

Figure 20.11 Accuracy assessment of the predictive model.

Dou et al. 2019; Roy et al. 2020). Here the AUC values of RF model for training and validation datasets are 92 430 and 94 571, respectively. This type of outcome is very authentic and can be applicable in this field for getting an optimal outcome (Figure 20.11).

20.4 Discussion

The RF machine learning method and GIS techniques were considered for estimating the GV in this region. The demand for dynamic water resources are increasing continuously due to rapid establishment of new settlements. As well. people are trying to maximize their agricultural production by incorporating greater infrastructure. For this reason, the irrigation practices also rapidly increase, and these are responsible for depletion of groundwater resources in this region. The growing tendency of impervious areas is a major problem in most of the developing countries and is responsible for decreasing of possible recharge.

Vulnerability means a condition of transformation in which there is a probability for damages toward society (Das and Pal 2020). GV is usually included to signify the potential for groundwater pollution from toxic elements; however, in many studies, the vulnerability in terms of sustainable development of changing water resources has still not been explored. The aquifer or the complex groundwater is in drought stress in several areas, so studying GV like a commodity is important nowadays (Das and Pal 2019a; Thangarajan 2007). This region is facing the acute problem of drought in most years due to extreme water crisis. Most of the region is associated with paddy cultivation, which is the most important crop of subsistence-based agriculture practices. The problem of acute water crisis is associated with diverse impact of agricultural production in most of the region. The uncertainty and uneven distribution of monsoonal rainfall is the main barrier to agricultural sustainability in major portions of this region, though the eastern portion facilitated the better irrigation system by various modes of irrigation. The role of groundwater to irrigation systems in the eastern part is rapidly increasing. Apart from the subsistence-based crops, the establishment of cash crops are increasing to sustain the livelihood of the local people.

An efficient protection against pollution is necessary to effectively preserve the groundwater resources. Vulnerability mapping plays a significant function in this mitigation challenge due to its ability to display diverse hydrogeological structures in a comprehensible way. GV may be categorized into two types: (i) intrinsic vulnerabilities, which rely solely on hydrogeological characteristics of the aquifer network, and (ii) particular vulnerability, which relates to vulnerability of different pollutants and includes the physiochemical characteristics of each pollutant in the assessment.

From the previous studies, it can be said that the infiltration is the main source of the pollutants from different sectors like agriculture, industry, etc. (Rahman 2008). The lineament zones are the most important arrears where there is greater probability of pollution in groundwater resources (Chakrabortty et al. 2018). The direction and orientation of major and minor lineaments is favourable for groundwater recharge on one hand, but on other hand it is one of the ways to polluting the groundwater (Chenini et al. 2010). The chemical fertilizer and pesticides are main sources of pollution in rural environments but in urban environments the industrial and solid waste are also responsible for large-scale pollution.

This may contribute to a rise in groundwater contamination risk due to the rising amount of water penetration in the cultivated lands due to farming practices. After all, the annual total recharge due to infiltrated rainfall and irrigated water is impacted by the associated surface texture of the harvested region; soil texture, crops, mulch cover, as well as other factors usually influence of the level of the infiltrated water. In this analysis, due to the large paddy cultivation in the study region, the moisture interceptions and crop absorbing were almost similar with the separate cultivating regions.

While farming practices can decrease the number of vertical infiltration rate comparative to the general form, use of chemical pesticides raises the possibility of toxicity of the underground water (Reichenberger et al. 2007). Throughout the course of penetration, rainfall and irrigation water bring contaminants that concentrate in the aquifer over time; hence, the deformation and accumulation of contaminants increases the real risk of contamination (Appelo and Postma 2004). Invariably, agricultural practices raise the probability of pollution of the underground water. Extensive prediction of the risk of contamination from groundwater is an important foundation for successful environmental safety and consumption.

20.5 Concluding Remarks

Assessment and estimation of the GV is essential because it is related to the sustainability of this precious resource and sustainability of this region. In most of the research, the vulnerability of this resource has been estimated considering various parameters related to the contamination from pollutants. The source of the pollutants and the possible places that can be crucial for contamination should be identified for the vulnerability assessment and to work out the suitable remedies. The GV assessment through machine learning and

GIS environment is time consuming, but also give us the optimal results with better accuracy. The main advantage of the RF model is that it can handle both the categorical and continuous variables in explanatory form. This type of outcome can be essential in different purposes because it can: (i) identify the present condition of water resources; (ii) identify the suitable areas for good condition of water resources by considering the quantitative and qualitative parameters; and (iii) be helpful in monitoring activities. The western and middle portion of this region are associated with very high to high vulnerable areas, which should be give proper emphasis to management and development strategies. There is acute crisis of water in most of the areas during the dry season. The monsoon season gets sufficient rainfall but most of the rainfall was confirmed in storm events. Due to the steep topography the rate of runoff is very high, so the storage of subsurface water resources is marginal in this region. The scope of the future researcher is to establish the most appropriate factors related to vulnerability and slightly modify this algorithm in different spatial scales to estimate the vulnerability in a more appropriate way.

References

Alagumuthu, G. and Rajan, M. (2008). Monitoring of fluoride concentration in ground water of kadayam block of Tirunelveli district, India: correlation with physico-chemical parameters. *Rasayan Journal of Chemistry* 1 (4): 757–765.

Ahmed, I., Al-Othman, A.A., and Umar, R. (2014). Is shrinking groundwater resources leading to socioeconomic and environmental degradation in Central Ganga Plain, India? *Arabian Journal of Geosciences* 7 (10): 4377–4385.

Appelo, C.A.J. and Postma, D. (2004). Geochemistry, Groundwater and Pollution. CRC Press.

Arabameri, A., Rezaei, K., Cerda, A. et al. (2019). GIS-based groundwater potential mapping in Shahroud plain, Iran. A comparison among statistical (bivariate and multivariate), data mining and MCDM approaches. *Science of the Total Environment* 658: 160–177.

Breiman, L. (2001). Random forests. *Machine Learning* 45 (1): 5–32.

Castaño, S., Sanz, D., and Gómez-Alday, J.J. (2010). Methodology for quantifying groundwater abstractions for agriculture via remote sensing and GIS. *Water Resources Management* 24 (4): 795–814.

Chae, G.-T., Kim, K., Yun, S.-T. et al. (2004). Hydrogeochemistry of alluvial groundwaters in an agricultural area: an implication for groundwater contamination susceptibility. *Chemosphere* 55 (3): 369–378.

Chakrabarti, S. and Bhattacharya, H.N. (2013). Inferring the hydro-geochemistry of fluoride contamination in Bankura district, West Bengal: a case study. *Journal of the Geological Society of India* 82 (4): 379–391.

Chakrabortty, R., Pal, S.C., Malik, S., and Das, B. (2018). Modeling and mapping of groundwater potentiality zones using AHP and GIS technique: a case study of Raniganj Block, Paschim Bardhaman, West Bengal. *Modeling Earth Systems and Environment* 4 (3): 1085–1110.

Chakrabortty, R., Pal, S.C., Chowdhuri, I. et al. (2020). Assessing the importance of static and dynamic causative factors on erosion potentiality using SWAT, EBF with uncertainty and plausibility, logistic regression and novel ensemble model in a sub-tropical environment. *Journal of the Indian Society of Remote Sensing* 48: 1–25.

Chenini, I., Mammou, A.B., and El May, M. (2010). Groundwater recharge zone mapping using GIS-based multi-criteria analysis: a case study in Central Tunisia (Maknassy Basin). *Water Resources Management* 24 (5): 921–939.

Chowdhuri, I., Pal, S.C., and Chakrabortty, R. (2020). Flood susceptibility mapping by ensemble evidential belief function and binomial logistic regression model on river basin of eastern India. *Advances in Space Research* 65 (5): 1466–1489.

Das, B. and Pal, S.C. (2019a). Combination of GIS and fuzzy-AHP for delineating groundwater recharge potential zones in the critical Goghat-II block of West Bengal, India. *HydroResearch* 2: 21–30.

Das, B. and Pal, S.C. (2019b). Assessment of Groundwater Recharge and Its Potential Zone Identification in Groundwater-Stressed Goghat-I Block of Hugli District, West Bengal, India, 1–19. Environment, Development and Sustainability.

Das, B. and Pal, S.C. (2020). Assessment of groundwater vulnerability to over-exploitation using MCDA, AHP, fuzzy logic and novel ensemble models: a case study of Goghat-I and II blocks of West Bengal, India. *Environmental Earth Sciences* 79 (5): 1–16.

Das, B., Pal, S.C., Malik, S., and Chakrabortty, R. (2019). Modeling groundwater potential zones of Puruliya district, West Bengal, India using remote sensing and GIS techniques. *Geology, Ecology, and Landscapes* 3 (3): 223–237.

Dou, J., Yunus, A.P., Tien Bui, D. et al. (2019). Assessment of advanced random forest and decision tree algorithms for modeling rainfall-induced landslide susceptibility in the Izu-Oshima Volcanic Island, Japan. *Science of the Total Environment* https://doi.org/10.1016/j.scitotenv.2019.01.221.

Dutta, S. (2018). Cause-effect analysis between irrigation and agricultural expansion on sub-surface water resources: a case study of Kanksa Block in Ajay–Damodar interfluve of Barddhaman District, West Bengal, India. *Sustainable Water Resources Management* 4 (3): 469–487.

El-Naqa, A., Hammouri, N., and Kuisi, M. (2006). GIS-based evaluation of groundwater vulnerability in the Russeifa area, Jordan. *Revista Mexicana de Ciencias Geológicas* 23 (3): 277–287.

Fusco, F., Allocca, V., Coda, S. et al. (2020). Quantitative assessment of specific vulnerability to nitrate pollution of shallow alluvial aquifers by process-based and empirical approaches. *Water* 12 (1): 269.

Hosseinalizadeh, M., Kariminejad, N., Chen, W. et al. (2019). Gully headcut susceptibility modeling using functional trees, naïve Bayes tree, and random forest models. *Geoderma* 342: 1–11.

Jenks, G.F. (1967). The data model concept in statistical mapping. *International Yearbook of Cartography* 7: 186–190.

Khosravi, K., Sartaj, M., Tsai, F.T.-C. et al. (2018). A comparison study of DRASTIC methods with various objective methods for groundwater vulnerability assessment. *Science of the Total Environment* 642: 1032–1049.

Kumar, P., Bansod, B.K., Debnath, S.K. et al. (2015). Index-based groundwater vulnerability mapping models using hydrogeological settings: a critical evaluation. *Environmental Impact Assessment Review* 51: 38–49.

Kumar, B., Singh, U.K., and Mukherjee, I. (2017). Hydrogeological influence on the transport and fate of contaminants in the groundwater, India. *JSM Biology* 2: 1–11.

Machiwal, D., Cloutier, V., Güler, C., and Kazakis, N. (2018). A review of GIS-integrated statistical techniques for groundwater quality evaluation and protection. *Environmental Earth Sciences* 77 (19): 681.

Matthess, G. (1982). The properties of ground-water, 406. New York: John Wiley and Sons.

McMaster, R. (1997). In Memoriam: George F. Jenks (1916–1996). *Cartography and Geographic Information Systems* 24 (1): 56–59.

Menon, S. V. (2007). *Ground water management: Need for sustainable approach* Unpublished paper.

Mukherjee, P., Singh, C.K., and Mukherjee, S. (2012). Delineation of groundwater potential zones in arid region of India—a remote sensing and GIS approach. *Water Resources Management* 26 (9): 2643–2672.

Pahuja, S., Tovey, C., Foster, S., and Garduno, H. (2010). Deep Wells and Prudence: Towards Pragmatic Action for Addressing Groundwater Overexploitation in India. The World Bank, Washington, DC.

Rahman, A. (2008). A GIS based DRASTIC model for assessing groundwater vulnerability in shallow aquifer in Aligarh, India. *Applied Geography* 28 (1): 32–53.

Rahmati, O., Pourghasemi, H.R., and Melesse, A.M. (2016). Application of GIS-based data driven random forest and maximum entropy models for groundwater potential mapping: a case study at Mehran Region, Iran. *Catena* 137: 360–372.

Reichenberger, S., Bach, M., Skitschak, A., and Frede, H.-G. (2007). Mitigation strategies to reduce pesticide inputs into ground-and surface water and their effectiveness; a review. *Science of the Total Environment* 384 (1–3): 1–35.

Rodriguez-Galiano, V., Mendes, M.P., Garcia-Soldado, M.J. et al. (2014). Predictive modeling of groundwater nitrate pollution using Random Forest and multisource variables related to intrinsic and specific vulnerability: a case study in an agricultural setting (Southern Spain). *Science of the Total Environment* 476: 189–206.

Rosegrant, M.W., Ringler, C., and Zhu, T. (2009). Water for agriculture: maintaining food security under growing scarcity. *Annual Review of Environment and Resources* 34: 205–222.

Roy, P., Chakrabortty, R., Chowdhuri, I. et al. (2020). Development of different machine learning ensemble classifier for gully erosion susceptibility in Gandheswari watershed of West Bengal, India. In: Machine Learning for Intelligent Decision Science, 1–26. Springer.

Saha, D. and Alam, F. (2014). Groundwater vulnerability assessment using DRASTIC and Pesticide DRASTIC models in intense agriculture area of the Gangetic plains, India. *Environmental Monitoring and Assessment* 186 (12): 8741–8763.

Saha, D., Dhar, Y.R., and Sikdar, P.K. (2008). Geochemical evolution of groundwater in the Pleistocene aquifers of south Ganga plain, Bihar. *Journal of the Geological Society of India* 71 (4): 473.

Sahoo, S., Dhar, A., Kar, A., and Chakraborty, D. (2016). Index-based groundwater vulnerability mapping using quantitative parameters. *Environmental Earth Sciences* 75 (6): 522.

Sankararamakrishnan, N., Sharma, A.K., and Sanghi, R. (2005). Organochlorine and organophosphorous pesticide residues in ground water and surface waters of Kanpur, Uttar Pradesh, India. *Environment International* 31 (1): 113–120.

Santos, I.R., Machado, M.I., Niencheski, L.F. et al. (2008). Major ion chemistry in a freshwater coastal lagoon from southern Brazil (Mangueira Lagoon): influence of groundwater inputs. *Aquatic Geochemistry* 14 (2): 133–146.

Sekhri, S. (2013). Sustaining groundwater: Role of policy reforms in promoting conservation in India. *Shekhar Shah Barry Bosworth Arvind Panagariya*, 149.

Shirazi, S.M., Imran, H.M., and Akib, S. (2012). GIS-based DRASTIC method for groundwater vulnerability assessment: a review. *Journal of Risk Research* 15 (8): 991–1011.

Shrestha, S., Kafle, R., and Pandey, V.P. (2017). Evaluation of index-overlay methods for groundwater vulnerability and risk assessment in Kathmandu Valley, Nepal. *Science of the Total Environment* 575: 779–790.

Suhag, R. (2019). Overview of Ground Water in India. PRS.

Thangarajan, M. (2007). Groundwater models and their role in assessment and management of groundwater resources and pollution. In: Groundwater (ed. M. Thangarajan), 189–236. Springer.

Tiktak, A., Boesten, J., Van der Linden, A.M.A., and Vanclooster, M. (2006). Mapping ground water vulnerability to pesticide leaching with a process-based metamodel of EuroPEARL. *Journal of Environmental Quality* 35 (4): 1213–1226.

Trigila, A., Iadanza, C., Esposito, C., and Scarascia-Mugnozza, G. (2015). Comparison of logistic regression and random forests techniques for shallow landslide susceptibility assessment in Giampilieri (NE Sicily, Italy). *Geomorphology* 249: 119–136.

Were, K., Bui, D.T., Dick, Ø.B., and Singh, B.R. (2015). A comparative assessment of support vector regression, artificial neural networks, and random forests for predicting and mapping soil organic carbon stocks across an Afromontane landscape. *Ecological Indicators* 52: 394–403.

Index

Groundwater Geochemistry: Pollution and Remediation Methods, First Edition.
Edited by Sughosh Madhav and Pardeep Singh.